Land surface processes in atmospheric general
circulation models

About this book

This book consists of papers presented at the World Climate
Research Programme study conference on land surface processes
held in Greenbelt, Maryland from 5 to 10 January 1981.
The papers cover the following: the state of knowledge of the
sensitivity of atmospheric general circulation models on
hydrology and other land surface processes; assessment of
the state of knowledge of numerical modelling of hydrology
and other land surface processes at the scale of atmospheric
general circulation models; recommendations for research
activities; establishment of data requirements for initialization, validation, and parameter evaluation. This book will be
of interest to atmospheric scientists, soil physicists,
hydrologists and climatologists.

Cambridge University Press is publishing the papers for the
World Meteorological Organization and the International Council
of Scientific Unions in order to achieve a wider distribution
for these important results.

Land surface processes in atmospheric general circulation models

Papers presented at
the World Climate Research Programme study conference,
held under the auspices of
the WMO/ICSO Joint Scientific Committee
in Greenbelt, Maryland
5 to 10 January, 1981

Edited by
P.S. Eagleson
Massachusetts Institute of Technology,
Cambridge, Massachusetts, USA

CAMBRIDGE UNIVERSITY PRESS
Cambridge
London New York New Rochelle
Melbourne Sydney

Published by the Press Syndicate of the University of Cambridge
The Pitt Building, Trumpington Street, Cambridge CB2 1RP
32 East 57th Street, New York, NY 10022, USA
296 Beaconsfield Parade, Middle Park, Melbourne 3206, Australia

© Cambridge University Press 1982

First published 1982

Printed in Great Britain at the University Press, Cambridge, UK

Library of Congress catalogue card number: 82-9740

British Library Cataloguing in Publication Data
Land surface processes in atmospheric general circulation models
1. Climatic geomorphology — Congresses
I. Eagleson, P.S.
551.4 GB406
ISBN 0 521 25222 9

CONTENTS

	Page
Foreword	vii
Introduction	ix

SESSION I - Atmospheric General Circulation Models and Climate Simulations

J. Smagorinsky	Large-scale climate modelling and small-scale physical processes	3
S. Manabe	Simulation of climate by general circulation models with hydrologic cycles	19
D.J. Carson	Current parameterizations of land-surface processes in atmospheric general circulation models	67
Y. Mintz	The sensitivity of numerically simulated climates to land surface conditions	109

SESSION II - The Microphysical Processes of Momentum, Heat and Water Transfers Across and Near the Surface of the Land

W.H. Brutsaert	Vertical flux of moisture and heat at a bare soil surface	115
L.J. Fritschen	The vertical fluxes of heat and moisture at a vegetated land surface	169
M. Kuhn	Vertical flux of heat and moisture in snow and ice	227

SESSION III - Mesoscale Parameterizations of the Transfer Processes

J.C.I. Dooge	Parameterization of hydrologic processes	243
P.S. Eagleson	Dynamic hydro-thermal balances at macroscale	289

SESSION IV - Land Surface Global Data Sets

M.J. Gardiner	Use of regional and global soils data for climate modelling	361
A. Perrier	Land surface processes: Vegetation	395
V.M. Kotliakov and A.N. Krenke	Data on snow cover and glaciers for the global climatic models	449

SESSION IV (Contd.)

		Page
K. Ya. Kondratyev V.I. Korzov V.V. Mukhenberg and L.N. Dyachenko	The shortwave albedo and the surface emissivity	463
A. Baumgartner	Water balance	515

SESSION V - <u>Acquisition of Land Surface Data</u>

| K.I. Itten | Possibilities for remote sensing of surface characteristics | 541 |

FOREWORD

Climate defines a part of the natural environment in which man has evolved and now exists. Climate can foster human activities or hinder them. The variability of climate can be beneficial or it can be violent and disastrous. Increasing world-wide concern and awareness of the possibility of climate changes, either natural or man-induced, has led to major international initiatives in studying and attempting to understand climate and its socio-economic impact in the form of the World Climate Programme. The objectives of this Programme are to improve the application of climatic data and knowledge to all aspects of man's activities, and to increase the present knowledge of climate and of those factors which influence it so that an attempt can be made to foresee possible changes of climate especially any man-made changes adverse to the well-being of humanity. A key component within the framework of the World Climate Programme is the World Climate Research Programme (WCRP), which will study why, how and where climate changes and variations occur, and assess thereby the feasibility of prediction and of their occurrence. The major objectives of the Programme are to determine:

- To what extent climate can be predicted
- The extent of man's influence on climate

It will be readily appreciated that the objectives that have been set are highly ambitious, and to achieve them requires a major extension of our knowledge and understanding of climate, climate processes, influences on climate, and how climate and climate variations can be modelled. A great deal of fundamental research and technical work is entailed, some of it in areas that have not been previously studied or even observed in detail. Overall, a tremendous depth of international and widespread scientific cooperation is necessary to answer the many complex, interdisciplinary questions involved. The World Climate Research Programme is conducted jointly by the World Meteorological Organization and the International Council of Scientific Unions and is an excellent example of where collaboration, in this case between a governmental international organization and a non-governmental international organization, is being directed towards the advancement of science to the achievement of ambitious objectives.

With the fundamental and innovative research and technical work going ahead as part of the WCRP, and the development of understanding that will take place, it is essential that as broad a distribution as possible be given to our extending knowledge to a wide community of interested persons and bodies. Accordingly, the

WMO/ICSU Joint Scientific Committee for the WCRP has agreed that the present volume should be published by the Cambridge University Press and distributed widely. The volume contains a series of papers presented at a Study Conference on Land Surface Processes held at Greenbelt, Maryland in January 1981. The cooperation and assistance of the Modelling and Simulation Facility of the Laboratory of Atmospheric Sciences (NASA) in convening and providing the facilities and technical support for conducting this Conference is greatly appreciated.

Improved understanding of climate can only be achieved, inter alia, by the study of climatologically significant processes, and the Joint Scientific Committee considers that the papers presented at the Conference contribute greatly to that understanding.

As Chairman of the WMO/ICSU Joint Scientific Committee, I would like to express my thanks to all those who contributed to the Conference. It is also a pleasure to give special acknowledgement to Professor P.S. Eagleson (Chairman of the Joint Scientific Committee Working Group on Land Surface Processes, previously the Joint Organizing Committee Working Group on the same subject) who has played such an important role in the organization of the Conference and gathering together such a distinguished set of papers.

John T. Houghton
Chairman
WMO/ICSU Joint Scientific Committee

Oxford, May 1982

INTRODUCTION

In pursuing the second objective of the Global Atmospheric Research Programme (GARP), namely the understanding of the physical processes determining climate, the (then) Joint Organizing Committee (JOC) established a Working Group on Land Surface Processes in 1977. The Working Group was charged with:

(i) Assessing the state of knowledge in incorporating such land surface features as snow cover, vegetation, albedo, surface roughness, soil moisture etc. in numerical models of climate

(ii) Identifying the problem areas and gaps in knowledge, and

(iii) Proposing a research programme to resolve these difficulties.

In pursuit of these objectives, the JOC approved the organization of a Study Conference on Land Surface Processes in Atmospheric General Circulation Models which was held in Greenbelt, Maryland from 5 to 10 January 1981. The Study Conference was convened to:

(i) Assess the state of knowledge of the sensitivity of atmospheric general circulation models to hydrology and other land surface processes

(ii) Assess the state of knowledge of numerical modelling of hydrology and other land surface processes at the scale of atmospheric general circulation models

(iii) Recommend research activities and numerical experiments designed to improve knowledge in the above areas

(iv) Establish data requirements for initialization, validation, and parameter evaluation.

The Conference deliberations confirmed that atmospheric general circulation models are highly sensitive to both the land surface albedo and to the concentration of near-surface soil moisture and it made recommendations regarding the further research needed to cope with these sensitivities through improved land surface parameterizations.

The Study Conference was notable for its initiation of a focussed scientific discussion among general circulation modellers, atmospheric and soil physicists, and hydrologists. The collection of review papers included in this volume should be a valuable addition to the scientific literature and a spur to interest in this important research area.

Peter S. Eagleson

SESSION I

ATMOSPHERIC GENERAL CIRCULATION MODELS

AND

CLIMATE SIMULATIONS

LARGE-SCALE CLIMATE MODELING
AND
SMALL-SCALE PHYSICAL PROCESSES

by

Joseph Smagorinsky
Geophysical Fluid Dynamics Laboratory/NOAA
Princeton University
Princeton, New Jersey 08540

Some antecedents

My object today is to give an overview of what role parameterizations play in the construction of climate models and of the problems in devising such parameterizations - all as background and context for this particular conference on the role of land surface processes in climate.

Subsequent talks by Manabe and Carson will deal increasingly with the particulars. Much of what I will have to say is covered at length in GPS 8 (Parameterization of Subgrid Scale Processes)* and GPS 16 (The Physical Basis of Climate and Climate Modeling)**. I am also borrowing heavily from my 1974 paper (Global Atmospheric Modeling and the Numerical Simulation of Climate)***.

The quest for a physical understanding of the terrestrial climate system is a venerable one. The early descriptions of the temporal and spatial structure of climate were an attempt to reconcile the more obvious seasonal and latitudinal radiative variations with local modifications resulting from topography, the proximity of bodies of water, soil properties and vegetation. Köppen's classification reflects such perceptions. As the complex geographical variations of climate become more evident, it becomes clearer that what man experienced at the earth's surface could not easily be explained in simple terms. The great subtropical deserts of the world are a cogent example. In some instances, the statistical ensemble of weather events constituting climate was simply a reflection of persistent situations, such as in the subtropics. In other instances, the climatic ensembles reflected ubiquitous but nevertheless sporadic transient episodes such as mid-latitude east coast cyclogenesis. The point is that the dynamical characteristics of the climate system could not be ignored in trying to assemble a comprehensive and self-consistent theory of climate that could explain the observed climatic structure near the earth's surface. The atmosphere's general circulation had to be understood as an inherent part of the problem and, as it turns out, that of the ocean as well. To these must be added the roles of the continental and oceanic cryosphere and the uppermost layer of the continents.

---

*   JOC Study Conference, Leningrad, 20-27 March 1972.

**   International Study Conference, Stockholm, 29 July-10 August 1974.

***   Weather and Climate Modification, ed., W. N. Hess, John Wiley & Sons, Inc., 1974, pp. 633-686.

Although a large body of work had been done the first half of this century in developing phenomenological theories for the dynamics of the atmosphere and the ocean, attempts to deal with them comprehensively and interactively did not really start until Normal Phillips' work on atmospheric general circulation modeling in the early 1950's. It may seem remote now, in retrospect, that Phillips' simple 2-level geostrophic general circulation model, which was confined to just a portion of the global atmosphere, was the beginning of comprehensive climate modeling - but it is a fact. His model did not have oceans and continents or mountains or complex interactions between the atmosphere and its lower boundary. Nevertheless, it provided the basic framework upon which virtually all advances in the past quarter century were made in 3-D climate modeling.

Phillips' model itself was a natural extension of earlier models designed to explain the transient characteristics of large-scale atmospheric motions. First the simple barotropic models of the late 1940's, in which the kinetic energy was conserved in the large, were remarkably able to account for mid-tropospheric wind variations over a 24-hour period. Such a model was derived from a simplification of the inviscid Navier-Stokes equation. A later advance, in the early 1950's, provided for conversions of potential to kinetic energy, and therefore for cyclogenesis, somewhat extending the accuracy and range of validity of forecasts. This was by means of a so-called baroclinic model. The modified vorticity equation, called the potential vorticity equation, took the first law of thermodynamics into account and yielded a mid-tropospheric temperature prediction of wind at several atmospheric levels. Both the barotropic and baroclinic models were derived from the conservation laws for momentum and heat. On such short-time scales, sources and sinks of energy could be ignored - a fortunate property of the large-scale terrestrial atmosphere which I do not have the time to dwell upon now. In other words, the large-scale atmosphere could be isolated and dealt with relatively simply. One could safely ignore phenomena that were happening on scales smaller than can be resolved by computational grids of the order of several hundred kilometers: radiative heat transfer, viscosity and phase changes of water substance.

Phillips' general circulation model, which aimed at dealing with longer time evolutions of the atmosphere, the general circulation, could no longer ignore external energy exchanges.

Permit me now to begin to enlarge on the nature of problems of constructing general circulation models and, in the ultimate, climate models. This will provide a context for the subject of this Study Conference.

The scale characteristics of atmospheric phenomena

The time-space domain for characteristic atmospheric phenomena is shown in Figure 1. This is typical for the free atmosphere (away from the lower boundary). Note that it covers a 10 order of magnitude span of horizontal dimension, but does not include the size of cloud water droplets, about 1 mm. There is a major energy peak at around 3000 to 6000 km corresponding to a zonal wave number of 5 to 8 (the number of waves in the wind or temperature field around a latitude circle). These extratropical cyclones are the result of "baroclinic instability", the major potential to kinetic energy converting mechanism in extratropical latitudes. There are secondary peaks corresponding to tropical cyclones (hurricanes), fronts, cumulus convection, tornadoes, cloud and clear air turbulence. They are generally intermittent and sparse phenomena and thus may not appear in a spectrum taken at any one time.

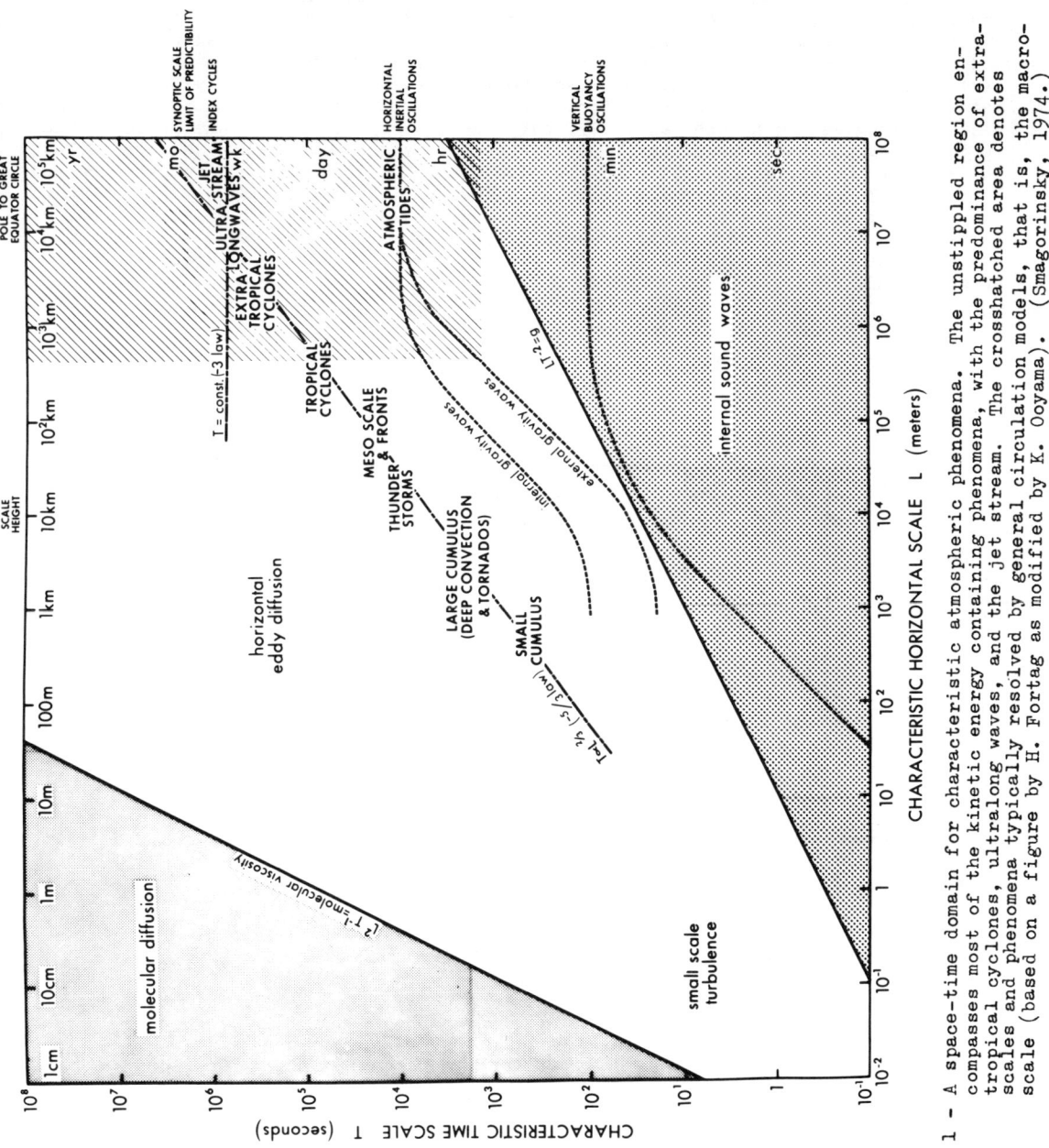

Figure 1 - A space-time domain for characteristic atmospheric phenomena. The unstippled region encompasses most of the kinetic energy containing phenomena, with the predominance of extra-tropical cyclones, ultralong waves, and the jet stream. The crosshatched area denotes scales and phenomena typically resolved by general circulation models, that is, the macro-scale (based on a figure by H. Fortag as modified by K. Ooyama). (Smagorinsky, 1974.)

Usually, because of computer limitations, a numerical model can at most resolve a two order of magnitude span of scales of this spectrum (Figure 2). Hence a global model with a horizontal grid size of 100 km can resolve wavelengths of 400 km and larger on to the planetary size (about 40,000 km). These are the explicitly treated macroscales. Interactions with smaller scales must be dealt with by other means. As I pointed out already, for some simple problems such as short-range prediction, they can be ignored entirely, but usually they must be <u>parameterized</u>. That is the integrated effects of the truncated part of the spectrum must be represented in terms of the variables explicitly resolved: the macroscale wind, temperature and humidity. In general, a parameterization allows for a two-way interaction between the explicit and parameterized parts of the spectrum. The parameterization usually introduces empirical constants (parameters) that enter into the prescription. An example of such a parameter is the eddy viscosity coefficient in a turbulent parameterization.

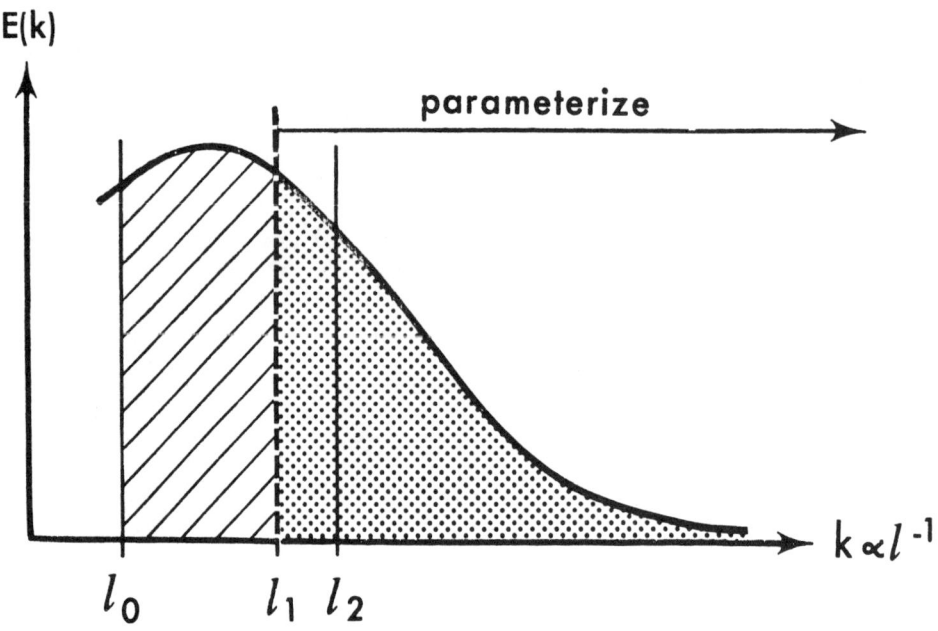

Figure 2 - The curve represents the variation of atmospheric energy density as a function of the wave number $\kappa \propto \ell^{-1}$. The hatched area is the explicitly resolved part of the spectrum, the macroscale. The stippled area indicates the portion of the energy spectrum within which the physical processes cannot be resolved by a domain with the dimension ℓ_o and the mesh width ℓ_1. (Smagorinsky, 1974.)

On the other hand, where greater detail is desired locally, an intermediate solution would be to deal with a subregion of higher but overlapping resolution. The calculation of the interaction between two domains of differing resolution, one "nested" in the other, is not yet a completely solved problem, technically. It has applicability to a number of important modeling problems, such as the hurricane embedded in the planetary flow.

I should point out that even the continuum Navier-Stokes equations of motion already represent a parameterization of the molecular motions in the representation of viscosity. Similarly the notion of temperature, as it enters into the continuum thermodynamic laws, already implies a parameterization of the molecular kinetic energy according to statistical mechanics.

These we take for granted. But as one looks upon the physics of increasingly macroscale volumes in practical <u>finitistic</u> representations formulated for numerical solution, the problem not only <u>remains but</u> amplifies as one needs to account for the interaction with more and more phenomena that are not explicitly resolved in the so-called <u>subgrid scales</u>.

Simply, the object of atmospheric modeling is to approach the problem of simulating the large-scale atmospheric behavior through a consistent application of the governing fundamental physical laws and boundary conditions; I should remind you that the macrostructure of these laws has been known for over a century. What had been lacking, and still is the main subject for research, is an understanding of the interactions of the macroscales of primary interest (i.e., greater than several hundred kilometers) with processes of lesser dimensions, such as radiative transfer, turbulent fluxes, and the processes within clouds and by clouds.

The governing physical laws

Let us qualitatively consider the system of physical laws that together constitute a fully consistent set of mathematical equations to predict the time variations of the primary dependent variables (the macroscale variables) at each point (really a finite volume) in the three-dimensional atmosphere (Figure 3). As we shall see, these variables are the two horizontal wind components (which define a two-dimensional vector $\mathbf{V}$), the temperature (θ) and the humidity (q). Furthermore, at the lower boundary of each finite column, one must also predict variations in the surface pressure ($p*$). From a knowledge of these primary dependent variables, at any one time it is then possible to determine such subsidiary variables as the density (ρ) and a measure of the vertical velocity (ω). Actually, ω is the change of pressure in an elemental volume moving with the total three-dimensional wind, and is approximately proportional to the downward component of the air motion.

For a system to be solvable in terms of these primary variables, it is first necessary to assume <u>a priori</u> that, for the macroscales of interest, the vertical particle accelerations are small compared to the acceleration of gravity. This <u>hydrostatic</u> approximation is valid for horizontal macroscales larger than a few kilometers, which certainly includes the scales we are concerned with, that is, of hundreds of kilometers and larger. Such a simplification of the differential (and difference) equations eliminates (or "filters") from the final solutions those corresponding to the vertical propagation of sound waves, which, relative to the other macroscale modes, carry relatively little energy and are of high frequency. This permits one to use time increments of the order of minutes in the numerical integration rather than seconds.

The horizontal momentum equations and those of state and hydrostatic balance are schematically combined in Figure 3. This set can predict changes in $\mathbf{V}$ provided we can also predict θ and $p*$, and determine ω. The law expressing the indestructibility of <u>mass</u>, the equation of continuity, provides the means of determining $p*$ and ω, if we know $\mathbf{V}$.

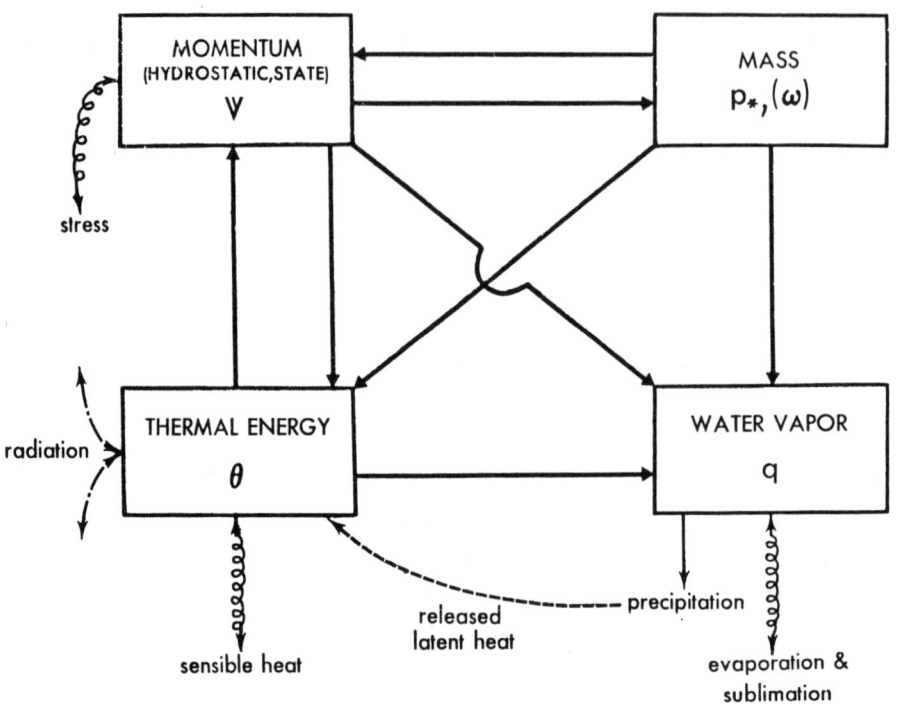

Figure 3 - A schematic diagram of the relationship of the physical conservation laws which together define the variations of the primary macroscale variables in an elemental volume. Also shown are external sources and sinks of momentum, heat, and water vapor. (Smagorinsky, 1974.)

To predict changes in θ we impose the thermal energy equation (i.e., the first law of thermodynamics). To calculate such changes, one first of all needs to know $\mathbf{V}$, p_*, and ω. Subgrid scale diffusive processes may also alter the energy in an elemental volume. If water vapor changes phase, for example, then the released latent heat must be accounted for. It may also be possible that energy may be radiated into or out of the volume. The scattering, absorption, or transmission will in general depend on the changing content of water vapor and of liquid water particles in the volume. Finally, an elemental volume next to the planetary surface could exchange sensible heat through the turbulence.

The remaining predictive law needed is that for q, the <u>water vapor</u> conservation equation. For this, we say that the water vapor content of an elemental volume moving with the macroscale motion can change (analogously to energy) only by internal subgrid scale diffusion, turbulent transfer from the planetary surface (evaporation or sublimation) or water phase changes (precipitation). To calculate these one must know $\mathbf{V}$, p_*, ω and θ.

In practice one does not deal explicitly with moving elemental volumes (a "Langrangian" framework), but one can more conveniently make the equivalent calculation for a geographically and temporally fixed volume (a "Eulerian" framework).

Let us now discuss some of the processes that must be represented in a climate model (Figure 4).

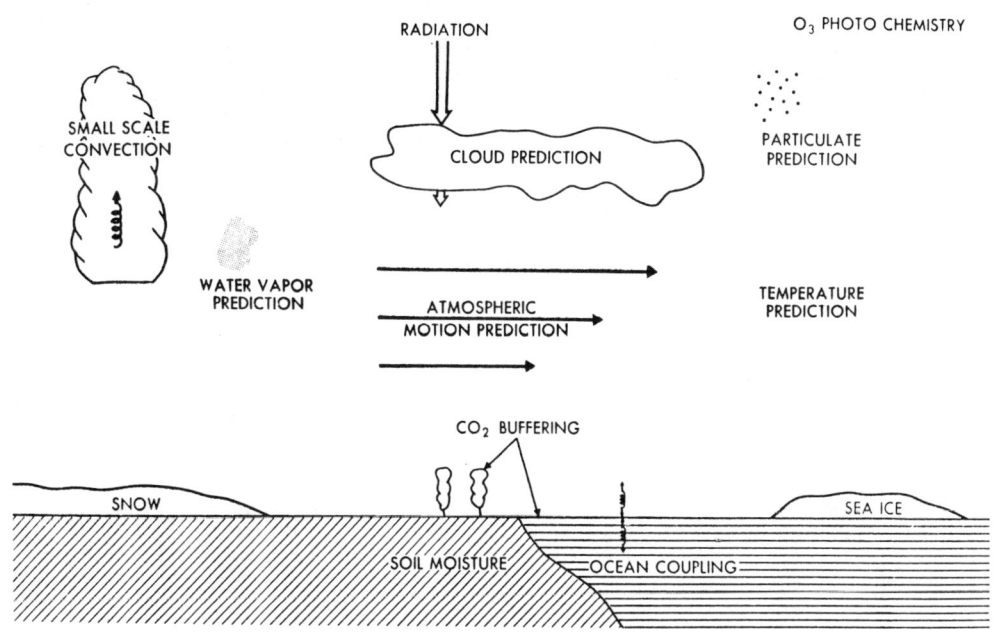

Figure 4 - Schematic representation of processes that are elements of climate or that influence climate.

Subgrid scale diffusion

The major transports of momentum, heat, and water vapor generally are accomplished by the macroscale motions, that is, those that are explicitly resolved. Diffusion by motion of scales smaller than an elemental volume may be important and must be parameterized. This then requires the specification of some additional empirical parameters.

Of special importance are the small-scale modes which can be excited when the macroscale becomes buoyantly unstable locally. In the atmosphere this is visually observed as cumulus clouds which have dimensions of several kilometers. Families of such clouds can occupy areas of hundreds or thousands of kilometers and, therefore, can have a profound effect back on the macroscale. The excitation of cumulus convection is relatively rapid, of the order of a few hours, and we know that, as a result, large vertical transport of heat, water vapor, and quite likely, momentum occur with the net result that the macroscale is restored to a buoyantly stable state. This process, as we know from observation, is at least of intermittent importance in extratropical latitudes, but is a primary mechanism in the tropics. The parameterization of cumulus convection is therefore a critical element in modeling the earth's atmosphere. Unfortunately, a generally accepted parameterization has yet to be devised; in part, the problem is that there are inadequate observations of the details of convection, so that conclusive verification of candidate parameterization hypotheses is not really possible. The results of the GARP Atlantic Tropical Experiment in 1974 (GATE) have provided some new insights.

Of similar importance is the role of small-scale diffusive processes which may occur within the atmosphere without any accompanying condensation or cloud. There is indirect evidence that such processes may be responsible for as much as half of the total energy dissipated by the macroscale motions. The remainder is lost within the planetary boundary layer, which is about 1 km in depth, being thicker at equatorial latitudes. Provisional parameterizations of internal clear air diffusion and dissipation have been used with some success. However, definitive determination in this also awaits the results of detailed observational experiment and theoretical studies.

Surface balance conditions

The interaction of the atmosphere with its lower boundary is of critical importance. For example, through surface stresses the atmosphere loses about half its energy. Furthermore, the atmosphere is largely transparent to the sun's radiation. Most of this radiation is absorbed by the planetary surface and then fed back up by infrared radiation and through turbulent transfer of heat in sensible and in latent form. The details of this upward return of heat and how it depends on characteristics of the planetary surface, which may be slowly or rapidly variable depending in part on the atmosphere itself, provide some of the most interesting and challenging modeling problems with particular relevance to questions of the stability of climatic regimes.

In the case of large oceanic expanses, the wind stress together with the heat transfer and freshwater exchange (precipitation and evaporation) are ultimately responsible for driving the ocean circulation. The uppermost layer of the ocean responds most rapidly to the interfacial exchanges with a relaxation time of the order of several weeks. On the other hand, the deep ocean circulation is the result of atmospheric coupling over hundreds or even thousands of years.

The oceans themselves are responsible for transporting a great deal of heat poleward. It may be as much as that of the atmospheric transport.

Therefore, the response of the earth's entire fluid envelope to the driving latitudinal heating gradient of the sun's radiation must be considered. A true climate model, therefore, cannot be bounded by the earth's surface but must extend into the ocean down to its bottom.

Let us now consider those elements that enter into the nature and degree of the interaction between the atmosphere and its lower boundary.

Momentum. The turbulent stress exerted on the lower boundary by the atmosphere would be expected to be of direct importance to a boundary which is mobile, that is, a large body of water.

The stress over both land and sea, however, will exchange momentum with the atmosphere locally, and the surface wind will also be a factor in determining the contribution of turbulent exchange processes in the heat and water vapor balances at the earth-atmosphere interface.

Heat (Figure 5). The limiting cases that one can conceive of are as follows: (a) a continental surface with zero heat capacity admitting no conduction of heat into the ground. The surface temperature would, therefore, be determined from a balance of the net radiation at the earth's surface and the turbulent flux of sensible and latent heat (due to evaporation). Since diurnal variability of surface temperature is in part determined by the daily cycle of solar radiation and by the molecular conduction of heat by the soil, the diurnal cycle could not be included in this limiting case;

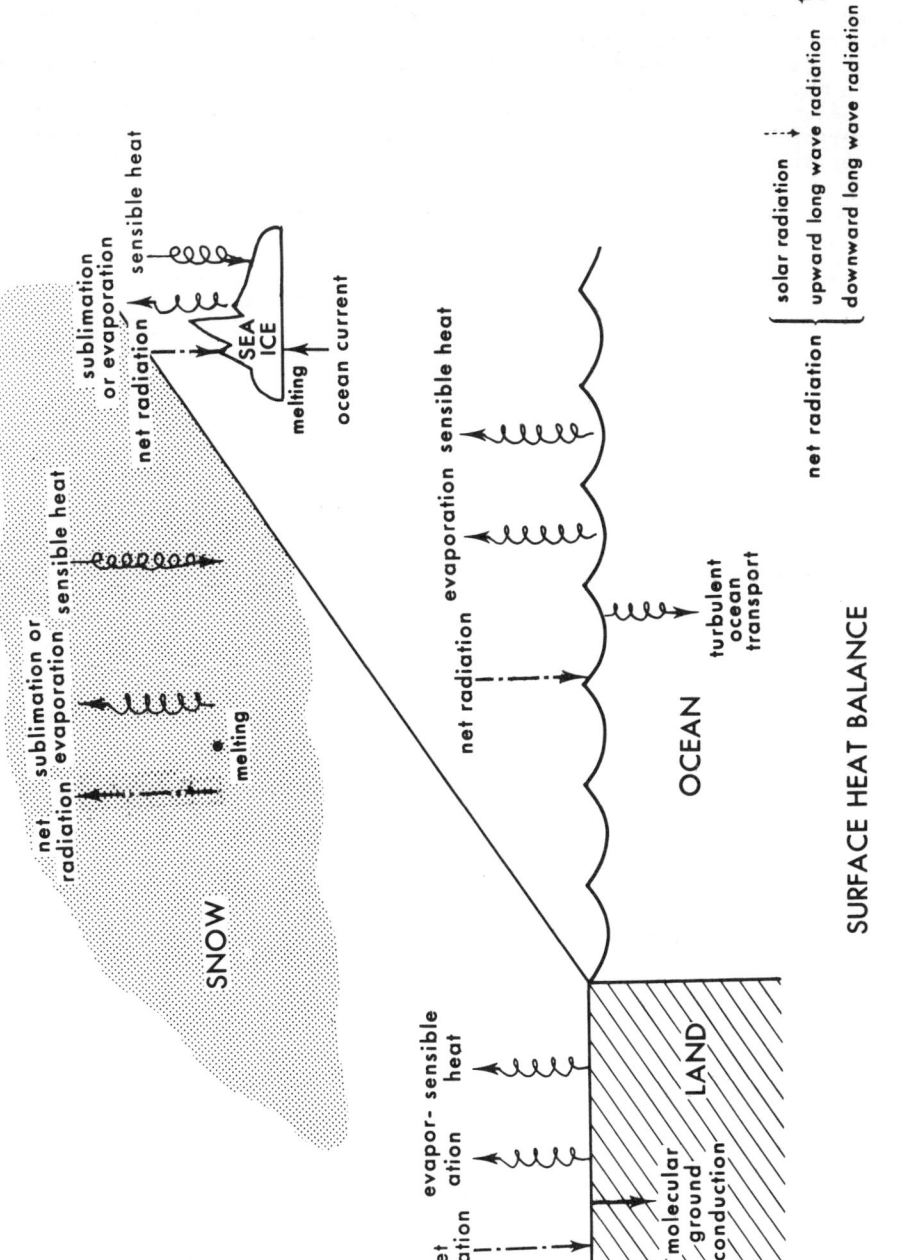

Figure 5 - A schematic of the physical processes entering into the heat balance at four types of earth's surface: bare land, snow-covered land, open sea, and ice-covered sea. (Smagorinsky, 1974.)

(b) an oceanic surface with infinite heat capacity is the other extreme. This then would be an ocean which transmits heat vertically so rapidly (presumably by turbulent processes) that its surface temperature is unaffected by radiative or turbulent heat fluxes at the interface. This is a reasonable approximation for periods of less than 1 or 2 weeks. In this limiting case the sea surface temperature must be specified as an external parameter. Many useful climatic simulation experiments of limited scope have thus been performed by using the observed distribution of sea surface temperature as a boundary condition.

Let us now proceed to the intermediate cases, namely, those of continents with small but finite heat capacity due to molecular conduction, and oceans with large but finite heat capacity because of their vertical and lateral mobility. This then requires a predictive framework for the surface temperature. Effectively, the boundary of the predictive system is moved to some point below the earth's surface.

First, note that the net radiation at the lower boundary is the sum of downward solar radiation and the upward and downward long-wave (or infrared) radiation.

A snowfree land surface has a temperature determined in the manner just described. The main difference for snow cover is that molecular conduction becomes negligible, but the heat that goes into the melting of snow becomes a factor in the heat balance.

The ocean surface heat balance in the presence of sea ice is controlled by the exchange of heat between ocean and atmosphere through molecular conduction in the ice. It is of particular importance that pack ice forms a significant shield which reduces the free interaction of the ocean and atmosphere. Cracks in the sea ice are a complicating factor.

It must already be apparent that to make the above calculations of the heat balance over the variety of boundary surfaces, the water content and snow or ice cover must also be predicted.

<u>Hydrology and cryology</u> (Figure 6). As in the case of the heat balance, one can conceive of limiting cases in the hydrologic interaction of the earth's surface with the atmosphere. In particular, one can assume that the lower boundary provides only an evaporative source and a precipitation sink: (a) for a continental boundary one must specify empirically the soil moisture available for evaporation as a boundary condition; (b) over an ocean boundary, consistent with the corresponding limiting condition for the heat balance, one need only specify the sea surface temperature which determines the amount of water available for evaporation at 100 per cent relative humidity.

Again, the real situation lies somewhere between these two limiting situations and for the intermediate cases one must predict surface variation of water substance.

Over a snowfree continental surface, one can prescribe the evaporation rate as a function of soil moisture, which must be predicted. For various soils one can assume a water storage capacity beyond which the excess of precipitation over evaporation will run off in the watershed determined by the topography, eventually reaching the sea. This runoff will contribute fresh coastal surface water to the oceans which, with the local ocean rain-evaporation imbalance, will determine the local alteration of the surface salinity. Since ocean salinity and temperature determine the density of seawater, the surface hydrologic imbalance contributes to determining the oceanic motion field.

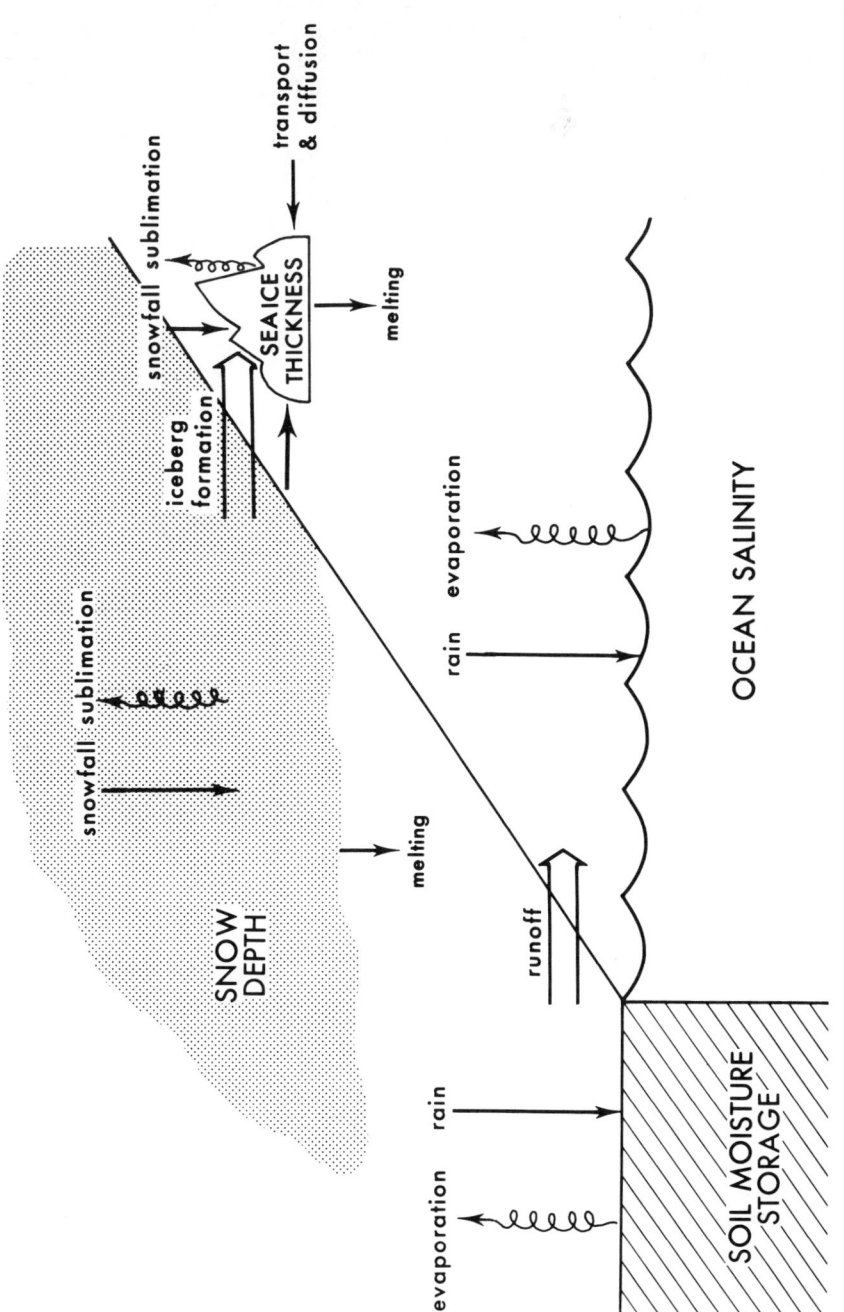

Figure 6 - A schematic of the physical processes entering into the hydrologic balance at four types of earth's surface: bare land, snow-covered land, open sea, and ice-covered sea. (Smagorinsky, 1974.)

Returning to continental surfaces, but now snow covered, the difference between snowfall and sublimation, as modified by snow melt, which alters the local soil moisture, determines the snow depth and, where it goes to zero, the snow perimeter. Icebergs form from coastal breakoffs of continental snow (and ice) cover. As we have seen, sea ice may materially alter the thermal characteristics of the ocean, and so its thickness must be predicted in a general model which one would hope could be applicable to the study of the onset of ice ages, namely, the systematic equatorward growth of the snow perimeter over continents, and of sea ice. The sea ice thickness will change as a net result of snowfall, sublimation, and melting. Where the sea ice is mobile, one must keep track of the ice floe's movements by the upper ocean currents of large scales (transport) and of small scales (diffusion). Melting will also alter the salinity and hence the density stratification of the upper layers of the ocean.

There are, then, some fundamental differences between the interaction of land surfaces with the atmosphere and that of the sea. For example:

- Because the sea is mobile, heat can be exchanged through greater depths as well as laterally. The effectively much larger heat capacity of the sea gives it a much longer time scale and a smaller amplitude of local reaction.

- In the case of the water balance, precipitation over the sea mainly contributes to changing the salinity and hence the ocean's density gradient. Over land it is the soil moisture.

Basic parameters of a model

In any model depending upon the parameterizations incorporated in the formulation - that is, the modeling approximations - a number of empirically determined numbers must be specified (Figure 7).

Planetary data

These are clearly external and assume the planet to be nearly spherical.

- a The main radius.
- g The effective acceleration of gravity, which depends on the mass of the planet, and takes account of the centrifugal acceleration and its effect on the slight oblateness of the planet.
- Ω The angular velocity of the planet will determine the Coriolis forces, that is, the apparent forces action on atmospheric motions relative to the planet. It will also determine the length of day and night, which is important for the way that solar radiation is absorbed by the atmosphere and by the planet's surface.
- Orbital   The geometry of the planet's annual orbit about the sun and the inclination of the planet's axis of rotation to the plane of the ecliptic.
- $S_\infty(\lambda)$ The quantity and quality of the sun's radiation reaching the outer limits of the planetary atmosphere. By quality is meant how much energy is received at different wavelengths, since various gases react differently to radiation which is multichromatic.

Here we assume that we already know the constituents of the atmosphere, by-passing, for the most part, the problem of accounting for its evolution.

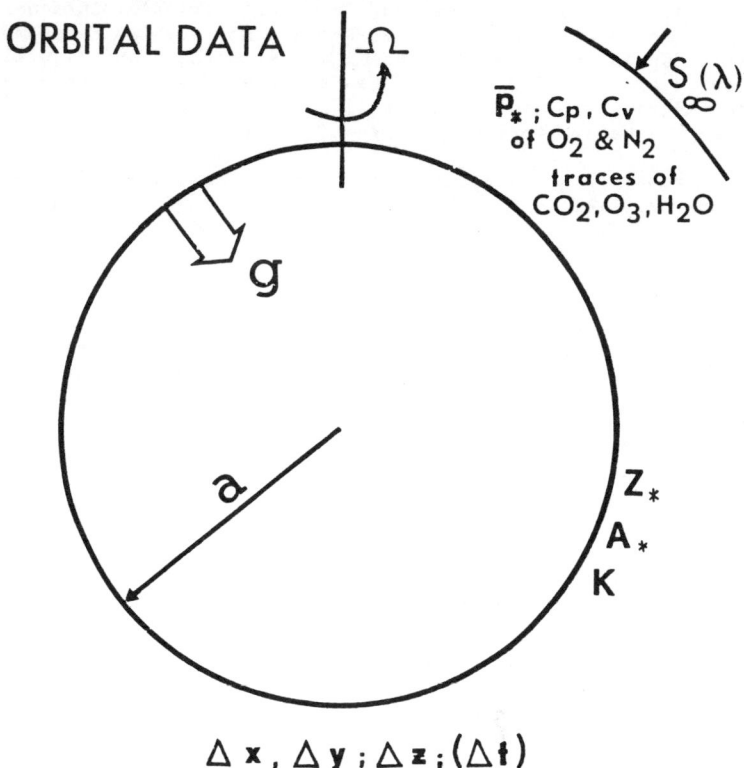

Figure 7 - Some of the basic parameters of a climate model (see text for symbol definitions).

$\bar{p}_*$ The total mass of all gaseous constituents expressed as the average force per unit area of the planet's surface - the mean surface pressure.

c_p and c_v The effective specific heats, at constant pressure and at constant volumes, of the mixture of gases.

The most massive constituents of the planetary atmosphere determine p_*, c_p, and c_v. For the earth, N_2 and O_2 account for about 98 per cent of the total mass (and over 99 per cent of the total volume). However, the trace gases may be quite important for their radiative impact. For our planet, the radiatively active gases are CO_2, O_3 and H_2O as are particulates such as clouds, haze, and dust. The other trace gases that have been found in our atmosphere - argon, neon, helium, and krypton - are relatively inert radiatively, just as are N_2 and O_2.

Boundary properties that depend on internal variations

Examples are the surface roughness (z_*), the reflectivity or albedo (A_*), and the thermal conductivity of the earth's surface (K_*). In the first approximation one can, as has been the custom in simple models, specify an average value determined from observed data. In the limit one can, for short-period evolutions, ignore them entirely, but not for problems of climatic scope. For sophisticated models z_* for a sea surface will depend on the wind in the lower atmosphere and A_* will be materially altered by snow cover. K_* will vary for sea because of wind stirring and even on land will depend on the water content of the soil which is the net result of precipitation, evaporation, and the water storage capacity of the soil.

I should remark at this point that there are relatively few free parameters available for prescription. It might appear that the more comprehensive the model the greater the number of parameters. On the other hand, one should also expect that as the model is more fundamentally formulated, the fewer the parameters available to be tweaked.

For example, in the case of simplified 2-dimensional climate models, the reduction in dimensionality requires a parameterization of the transport properties of north-south atmospheric motions resulting from east-west variations.

Under any circumstances, good modeling practice demands an awareness of the sensitivity of the model simulations to the choice of each parameterization, including of course, the numerical constants that are introduced. This is not always simple to do. Because of the inherent non-linearities of complex climate models, the nature of the sensitivities can be quite subtle.

It is often a problem in devising parameterizations to be certain that a closure scheme should in principle exist, and under what conditions.

Furthermore, there are still some instances where hardly any progress has been made. For example, many line phenomena defy attempts to represent their large-scale effects in terms of the macroscale variables: the effects of coasts and steep mountains (especially in the presence of irreversible processes involving water phase changes) and heat transfer through leads in sea ice.

In conclusion, let me now cite some points made in the report "Parameterization of Sub-grid Scale Processes " (GPS 8).

One of the most important and difficult problems in large-scale modeling is to identify which scale interactions are of importance and then to determine the closure forms - the parameterizations themselves. These are not unique and can be approached in a variety of ways.

Successful parameterization requires a sequence of steps:

1. The process involved must be precisely identified and its role distinguished meaningfully from other processes which may be simultaneous.

2. Its importance for the explicit scale must be determined.

3. An intensive study of individual cases must establish the fact that the relevant physics and dynamics are adequately understood.

4. Quantitative rules must be found for expressing the frequency of occurrence and intensity of the sub-grid scale process in terms of the explicit scale variables.

5. Rules for determining the grid scale average of the associated transports of mass, momentum, heat and moisture must be inferred from theory and verified by direct observation.

6. These rules must be translated as accurately as required into practical algorithms for specific numerical models, care being taken to ensure that all important processes are included but that duplication is avoided.

All these steps are essential and are to some degree interdependent. Thus the meaningful isolation for study of a possible transfer process (e.g., clear air turbulence) pre-supposes that the factors controlling its occurrence are embedded in the synoptic scale variables rather than in other intermediate scale phenomena such as gravity waves or frontal surfaces, i.e., that the fourth step is in principle possible. Also, judgements of the importance of some processes may perforce remain fairly intuitive until at least a crude parameterization has actually been tested in a general circulation model, and comparative studies made with it, included and excluded. Case studies must be oriented towards the over-all goal and must not degenerate into the pursuit of detail for its own sake unless fundamentals are at stake. For each step different expertise is required and several specialists must co-operate to achieve the over-all solution with an approach which is at the same time broad and flexible yet precisely planned and executed.

Probably the most difficult and critical step is the first - the precise definition of the problem to be tackled in a tractable form in which the relevant variables are identified and the less important excluded. Also, several different processes may be present simultaneously (e.g., gravity waves and moist convection) and may interact among themselves. Judgement is needed as to whether - in so far as they affect the synoptic scale motion - such interactions are secondary (so that the waves and convection may be parameterized separately) or whether they are so strong that both effects must be considered together. Problems also arise when a continuum of scales is involved. For example, mountains are important both directly on synoptic scales and indirectly on smaller scales because of locally induced precipitation, waves and surface friction. Care is needed to ensure a rational division in the model between those parts of the transfers which are on a large enough scale to be resolved directly in the finite difference scheme and those which are accounted for by parameterization. Otherwise redundancies or complete omissions will result.

In most cases observational programmes are an essential element for gathering fundamental data upon which to base parameterization hypotheses, to calibrate them and to test them independently. Also required in all cases are large-scale models which can simulate the parameterized interaction to be judged against hypothetical or actual initial and verifying conditions.

Another numerical aid comes from explicit small-scale models for the specific phenomenon itself, such as for cumulus convection or fronts, which should prove an invaluable aid in many cases for providing parameterization insights.

In conclusion, although a classification and sub-division of the various subgrid scale processes to be parameterized is clearly essential if progress is to be made, it is vital that such a sub-division should not be arrived at hastily and should be reviewed continuously as our knowledge increases.

SIMULATION OF CLIMATE BY GENERAL CIRCULATION MODELS
WITH HYDROLOGIC CYCLES

S. Manabe

Geophysical Fluid Dynamics Laboratory/NOAA
Princeton University
Princeton, New Jersey 08540

Abstract

This paper describes the basic structure of a typical general circulation model of the atmosphere and that of the joint ocean atmosphere system. It discusses some alternate approaches to the construction of key components of a model. In addition, the present state of the art in climate simulation is discussed with reference to the climates produced by some of the models developed at the Geophysical Fluid Dynamics Laboratory. Special emphasis is placed upon the description of the hydrologic aspect of the simulation. Finally, the future strategies for model improvement are suggested.

1. Introduction.

The study of climate with a general circulation model of the atmosphere began with the pioneering work of Phillips (1956). Using a simple 2-layer weather prediction model of the atmosphere and assuming a latitudinal distribution of zonally uniform heating, he succeeded in simulating the jet stream and the zonal mean cells of the meridional circulation (i.e., Hadley cell, Ferrel cell, and polar cell). His model, however, was not suitable for dealing with the hydrologic cycle because the quasi-geostrophic approximation used in his model is not valid in regions of strong condensational heating, particularly in low latitudes.

Smagorinsky (1963) succeeded in making a long-term integration of a general circulation model of the atmosphere with the so-called primitive equations of motion. Thus, he was able to avoid the use of the quasi-geostrophic approximation and opened the way for the incorporation of the hydrologic cycle into an atmospheric model.

Mintz (1965, 1968) constructed a general circulation model of the atmosphere with a global computational domain and realistic geography. He succeeded in simulating gross features of the horizontal distribution of sea-level pressure.

Manabe et al. (1965) incorporated the processes of the hydrologic cycle (i.e., evaporation, moisture transport and precipitation) into their hemispheric model with a uniformly wet surface. They succeeded in reproducing the zonal mean features of precipitation such as the tropical rainbelt, the subtropical belt of meager precipitation and the middle latitude rainbelt.

The process of surface hydrology was taken into account by Manabe (1969) when he investigated the climatic influence of the land-sea contrast in wetness by use of a model with limited computational domain and idealized geography.

Recently, many investigators have incorporated the hydrologic processes into their general circulation models with a global computational domain and realistic geography. (Holloway and Manabe, 1971: Kasahara and Washington, 1971: Arakawa, 1972: Gates and Schlesinger, 1977: Somerville et al., 1974: Washington and Williamson, 1977: Corby et al., 1977). They succeeded in simulating many large-scale characteristics in the geographical distributions of climate and hydrologic variables.

Manabe and Holloway (1975) extended the time integration of their global model with hydrologic cycle beyond one annual cycle and succeeded in reproducing the gross features of the seasonal variations in the geographical distributions of hydrologic variables i.e., rates of precipitation, evaporation, and run off.

Some of the most recent results of climate simulation were presented at the JOC Study Conference on Climate Models. It is recommended that the reader refer to the proceeding of the conference (GARP, 1979) for these results.

One of the motivations for the development of a general circulation model of the atmosphere is the usage of the model for the study of climate and its variation. For example, Mintz (1981) discussed the sensitivity of climate to the wetness of soil based upon the results from various numerical experiments. Manabe and Stouffer (1980) used an ocean-atmosphere model for the evaluation of the climate changes resulting from the future anthropogenic increase of atmospheric CO_2-concentration. Recently, Charney and Shukla (1980) made a preliminary attempt to investigate the predictability of climate variation with intermonthly time scale by use of the results from a general circulation model of the atmosphere.

In order to obtain a reliable result from numerical experiments with a general circulation model of the atmosphere, it is necessary to confirm that the model behaves realistically. The similitude of model climates to the actual climate is the topic of the extensive discussion in this review paper.

3. Model Construction.

This section describes the basic structure of a general circulation model of the atmosphere. In addition, it briefly describes various approaches to constructing key components of a model. Refer to the book edited by Chang (1977) and the GARP Publication No. 14 (JOC, 1974) for detailed descriptions of the structures of various models.

As indicated by Fig. 1, a general circulation model of the atmosphere contains prognostic equations for horizontal momentum components, surface pressure, temperature and water vapor. These are the equations of motion, the equation of mass continuity, the thermodynamical equation and the continuity equation for water vapor. In addition, temperature and wetness of continental surfaces are computed from the budget requirements of heat and moisture, respectively.

a. Basic prognostic equations.

Using the so-called hydrostatic approximation, one can write the equations of motion in the form in which only horizontal wind components are prognostic variables. The vertical component may be computed from the requirement of mass-continuity. As suggested by Phillips (1957), the vertical coordinate is usually chosen to be sigma, i.e., pressure normalized by surface pressure. This sigma-coordinate system is introduced because it enables one to incorporate the dynamical effect of an uneven lower boundary surface (i.e., mountains) in a straightforward manner. The usage of the hydrostatic approximation mentioned above is justified because the horizontal scale of large scale atmospheric disturbances is much larger than the scale-height of the atmosphere.

The equations of motion explicitly compute the change of momentum resulting from the large scale flow field, whereas it evaluates the

Model Equations

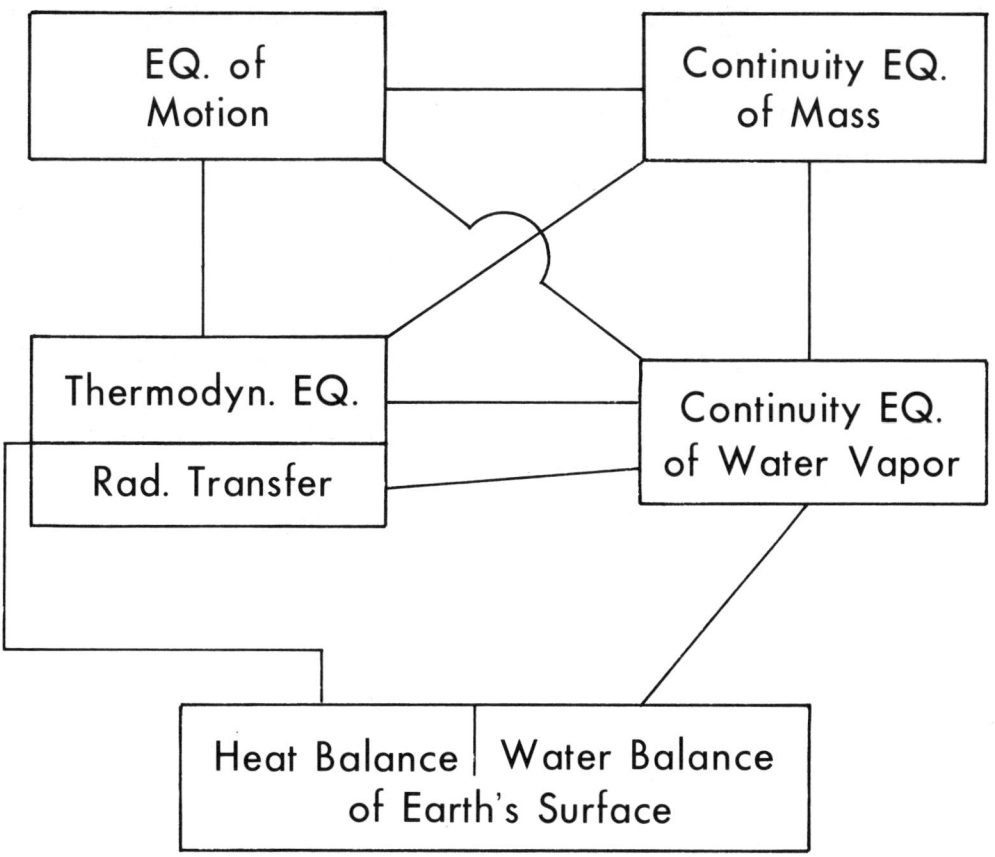

Figure 1 - Box diagram of a general circulation model of the atmosphere.

momentum change which is due to subgrid scale eddies from a parameterization based upon a closure hypothesis (e.g., Smagorinsky, 1963: Leith, 1969).

The thermodynamical equation computes the rate of temperature change which is due to the three dimensional heat transports by the large- scale flow and subgrid scale eddies, adiabatic compression, radiative heating or cooling, dry and moist convection, and heat released by convective and non-convective condensation.

The continuity equation of water vapor predicts the rate of change of the mixing ratio of water vapor in air, taking into consideration the three-dimensional advection of moisture by large-scale flow, mixing by subgrid scale eddies and by moist and dry convection, and convective and non-convective condensation.

b. Radiative transfer.

The fluxes of solar and terrestrial radiation are usually computed separately by exploiting the spectral separation of these two kinds of radiations. The heating due to the convergence of these radiative fluxes accounts for a part of the heat source term in the thermodynamical equation.

The upward and downward fluxes of terrestrial radiation are computed from the equation of radiative transfer by taking into consideration the selective absorption and emission by some atmospheric gases and cloud cover. Some of the most refined schemes suitable for incorporation into an atmospheric model have been developed by Rodgers and Walshaw (1966), Kaplan (1959) and Fels and Schwarzkopf (1975). The schemes of intermediate complexity, which are essentially equivalent to the so-called radiation diagram method (e.g., Möller, 1943; Elsasser, 1942; and Yamamoto, 1952) have been constructed, for example, by Manabe and Strickler (1964),

Sasamori (1968) and Rodgers (1967). On the other hand, Takahashi et al. (1960) and Smagorinsky (1963) developed some of the simplest schemes incorporated into atmospheric models. In these schemes, the fluxes of terrestrial radiation are determined as a function of temperature at a few selected levels in the atmosphere.

For the computation of the fluxes of solar radiation, it is desirable to take into consideration the absorption by atmospheric gases, reflection and absorption by cloud cover, and the scattering of radiation by air molecules and aerosols. Most of the schemes incorporated into atmospheric models deal with all of these effects with the exception of the scattering by aerosols. One of the most refined schemes developed so far is the algorithm which was proposed by Lacis and Hansen (1974) and incorporated into the general circulation model developed by Sommerville et al. (1974) at the Goddard Institute for Space Studies. By use of a multiple scattering computation technique, the influence of cloud cover is evaluated in a very straightforward manner.

The incoming solar radiation at the top of a model atmosphere is prescribed as a function of latitude and season. The gaseous absorbers, which are usually considered in the computations of both solar and terrestrial fluxes, are water vapor, carbon dioxide and ozone. The distribution of water vapor is determined by the prognostic system of water vapor. Most models assume a prescribed distribution of ozone although a few models predict ozone concentration from the equations of photochemical processes (Hunt, 1969; Cunnold et al., 1975). The mixing ratio of carbon dioxide in air is assumed to be constant everywhere. Some models predict the distribution of cloud cover referring to a relevant atmospheric variable such as relative humidity (Kasahara and Washington, 1971; Sommerville et al., 1974; Manabe and Wetherald, 1980a;

Wetherald and Manabe, 1980). Other models use a prescribed distribution of cloud (Manabe et al., 1965; Corby et al., 1977).

c.   Moist convection and condensation.

The moist convection process is responsible for the upward transports of heat and moisture, condensation of water vapor and the release of latent heat. Therefore, its effects are taken into consideration in the formulations of both the continuity equation of water vapor and the thermodynamical equation. It plays a major role in determining the static stability and vertical moisture distribution in the atmosphere, particularly in lower latitudes and accordingly has strong influence upon the intensities of the atmospheric circulation and the hydrologic cycle. The development of the parameterization representing the moist convective process has been one of the most difficult tasks in the construction of general circulation models of the atmosphere.

Atmospheric modelers have developed many parameterizations with various degrees of complexity. One of the simplest parameterization schemes proposed is the so-called moist convective adjustment which involves the adjustment of static stability and relative humidity when certain conditions for free moist convection are satisfied (Manabe et al., 1965). Recently, Arakawa and Shubert (1974) formulated a highly sophisticated parameterization scheme of moist convection. For this purpose, they considered the physical and dynamical structure of a moist convective ensemble and introduced a closure hypothesis on the statistical dynamics of cumulus convection. A scheme of intermediate complexity was proposed by Kuo (1965) and has been extensively used by many modelers after some modifications.

One can test the performance of various parameterization schemes by incorporating them into an atmospheric model and evaluating the climate as simulated by the model. So far, very few tests of this kind have been conducted. Refer to Miyakoda and Sirutus (1977) for the results from their preliminary evaluation.

In addition to convective precipitation, a model predicts non-convective precipitation when the value of relative humidity becomes supercritical. The heat thus released from condensation becomes a part of the heat source term in the thermodynamical equation.

Both convective and non-convective precipitations are usually apportioned into rainfall and snowfall depending upon the temperature distribution near the earth's surface in the model atmosphere.

d. Mixing through boundary layer.

Some of the key processes in an atmospheric circulation model are the exchanges of heat, moisture and momentum between the earth's surface and the free atmosphere through the planetary boundary layer. Although many schemes parameterizing the boundary layer exchanges have been proposed, one can classify them into two broad categories. The schemes, which belong to the first category, have been proposed, for example, by Lettau (1959), Blackadar (1962), and Monin and Zilitinkievich (1967). They developed the similarity approach in which the drag coefficient and the bulk heat transfer coefficient depend upon the bulk Richardson number defined for the entire planetary boundary layer. Thus, it is not necessary to have many finite difference levels inside the planetary boundary layer. This approach has been adopted by Arakawa (1972), Corby et al., (1972), Washington et al., (1979) and Deardorf (1972) in constructing their atmospheric models.

The schemes of the second category predict explicitly the temporal variations of the structure of the planetary boundary layer resulting from the interaction between turbulence and the boundary layer environment. This approach has been proposed, for example, by Rossby and Montgomery (1935) and has been incorporated in GFDL models. Recently, Mellor and Yamada (1974) developed a hierarchy of turbulence closure models of the planetary boundary layer with various degrees of complexity and discussed the performance of each version of their model. Their study enables atmospheric modellers to choose a scheme having optimum complexity depending upon the specific objective of their investigation. One significant disadvantage of the schemes of this category is that they require at least several finite-difference levels in the planetary boundary layer to achieve tolerable computational accuracy. On the other hand, this approach enables a model to resolve the vertical distribution of the large scale flow in the planetary boundary layer explicitly. One can appreciate the advantage of this approach if one recalls that the major portion of the meridional moisture transport by the tropical Hadley cell occurs _inside_ the planetary boundary layer. (Also, note that both the speed of mean flow and moisture content of air change markedly with height in the planetary boundary layer.)

e. Surface heat budget.

The heat budget of continental surfaces is maintained among the following components: net fluxes of solar and terrestrial radiation, turbulence-fluxes of sensible and latent heat, and conductive heat flux into (or out of) the soil. With the exception of the net flux of solar radiation, all other fluxes depend upon the surface temperature. In order to satisfy the requirement of the heat balance at the earth's surface, the sum of all contributions from the fluxes identified above should be zero. It is therefore possible to determine the temperature of continental surfaces by use of this heat balance condition. For the sake of simplicity, some modellers (i.e., Manabe et al., 1965) neglect the conductive heat flux into the soil assuming that the heat capacity of continental surfaces is equal to zero. Since this assumption naturally results in an exaggeration of the amplitude of the diurnal surface temperature variation, it is desirable to remove diurnal variation from the insolation imposed at the top of the model atmosphere when this assumption is made.

A more straightforward method for the prediction of surface temperature includes a numerical time integration of the equation of heat conduction in soil with a boundary condition of heat flux at the soil surface. Many atmospheric modellers have simplified this procedure by introducing a parameter, i.e., equivalent heat capacity of soil for diurnal temperature variation (Arakawa, 1972; Bhumralkar, 1975). For further details of this approach, see the review articles by Carson (1981) and Eagleson (1981).

Over the oceanic regions, most models have the observed distribution of sea-surface temperature as a lower boundary condition. For a description of the models, in which sea surface temperature is determined from the interaction between ocean and atmosphere, see Section 4.

f. Surface hydrology.

It is well known that soil moisture has a profound effect on the climate by enhancing the ventilation of the earth's surface through evaporation. In most general circulation models of the atmosphere, soil moisture is predicted by use of a simple box method. This approach is designed to incorporate qualitatively the most critical features of soil behavior without requiring the detailed information demanded by a more rigorous formulation, such as those of Philips and de Vries (1952), and Philip (1957) and Sasamori (1970). In this box method, a box represents a moisture holding capacity of soil. The change of water amount contained in the box is computed as a residual of contributions from processes that increase soil moisture (i.e., rainfall and snowmelt) and those that decrease it (i.e., evaporation and runoff). In some models (i.e., Manabe, 1969; and Washington and Williamson, 1977), runoff is predicted at a grid point only if a computed change in water in a soil-box would result in a water depth exceeding the total depth of the box i.e., field capacity of soil. Other modellers use somewhat different methods. See the review paper by Carson (1981) for further details of these methods.

The effect of soil moisture on evaporation is incorporated into a model by a simple scheme used by Budyko (1956) and Thorthwarte and Mather (1955). When the soil does not contain a sufficient amount of water, it is assumed that evaporation rate is smaller than that of perfectly wet surface, i.e., potential evaporation rate. If the soil moisture is greater than a certain critical percentage of the maximum soil capacity, the evaporation rate is assumed to be equal the maximum rate. Otherwise, the rate of evaporation from land is computed to be a monotonically increasing function of soil moisture up to this critical value. (See Eagleson (1978) for the evaluation of this relationship.)

The rate of change of water equivalent depth of snow is computed as the residual of the rate of snowfall minus the sum of snowmelt and sublimation rates. Precipitation, which is deduced from the prognostic system of water vapor described earlier, is regarded as snowfall if the temperature near the earth's surface is below freezing. The snowmelt rate is calculated from a surface heat budget under the assumption that the temperature of the snow surface does not exceed the freezing point and that the conductivity of snow is zero. The rate of sublimation from a surface of snow or ice is computed under the assumption of water vapor saturation at the snow surface. The method of snow prediction described here is incorporated into a general circulation model of Manabe (1969). Most other models have essentially similar schemes of snow prediction.

g.   Numerical computation.

Each term on the right hand side of the prognostic equation identified in the subsection "a" are usually computed by use of the finite-difference technique. The finite-difference equivalents of these equations are formulated such that the mass-integrals of the prognostic variables (i.e., wind components, temperature) and their second order moments are unaltered by the contribution of the non-linear advection term in these equations. This is done in order to prevent a systematic error in the numerical computation and to prevent the so-called non-linear instability discussed by Phillips (1957 and Arakawa (1966). The horizontal grid sizes of various models range from 200 to 600 km. In the vertical direction, several finite-difference levels are chosen such that the thermal and dynamical structures of the troposphere and lower stratosphere is properly resolved by the finite-difference computation.

An attractive alternative to a finite-difference model described above is a spectral model of the atmosphere. In a spectral model, the prognostic variables are represented by a finite sum of spherical harmonics rather than by values at discrete points. The model equations predict spectral components rather than grid point values. Since the horizontal derivatives in the prognostic equations are computed analytically, this approach usually yields high computational accuracy. The shortcoming of the spectral approach as originally proposed by Platzman and Baers, (1960) was that the computer time required for evaluating the non-linear advection terms increases rapidly with increasing spectral resolution of the model. To avoid this difficulty, Orsag (1970) and Eliassen et al. (1970) recently proposed a spectral method with the so-called transform technique. From the spectral representation of the fields of relevant variables, non-linear terms are evaluated on grid points and the results of these computations are transformed back into the spectral domains. This transform technique makes it unnecessary to evaluate a large number of the non-linear, triad interaction terms and markedly reduces the computational requirement of a spectral model with relatively high resolution. Because of their computational accuracy, many groups have recently developed spectral atmospheric models (Bourke, 1974; Hoskins and Simmons, 1975 and Gordon, 1974). The performance of spectral models in simulating climate is competitive with grid models (See, for example, Bourke et al., 1977 and Manabe et al., 1979).

h. Time integration.

A climate is obtained as the statistically stable state of the model atmosphere which emerges from the time integrations of the prognostic equations of the model over the period of several hundred days. The initial condition for a time integration is usually an isothermal and dry atmosphere at rest. As the time integration proceeds, a model atmosphere gradually adjusts to the underlying sea-surface temperature and the insolation imposed at the top of the model atmosphere and transforms into a seasonally varying climate such as that described in the following section.

3. Climate of an Atmospheric Model.

To convey a general impression on the present state of the art in climate modeling, this section presents a climate as simulated by a general circulation model of the atmosphere developed at the Geophysical Fluid Dynamics Laboratory of NOAA. The model was developed by Manabe and Holloway several years ago (Holloway and Manabe, 1971 and Manabe and Holloway, 1975). It is a finite-difference model with grid size of approximately 250 km and has a global computational domain and realistic geography. The insolation, which varies seasonally and latitudinally but not diurnally, is prescribed at the top of the model atmosphere. The model assumes the observed distribution of sea-surface temperature which varies with respect to season. The simulated climate described below was obtained from a time integration of the model over a period of about 3 model years. For specific details of the model structure, refer to the paper by Manabe and Holloway (1975).

Fig. 2 illustrates both the simulated and observed monthly mean flow fields near the earth's surface for January and July. This figure clearly indicates that the model successfully reproduces large-scale features of the surface flow field. For example, it simulates reasonably well the seasonal and longitudinal variation in the position of the intertropical convergence zone, the seasonal reversal in the direction of the South Asian Monsoon flow and the winter outbreak of cold air along the east coast of the Eurasian continent extending all the way to Indonesia. Some of the failures of the simulation include the unrealistically weak and narrow oceanic anticyclone and the excessive development of cyclonic vortices along the east coast of the Eurasian continent in the July distribution of the flow field.

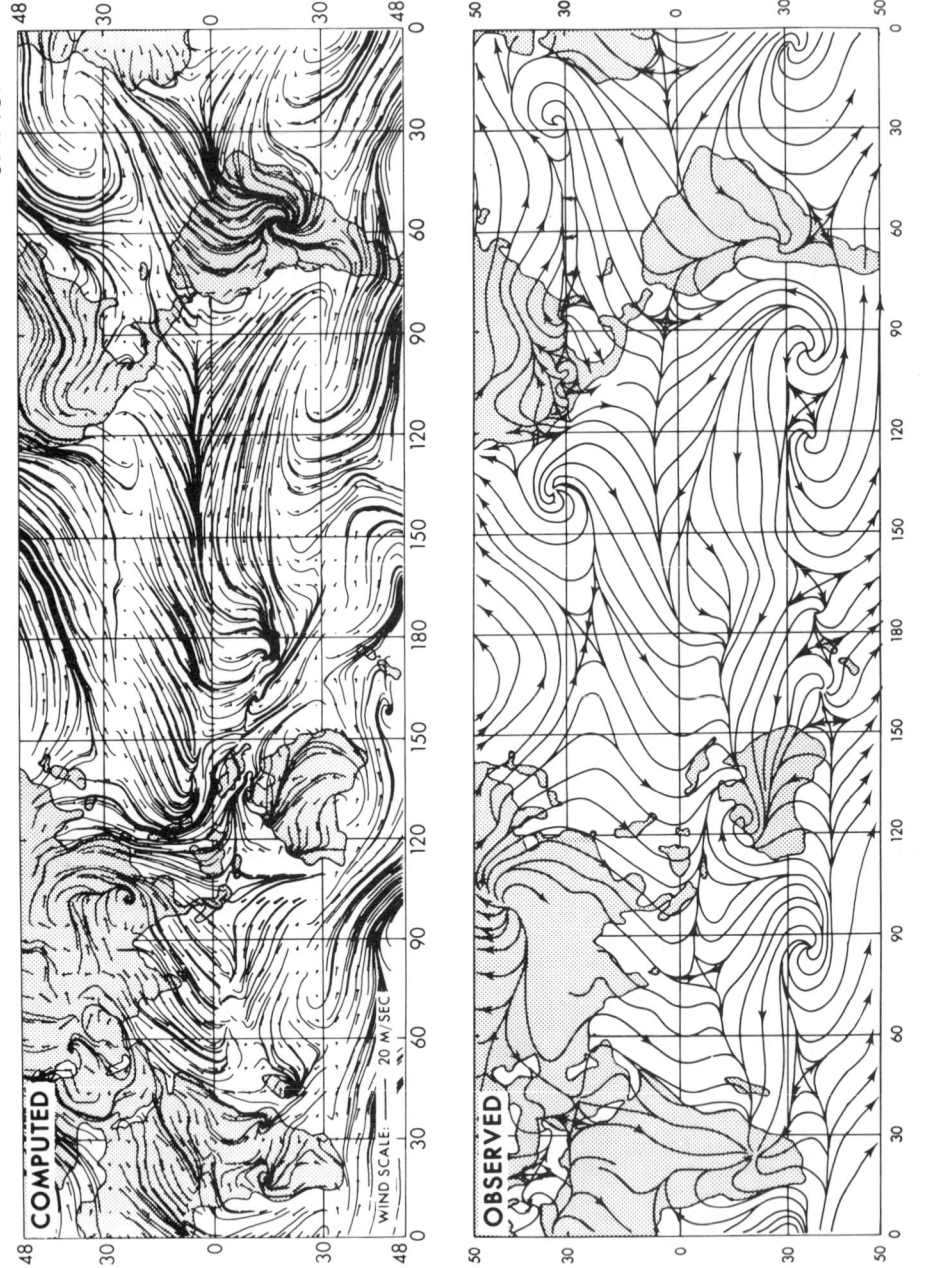

Figure 2a - Top: One-month mean wind vectors and streamlines near the earth's surface of the global model in January. Bottom: Observed one-month mean streamlines near the earth's surface in January (Mintz and Dean, 1952).

-35-

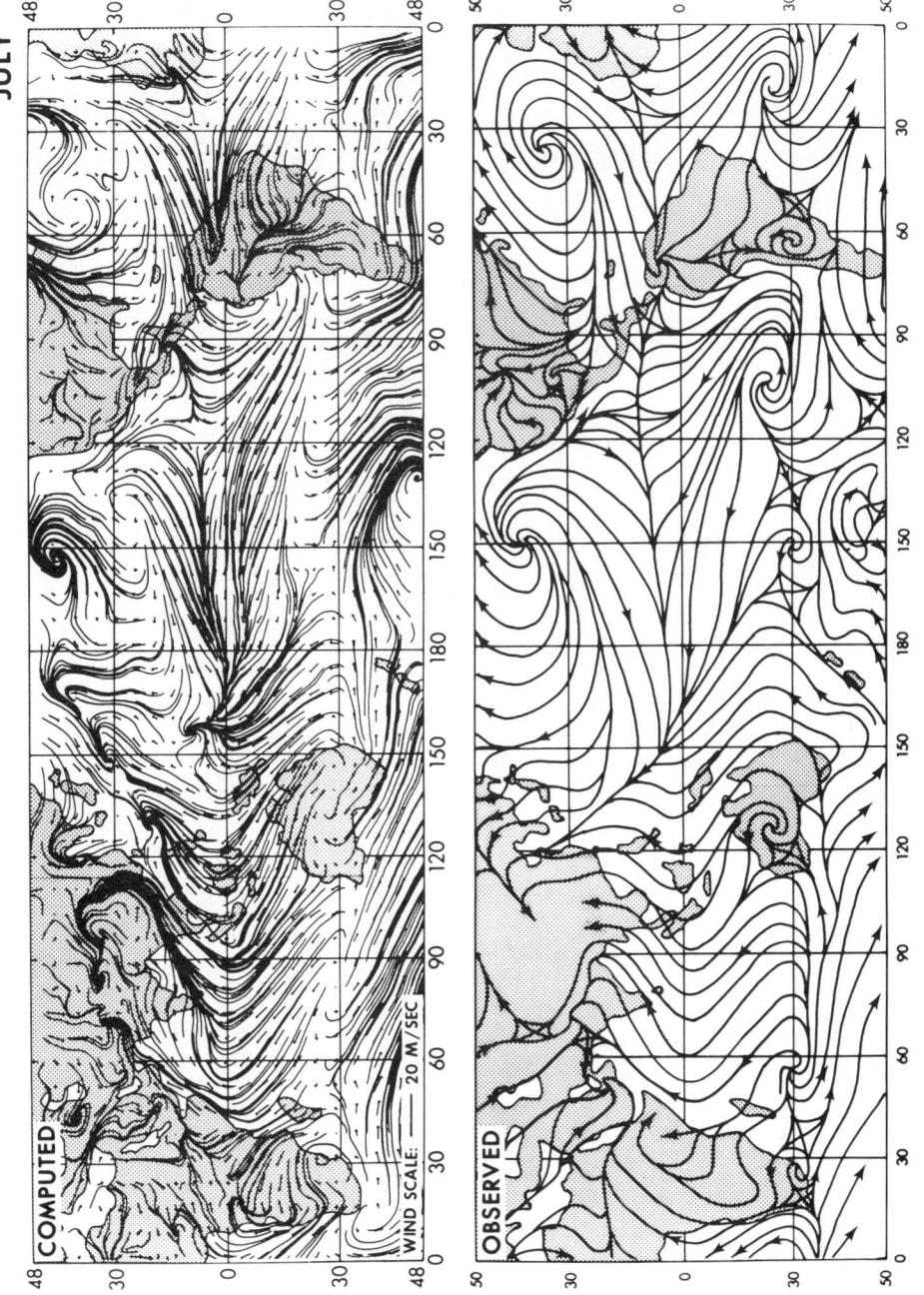

Figure 2b - Same as 2a but for July.

In Fig. 3, the global distributions of precipitation rate derived from the model simulation is compared with the corresponding distributions of the observed rate for two seasons of the year. This figure reveals that the model reproduces very well the seasonal movement of the tropical rainbelt. This result appears to be consistent with the successful simulation of the seasonal excursion of the intertropical convergence zone.

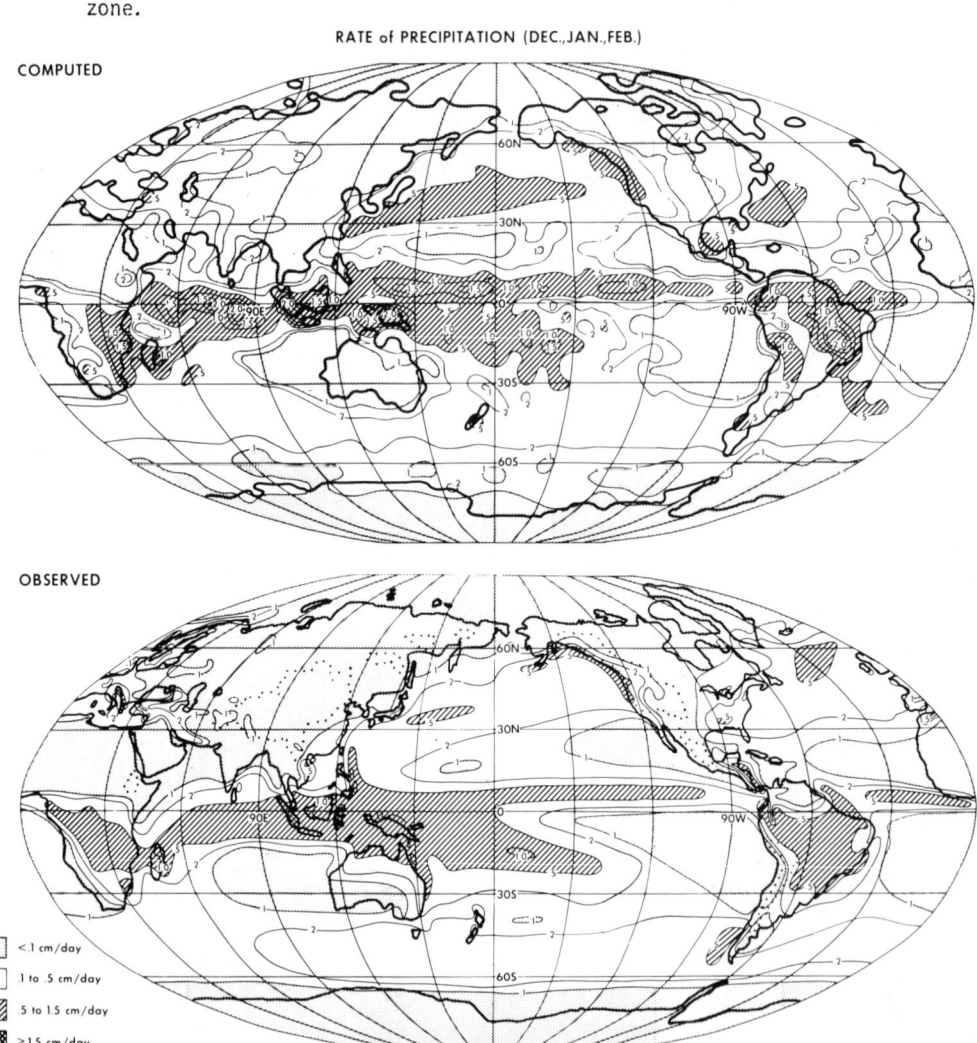

Figure 3a - Global distribution of the mean rate of precipitation (cm day^{-1}) computed by the global model (top) compared with that estimated by Möller (1951) from observed data (bottom) for December, January and February.

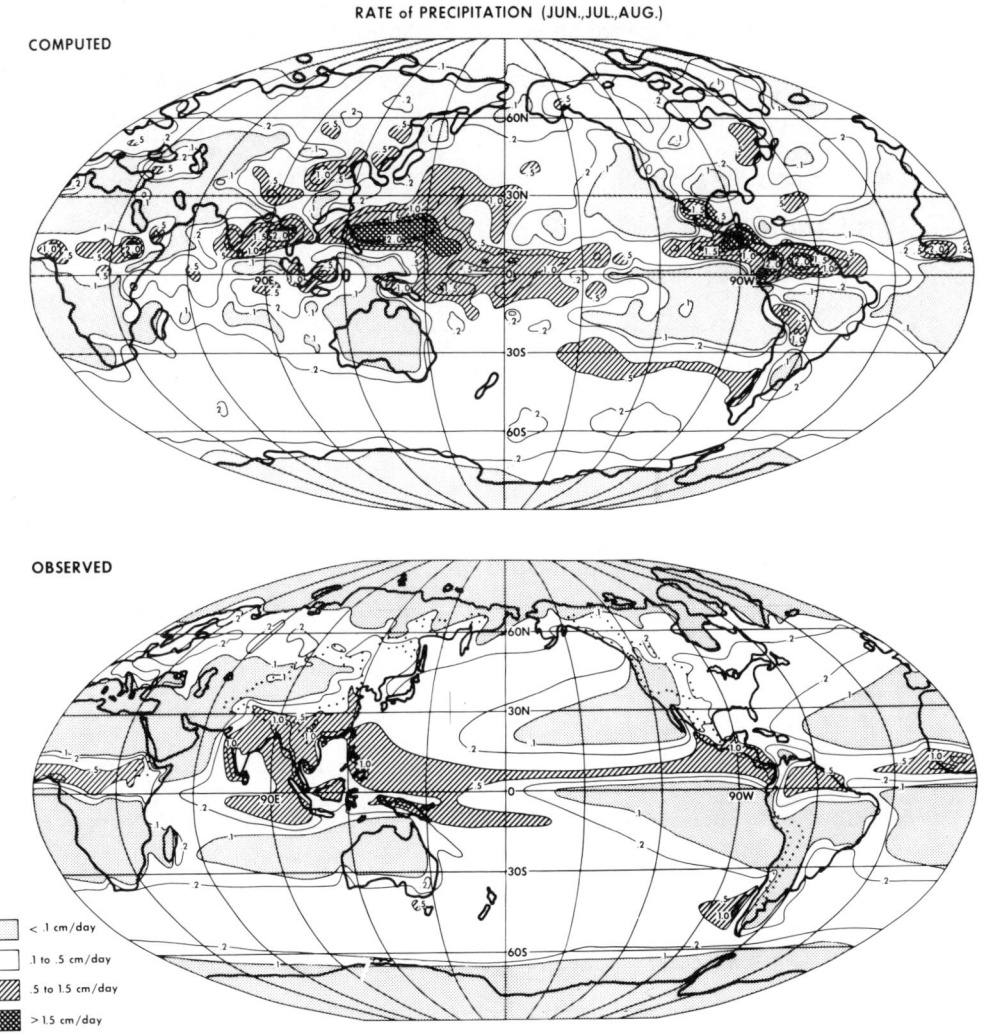

Figure 3b - Same as Figure 3a but for June, July and August.

In Australia, the Sahara, and Central Asia of the model, the rate of precipitation is small throughout the year in agreement with the features of the observed precipitation rate. In general, the seasonal variation of the precipitation rate over Africa, Eurasia and North America is simulated well although the precipitation rate over Texas and Southwest Asia is grossly underestimated during the summer season.

It is of interest to compare the distributions of the simulated precipitation rate described above with those of eddy kinetic energy at the 850 mb-level of the model atmosphere shown in Fig. 4. According to this comparison, the precipitation rate tends to be higher where eddy kinetic energy is greater. For example, in January, both the precipitation rate and eddy kinetic energy are relatively large along the east coast of the North American continent and over the Pacific belt extending from the east coast of the Eurasian continent to the northwest coast of the North American continent. In July, both of these quantities are large over the zonal belt located in middle latitudes of the Southern Hemisphere. The correspondence between the distribution of these two quantities is indicative of the effectiveness of large-scale eddies in inducing precipitation in the model atmosphere. Over the model tropics, the region of large eddy kinetic energy again coincides with tropical rainbelts indicating the mutually-enhancing relationship between heat of condensation and tropical disturbances. It is encouraging that the horizontal distribution of eddy kinetic energy in the model atmosphere described above well resembles the corresponding distribution in the actual atmosphere.

The geographical distribution of the annual mean rate of runoff from the model is compared in Fig. 5 with the distribution of the actual runoff rate estimated by Lvovitch and Obtchinikov (1964). In this comparison, the regions of large runoff rate correspond to the areas of heavy precipitation. For example, the model computes a high rate of runoff over the tropical portions of Africa and South America, Indonesia, and the northern part of the west coast of North America in qualitative agreement with the characteristics of the observed distribution. It, however, fails to simulate the regions of abundant runoff over South China and Northern India. Similar underestimation is notable in the computed distribution of precipitation rate described earlier.

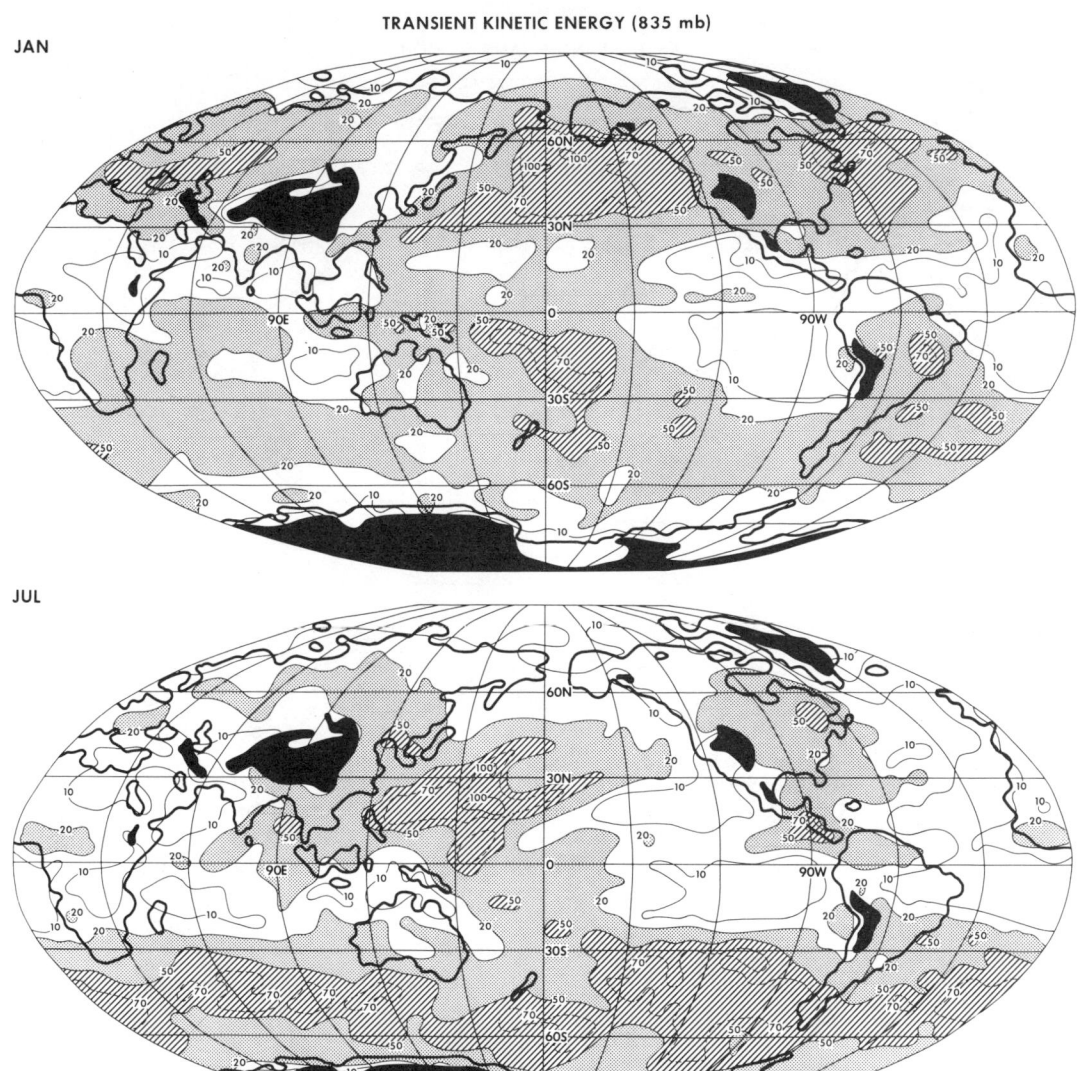

Figure 4 - Geographical distribution of eddy kinetic energy of the model-simulated transient eddies at the 850 mb-level in January (top) and July (bottom). Units are in square meters per second.

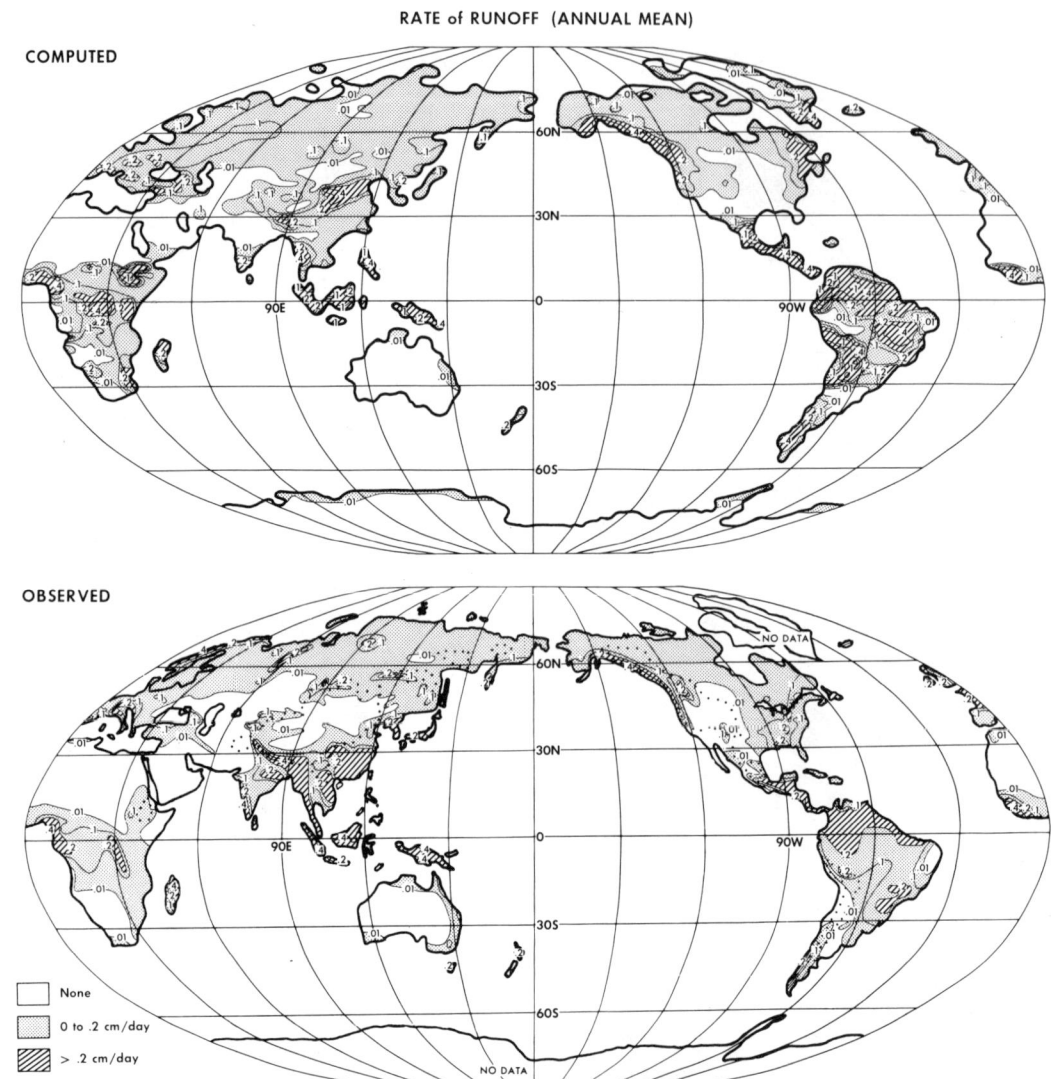

Figure 5 - Global distribution of annual mean rate of runoff (cm/day) simulated by the global model (top) compared with an observed distribution (bottom) based on data derived from Levovitch and Obtchinikov (1964). Small-scale features of contours are removed subjectively by the present author.

Most of the regions specified by the model as having no significant runoff approximately coincide with the areas of meager precipitation. For example, runoff is completely missing from the regions of the Sahara desert and the desert of the Arabian peninsula, as well as from most of Australia. On the other hand, the large areas of no runoff in the deserts and steppes of Central Asia are significantly narrowed in the model results.

The model succeeds in reproducing moderately large runoff over Canada, Siberia and Europe. Although the rate of precipitation is not very large in these high latitude regions, the rate of evaporation (or sublimation) is very small because of the weak insolation available.

It is of interest that the runoff is very small in the region of the Gobi Desert which is situated in middle latitudes instead of in the subtropics where most of the dry regions are located. The results from recent numerical experiments indicate that the Tibetan Plateau plays a major role in maintaining the arid region in Central Asia. For example, Mintz (1965) and Manabe and Terpstra (1974) have noted that, in models without mountains, the Siberian high is weak and is located in the southern part of the Eurasian continent in winter. Accordingly, a significant amount of precipitation occurs in this no-mountain model over the otherwise arid region of middle latitudes. Hahn and Manabe (1975) have shown that the Southern Asian low, which induces monsoon precipitation in summer, is shifted to the northern part of China in a mountainless model. In short, these results suggest that it is necessary to take into consideration the effect of mountains in order to explain the existence of the arid regions of Central Asia in middle latitudes.

So far, the geographical distributions of some of the key hydrologic variables have been described. In Fig. 6, the latitudinal distribution of the zonal mean, annual mean precipitation rate from the model is

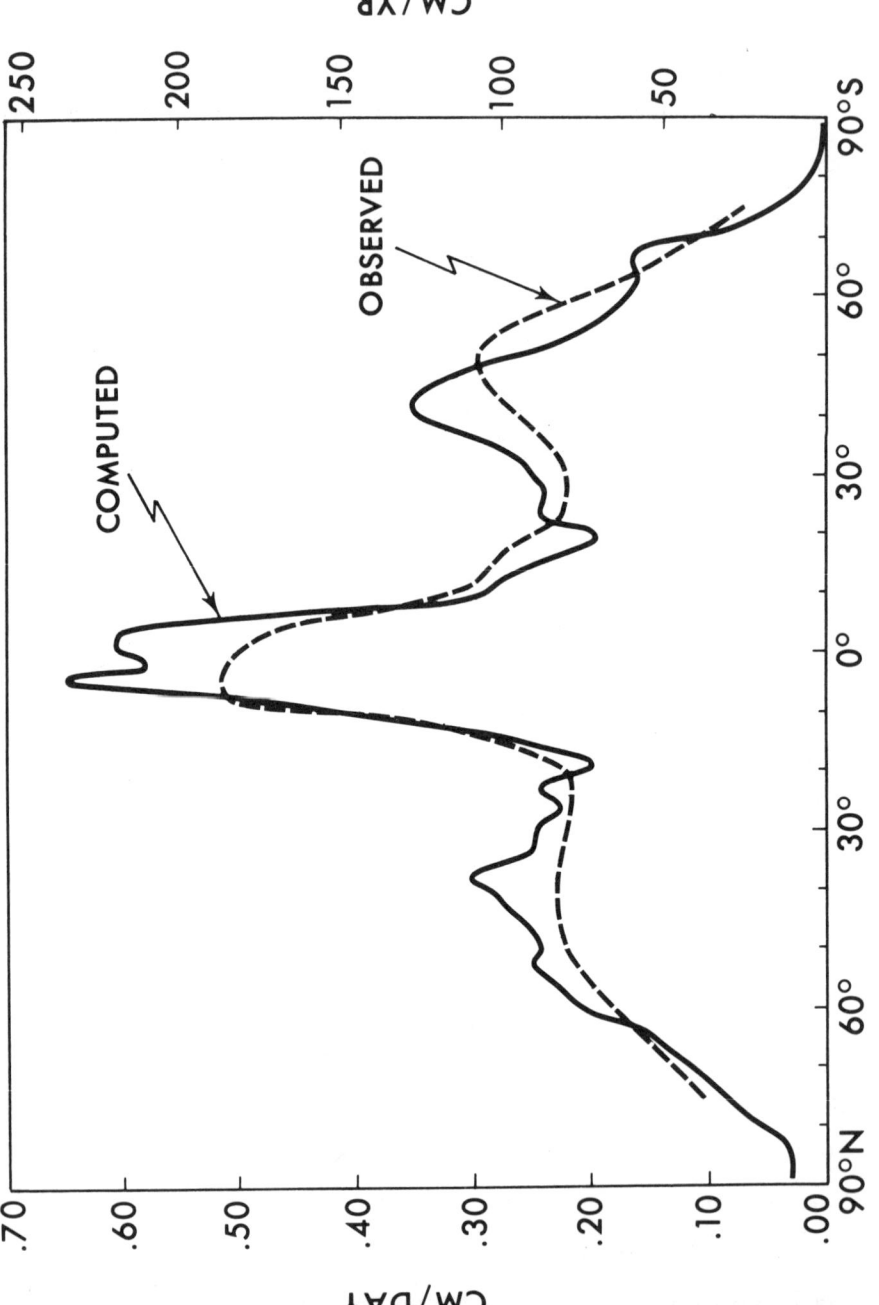

Figure 6 - Latitudinal distribution of the annual mean rate of precipitation computed by the model (solid line) and observed (Budyko, 1956) (dashed line).

compared with the corresponding distribution of the observed rate of precipitation. As this figure indicates, the model reproduces the precipitation maximums in the tropics and middle latitudes and the minimums in the subtropics although the latitudes of these maximums and minimums tend to be too low by several degrees.

The latitudinal distributions of the zonal mean precipitation rate described above may be examined by referring to the distribution of the annual mean meridional circulation in the model atmosphere shown in Fig. 7. (For the reader's information, the observed distribution of meridional circulation obtained by Oort and Rasmusson (1970) is added to the same figure.) This comparison indicates that the tropical rainbelt of the model is located near the center of the upward branch of the Hadley cell where the activities of tropical distrubances are also at a local maximum. This conincidence is a manifestation of a mutually-enhancing relationship among the Hadley circulation, the tropical disturbances and condensational heating. (Manabe et al., 1974). In middle latitudes, the rainbelt of the model is not located at the center of the upward motion branch of the Ferrel cell. Instead, it is located at the latitude where the kinetic energy of rain-producing baroclinic waves is at a maximum (see Fig. 4). As one may expect, the subtropical minimum in the simulated precipitation rate is located near the center of the downward motion branch of the meridional circulation.

The computed distribution of precipitation rate shown in Fig. 6 is significantly different from the latitudinal distribution of annual mean evaporation rate from the model illustrated in Fig. 8. In the model subtropics, the rate of evaporation exceeds that of precipitation, whereas the reverse is the case in the tropics and middle latitudes of the model.

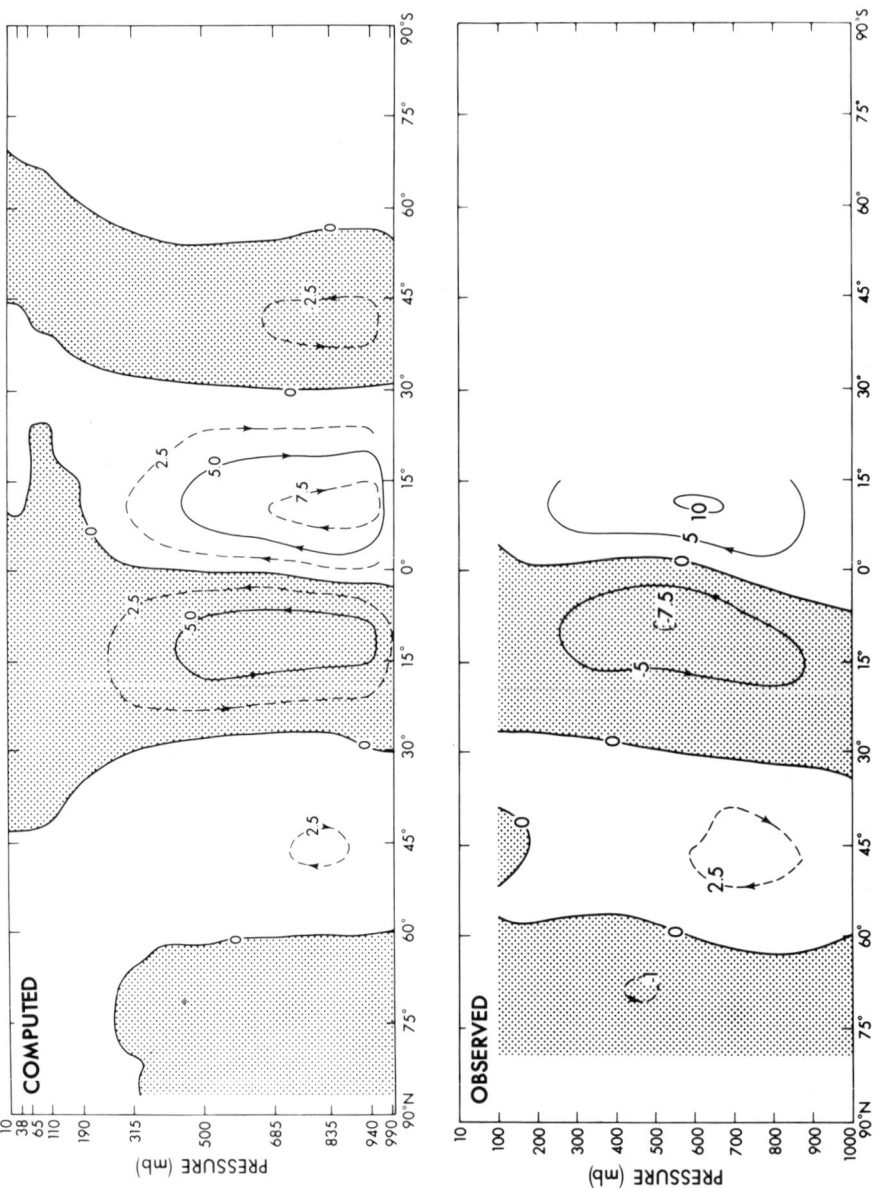

Figure 7 - Stream function (10^{13} g s^{-1}) of the annual mean meridional circulation; computed (top), observed in Northern Hemisphere only (bottom) (Oort and Rasmusson, 1971).

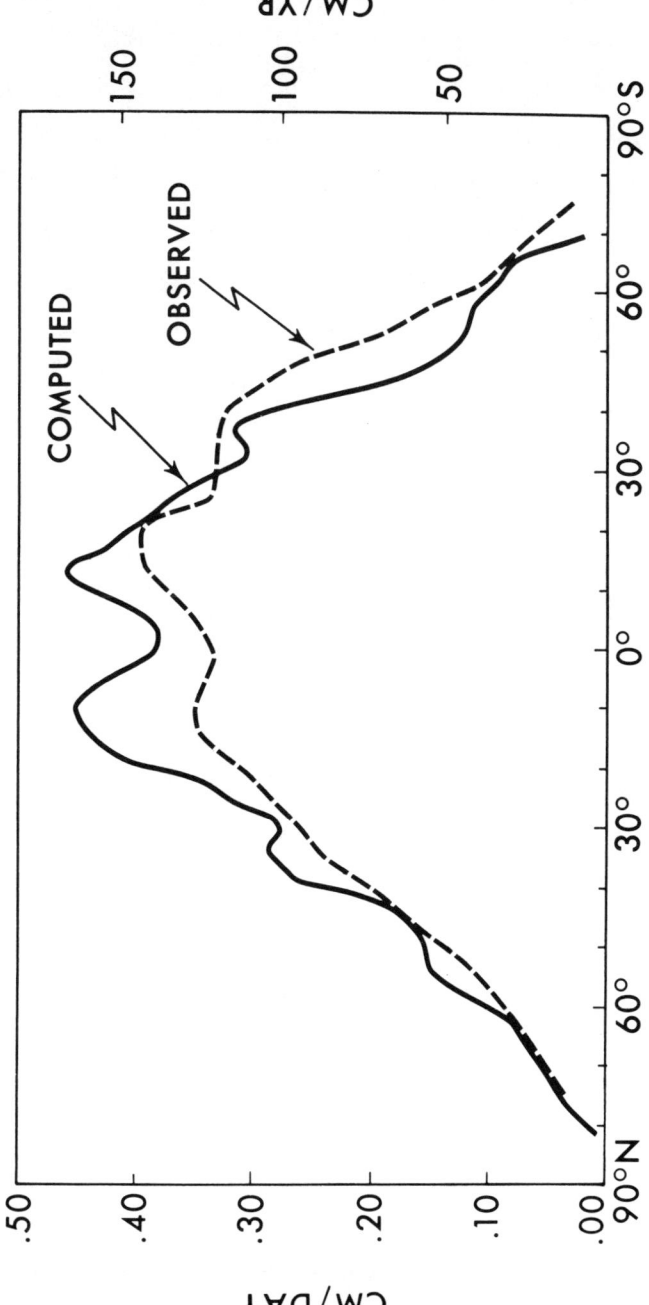

Figure 8 - Latitudinal distribution of annual mean rate of evaporation computed by the model (solid line) and derived from observed data (Budyko, 1963) (dashed line).

This difference constitutes a source or sink of moisture for the model atmosphere and is compensated by the divergence or convergence of meridional moisture transport by the atmospheric circulation. Fig. 9 presents curves of the zonal mean values of the model-simulated annual mean northward transport of water vapor by the meridional circulation, by large scale eddies and by parameterized subgrid scale mixing. Plotted on this same graph are values of the first two transports as obtained by Oort and Rasmusson (1970) from the observed data. According to this comparison, the computed and the observed distributions of eddy transport agree very well with each other. However, the distribution of transport by the meridional circulation in the model atmosphere differs somewhat from that estimated by Oort and Rasmusson (1970).

Fig. 9 shows that the meridional circulation (the Hadley cell in this case) brings moisture into the region of the tropical rainbelt from the subtropical belts where evaporation far exceeds precipitation. On the other hand, the large-scale eddies are principally responsible for transporting water vapor from the subtropics into the regions of the middle latitude rainbelts. Finally, the contribution of subgrid scale mixing is negligible.

A general impression of the skill of a model in simulating the climate may be gained by constructing an atlas of model climate as identified by Koppen's classification system and comparing it with the corresponding atlas of the actual climate. Fig. 10 presents such a comparison. This figure clearly indicates that the simulated climate resembles the actual climate with the exception of regions near major mountain ranges such as the west coast of South America, southwestern portions of North America and southern China where unrealistic distribution of deserts are indicated.

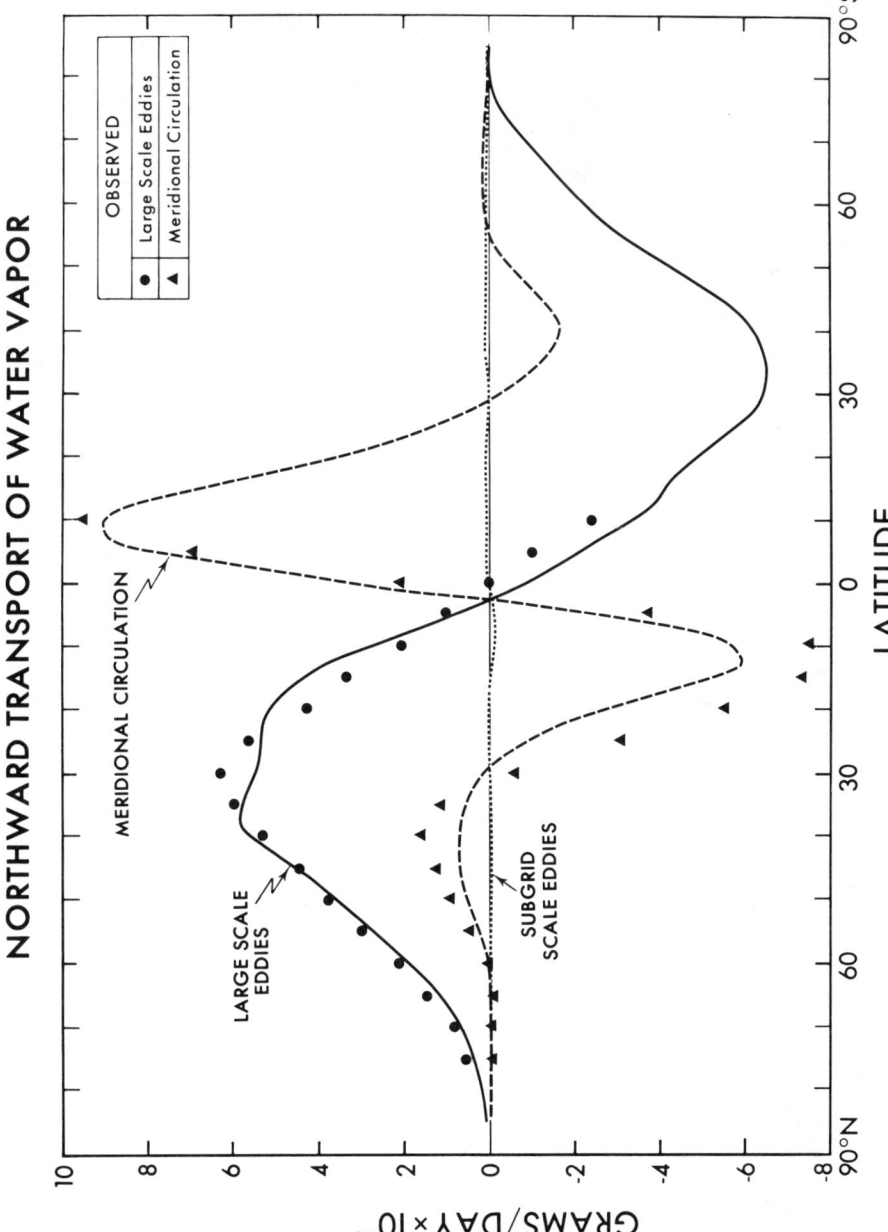

Figure 9 - Latitudinal distribution of the annual mean, simulated northward transport of water vapour by the meridional circulation (dashed line), by large-scale eddies (solid line), and by subgrid scale eddies (dotted line). Observed values of the first two transports derived by Oort and Rasmusson (1970) are plotted on this graph as triangles and dots, respectively.

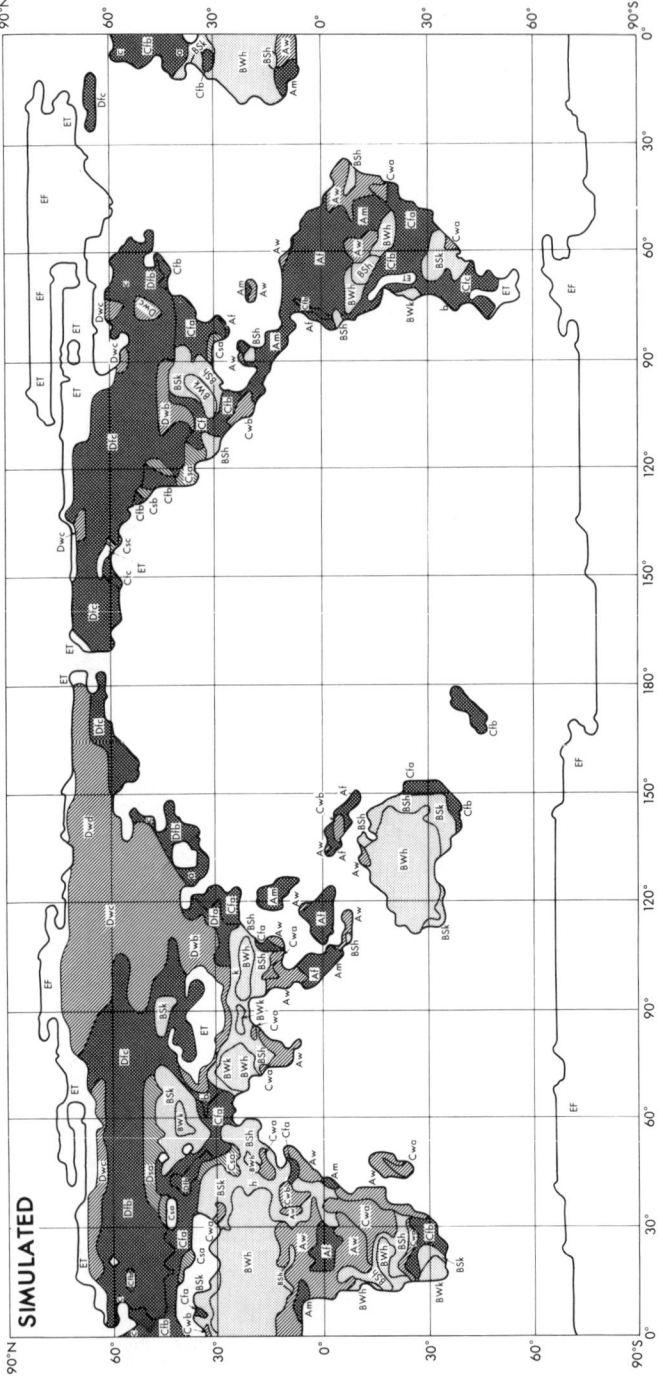

Figure 10a - Global distribution of Köppen climate types based on model-simulated data. The first capital letters indicate as follows: A, tropical climate; B, dry climate; C, warm temperate rainy climate; D, cool snowforest climate; E, polar climate. The second lower case letters indicate as follows: f, absence of dry season; w, dry season in winter; s, dry season in summer.

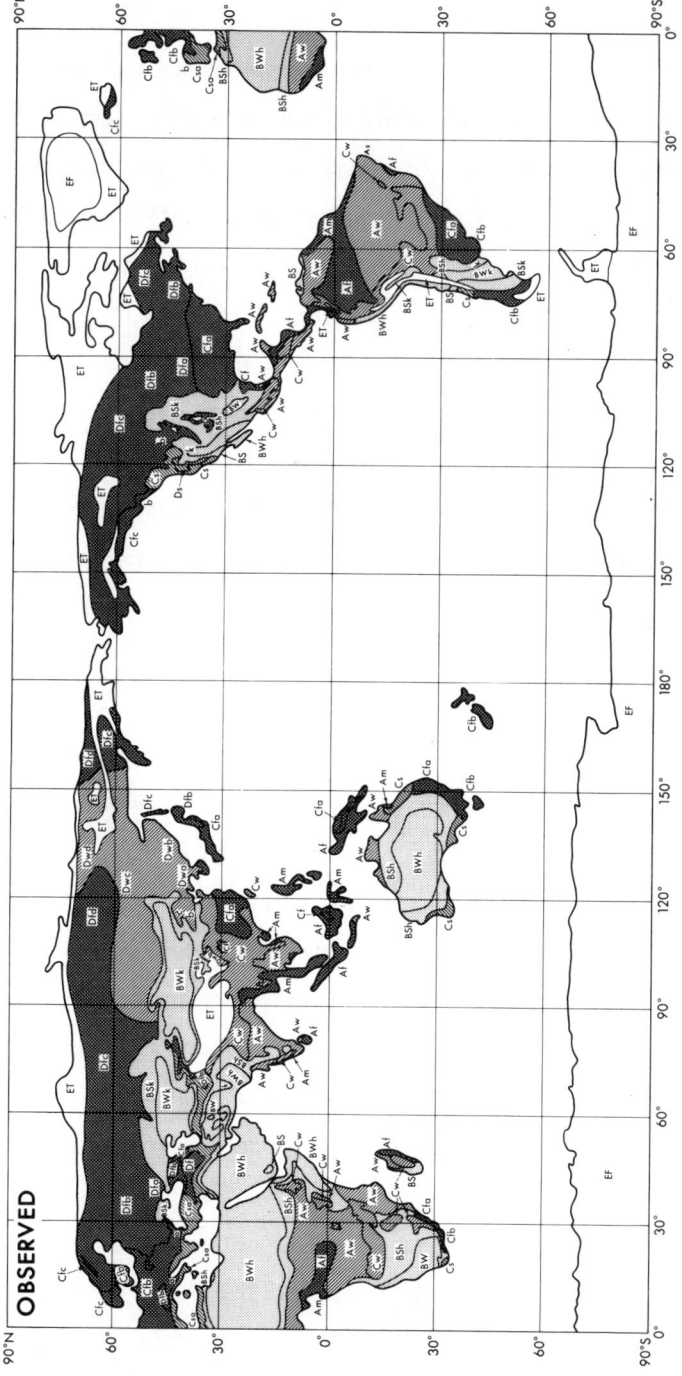

Figure 10b - Global distribution of Köppen climate type based on observed data (Trewartha, 1968).

4. Joint Ocean-Atmosphere Model.

The general circulation model of the atmosphere has a prescribed distribution of sea-surface temperature as a lower boundary condition. However, it is well known that climate and its variation is strongly controlled by the interaction among the atmosphere, continental surfaces and oceans. Therefore, in order to investigate a long-term climate variability or sensitivity of climate to an external forcing, it is necessary to construct a combined ocean-atmosphere model, in which the oceanic and atmospheric part of the model are allowed to interact with each other freely. The development of such a model has been in progress for some time. (See, for example, Manabe and Bryan, 1969). This section contains the brief description of the structure of a joint ocean-atmosphere model developed by Manabe et al. (1979) and its performance in simulating climate.

a. Model structure.

Fig. 11 shows a box diagram illustrating the basic structure of a joint ocean-atmosphere model. As indicated by this figure, a joint model consists of a general circulation model of the atmosphere described in the preceding sections and an ocean circulation model.

The ocean model is similar to that developed by Bryan and Lewis (1979). It predicts fields of currents, temperature and salinity by use of the equations of motion, the thermodynamical equation and the continuity equation of salinity. The density of water is calculated from the equation of state. In addition, the oceanic model incorporates a simplified method for calculating the growth and the movement of pack ice in polar latitudes. For further information on the various versions of the ocean circulation model, see, for example, Holland (1979).

STRUCTURE of JOINT OCEAN-ATMOSPHERE MODEL

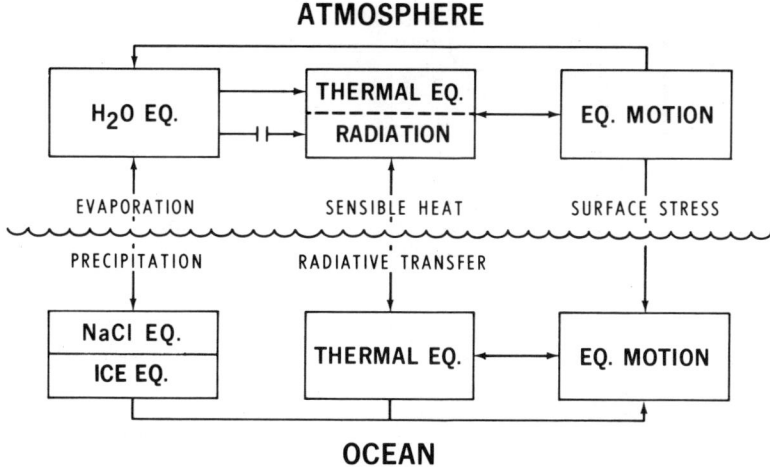

Figure 11 - Box diagram of the joint atmosphere-ocean model.

As Fig. 11 illustrates, the atmospheric and oceanic parts of the joint model exchange momentum, heat and moisture. The hydrologic processes incorporated into the model are illustrated by the box diagram shown in Fig. 12. (See the papers by Manabe (1969b) and Bryan (1969) for further details of this diagram.) For the computation of radiative transfer, seasonal variation of insolation is prescribed at the top of the model atmosphere.

Again, the model climate described below is obtained as the stable state which emerges from the long term integration of the joint model. The initial condition for this time integration is an isothermal ocean and an isothermal atmosphere at rest. Since it takes an extremely long time for the joint system to reach stable states, an economical method of time integration is developed. This method synchronizes the long-term integration of the atmospheric part of the model with the relatively short term integration of the oceanic part. (See Manabe et al., 1979 and Schlesinger, 1979 for further details and discussion of this economical method.)

HYDROLOGIC CYCLE

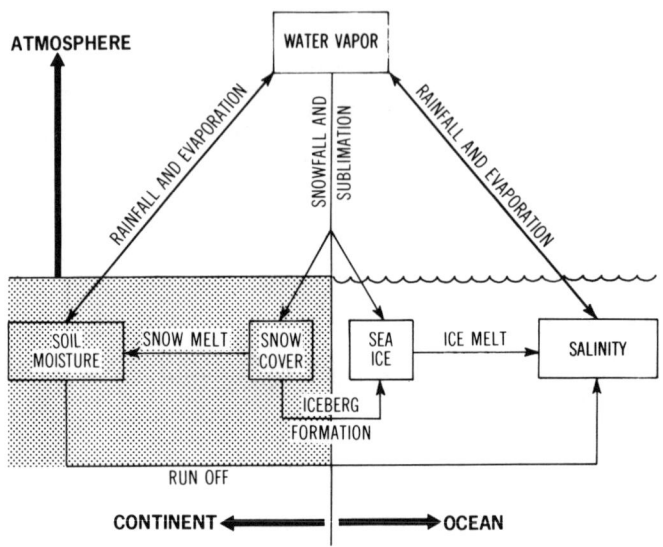

Figure 12 - Box diagram of the hydrologic processes in the joint model. (From Manabe, 1969a.)

b. **Simulated climate.**

In Fig. 13, the geographical distributions of monthly mean surface air temperature in the model atmosphere are compared with the corresponding distributions in the actual atmosphere. According to this comparison, the model substantially overestimates the surface air temperature of the circum-antarctic ocean and underestimates the surface air temperature of the Greenland Sea. Nevertheless, it successfully reproduces the basic characteristics of the seasonal and longitudinal variations of the observed surface air temperature.

The geographical distribution of precipitation rate obtained from the joint model is illustrated in Fig. 14 for two seasons of the year.

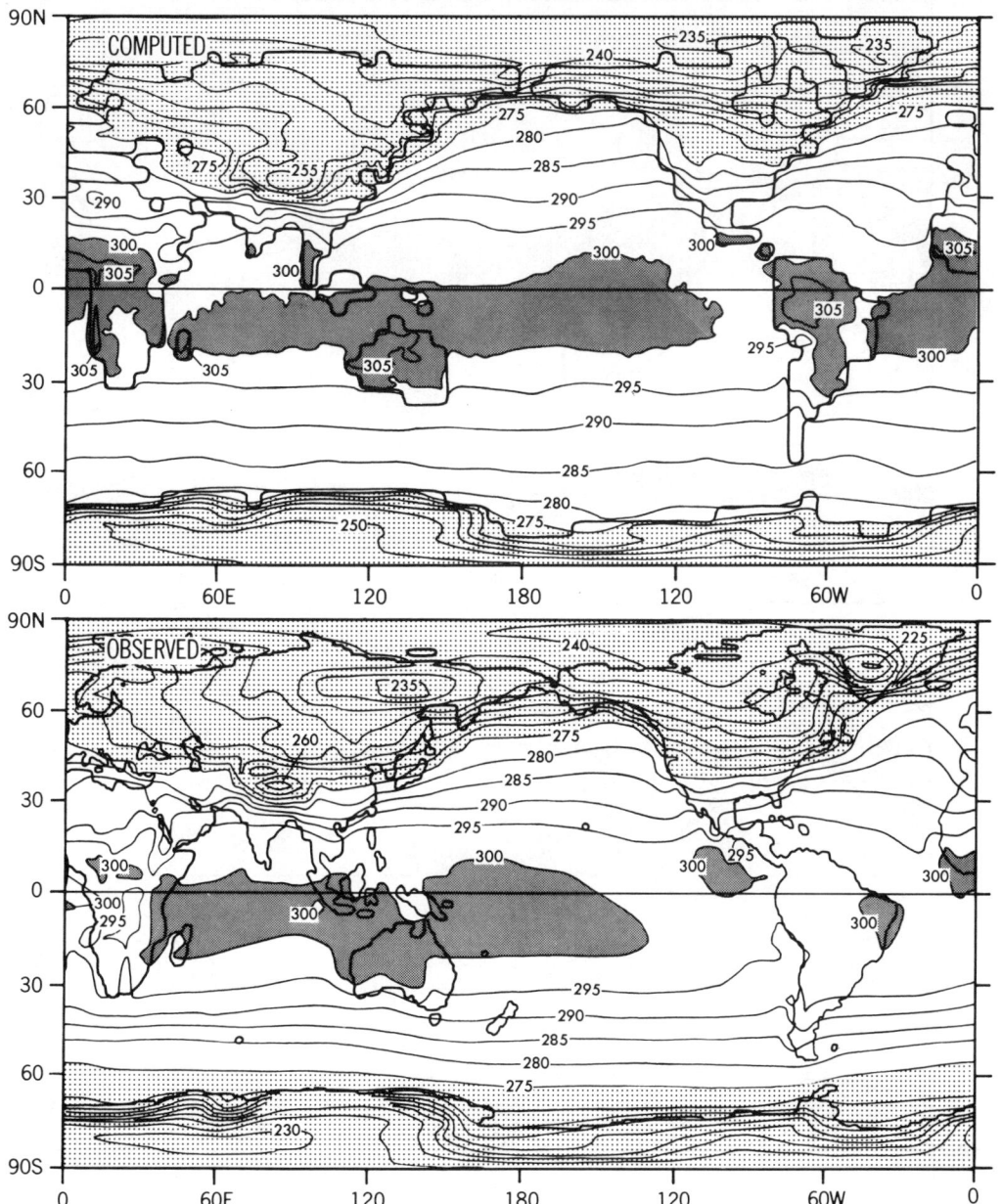

Figure 13a - Geographical distribution of the monthly mean surface air temperature (°K) in February. Top: computed. Bottom: observed.

-54-

SURFACE AMBIENT TEMPERATURE (AUG.)

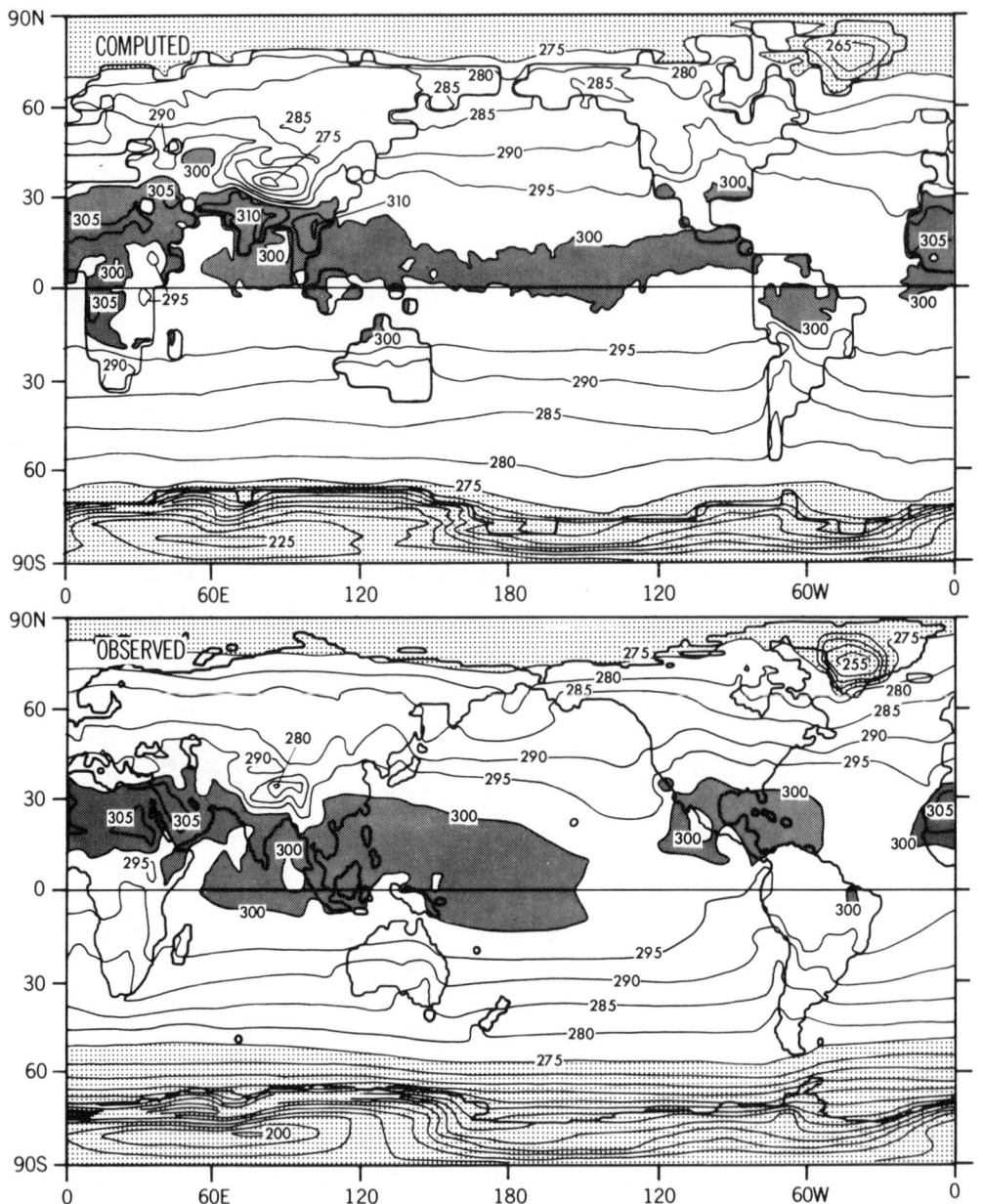

Figure 13b - Same as in Figure 13a but for August.

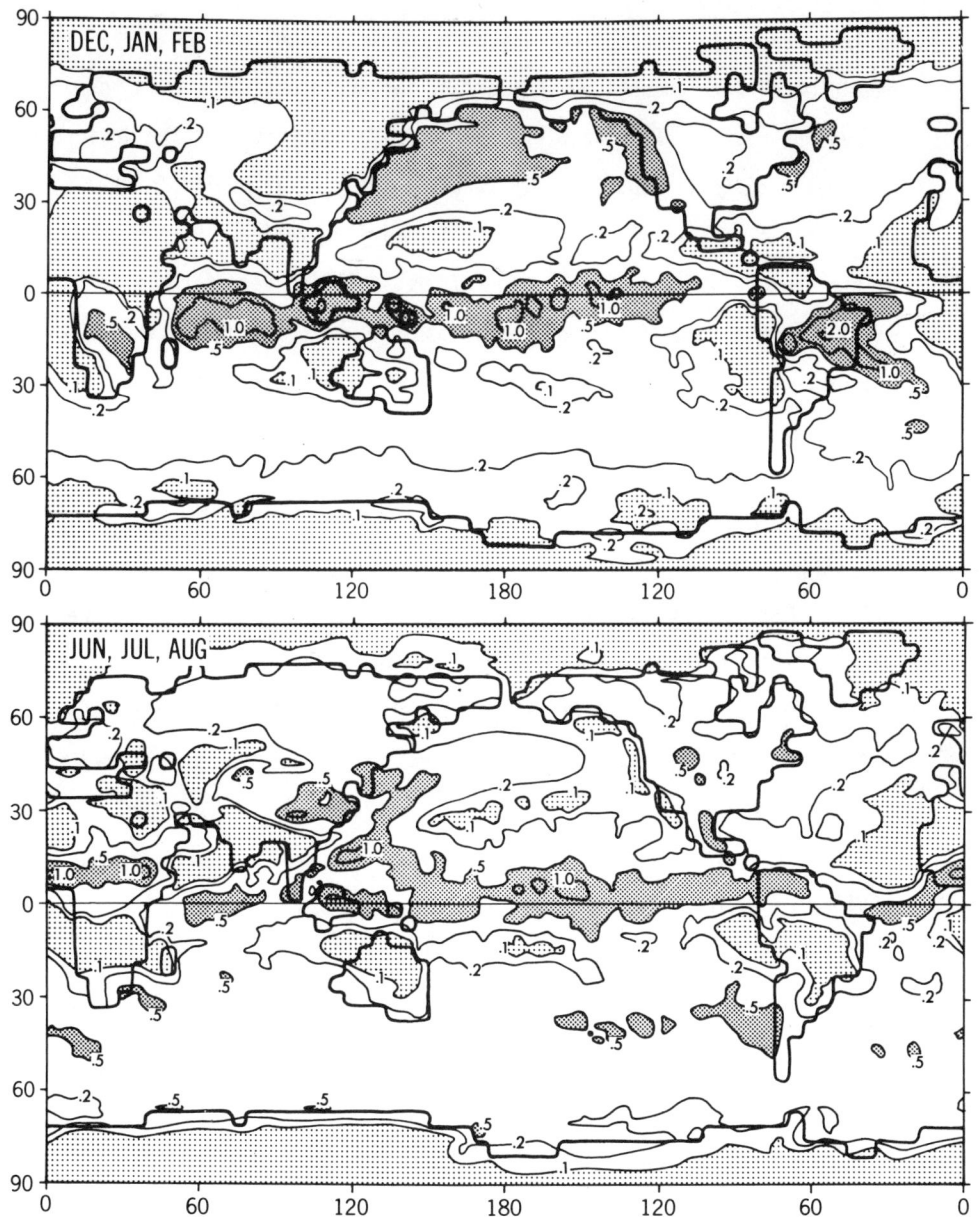

Figure 14 - Geographical distribution of precipitation rate (cm/day) simulated by the joint model. Top: Dec-Jan-Feb. Bottom: Jun-Jul-Aug. Refer to Figure 3 for observed distributions.

In general, the distribution of precipitation rate from the joint model is less realistic than the corresponding distribution from the atmospheric model with prescribed sea-surface temperature described in the preceding section. For example, the precipitation rate over the western tropical Pacific is substantially underpredicted as a result of the underestimation of sea surface temperature in the region. However, many of the features of the distribution of precipitation rate for the joint model resemble those of the actual atmosphere which encourages the use of such a model for a preliminary study of climate response to various forcing, such as an anthropogenic increase of CO_2 in the atmosphere.

5. Concluding Remarks.

It has been shown that the global model of climate described here successfully reproduces some of the large-scale characteristics of the seasonal and geographical variation of climate. However, the simulated climate also contains unrealistic features in the neighborhood of major mountain ranges such as the Andes, Rockies and Himalyas. The technique for computing the air flow over and around large scale mountain ranges obviously requires further improvement.

In this presentation, the preformances of general circulation models developed at the Geophysical Fluid Dynamics Laboratory have been described. Similar skill in climate simulation has been indicated by models developed by other groups. Some models perform better than other models in simulating certain characteristics of climate, whereas the opposite is the case with respect to other characteristics. Unfortunately, it has not been possible to determine the main causes of the differences in performance among various models.

The hydrologic processes incorporated into the climate models described here include many simplifications and idealization. It is therfore very probable that improvement in the hydrologic parameterization in models may yield a more realistic climate. The present state of art of the hydrologic parameterization and the future prospect for improvement are discussed by Carson (1981) in his review paper and are not the topics for this presentation. However, it is recommended to develop a hierarchy of hydrologic parameterizations with various degrees of complexity and to evaluate how the simulated climate is altered by each additional sophistication of the hydrologic parameterization. Based upon the results from this sensitivity study, one can derive a hydrologic parameterization with optimum sophistication and thus avoid the unnecessary complexity which prevents the effective analysis of the results from numerical model experiments. Furthermore, the parameterization experiments described above are very useful for identifying the causes of failures in climate simulation.

A similar approach is required for the improvement of other key components of a model such as the parameterized processes of radiative transfer, moist convection, boundary layer turbulence, condensation, and ocean-atmosphere interaction.

It is important to recognize that success in simulating the observed climate by a mathematical model of the atmosphere does not necessarily imply satisfoctory understanding of the mechanisms which maintain the climate. So far, relatively few works addressing this problem have been published. For example, Manabe and Bryan (1969) investigated the climatic effect of oceanic heat transport by comparing two climates simulated by their models with and without the effect of heat transport by ocean currents. To study how the large-scale mountains affect the general circulation of the atmosphere, several numerical experiments have been conducted by Mintz (1965), Kasahara and Washington (1971), Manabe and Terpstra (1974), and Nakamura (1979). Some of the results from these experiments were discussed in this presentation. These experiments have been useful in obtaining a preliminary idea of the relative imporantce of various climate-controlling processes. It is therefore recommended to perform numerical experiments of a similar kind designed to identify the role of various physical processes in maintaining the existing climate. Such experiments will be useful not only for improving the understanding of climate but also for developing a strategy for improving the capability of a model for simulating climate.

References

Arakawa, A., 1966: Computational design for long-term meridional integration of the equation of fluid motion: Two dimensional incompressible flow. Part I. J. Comp. Phys., 1, 119-143.

_____, 1972: Design of the UCLA atmospheric general circulation model. Tech. Rept. No. 7., Dept. of Meteorology, U.C.L.A.

_____ and W. Shubert, 1974: Interaction of a cumulus cloud ensemble with the large-scale enviornment, Part I. J. Atmos. Sci., 31, 674-701.

Bhumralkar, C. M., 1975: Numerical experiments on the computation of ground surface temperature in an atmospheric general circulation model. J. Appl. Meteorol., 14, 1246-1258.

Blackadar, A. K., 1962: The vertical distribution of wind and turbulence exchange in a neutral atmosphere. J. Geophys. Res., 67, 3095-3102.

Bourke, W., 1974: A multi-level model. I. Formulation and hemispheric integrations. Mon. Wea. Rev., 102, 688-701.

_____, B. McAvaney, K. Puri and R. Thurling, 1977: Global modeling of atmospheric flow by spectral methods. Methods in Computational Physics., 17, 267-324.

Bryan, K., 1969: Climate and ocean circulation: III The ocean model. Mon. Wea. Rev., 97, 806-827.

_____ and L. J. Lewis, 1979: A water mass model of the world ocean. J. Geophys. Res., 84, 2503-2517.

Budyko, M. I., 1956: Teplovoi Balans Zemnoi Poverkhnosti, Gidrometeorologicheskoe Izdatel'stvo, Leningrad (English translation by N. A. Stepanova, 1958, The heat balance of the earth's surface. Office of Technical Service, U.S. Dept. of Commerce, Wash., D.C., 1958).

Carson, D. J., 1981: Current parameterization of land surface processes. Proceedings of Study Conference on Land Surface Processes in Atmospheric General Circulation Models, Greenbelt, Md., 5-10 Jan. 1981. This volume.

Chang, J., 1977: General circulation models of the atmosphere. Edited by J. Chang. Method in Computational Physics, Vol. 17, Academic Press, New York.

Charney J., and J. Shukla, 1981: Predictability of Monsoons. Monsoon Dynamics. Eds. Sir James Lighthill and R. P. Pearce, p. 99-110, Cambridge University Press, London, England.

Corby, G. A., A. Gilchrist and P. R. Rowntree, 1977: United Kingdom Meteorological Office Five-level general circulation model. Method in Computational Physics, 17, (Academic Press, New York), pp. 67-110.

Cunnold, D., F. Alyea, N. Phillips and R. Prinn, 1975: A three-dimensional dynamical-chemical model of atmospheric ozone. J. Atmos. Sci., 32, 170-194.

Deardorf, J. W., 1972: Parameterization of the planetary boundary layer for use in general circulation models. Mon. Wea. Rev., 100, 93-106.

Döös, B. R., 1969: The influence of the large-scale heat sources on the dynamics of the ultra-long waves. Tellus, 21, 25-39.

Eagleson, P. S., 1978: Climate, soil, vegetation. Parts 1-7. Water Resources Research, 14, 705-776.

Eagleson, P. S., 1981: Dynamic hydrothermal balance at macroscale. Proceedings of Study Conference on Land Surface Processes in Atmospheric General Circulation Models, Greenbelt, Md., 5-10 Jan. 1981. This volume.

Eliassen, E., B. Machenhauer and E. Rasmussen, 1970: On a numerical method for integration of the hydrodynamical equations with a spectral representation of the horizontal fields. Rep. No. 2, Institut for Teoretisk Meteorologi, Kobenhavns Universitet, Denmark.

Elsasser, W. M., 1942: Heat transfer by infrared radiation in the atmosphere. Harvard Meteor. Stud. No. 6.

Fels, S. B., and M. D. Schwarzkopf, 1975: The simplified exchange approximation: a new method for radiative transfer calculations. J. Atmos. Sci., 32, 1475-1488.

GARP Joint Organizing Committee, 1979: Report of the JOC study conference on climate models: Performance, Intercomparison, and Sensitivity Studies. GARP Publication Series, No. 22, 1049 p.

Gates, W. L., and M. E. Schlesinger, 1977: Numerical simulation of the January and July global climate with a two-level atmospheric model. J. Atmos. Sci., 34, 36-76.

Gordon, T., and B. Stern, 1974: The GARP program on numerical experimentation, Report No. 7, 46-87.

Hahn, D. G., and S. Manabe, 1975: The role of mountains in the south Asian monsoon circulation - A comparison of two global circulation experiments. J. Atmos. Sci., 32, 1515-1541.

Holland, W. R., 1979: The general circulation of the ocean and its modeling. Dyn. Atmos. Oceans, 3, 111-142.

Holloway, J. L., Jr., and S. Manabe, 1971: Simulation of climate by a global general circulation model. Mon. Wea. Rev. 99, 335-370.

Hoskins, J. J. and X. Simmons, 1975: A multi-layer spectral model and the semi-implicit method. Quart. J. Roy Met. Soc., 101, 637-655.

Hunt, B. G., 1969: Experiments with a stratospheric general circulation model: III Large scale diffusion of ozone including photochemistry Mon. Wea. Rev., 97, 287-306.

J.O.C., 1974: Modeling for the first GARP global experiment. GARP Publication Series No. 14, 261 p., World Meteorological Organization.

Kaplan, L. D., 1959: A method for the calculation of infrared flux for use in numerical models of atmospheric motion. The Atmosphere and the Sea in Motion, B. Bolin, Ed., Rockefeller Institute Press and Oxford University Press, 170-177.

Kasahara, A. and W. M. Washington, 1971: General circulation experiments in the six-layer NCAR Model, including orography, cloudiness and surface temperature calculations. J. Atmos. Sci., 28, 657-701.

Kuo, H. L., 1965: On formation and intensification of tropical cyclones through latent heat released by cumulus convection. J. Atmos. Sci., 22, 40-63.

Leith, C. E., 1969: Numerical simulation of turbulent flow. Properties of Matter Under Unusual Conditions . Edited by H. Mark and S. Fernback. pp. 267-271. Interscience Publishers, New York.

Lettau, H. H., 1959: Wind profile, surface stress and geostrophic drag coefficient in the atmospheric surface layer. Adv. Geophys., 6, 241-257.

Lvovich, M. I., and S. P. Ovtchinnikov, 1964: Physical-geographical atlas of the world (in Russian). Academy of Sciences, U.S.S.R. and Department of Geodesy and Cartography, State Geodetic Committee, Moscow.

Manabe, S., and R. F. Strickler, 1964: Thermal equilibrium of the atmosphere with a convective adjustment. J. Atmos. Sci., 21, 361-385.

_____, J. Smagorinsky and R. F. Strickler, 1965: Simulated climatology of a general circulation model with a hydrologic cycle. Mon. Wea. Rev., 93, 769-798.

_____, 1969: Climate and the ocean circulation. I. The atmospheric circulation and the hydrology of the earth's surface. Mon. Wea. Rev., 97, 739-774.

_____, and K. Bryan, 1969: Climate calculations with a combined ocean-atmosphere model. J. Atmos. Sci., 26, 786-789.

_____ and T. B. Terpstra, 1964: The effects of mountains on the general circulation of the atmosphere as identified by numerical experiments. J. Atmos. Sci., 31, 3-42.

_____ and J. L. Holloway, Jr., 1975: Seasonal variation of hydrologic cycle as simulated by a global model of the atmosphere. J. Geophy. Res., 80, 1617-1649.

_____, D. G. Hahn, and J. L. Holloway, Jr., 1979: Climate simulation with a GFDL spectral model of the atmosphere: Effect of spectral truncation. Report of the JOC study conference on climate models: Performance, intercomparison, and sensitivity studies. GARP Publication Series No. 22, 41-94.

_____ and R. T. Wetherald, 1980a: On the distribution of climate change resulting from an increase in CO_2 content of the atmosphere. J. Atmos. Sci., 37, 99-118.

Mellor, G. L., and T. Yamada, 1974: A hierarchy of turbulence closure models for planetary boundary layers. J. Atmos. Sci., 31, 1791-1806.

Mintz, Y., 1965: Very long term global integration of the primitive equations of atmospheric motion. Proceedings of WMO-IUGG Symposium on Research and Development Aspect of Long Range Forecasting. WMO Technical Note, 66, 141-167, World Meteorological Organization, Geneva.

_____, 1968: Very long-term global integration of the primitive equation of atmospheric motion: An experiment in climate simulation. Meteorological Monograph, Vol. 8, No. 30, 20-36, American Meteorological Society, Boston, Mass.

_____, 1981: Climate sensitivity to land surface state. Proceedings of Study Conference on Land Surface Processes in Atmospheric General Circulation Models, Greenbelt, Md., 5-10 Jan. 1981. This volume.

_____, and G. Dean, 1952: The observed mean field of motion of the atmosphere, Geophys. Res. Papers, No.17, 11-65.

Miyakoda, K. and J. Sirutis, 1977: Comparative integrations of global models with various parameterized processes of subgrid-scale vertical transport: Description of the parameterizations. Beitrage zur Physik der Atmosphäre, 50, 445-447.

Möller, F., 1943: Das Strahlungs diagramm. Wiss. Abh. D.R. Reich. Wetterd.

_____, 1951: Quarterly charts of rainfall for the whole earth (in German) Petermans Geograph Mitt., 95, 1-7.

Monin, A. S. and S. S. Zilitinkievich, 1967: Planetary boundary layer and large scale atmospheric dynamics. Study conference on GARP (ICSU/IUGG, WMO and COSPAR).

Oort, A. H. and E. M. Rasmusson, 1970: On the annual variation of the monthly mean meridional circulation. Mon. Wea. Rev., 98, 423-442.

_____ and _____, 1971: Atmospheric circulation statistics. NOAA Professional Paper, No. 5, 323 pp., U.S. Govt. Print. Office, Washington, D.C.

Orsag, S. A., 1970: Transform method for the calculation of vector-coupled sums: Application to the spectral form of the vorticity equation. Mon. Wea. Rev., 102, 688-701.

Philip, J. R., 1957: Evaporation and moisture and heat fields in the soil. J. of Meteorology, 14, 354-366.

_____ and D. A. de Vries, 1957: Moisture movement in porous materials under temperature gradients. Trans. Amer. Geophys. Union, 38, 222-232.

Phillips, N. A., 1956: The general circulation of the atmosphere: A numerical experiment. Quar. J. of the Roy. Meteor. Soc., 82, 123-164.

_____, 1956: An example of non-linear computational instability. The Atmosphere and the Sea in Motion, Rockefeller Institute Press in Association with Oxford University Press, New York. pp. 501-504.

_____, 1957: A coordinate system having some special advantages for numerical forecasting. J. Meteorol. 14, 184-185.

Platzman, G. W., 1960: The spectral forms of the vorticity equation. J. Met., 17, 635-644.

Rodgers, C. D., and C. D. Walshaw, 1966: The computation of infrared cooling rate in planetary atmospheres. Quart. J. Roy. Met. Soc., 92, 67-92.

_____, 1967: The use of emissivity in atmospheric radiation calculation. Quart. J. Roy. Met. Soc., 92, 43-54.

Rossby, C. G. and R. B. Montgomery, 1935: The layer of frictional influence in wind and ocean currents. Papers in Physical Oceanography and Meteorology, Vol. 3, No. 3, Massachusetts Institute of Technology and Woods Hole Oceanographic Institution, Apr. 1935, 101 pp.

Samamori, T., 1968: The radiative cooling calculation for application to general circulation experiments. J. Appl. Meteor., 7, 721-729.

_____, 1970: A numerical study of atmospheric and soil boundary layers. J. Atmos. Sci., 27, 1122-1137.

Smagorinsky, J., 1963: General circulation experiments with primitive equations, I. The basic experiment. Mon. Wea. Rev, 93, 99-164.

Somerville, R. C., P. H. Stone, M. Halem, J. E. Hansen, J. S. Hogan, L. M. Druyan, G. Russel, A. A. Lacis, W. J. Quirk, and J. Tennenbaum, 1974: The GISS model of the global atmosphere. J. Atmos. Sci., 31, 84-117.

Sverdrup, H. U., M. W. Johnson and R. H. Fleming, 1942: The Oceans. New York, Prentice-Hall, 1087 pp.

Takahashi, K., A. Katayama and T. Asakura, 1960: A numerical experiment of the atmospheric radiation. J. Met. Soc. Japan, 38, 175-181.

Thornthwaite, C. W. and J. R. Mather, 1955: The water balance. Publications in climatology, 8, 1, Laboratory of Climatology, Centerton, N.J. 86 pp.

Trewartha, G. T., 1968: An introduction to climate. 4th ed. McGraw-Hill New York.

Washington, W. M. and D. L. Williamson, 1977: A description of the NCAR global circulation models, Method in Computational Physics, 17, (Academic Press, New York), pp. 111-172.

_____, R. Dickinson, V. Ramanathan, T. Mayer, D. Williamson, G. Williamson, and R. Wolski, 1979: Preliminary atmospheric simulation with the third generation NCAR general circulation model. January and July. Report of the JOC study conference on climate models: Performance, intercomparison and sensitivity studies. GARP Publication Series, No. 22, 95-138.

Wetherald, R. T. and S. Manabe, 1980b: Cloud Cover and Climate Sensitivity. J. Atmos. Sci., 37, 1485-1510.

Yamamoto, G., 1952: On a radiation chart, Science Reports of the Tohoku University, Ser. 5, Geophysics, Vol. 4, No. 1, June 1952, pp. 9-23.

CURRENT PARAMETERIZATIONS OF LAND-SURFACE PROCESSES IN
ATMOSPHERIC GENERAL CIRCULATION MODELS

by

D. J. Carson

Meteorological Office, Bracknell, England

1.   INTRODUCTION

In the quest for a better understanding of climate and climatic change many types of so-called 'climate model' are being developed. The JSC Working Group on Land-surface Processes has taken the conscious decision to focus its attention at present on the practices and requirements of those climate models being developed from the most advanced, three-dimensional atmospheric general circulation models. The aim of this paper is to review current practices relating to the parameterization of the land-surface properties and processes in such models. The structure of a typical general circulation model and current quality of simulations of known characteristics of the global atmospheric circulation are described in the companion paper of Manabe (Review Paper 2). Experimental evidence of the sensitivity of certain facets of such simulations to changes in the land-surface boundary conditions is presented by Mintz (Review Paper 4).

It is virtually impossible for a review of this nature to be complete and comprehensive and its scope and limitations need to be identified. The parameterization of land-surface processes is currently a very active field of research and model development, and methods cited in the readily available literature may quickly become outdated. Viable techniques are continually being tested in sensitivity studies with global models or are being used in three-dimensional meso-scale and numerical forecasting models. In principle, some of these could be incorporated ultimately into full three-dimensional climate models and it is anticipated that reports on such progress will be forthcoming at the Greenbelt Study.Conference in 1981.

A reasonable objective criterion for selecting global models to be included in the survey would be that they have already demonstrated an ability to produce acceptable atmospheric circulation simulations for January and/or July; this is effectively what has been done. Table 1 lists the major climate modelling centres with a pertinent reference to the most recent version of each model used for general circulation studies with some indication of type and spatial resolution. This choice was made easier by the timely appearance of the Report of the JOC Study Conference on Climate Models: Performance, Intercomparison and Sensitivity Studies (1979), Volume I of which contains most of the references given in Table 1. It is important to stress for those not involved directly with general circulation modelling that there is as yet no generally accepted 'best way' of designing even the basic components of such a model. The models listed in Table 1 represent a wide range of approaches including different numerical schemes (both spatial and temporal), vertical and horizontal resolutions, horizontal grid structures and representations of the physical and other sub grid-scale processes.

Table 1 – The primary atmospheric general circulation models considered in the review.

$$\sigma = (p - p_T)/(p_0 - p_T)$$ where p is pressure, p_0 is surface pressure and p_T is specified in the table. N is the numer of model rows between pole and equator.

LABEL	CENTRE	VERTICAL REPRESENTATION	HORIZONTAL REPRESENTATION	REFERENCES
AES	Atmospheric Environment Service, Canadian Climate Centre, Downsview, Ontario.	$\sigma \, (p_T=0)$ 5 levels	Spectral 20 waves rhomboidal truncation	Boer and McFarlane (1979)
ANMRC	Australian Numerical Meteorology Research Centre, Melbourne.	$\sigma \, (p_T=0)$ 9 levels	Spectral 15 waves rhomboidal truncation	McAvaney et al (1979, 1978) Bourke et al (1977)
CCSAS	Computing Centre, Siberian Academy of Sciences, Novosibirsk.	$\sigma \, (p_T=0)$ 3 or 5 levels	Finite difference 5° or 10° latitude-longitude	Marchuk et al (1979)
GFDL	Geophysical Fluid Dynamics Laboratory/NOAA, Princeton, New Jersey.	$\sigma \, (p_T=0)$ 9 levels	Spectral 15, 21 or 30 waves rhomboidal truncation	Manabe et al (1979)
		$\sigma \, (p_T=0)$ 9 levels	Finite difference irregular lat-long approx 500 km (N24) or 250 km (N48)	Holloway and Manabe (1971) Manabe and Terpstra (1974)
		$\sigma \, (p_T=0)$ 11 levels	Finite difference irregular lat-long approx 250 km (N48)	Manabe, Hahn and Holloway (1974) Manabe and Holloway (1975)
		$\sigma \, (p_T=0)$ 18 levels	Finite difference irregular lat-long approx 500 km (N24)	Miyakoda and Sirutis (1977)
GLAS	NASA/Goddard Space Flight Centre, Laboratory for Atmospheric Sciences, Greenbelt, Maryland.	$\sigma \, (p_T=10mb)$ 9 levels	Finite difference 4° latitude by 5° longitude	Halem et al (1979)
GISS	Institute for Space Studies, Goddard Space Flight Centre, NASA, New York.			Somerville et al (1974) Stone et al (1977)

Table 1 (Continued)

LABEL	CENTRE	VERTICAL REPRESENTATION	HORIZONTAL REPRESENTATION	REFERENCES
MO	Meteorological Office, Bracknell, England.	σ ($p_T = 0$) 5 levels 11 levels	Finite difference irregular lat-long approx 330 km (N30) approx 220 km (N45)	Corby et al (1977) Gilchrist (1979) Saker (1975) Carson and Cunnington (1979)
NCAR	National Centre for Atmospheric Research, Boulder, Colorado.	z Third-generation model: 8 layers approx 3 km thick (Top of model 24km) Second-generation model: 6 layers approx 3 km thick (Top of model 18km)	Finite difference 5° lat-long	Washington et al (1979) Washington and Williamson (1977) Kasahara and Washington (1971)
OSU	Department of Atmospheric Sciences and Climatic Research Institute, Oregon State University, Corvallis, Oregon.	σ ($p_T = 200$ mb) 2 levels	Finite difference 4° latitude by 5° longitude	Schlesinger and Gates (1979)
RAND	The Rand Corporation Santa Monica, California.			Gates and Schlesinger (1977)
UCLA	Department of Atmospheric Sciences, University of California, Los Angeles, California.	Generalised σ Combination of σ for lower atmosphere and p for upper atmosphere. 12 levels or 6 levels. 3- and 6-level models	Finite difference 4° latitude by 5° longitude	Arakawa and Lamb (1977) Arakawa (1972)

Many physical processes which have to be modelled have a direct and immediate effect on the surface and sub-surface but are not directly governed by, nor specified in terms of, the land-surface and sub-surface characteristics and processes. For example:

(a) The net radiative flux at the surface depends very critically on atmospheric structure and composition and the presence of clouds, as well as on the surface properties of albedo, emissivity, temperature, etc.;

(b) Soil moisture budgets depend critically on the atmospheric processes governing the model's rainfall and snowfall;

(c) Surface turbulent fluxes of momentum, heat and moisture are commonly deemed to depend not only on surface conditions and characteristics but also on the internal structure of the boundary layer which in turn is determined by turbulent mixing throughout the whole boundary layer, convection throughout a deeper layer, radiative cooling, etc.

There are clearly very complicated, highly non-linear, feed-back mechanisms operating which link the surface processes and state to the internal atmospheric structure and other meteorological processes and which ultimately determine the lower boundary conditions. This paper is limited to a description of the general principles employed to determine the lower boundary conditions in atmospheric general circulation models with details only of the formulations used for surface albedo and emissivity, the surface turbulent fluxes of momentum, heat and water vapour and the soil thermal and hydrological processes.

It is perhaps worth reminding the reader that the problem of assessing a new parameterization within the framework of a general circulation model can be a very difficult one and that one must be particularly wary of passing judgement on a specific parameterization in isolation from the rest of the modelled structure and processes. It is also worth noting that there are other important surface factors which determine the lower boundary condition in an atmospheric general circulation model but which are beyond the compass of this paper. These include the particular problems associated with the ocean surface and air-sea interaction, the distribution and formation of sea-ice, the specification of realistic distributions of land, sea and ice and orographic heights and the parameterization of sub grid-scale orographic effects. A further potentially important aspect of surface modelling not catered for sufficiently here concerns the initialization of surface and soil properties such as the soil moisture content, snow and ice cover and the sea-surface temperatures. The importance of such considerations should be apparent from the sensitivity studies discussed by Mintz (Review Paper 4).

The remainder of the paper considers the land-surface processes very much from the viewpoint of the general circulation modeller. Although the discussions and descriptions are intended to be objective and reasonably comprehensive it has been convenient or even necessary when illustrating the details of some particular methods to use the U.K. Meteorlogical Office (MO) models with which the reviewer is most familiar and to which he has direct access.

2. SURFACE RADIATIVE PROPERTIES AND FLUXES

Since solar radiation provides most of the energy needed to maintain the general circulation of the atmosphere and since the major input of this energy to the Earth-atmosphere system occurs at the surface, it seems natural that the first section on parameterization should address the surface radiative properties and fluxes. It is recognized that fluxes from the surface have to be modelled quite differently over the sea from the way they are over the land. For most modelling purposes to date the sea has been considered to operate essentially as an infinite reservoir of heat and moisture with only very gradual changes in the sea-surface temperatures (usually specified and imposed rather than predicted by the model). Land on the other hand has a much more limited capacity for storage and the fluxes and state of the surface respond much more quickly to external influences such as incoming solar radiation, rainfall, snowfall, etc. The general circulation modeller acknowledges therefore that it is particularly important to determine the correct net imbalance of radiative fluxes to and from the land surface.

The net downward radiative flux at the surface, R_N, is expressed as

$$R_N = R_{SN} + R_{LN}$$
$$= (1-\alpha) R_{S\downarrow} + \varepsilon (R_{L\downarrow} - \sigma T_o^4) \qquad (1)$$

where R_{SN} is the net downward solar radiative flux at the surface,
R_{LN} is the net downward long-wave radiative flux at the surface,
$R_{S\downarrow}$ is the downward solar radiative flux at the surface,
$R_{L\downarrow}$ is the downward long-wave radiative flux at the surface,
T_o is the surface temperature, α is the surface albedo,
ε is the surface long-wave emissivity or absorptivity and
σ is the Stefan-Boltzmann constant.

The evaluation of the fluxes $R_{S\downarrow}$ and $R_{L\downarrow}$ within general circulation models is an aspect of the determination of the surface boundary conditions which is beyond the scope of this particular review. The remainder of this section concentrates on the modelling of the surface radiative characteristics other than T_o which are most important in determining R_N, viz., ε and α.

Surface emissivity (ε).   Consider first the surface emissivity which is known to have a wavelength dependence and to vary according to the character of the surface as discussed for example by Kondratyev (1972) and Paltridge and Plat (1976). Values of the vertical emissivity listed there but ascribed in the main to Beuttner and Kern (1965) range from 0.997 for wet snow to 0.71 for quartz with the added caution that the meteorologically relevant global emissivity for natural surfaces may be much less.

The present modelling of ε can be dismissed almost summarily by noting that virtually every general circulation model assumes that all surfaces act like black bodies for long-wave radiation with ε = 1. The only exception to this of the models in Table 1 is the so-called third-generation NCAR general circulation model (Washington et al., 1979) in which the treatment of radiation has been extensively revised since the earlier description in Washington and Williamson (1977) to include (among many other changes) surface emissivities which are both spectrally dependent and vary with the type of the surface. The details of the approach are not known to the reviewer but it is apparent that the radiative properties including the long-wave emissivities are different for vegetated and non-vegetated land surfaces. This distinction is undoubtedly related to the incorporation in the NCAR model of Deardorff's (1978) recent proposals for the explicit inclusion of a layer of vegetation at the surface. The net radiation at the ground surface obeys relation (1) where $R_S\downarrow$ is expressed in terms of the downward solar radiative flux at the top of the vegetative canopy and a foliage shielding factor of the ground from short-wave radiation and $R_L\downarrow$ is a specified function of both ground and foliage emissivities as well as the temperatures of the ground surface and the foliage or leaf surface, respectively, the downward long-wave radiative flux at the top of the canopy and again the foliage shielding factor.

The treatment of atmospheric radiation in general circulation models is currently a very active area of research and development in its own right and the models listed in Table 1 employ a wide range of radiation and cloud parameterizations. In some of the simpler approaches, such as that used in the version of the MO 5-layer model described by Corby et al. (1977), the net downward long-wave flux, R_{LN}, is represented rather by $-e\sigma T_o^4$, thus defining e the effective surface emissivity which is determined empirically and is likely to have some geographical dependence at least. In such formulations it is not always immediately obvious that the true surface emissivity $\mathcal{E}$ has been explicitly or implicitly assumed to be unity.

Surface albedo (α). In contrast to the almost universal assumption that $\mathcal{E} = 1$ in current general circulation models there exists a greater variety of approaches to the specification of surface albedo as indicated in Table 2. In reality α depends on the solar zenith angle, the spectral distribution of the solar radiation incident on the surface and whether that radiation is direct on diffuse, as well as on the character of the surface as determined by the vegetation (its type, density and state), the soil type, the soil moisture and whether the surface is snow or ice covered. Although generally a long way removed from representing the full complexity of the surface albedo's functional dependence on all such quantities α is usually accorded some variation with the broad character of the surface and indeed in the majority of current models has a specified geographical dependence.

The three basic existing approaches for snow- and ice-free surfaces may be summarized in order of increasing complexity:

(1) A single fixed value for so-called bare land; for example, 0.2 in the CCSAS model and in the MO 11-layer model and 0.14 in versions of the GISS, GLAS and UCLA models. The prime surface distinction other than allowing for snow- and ice-covered surfaces is ascribing an ocean surface albedo of 0.06 at MO, 0.07 at GISS, GLAS and UCLA and 0.1 at CCSAS. Since it is rare at present to employ a fully interactive ocean surface in a general circulation model and constrain the ocean surface temperature to obey a surface energy balance equation it is impossible to determine how the specification of any particular value of the sea-surface albedo affects the radiative balance of the model's atmosphere. This will depend very much on what provision is made in the radiation scheme to deal with solar insolation reflected from the surface. In many cases the actual value allotted to ocean albedo is immaterial and it is other aspects of the distinction between land and sea such as the ocean's infinite thermal capacity and its very different surface roughness length which are the important distinctions.

A slight extension to this approach was used in the August simulation with the GLAS model reported by Halem et al. (1979). Following Charney et al. (1977) vegetated land and deserts have been delineated with albedos of 0.14 and 0.35 respectively.

(2) Albedo specified as a function of latitude only. A good example is the use of values in the MO 5-layer model (see Table 3) which were obtained by linear interpolation from the average values at $10°$ latitude intervals listed by Corby et al. (1977). A full account of the derivation of the latter values using geographical surveys of vegetation and published values of the albedo of different types of surface is given by Rowntree (1975). The ANMRC spectral model is another which currently employs latitudinally varying albedo over land (and sea) derived in this case from the global distributions of surface albedo prepared by Posey and Clapp (1964). It may be noted that Rowntree (1975) considered the popular Posey and Clapp values along with several others in his derivation of the latitudinal dependence of α but adopted values generally higher than would be obtained directly from their maps mainly because he used a higher albedo for forests, grassland, etc. equatorward of $40°$ latitude.

Table 2 — Summary of surface albedos used in atmospheric general circulation models

MODEL	REFERENCES	SNOW- AND ICE-FREE SURFACES	SNOW- AND ICE-COVERED SURFACES
AES	Boer and McFarlane (1979)	No specific details but implied geographical distribution based on Posey and Clapp (1964).	Follows Holloway and Manabe (1971) - see GFDL. Equation (2)
ANMRC	McAvaney et al (1978)	Latitudinal variation based on Posey and Clapp (1964).	Snow albedo prescribed in latitudinal variation of α. α (sea-ice) = 0.7.
CCSAS	Marchuk et al (1979)	$\alpha = \begin{cases} 0.2 & \text{bare ground} \\ 0.1 & \text{ocean} \end{cases}$	Snow: $\alpha = 0.2 + 0.4\, d_{sw}$ $\alpha \leq 0.6$ Equation (3) - same as NCAR Ice: $\alpha = 0.6$
GFDL	Holloway and Manabe (1971)	Geographical distribution based on Posey and Clapp (1964).	Snow: $\alpha = \begin{cases} \alpha_\ell + (0.6-\alpha_\ell)d_{sw}^{1/2} & d_{sw} < 1\,cm \\ 0.6 & d_{sw} \geq 1\,cm \end{cases}$ Equation (2) Poleward of 75° lat., albedo for land and pack-ice 0.75
GLAS	Halem et al (1979)	Feb Geographical distribution based on Posey and Clapp (1964) Aug Charney et al (1977) Vegetated land 0.14 Desert 0.35 Ocean 0.07	Snow/Ice 0.70 Holloway and Manabe (1971) Equation (2)
GISS	Somerville et al (1974)	Land 0.14 Sea 0.07	Ice and Snow 0.7

Table 2 (Continued)

MODEL	REFERENCES	SNOW- AND ICE-FREE SURFACES	SNOW- AND ICE-COVERED SURFACES
MO	Corby et al (1977) Slingo (1979)	5-level model: Snow-free land values vary with latitude – see Table 3 Range 0.150–0.223.	Snow: $\alpha = \alpha_\ell + 0.38 \, d_{sw}^{1/2}$ d_{sw} in cm $\alpha \leq 0.6$ Similar to Holloway and Manabe (1971). Equation (2). Sea-ice and permanent snow cover: $\alpha = \begin{cases} 0.5 & T_0 < 271.2 \, K \\ 0.5 & T_0 \geq 271.2 \, K \end{cases}$
	Saker (1975)	11-level model: Land 0.2 Sea (where effective) 0.06	Transient snow cover 0.5 Permanent snow cover, land- and sea-ice 0.8.
NCAR	Washington and Williamson (1977)	Geographical distributions based on Posey and Clapp (1964).	Snow or ice: Equation (3) $\alpha = 0.2 + 0.4 \, d_{sw}$ $\alpha \leq 0.6$
	Washington et al (1979)	Third-generation model: dependence on vegetation.	
OSU	Schlesinger and Gates (1979)	Geographical distribution based on Posey and Clapp (1964) and models 9 surface types.	Fixed value.
RAND	Gates and Schlesinger (1977)	Geographical distribution based on Posey and Clapp (1964).	
UCLA	Arakawa (1972)	Bare soil 0.14 Ocean 0.07	Snow covered 0.7 Ice-covered soil or sea water 0.4

Table 3 - Land surface snow-free albedo for each row of the
MO 5-level model (Corby et al., 1977). These values
apply for all times of the year.

LATITUDE	ALBEDO	
	Northern Hemisphere	Southern Hemisphere
88.5		
85.5		
82.5		
79.5	0.18	
76.5		
73.5		
70.5		
67.5	0.172	
64.5	0.163	0.19
61.5	0.154	
58.5	0.152	
55.5	0.158	
52.5	0.165	
49.5	0.172	
46.5	0.183	
43.5	0.194	
40.5	0.205	
37.5	0.208	0.198
34.5	0.210	0.209
31.5	0.211	0.219
28.5	0.214	0.221
25.5	0.218	0.214
22.5	0.222	0.207
19.5	0.223	0.200
16.5	0.212	0.191
13.5	0.201	0.183
10.5	0.190	0.174
7.5	0.177	0.166
4.5	0.165	0.158
1.5	0.152	0.150

(3)   A specified geographical distribution of albedo. Here the common approach follows Holloway and Manabe (1971) who used Posey and Clapp (1964) to introduce a geographical distribution of surface albedo for snow-free continental and oceanic surfaces into the GFDL general circulation models. Posey and Clapp derived global maps of climatological average values of α for the months of January, April, July and October (NB. Their Northern Hemisphere maps are clearly mislabelled and should read July, April, October and January for Figures 2-5 respectively) based on the twelve characteristic albedo values for natural surfaces given in Table 4 and using what is known about the different types of surface vegetation and ice and snow cover in all regions of the globe. Regional surface albedos based on the Posey and Clapp maps have also been used at OSU, RAND, NCAR, most probably at AES and also in the February simulation with the GLAS model reported by Halem et al. (1979). The actual fields used in January and July simulations with the RAND 2-layer model have been discussed and displayed by Gates and Schlesinger (1977). The separate and generally different values assigned to each grid point are summarized in their figures in five

Table 4 - Characteristic albedo values for natural surfaces used by Posey and Clapp (1964). The values were derived specifically for the Northern Hemisphere but similar values appear to have been used in the construction of the Southern Hemispheric maps.

	Surface	Albedo
1.	Ice in water navigable by icebreakers	0.35
2.	Ice in water navigable by heavy ships	0.25
3.	'Solid' pack-ice covered by snow	0.45 - 0.80 with variations made to agree with Larsson and Orvig.
4.	Snow in open fields, tundra, prairie	
	(a) $60°$ latitude and northward	0.70 - 0.80
	(b) $50-60°N$	0.60
	(c) Less than $50°$ lat.	0.55
5.	Snow in coniferous forest	0.20
6.	Snow in deciduous forest	0.30
7.	Deciduous forest without snow	
	(a) $60°$ lat. and northward	0.18
	(b) $50-60°N$	0.15
	(c) $40-50°N$	0.12
	(d) Less than $40°N$	0.07
8.	Coniferous forest without snow	
	(a) $60°$ lat. and northward	0.16
	(b) $50-60°N$	0.14
	(c) $40-50°N$	0.10
	(d) Less than $40°N$	0.07
9.	Fields, grassland, tundra and steppe	
	(a) $60°$ lat. and northward	0.19
	(b) $50-60°N$	0.17
	(c) $40-50°N$	0.14
	(d) Less than $40°N$	0.10
10.	Dry savannahs and semi-deserts	0.20
11.	Deserts	0.30
12.	Tropical rain forest	0.07

different categories, viz., dense forests (0.07-0.15); tundra in high latitudes or scattered forests in middle and low latitudes (0.16-0.19); patchy snow in high latitudes or desert, steppe or grasslands in middle and low latitudes (0.20-0.39); partial snow or ice cover (0.40-0.69); permanent ice and snow cover (0.70-0.90). The albedo over the oceans is a function of latitude only and corresponds to the average seasonal zenith angle of the sun. As a result of more recent development of this model at OSU α has been made a function of the models' nine prescribed surface types but with the values still based on the data of Posey and Clapp (see Schlesinger and Gates (1979)).

The specification of α in the NCAR models is very much a function of the state of development of the model. In the second-generation model discussed by Kasahara and Washington (1971) α is purported to be a function of the character of the surface and has a geographical distribution based upon the classification system of Budyko (1956) which takes into consideration the type of soil condition, the type of predominant vegetation and whether the surface is ice- or snow-covered. According to Washington and Williamson (1977) more recent versions have derived distributions from Posey and Clapp and it is likely that further developments are incorporated into the third-generation model described briefly by Washington et al. (1979) where different radiative properties are acknowledged for, for example, vegetated and non-vegetated land and where the ocean albedo has a zenith-angle dependence.

The geographical distribution of surface albedos used in the recently developed operational, 15-level, global model at the European Centre for Medium-range Weather Forecasts (ECMWF) (see, for example, Burridge and Haseler (1977) and Tiedtke et al. (1979)) was derived from Nimbus 3 satellite data by Preuss and Geleyn (1980).

Albedo of snow- and ice-covered surfaces. The most important feedback mechanism involving surface albedo allowed for in the current series of general circulations models is the coupling between snow and ice cover, albedo and the surface temperature. An important characteristic of snow- and ice-covered surfaces is their high reflectivity compared with other natural surfaces such that even a thin covering of fresh snow can drastically alter the albedo of a landscape. The local albedo of a snow-covered surface is very variable and a complicated function of many factors including the age of the snow pack (α decreases markedly as the snow becomes compacted and soiled), the wavelength and angle of the incident radiation and even diurnal cycles in the state of the snow surface, particularly when conditions are right for surface melting. The albedo may lie anywhere in the range from 0.95 for freshly fallen snow to about 0.35 for old, slushy snow.

At present the coupling between snow and ice and the surface albedo is generally prescribed very simply. Three types of snow- or ice-covered surfaces are generally acknowledged:

(1) Surfaces with instantaneously variable depth of snow either predicted or implied.

(2) Permanent or seasonally prescribed snow- and ice-covered land surfaces.

(3) Permanent or seasonally prescribed areas of sea-ice.

Holloway and Manabe (1971) used the results of a study by Kung et al. (1964) to introduce the following simple dependence of albedo on snow depth into the GFDL model:

$$\alpha = \begin{cases} \alpha_\ell + (\alpha_s - \alpha_\ell) d_{sw}^{1/2} & d_{sw} < 1 \text{ cm} \\ \alpha_s & d_{sw} \geq 1 \text{ cm} \end{cases} \quad (2)$$

where α_l is the snow-free land surface albedo,

α_s is the albedo of a deep-snow surface, assumed to be 0.60

and d_{sw} is the water equivalent depth of snow (here expressed in cm) assumed to be 1/10 of the actual snow depth. No allowance is made at present for the varying density of a snow pack. Therefore the assumption is that when the grid-point snow depth is greater than or equal to 10 cm, the value of the surface albedo is independent of snow depth and equal to 0.60. Also following Kung et al. (1964) the albedo over land and pack-ice poleward of $75°$ was assumed to be 0.75 everywhere. (N.B.: A slightly different version of the last statement appears in Manabe, Holloway and Spelman (1974) who in describing the same model state that 'when pack-ice is present over the sea, an albedo of 0.35 is used. Poleward of latitude $70°$, however, the albedo over snow-covered land and over pack-ice is assumed to be 0.70 everywhere.') The same scheme has been used at AES and GLAS. A very similar dependence of α on the square root of the snow depth has been adopted in the most recent version of the MO 5-layer model with an upper limit of 0.60. In specified areas of sea-ice and permanent snow cover where T_o is not allowed to exceed 273 K a value of 0.8 is generally used but with 0.5 if T_o is near 273 K. All land is assumed ice-covered south of $60°S$ and over ground above 1000m above mean sea level north of $66°N$. In earlier versions of this model (see for example Gilchrist (1979) and Corby et al. (1977)) snow depth was not accumulated in the model, but land surfaces (other than those covered permanently by snow and/or ice) were deemed to be snow-covered if $T_o < 269$ K and allotted a fixed albedo of 0.5.

In both the NCAR and CCSAS models where the surface albedo is modified by snow it is given by

$$\alpha = 0.2 + 0.4 \, d_{sw} \quad (d_{sw} \text{ in cm}) \qquad (3)$$

again with the restriction that $\alpha \leq 0.6$, the maximum value being first achieved when $d_{sw} = 1$ cm, approximately equal to 10 cm snow depth. The only essential difference between this formula and that of Holloway and Manabe (1971) is that α depends linearly on d_{sw} rather than on $d_{sw}^{1/2}$.

In other models which predict and monitor the mass of snowfall, such as the MO 11-layer model, the OSU model and the UCLA model of Arakawa (1972), a single albedo value is still used for any non-zero depth of snow. The first snowfall on a previously snow-free surface results in an immediate increase in surface albedo which will tend, at least initially, to accelerate the positive feedback of a further lowering of the surface temperature with an enhanced probability of further snow accumulation.

McAvaney et al. (1978) report that the albedo of land and sea in the ANMRC model varies latitudinally only as derived from the charts of Posey and Clapp with the rider that sea-ice albedo is fixed at 0.7. The implication is that the effect of snow on the surface albedo is prescribed with a fixed latitudinal dependence only. In the January simulation with the GISS model of Somerville et al. (1974) $\alpha = 0.7$ was used for ice and snow surfaces. The ice locations were fixed whereas the snow line in each hemisphere varied throughout the integrations as prescribed by a simple function of the time of year.

As noted above when ε was discussed, the surface radiative properties are bound up with other aspects of the radiation and cloud treatments and in some instances for example, an 'effective surface albedo' may be defined which is a modification of the clear-sky surface albedo to allow for multiple reflections of the solar radiation between clouds and the ground (e.g. Slingo (1979)).

3. SURFACE TURBULENT EXCHANGES

The surface turbulent exchanges of momentum, heat and moisture are parameterized such that they can be determined from values of wind, temperature and humidity evaluated within the model and from values of surface properties such as surface roughness, temperature and wetness. It is standard practice to represent the mean vertical surface turbulent flux, F_x, of some conservative quantity whose mean value is x, by

$$F_x = \overline{(\omega' x')}_o = - C_x \cdot V(z_\ell) \, \Delta x(z_\ell) \qquad (4)$$

where $\overline{(\omega' x')}_o$ defines F_x in conventional turbulence notation (i.e. it is the eddy covariance of x and the vertical velocity component ω),

$V(z_\ell)$ is the mean wind speed at some specified height within the boundary layer, z_ℓ above the surface,

$\Delta x(z_\ell) = x(z_\ell) - x_o$ is the difference between the value of x at z_ℓ and its surface value, x_o,

and C_x is the so-called bulk transfer coefficient.

Formally, equation (4) defines C_x in a strictly mathematical sense, independent of any physical modelling of the turbulent transport, but gives no information about its character. Theoretical and field observational studies indicate that, in general, C_x is different for each x and is a complicated function of height, atmospheric stability, surface roughness and other physical and physiological characteristics of the surface vegetation. It is appropriate at this stage to introduce a further, commonly used expression for F_x, that is,

$$F_x = - u_* x_* \qquad (5)$$

where u_* is the surface friction velocity defined by $u_* = \sqrt{\tau/\rho}$, where τ is the magnitude of the surface eddy shearing stress (or the surface turbulent flux of momentum) and ρ is a representative mean density of the air near the surface. Equation (5) defines x_* which is used as a scaling value for x in the surface layer of the atmospheric boundary layer. From relations (4) and (5) note that

$$C_x = \left(\frac{u_*}{V(z_\ell)}\right) \left(\frac{x_*}{\Delta x(z_\ell)}\right) \qquad (6)$$

The complete set of turbulent fluxes expressed in a form appropriate for numerical modelling is:

Momentum Flux (τ)
$$-\frac{\tau}{\rho} \equiv F_v = \overline{(\omega' v')}_o = - u_*^2 = - C_D V^2(z_\ell). \qquad (7)$$

Note that the horizontal shearing stress at the surface $\underset{\sim}{\tau}$ is conventionally a vector measure of the downward flux of horizontal momentum directed parallel to the limiting wind direction as the surface is approached. Formula (7) relates only the magnitude of $\underset{\sim}{\tau}$ to the wind speed $V(z_\ell)$ through the drag coefficient C_D. The vectorial character of $\underset{\sim}{\tau}$ remains to be determined but it can be represented conveniently by

$$\underset{\sim}{\tau} = \rho\, C_D\, V(z_\ell)\, \underset{\sim}{V}^s \qquad (8)$$

where V^s is a velocity of magnitude $V(z_\ell)$ and direction parallel to the surface wind. The various techniques used to determine the components of $\underset{\sim}{\tau}$ will be discussed below.

Sensible Heat Flux (H)

$$\frac{H}{\rho c_p} \equiv F_\theta = \overline{(w'\theta')}_o = -u_* \theta_* = -C_H V(z_\ell) \Delta\theta(z_\ell) \quad (9)$$

where θ is mean potential temperature and c_p is the specific heat of air at constant pressure.

Water Vapour Flux (E)

$$\frac{E}{\rho} \equiv F_q = \overline{(w'q')}_o = -u_* q_* = -C_E V(z_\ell) \Delta q(z_\ell) \quad (10)$$

where q is mean specific humidity.

The modelling problem is to achieve realistic estimates of the above surface fluxes at all parts of the globe constrained at present, not primarily by the inadequacies of available theoretical, experimental and observational studies, but by limitations such as too coarse vertical and horizontal resolutions imposed by the general structure of the general circulation models. To determine the fluxes the C_x (or the $x_*/\Delta x(z_\ell)$) must be prescribed or expressed in terms of modelled variables and parameters, the level z_ℓ must be identified, $V(z_\ell)$, $\theta(z_\ell)$ and $q(z_\ell)$ must be explicitly predicted or diagnosed within the model, the direction of the surface shearing stress must be estimated or prescribed and the surface temperature and humidity need to be known. The provision of a surface temperature, T_o, will be discussed in Section 4. The surface specific humidity, q_o, is not easy to predict explicitly and is therefore a measure not readily available from numerical models. In most cases its implied value is inextricably linked to the parameterization of the surface hydrology and its discussion is more appropriately left until Section 5. It may be assumed for the remainder of this section that q_o takes the surface saturation value $q_{o,sat}$ which is readily obtained as a function of the surface temperature and pressure from the Clausius-Clapeyron equation. The corresponding water vapour flux, E_p, thus evaluated is defined as the potential evaporation rate.

Simple specifications of C_x

A large variety of specifications of C_x is displayed in the general circulation models of Table 1 ranging from single fixed values to more advanced and complicated formulations based on boundary-layer theories of the past decade or so. The simplest approach is to be found, rather unexpectedly, in the GFDL models used for general circulation and climate studies. Manabe (1969) prescribes a universal surface roughness length, $z_o = 1$ cm, selects z_ℓ as the height of the lowest prognostic level (approximately 75m in the 9-layer model) and evaluates a drag coefficient appropriate for the logarithmic wind profile of a neutrally stable surface layer, i.e.,

$$C_D = \frac{k^2}{[\ln(z_\ell/z_o)]^2} \simeq 2 \times 10^{-3} \quad (11)$$

where k is the von Kármán constant (taken as 0.4).

The lowest prognostic level at z_ℓ is assumed to be close enough to the surface to use $\underline{V}^s = \underline{V}(z_\ell)$ in equation (8) to determine both the magnitude and the direction of the surface stress. It is further assumed that $C_H = C_E = C_D$ in all circumstances. Not only does this GFDL formulation fail to differentiate between C_D, C_H and C_E and ignore any dependence of these bulk transfer coefficients on boundary-layer stability but, perhaps more strikingly, it does not allocate different values over land and sea. It is particularly difficult to appreciate why it has not been deemed important by this stage to discriminate at least between the roughness characteristics of land and sea considering the results of the thorough testing of a hierarchy of surface flux parameterizations by, for example, Delsol et al. (1971) and Miyakoda and Sirutis (1977) in various versions of 9- and 18-level GFDL models. The most substantial effects noted by Delsol et al. (1971) (in 14-day, January, hemispheric forecasting integrations) were caused by varying the surface drag coefficient between land and sea.

Until the most recent developments in the third-generation model alluded to by Washington et al. (1979) the NCAR models have also relied on a simple specification of a single drag coefficient. In both the 6-level, second-generation model (Kasahara and Washington, 1971) and the 8-level, third-generation model (Washington et al. (1979) and Washington and Williamson (1977)) the lowest layer of 3 km thickness (in the absence of mountains) is too deep to capture any of the boundary-layer structure and so a shallow surface layer, in which the vertical fluxes of momentum, heat and moisture are supposed constant with height is simulated in order to evaluate the surface fluxes. The thickness of this layer is not defined explicitly but is assumed to be typically within the range 20-200m to include the anemometer level at about 10m. The surface fluxes are derived from the formulae (7)-(10) where the variables and the bulk transfer coefficients at z_ℓ are appropriate to the 10m-level and $\underline{V}^s = \underline{V}(z_\ell)$. Washington and Williamson (1977) give $C_D = C_H = 3 \times 10^{-3}$ with the coefficient for potential evaporation, $C_E = 0.7\, C_D$. In some earlier versions of the model in which soil moisture was not computed explicitly (see, for example, Kasahara and Washington (1971)) the surface evaporation was not computed in this way but was defined by $L_e E = H/B$ where L_e is the latent heat of vaporization (or sublimation) and B is the Bowen ratio, assumed to be unity, thereby decoupling the evaporative flux from the hydrological cycle. The anemometer level quantities $\underline{V}(z_\ell)$, $\theta(z_\ell)$ and $q(z_\ell)$ are not determined like the normal prognostic variables at the model's explicit levels but are obtained by an assumption of continuity of turbulent fluxes at the interface of the implied surface layer and the Ekman layer. Ekman-layer fluxes derived from flux-gradient relationships requiring the specification of boundary-layer eddy diffusivities are equated to the surface layer formulae (7)-(10) to provide the anemometer level quantities. Boer and McFarlane (1979) use essentially the same method in the AES model with the exception that $C_E = C_D$ in equation (10) for the potential evaporation. In order to ensure reasonable amounts of heat and moisture transfer across the surface in light-wind conditions it is necessary in such formulations to set a minimum value for $V(z_\ell)$. Holloway and Manabe (1971) used a minimum value of 1 ms^{-1} in the GFDL 9-layer model and Washington and Williamson (1977) report a value of 5 ms^{-1} used in the NCAR formulation.

An equally simple scheme is used in the 9-level, ANMRC model (Bourke et al., 1977) which allows a marked distinction between the drag coefficients of land and sea:

$$C_D = C_H = C_E = \begin{cases} 4 \times 10^{-3} & \text{over land} \\ 1 \times 10^{-3} & \text{over sea and sea-ice.} \end{cases} \quad (12)$$

As in the GFDL formulation z_ℓ corresponds to the lowest model level and $\underline{V}^s = \underline{V}(z_\ell)$.

The surface flux formulations of several models stem originally from the parameterizations used in the 2- and 3-level UCLA models designed by Mintz and Arakawa. The current GLAS 9-level model (Halem et al., 1979) is the latest in the line of model development which began at UCLA, particularly with the 3-level model described by Arakawa (1972), and continued with the 9-level model at GISS (Somerville et al., 1974). The current OSU 2-level model (Schlesinger and Gates, 1979), derived from the RAND 2-level model discussed by Gates and Schlesinger (1977), has evolved much more obviously from the UCLA, Mintz-Arakawa 2-level model documented comprehensively by Gates et al. (1971). It should again be acknowledged that the boundary-layer and surface-flux treatments in the latest versions of these models have undergone and are still undergoing continuous modification from the original formulations. However, it is still instructive, in the context of the evolution of the parameterization of the surface fluxes and perhaps for appreciating some of the results abundant in the literature from earlier versions of these models, to describe briefly their surface-flux formulations. In each case, as with the NCAR formulation, the bulk formulae (7)-(10) determining the surface fluxes are expressed in terms of near-surface variables not predicted explicitly within models of such coarse vertical resolution. The near-surface wind is in each case determined from extrapolation of the winds at higher levels towards the surface and the near-surface temperatures and humidities are obtained as described previously by equating the surface fluxes of sensible heat and water vapour to the same fluxes at the bottom of the Ekman layer.

In the Mintz-Arakawa model, as described by Gates et al. (1971), C_H and C_E were equated to the surface drag coefficient C_D, defined independently of stability by

$$C_D = \begin{cases} 2 \cdot 10^{-3} \left(1 + \dfrac{3 z_g}{5000}\right) & \text{over land, ice, snow} \\ \min\left[10^{-3}(1 + 0.07 V(z_\ell)),\; 2.5 \cdot 10^{-3}\right] & \text{over ocean} \end{cases} \quad (13)$$

where z_g is the surface elevation in metres above the mean sea level.

Note here the distinction between ocean and other surfaces. The range of C_D over the ocean from about 10^{-3} in light winds to a maximum of 2.5×10^{-3} for winds greater than about 20 ms^{-1} is intended to reflect the changing character of the surface roughness as the sea surface responds to increasing near-surface winds. Although there remains some debate about the exact nature of the relationship between z_o and the near-surface wind speed (see, for example, Smith and Carson (1977)) the relationship proposed here is roughly supported by some ocean data. It is not the purpose of this paper to dwell upon the details of the parameterization in general circulation models of sea-surface processes but it is pertinent to note that the proper distinction and contrast of the surface characteristics of the ocean from those of the land should be an important consideration in the design of a climate model.

Note that the land-surface drag coefficient in (13) varies between 2×10^{-3} over lowlands and sea-ice to about 7×10^{-3} over the higher mountains. This increase of C_D with z_g is a naive attempt to simulate the increased roughness of mountainous terrain as suggested by Cressman (1960). The specification of the effective drag coefficients of mountains is a particularly difficult problem about which comparatively little is known (see, for example, Smith and Carson (1977)) and formulations such as (13) which relate C_D to surface elevation are not only bound to be a gross oversimplification but are clearly quite inappropriate over certain elevated but extensive flat areas. A not extreme example is the prairies at O'Neill (elevation 614m), Nebraska chosen as the site of a major boundary-layer field experiment (Lettau and Davidson, 1957) for their uniformly and extensively flat terrain.

Formulation (13) was used in the RAND version of the Mintz-Arakawa model discussed by Gates and Schlesinger (1977) but according to Schlesinger and Gates (1979) the treatment of the boundary layer has been modified substantially in the most recent OSU version of the model and it is implied that the bulk transfer coefficients now have a stability dependence. A similar approach was adopted by Arakawa (1972) with the addition of a dependence of C_D on the stability of the surface layer. Let $(C_D)_N$ defined by (13) be the value of C_D for neutral stability; then Arakawa's (1972) formulation is

$$C_D = \begin{cases} (C_D)_N \left(1 + \dfrac{7\,\Delta T(z_\ell)}{V^2(z_\ell)}\right)^{-1} & \Delta T \geq 0 \\ & \text{(stable conditions)} \\ \\ (C_D)_N \left(1 + \sqrt{\dfrac{|\Delta T(z_\ell)|}{V^2(z_\ell)}}\right) & \Delta T \leq 0 \\ & \text{(unstable conditions)} \end{cases} \quad (14)$$

Note that the actual temperature, T, appears to have been used here and not the potential temperature as used in (9). In this instance the 'surface layer' has been defined as being neutrally stable when the near-surface air temperature $T(z_\ell)$ equals the surface temperature T_0. Equality is still assumed for C_D, C_H and C_E but now the bulk transfer coefficients for each type of surface are expressed effectively in terms of the surface-layer bulk Richardson number R_{i_B} $\left(\propto \Delta T(z_\ell)/V^2(z_\ell)\right)$ such that C_D decreases with increasing stability in accordance with observational studies and boundary-layer theories. (Note that the ocean-surface neutral C_D defined in Arakawa (1972) Chapter X is in error and is intended to be as defined above in equation (13).) This particular specification was adopted in the GISS 9-level model of Somerville et al. (1974) and certain aspects may still be used in the GLAS 9-level model of Halem et al. (1979).

A further scheme influenced very strongly by Arakawa's (1972) formulation is that in the CCSAS 3- and 5-level models (Marchuk et al., 1979). In neutral or stable conditions the transfer coefficients $(C_x)_{N,S}$ for land and ice are given exactly by the corresponding neutral value of Arakawa (1972) given in (13). However, over an ocean surface the formulae become

$$(C_D)_{N,S} = \begin{cases} 0.5 \cdot 10^{-3} \sqrt{V(z_\ell)} & \text{if } V(z_\ell) \leq 15 \text{ ms}^{-1} \\ \\ 2.6 \times 10^{-3} & \text{if } V(z_\ell) > 15 \text{ ms}^{-1} \end{cases} \quad (15)$$

and

$$(C_H)_{N,S} = (C_E)_{N,S} = \begin{cases} (C_D)_{N,S}/1.2 & \text{if } V(z_\ell) < 5 \text{ ms}^{-1} \\ (C_D)_{N,S} & \text{if } 5 \text{ ms}^{-1} \leq V(z_\ell) \leq 10 \text{ ms}^{-1} \\ (C_D)_{N,S}/0.7 & \text{if } V(z_\ell) > 10 \text{ ms}^{-1}. \end{cases} \quad (16)$$

A stability factor, exactly as described in Arakawa (1972), is applied in unstable conditions only such that for all types of surface,

$$C_D = (C_D)_{N,S} \left(1 + \sqrt{\frac{|\Delta T(z_\ell)|}{V^2(z_\ell)}} \right) \qquad (17)$$

and

$$C_H = C_E = (C_H)_{N,S} \left(1 + \sqrt{\frac{|\Delta T(z_\ell)|}{V^2(z_\ell)}} \right) \qquad (18)$$

It is intriguing that Marchuk et al. (1979) should choose to adopt Arakawa's (1972) stability modification over land in unstable conditions and yet dispense with it in stable conditions when the coefficients are known to fall rapidly from their near-neutral values to the very small values characteristic of stably stratified turbulent flows. It is also interesting to note the decision to differentiate between C_D and C_H (and C_E) over the sea in both light- and strong-wind conditions. The reasons for these decisions are not given in Marchuk et al. (1979).

C_x from boundary-layer similarity theories

It remains to discuss two distinct approaches to formulating the bulk transfer coefficients which stem from recent, widely accepted similarity theories pertaining to the atmospheric boundary layer. One, the so-called Rossby number similarity theory is applicable throughout the whole depth of the boundary layer and is most appropriate for use in numerical models with relatively coarse vertical resolution and is therefore used as the basis of the formulation used in the MO 5-layer model (Corby et al., 1977). The other, the Monin-Obukhov similarity theory, is suitable for models which either imply the existence of the atmospheric boundary layer's shallow surface layer as described above or have sufficient vertical resolution to represent explicitly the main atmospheric variables at a level within this layer. The latter is assumed to be the case in the MO 11-layer model whose lowest level is assumed to be about 100m above the surface. There exists fairly extensive literature on both similarity techniques and for further general discussions the reader is referred to, for example, Sheppard (1969) or Haugen (1973) or to the very recent comprehensive review of the boundary layer by the CAS Working Group on Atmospheric Boundary-layer Problems (McBean et al., 1979). Both approaches were examined very fully for the specific benefit of the numerical modeller by Clarke (1970).

C_x from Rossby number similarity theory

In its simplest form the Rossby number similarity hypothesis (Method II of Clarke (1970)) states in particular that u_*/V_G, $\theta_*/\Delta_T\theta$, $q_*/\Delta_T q$ and α_o, the angle of turning of the surface wind from the surface geostrophic wind direction, are functions of the surface Rossby number,

$$R_o = \frac{V_G}{f z_o}, \qquad (19)$$

and an external stability parameter,

$$S = \frac{g \, \Delta_T \theta}{\bar{\theta} f V_G}, \qquad (20)$$

only, where V_G is the magnitude of the surface geostrophic wind, Δ_T denotes the difference in a property of its value at the top of the boundary layer from its surface value, f is the Coriolis parameter, g is the acceleration due to gravity and $\bar{\theta}$ is a representative value of potential temperature in the atmospheric boundary layer. Therefore it follows from (6) that the particular bulk transfer coefficients, conventionally referred to as geostrophic transfer coefficients, viz.,

$$(C_x)_G = \left(\frac{u_*}{V_G}\right)\left(\frac{x_*}{\Delta_T x}\right) \qquad (21)$$

are also functions of R_o and S only. It can also be argued and it is commonly written that u_*/V_g, $\theta_*/\Delta_T \theta$, $q_*/\Delta_T q$, α_o and hence $(C_x)_G$ are functions of R_o and μ_o only, where μ_o is another non-dimensional stability parameter defined by

$$\mu_o = \frac{k u_*}{f L} \qquad (22)$$

where L is the so-called Monin-Obukhov length for the surface layer.

Several authors have displayed the semi-empirically derived continuous functions $(C_x)_G (R_o, S \text{ or } \mu_o)$ and a good practical review and discussion were provided by Arya (1975). In a gross oversimplification of the full results of the application of Rossby number similarity theory, which is nevertheless compatible with its implications, Corby et al. (1977) have incorporated into the MO 5-layer model the values for $(C_D)_G$ and α_o given in Table 5. Assuming Rossby numbers of 10^6 and 10^8 for land and sea respectively (implying typical surface roughness lengths of about 10 cm and 0.1 cm respectively) these values were extracted as being representative of stable and unstable conditions from the curves for $(C_D)_G$ and α_o produced by Monin and Zilitinkevich (1967). The magnitude of the surface stress is estimated by using the appropriate $(C_D)_G$ in (7) where $V(z_l)$ represents the wind speed at the top of the boundary layer assumed equivalent to the geostrophic or gradient wind) and is approximated by either the wind at the lowest level ($\sigma = 0.9$) or, in some versions, by the wind diagnosed at the height of the explicitly carried boundary-layer top. The novel feature of this approach which distinguishes it from the others described above is that the direction of the surface stress is prescribed by the values of α_o in Table 5, i.e. $\underset{\sim}{V}^s$ has magnitude $V(z_l)$ but is rotated through an angle α_o to be parallel to the surface wind.

The corresponding geostrophic transfer coefficients for sensible heat and moisture transfer are discussed fully by Corby et al. (1977) and defined by

$$(C_H)_G = (C_E)_G = (C_D)_G \left[1 + \frac{A}{V(z_l)} \left(\frac{|\Delta_T \theta_v|}{\bar{\theta}_v} \right)^{1/2} \right] \qquad (23)$$

where $\theta_v = \theta(1 + 0.61q)$ is virtual potential temperature, $\bar{\theta}_v$ is a representative value of θ_v throughout the boundary layer and

$$A = \begin{cases} 50 \text{ ms}^{-1} & \text{if } \Delta_T \theta_v < 0 \quad \text{(unstable conditions)} \\ 0 & \text{if } \Delta_T \theta_v \geq 0 \quad \text{(stable conditions)} \end{cases}$$

Table 5. Values of the geostrophic drag coefficient, $(C_D)_G$ and the angle of turning of the surface wind from the geostrophic wind direction, α_0 used in the MO 5-level general circulation model. (See Corby et al. (1977).)

	LAND		SEA	
	$10^3 (C_D)_G$	α_0 (DEG)	$10^3 (C_D)_G$	α_0 (DEG)
STABLE	0.3	30	0.2	20
UNSTABLE	4.0	10	2.0	0

The geostrophic transfer coefficients for momentum, heat and water vapour transfer are thus equal in neutral and stable conditions but the coefficients for heat and water vapour transfer are greater than that for momentum in unstable conditions. The additional enhancing term in (23) defining $(C_H)_G$ and $(C_E)_G$ in unstable conditions has been introduced to ensure adequate fluxes of heat and water vapour from the surface in the approach to pure free or windless convection when dimensional arguments suggest that the magnitudes of these fluxes are independent of wind speed and are proportional to $|\Delta \theta_v|^{1/2} |\Delta \theta|$ and $|\Delta \theta_v|^{1/2} |\Delta q|$, respectively.

C_x from Monin-Obukhov similarity theory

The Monin-Obukhov similarity hypothesis for the fully turbulent surface layer (strictly under stationary and horizontally homogeneous conditions) is now well established and is the most widely accepted approach amongst boundary-layer workers for relating surface-layer vertical gradients of mean wind speed, temperature and specific humidity to the corresponding surface turbulent fluxes. It is the basis of Method I of Clarke (1970) and the formulation for the bulk transfer coefficients used in the MO 11-layer model. For the fully turbulent surface layer this similarity hypothesis suggests the flux-gradient relationships

$$\frac{\partial x}{\partial z} = \frac{x_*}{kz} \varphi_x (z/L) \tag{24}$$

where φ_x is a universal similarity function of z/L only which may be different for each mean transferable property, x. The φ_x have to be established empirically from analysis of surface-layer observations. From (6) and (24) it is easy to show that, having specified the $\varphi_x(z/L)$, then C_x is in general a function of z/L and z/z_0.

Note that in making this statement it has been assumed, as is common practice, that the characteristic 'surface roughness lengths' for heat and water vapour transfer are equal to that for momentum transfer, i.e., the aerodynamic surface roughness length, z_0. C_x can be expressed more suitably for modelling purposes as a direct function of the modelled variables $V(z_\ell), \Delta x(z_\ell)$ by introducing the bulk Richardson number for the layer from the surface to height z_ℓ:

$$Ri_B = \frac{g z_\ell [\Delta\theta(z_\ell) + 0.61 \bar{T} \Delta q(z_\ell)]}{\bar{T} \, V^2(z_\ell)} \qquad (25)$$

where $\bar{T}$ is a representative temperature for the layer. For specified φ_x and z_ℓ the C_x can be expressed as functions of z_0 and the stability measure Ri_B only. For a full description of the method and the assumptions made see, for example, Carson and Richards (1978).

In the MO 11-layer model z_ℓ is the height of the lowest model level (approximately 100m above the surface) and assumed to be in the surface layer. Two categories of surface roughness are admitted,

$$z_0 = \begin{cases} 10 \text{ cm} & \text{for land, snow and sea-ice} \\ 10^{-2} \text{ cm} & \text{for sea.} \end{cases}$$

In earlier versions of the model z_0 over the sea was allowed to vary essentially with wind speed according to the so-called Charnock (1955) formula,

$$z_0 = M \frac{u_*^2}{g} \qquad (26)$$

where M is a 'constant' estimated as about 0.012 by Charnock but later by others to be in the higher range 0.035 - 0.05. In view of the large uncertainty in reducing the complex behaviour of z_0 for a sea surface to such a simple formula it was deemed unwarranted at this stage to use (26) and for general circulation modelling it should suffice to take $z_0 = 10^{-2}$ cm everywhere over the sea. For further discussions and references see, for example, Smith and Carson (1977) and Delsol et al. (1971).

The main problem facing the modeller who adopts this approach is the selection of the φ_x. The general character of the similarity functions is fairly well established over a limited range of stability conditions, centred on neutral, but their specification for more extreme stability conditions (both stable and unstable) is much more debatable and uncertain. The MO 11-layer model generally uses the following formulae:

Unstable and neutral conditions:

$$\varphi_H = \varphi_E = \varphi_M^2 = (1 - 16 z/L)^{-1/2} \qquad 0 \geq z/L \geq -1 \qquad (27)$$

(Dyer and Hicks, 1970)

Stable conditions:

$$\varphi_H = \varphi_E = \varphi_M = \begin{cases} 1 + 5 z/L & 0 < z/L \leq 1 \\ 6 & 1 < z/L < 6 \end{cases} \qquad (28)$$

(Webb, 1970)

Note that the Dyer and Hicks (1970) formulae are strictly limited to $|z/L| < 1$ and so for more unstable conditions another approach was invoked. Consider the buoyancy flux

$$\Psi = \frac{H}{\rho c_p} + 0.61 \overline{T} \frac{E}{\rho} = -u_* (\theta_* + 0.61 \overline{T} q_*). \qquad (29)$$

On the assumption that $C_H \equiv C_E$ this can be written

$$\Psi = -C_H \cdot V(z_\ell) (\Delta\theta(z_\ell) + 0.61 \overline{T} \Delta q (z_\ell))$$

$$= -C_H V(z_\ell) \Delta\psi(z_\ell) \qquad , \text{say}, \qquad (30)$$

where Ψ is very akin to the virtual heat flux and $\Delta\Psi$ to the virtual potential temperature difference. The Monin-Obukhov length, L, is defined by

$$L = -\frac{u_*^3 \overline{T}}{k g \Psi} \qquad (31)$$

and the bulk Richardson number (25) can be expressed as

$$Ri_B = \frac{g z_\ell \Delta\psi(z_\ell)}{\overline{T} V^2(z_\ell)}. \qquad (32)$$

In the free convection limit defined for $|L| \to 0$ when $\Psi > 0$ and $u_* \to 0$, dimensional analysis implies

$$\Psi = A' \left(\frac{g z_\ell}{\overline{T}}\right)^{1/2} |\Delta\psi|^{3/2} \qquad (33)$$

where $\qquad A' = \left\{ 3c \left[(z_\ell/z_o)^{1/3} - 1 \right] \right\}^{-3/2} \qquad (34)$

and c is a constant, determined empirically as 0.84 (in fact $c \equiv H_*^{-2/3}$ where H_* was taken as 1.3 from Clarke (1970) (Appendix 1)). Replacing $V(z_\ell)$ in (30) from (32) and comparing with (33) indicates that in freely convecting conditions

$$C_H = A' |Ri_B|^{1/2} \qquad (35)$$

The technique employed in the MO 11-layer model is to empirically and smoothly extrapolate the functional dependence of the C_x on Ri_B from the moderately unstable values given by the Dyer and Hicks (1970) formulae of (27) to attain the free convection form (35) for large negative Ri_B. The proper behaviour of C_D in very unstable conditions is much more difficult to determine but since the surface stresses are normally quite small in highly unstable conditions C_D is simply assumed constant beyond some empirically determined value of Ri_B. This method leads to C_H (and C_E) $> C_D$ in unstable conditions.

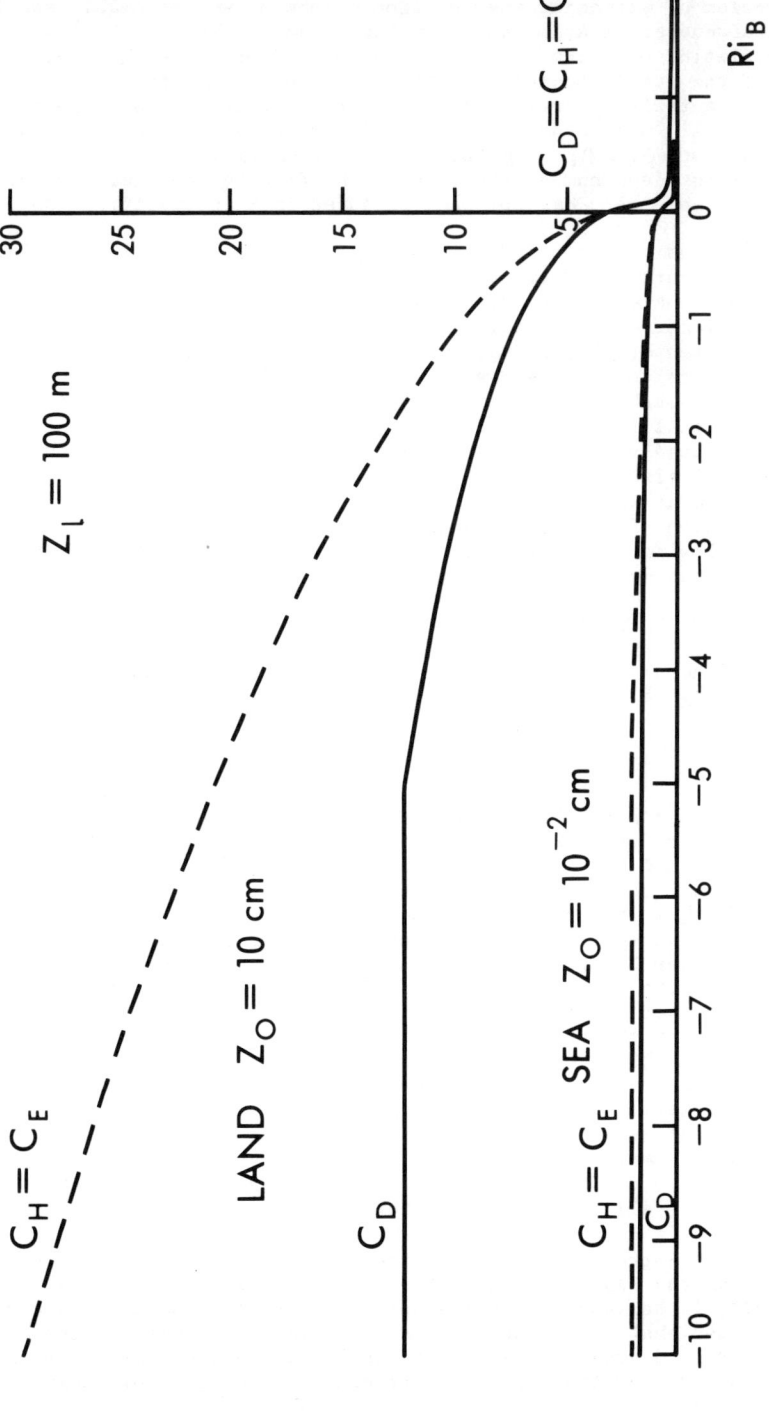

Figure 1. Surface layer bulk transfer coefficients derived from Monin-Obukhov similarity theory and used in the MO 11-level model

The problem of extending the functional form of φ_x to highly stable conditions has been discussed at some length by Carson and Richards (1978) who argued that the recent formulation of Hicks (1976) is that most compatible with both the current understanding of the stable boundary layer and the need in a numerical model to estimate fluxes representative over the area of a grid-box or element. However, in most versions of the MO 11-layer model Webb's (1970) formulae have been used with a lower limiting value set on $C_D = C_H = C_E$ for all Ri_g greater than a certain value. A critical stable Richardson number above which the flow becomes laminar and the surface turbulent fluxes become zero has not been included in this specification.

The continuous variation of C_D and $C_H = C_E$ employed in the MO 11-layer model is illustrated in Figure 1. Formulations of the last type are likely to become much more prevalent in climate models in the near future and it may be that similar schemes have already been incorporated into, for example, the new third-generation, 8-level NCAR model (Washington et al., 1979), the new UCLA, 12-level model (Arakawa and Lamb, 1977) and the OSU model of Schlesinger and Gates (1979). As stated earlier, formulations based on Monin-Obukhov similarity theory have of course been used in medium- to long-range numerical forecasting studies by Delsol et al. (1971) and Miyakoda and Sirutis (1977) who used various 9- and 18-level versions of the GFDL models. The ECMWF 15-level model also employs such a scheme as described by Louis (1979). A more recent development there (see Tiedtke et al. (1979) is the derivation of a geographical distribution of surface roughness lengths which have an implied dependence on vegetation, following Baumgartner et al. (1977) and on a parameter which represents the 'roughness' of the model's orography.

4. LAND-SURFACE TEMPERATURE AND HEAT BALANCE

The most general balance of the energy fluxes used at non-oceanic surfaces in climate models may be written

$$G_o = R_N - H - Q - Q_f + H_I \qquad (36)$$

where $Q = L_i E$ represents the latent heat flux due to surface evaporation or sublimation by turbulent transfer (L_i is the latent heat of evaporation, L_e, or sublimation, $L_s = L_e + L_f$, where L_f is the latent heat of fusion), $Q_f = L_f M_s$ represents the latent heat flux required to effect phase changes associated with melting or freezing at the 'surface' (M_s is the rate of snowmelt and/or icemelt) and H_I is a term included in many models specifically to represent the conduction of heat through sea-ice from unfrozen water below (generally following Holloway and Manabe (1971)). G_o, the net imbalance of the terms on the RHS of (36), represents a flux of sensible heat in the soil (sea-ice) at the surface which, conventionally, is assumed to be positive when directed away from the surface into the soil (sea-ice). All the terms in (36) are, in principle, functions of the surface temperature T_o. With G_o prescribed or evaluated, the common practice in atmospheric general circulation models is to determine T_o such that it satisfies the surface energy balance equation. The variety of techniques used in current general circulation models to specify or predict the soil heat flux term have only recently been reviewed, discussed and compared fairly comprehensively by Bhumralkar (1975) and Deardorff (1978) and so a brief acknowledgement of most of the methods is all that is needed here.

The simplest approach employed in the general circulation models at GFDL, ANMRC, CCSAS and in the RAND version of the Mintz-Arakawa 2-level model (Gates and Schlesinger, 1977) is to assume that the soil has zero heat capacity and set $G_o \equiv 0$. Manabe (1969) argued that it is justifiable to neglect the heat conduction into the soil when there is no diurnal variation of insolation in the model. Other simple parameterizations avoiding the explicit representation of the soil layers express G_o

as a simple function of, for example, H or R_N. For example earlier models at NCAR (see, for example, Washington and Williamson (1977)) have used $G_0 = H/3$ where the constant of proportionality, $\frac{1}{3}$, was chosen by Kasahara and Washington (1971) based on the results of Sasamori (1970).

$$G_0 = \begin{cases} \delta_1 R_N & R_N \geqslant 0 \\ \delta_2 R_N & R_N < 0 \end{cases} \tag{37}$$

is used over snow-free land surfaces in the MO 10-level operational forecasting model with $\delta_1 = 0.1$ and $\delta_2 = 0.5$ as proposed by Gadd and Keers (1970) and has been suggested similarly by Nickerson and Smiley (1975) with $\delta_1 = 0.19$ and $\delta_2 = 0.32$.

More advanced schemes evaluate G_0 from consideration of heat conduction and conservation in the soil. In the absence of horizontal heat transfer and melting or freezing within the soil the heat conservation equation for a normally assumed homogeneous soil reduces to

$$\frac{\partial T_g}{\partial t} = -\frac{1}{C} \frac{\partial G}{\partial y} \tag{38}$$

where T_g is the soil temperature, G is the soil heat flux, C is the volumetric heat capacity of the soil, $y = -z$ is the vertical co-ordinate in the soil and t is time. Heat conduction in the soil is governed by

$$G = -\lambda \frac{\partial T_g}{\partial y} \tag{39}$$

where λ is the thermal conductivity of the soil. Equations (38) and (39) yield the equation of conduction of heat in the soil

$$\frac{\partial T_g}{\partial t} = k \frac{\partial^2 T_g}{\partial y^2} \tag{40}$$

where $k = \lambda/C$ is the thermal diffusivity of the soil.

Several general circulation models use equations (38), (39) and the surface energy balance equation to produce a prognostic equation for evaluating T_0. The simplest approach of this kind requires the introduction of an effective depth D and an effective surface thermal capacity $C^* = CD = \rho_g c_g D$ where c_g is the specific heat capacity of the soil and ρ_g is its density, such that

$$G_0 = C^* \frac{\partial T_0}{\partial t} = CD \frac{\partial T_0}{\partial t} = \rho_g c_g D \frac{\partial T_0}{\partial t} \tag{41}$$

Note that integrating (38) with the assumptions that $T_0 \equiv (T_g)_{y=0}$ and that $G \to 0$ as $y \to \infty$ produces

$$G_0 = C \int_0^\infty \frac{\partial T_g}{\partial t} dy \tag{42}$$

which, if equated to G_o in (41) defines D in a strictly mathematical sense as

$$D = \left(\frac{\partial T_o}{\partial t}\right)^{-1} \int_0^\infty \frac{\partial T_g}{\partial t} d\gamma. \tag{43}$$

Note also that expressing G_o as in (41) does not imply, a priori (as suggested in Bhumralkar (1975)) that the surface temperature T_o has been replaced by and equated to the mean temperature throughout the shallow surface soil layer of finite depth, D, with the neglect of conduction of heat to or from lower soil layers. Rather it is an attempt, if the effective depth D can be determined sensibly, to parameterize the changing heat content of the whole soil layer by using the analogue of a single shallow layer of known thermal capacity C^* which is fully insulated at its lower boundary and whose uniform temperature, T_o, increases or decreases uniformly in response to the surface soil heat flux, G_o. It remains to determine D which appears to have been achieved in most cases by appealing to the theory of heat transfer in a semi-infinite homogeneous medium whose surface is heated in a simple periodic manner (see, for example, Sellers (1965)). If it is assumed that the surface temperature

$$T_o \equiv T_g(0,t) = \hat{T}_g + a_o \sin \omega t \tag{44}$$

where ω is the angular frequency of oscillation, $\hat{T}_g$ is the mean (over the period $P = 2\pi/\omega$) soil temperature, assumed to be the same at all depths, and a_o is the amplitude of the surface temperature wave then the solution of (40) is

$$T_g(\gamma,t) = \hat{T}_g + a(\gamma) \sin(\omega t - \gamma/\delta)$$

$$= \hat{T}_g + a_o \exp(-\gamma/\delta) \sin(\omega t - \gamma/\delta) \tag{45}$$

where $\delta = (kP/\pi)^{1/2}$ is the e-folding depth of the temperature wave of period P, i.e. it is the depth where the amplitude of the oscillation is reduced to e^{-1} times its surface value. The effective depth D corresponding to the soil temperature profile (45) is, from (43),

$$D = \frac{1}{a_o \cos \omega t} \int_0^\infty a(\gamma) \cos(\omega t - \gamma/\delta) d\gamma \tag{46}$$

$$= \frac{1}{\cos \omega t} \int_0^\infty \exp(-\gamma/\delta) \cos(\omega t - \gamma/\delta) d\gamma$$

$$= \frac{\delta}{\sqrt{2}} \cdot \frac{\sin(\omega t + \pi/4)}{\cos \omega t}. \tag{47}$$

Therefore the correctly determined effective depth satisfying (41) varies with time.

In the UCLA model of Arakawa (1972) and the AES model of Boer and McFarlane (1979) G_0 is represented as in (41) with the effective thermal capacity C^* defined as

$$C^* = \left(\frac{\lambda C}{\omega}\right)^{1/2} = \frac{C\delta}{\sqrt{2}}, \qquad (48)$$

i.e. the effective depth D is implied to be simply $\delta/\sqrt{2}$. Arakawa's (1972) scheme has also been adopted in the models at GISS and GLAS (see Somerville et al. (1974) and Halem et al. (1979)).

It is interesting to note that Rowntree (1975) defined the effective depth not as in (43) but in terms of the amplitude of the temperature wave such that

$$D = \frac{1}{a_0} \int_0^\infty a(\mathfrak{z}) \, d\mathfrak{z} = \delta \qquad (49)$$

which from (46) is tantamount to neglecting the changing phase of the temperature wave with depth. Schemes with G_0 defined by (41) and with D equated simply to the e-folding depth δ are used in the MO 5- and 11-layer models.

The predictive equation for T_0 proposed recently by Bhumralker (1975) and Deardorff (1978) (his so-called 'force-restore' method) can be expressed quite naturally and conveniently by replacing D in (41) with its value determined properly by (47), i.e.,

$$\frac{\partial T_0}{\partial t} = \frac{\sqrt{2}\, G_0 \cos \omega t}{C\delta \sin(\omega t + \pi/4)} \qquad (50)$$

This in turn can be expanded to give

$$\frac{\partial T_0}{\partial t} = \frac{2 G_0}{C\delta} - \frac{2\pi}{P}(T_0 - \hat{T}_g) \qquad (51)$$

which is Deardorff's (1978) equation (8a). Following Deardorff (1978), equation (51) and the surface energy balance equation (36) are being used to predict the changes in T_0 in the third-generation NCAR model (Washington et al., 1979) and also in the medium-range forecasting model at ECMWF (private communication). Bhumralkar (1975) has tested a very similar scheme in the RAND model and Boer and McFarlane (1979) have stated their intention to implement such a scheme in the AES model in the near future. Schlesinger and Gates (1979) report that the ground surface temperature has been made a prognostic variable in the OSU model and it seems likely that one of the versions of (41) discussed above is being used. In all the schemes so far the period of the diurnal temperature oscillation has been used as that appropriate for determining the thermal capacity of the effective surface layer.

The Deardorff (1978) method, equation (51), requires additional information about $\hat{T}_g$ which may be fixed or diagnosed over short periods of a few days but would need to be determined prognostically over the much longer periods of integration involved in climate modelling. Deardorff (1978) has suggested a second prognostic equation for $\hat{T}_g$ analogous to equation (41) but with the appropriate effective depth D_a, say, equated to the e-folding depth of the annual temperature wave. The reviewer is currently examining extensions of Deardorff's proposals which make a more deliberate attempt to capture the characteristics of both the diurnal and the annual variations of surface temperature compatible with the simple picture of periodic forcing at the surface. In one such system a third temperature $\hat{T}_g$ (representing the annual mean soil temperature at any level) is introduced and treated as the mean temperature of a very deep soil layer which varies according to an equation like (41) with an appropriately specified thermal capacity. Such parameterizations are then becoming analogous to the more elaborate schemes which explicitly model the heat conduction through several soil layers. No general circulation model yet intended for climate modelling normally incorporates a multi-level soil model but Delsol et al. (1971) and Miyakoda and Sirutis (1977) have used effectively a 2-level soil model in 9- and 18-level versions of the GFDL model and Benoit (1976) has experimented with six soil layers in the 6-level NCAR model. In such schemes G_0 from (39) is determined explicitly from the predicted soil temperature profile by

$$G_0 = \left(-\lambda \frac{\partial T_g}{\partial y}\right)_{y=0} \tag{52}$$

and then equation (36) is solved for T_0.

It remains to describe the thermal characteristics of the soils implied in models which use a finite soil thermal capacity. In the MO 5-layer model (Corby et al., 1977) no attempt has been made to model a variety of surface types and C^* has been given the universal value of 3 cal cm^{-2} K^{-1} (3 cm water equivalent) = 1.25 x 10^5 J m^{-2} K^{-1} representing a compromise for average conditions, for all land and ice surfaces (see Rowntree (1975) for details of this choice). A similarly universal value of 5 cal cm^{-2} K^{-1} = 2.09 x 10^5 J m^{-2} K^{-1} is used in the MO 11-layer model. Arakawa (1972) allows for a variation of C^* with type and state of the surface. He specifies the fixed values

$$C^* = \begin{cases} 5.1 & \text{cal cm}^{-2} \text{ K}^{-1} \quad \text{Over ice surfaces} \\ 2.3 & \text{cal cm}^{-2} \text{ K}^{-1} \quad \text{Over snow surfaces} \end{cases}$$

but merely states without providing the details that C^* over bare-soil surfaces is a function of the surface 'wetness'. A similar approach is adopted in the AES model (Boer and McFarlane, 1979) where Arakawa's (1972) values of C^* are used over ice and snow surfaces and over other land surfaces the quantities λ and C which determine C^* as defined in (48) are expressed as functions of soil wetness:

$$C = 0.20 + \frac{d_{w,max}}{100}\left(\mu_\ell + 0.5\mu_i\right) \quad \text{cal cm}^{-3} \text{K}^{-1} \tag{53}$$

and

$$\lambda = 0.10\left[0.276 + (0.11 + 0.15\mu)(1 - 0.5\frac{\mu_i}{\mu})\right] \quad \text{cal cm}^{-1}\text{s}^{-1}\text{K}^{-1} \tag{54}$$

where $d_{w,max}$ is the field capacity (probably expressed in cm of water) of a 1m-deep surface soil layer and μ, μ_ℓ and μ_i are, respectively, the total depth of soil moisture, the liquid component and the frozen component, normalized by that field capacity. The formula for C is claimed to be based on a formulation in Sellers (1965) adapted for a sandy loam and that for λ is attributed to Arakawa and his colleagues. Both C and λ and hence C^* are therefore allowed to increase with soil wetness in accordance with observations.

The NCAR third-generation model distinguishes between ocean, sea-ice and vegetated and non-vegetated land surfaces and incorporates Deardorff's (1978) proposals for surface temperature and moisture. Implicit in this approach is a dependence of the surface thermal properties on the type of surface, whether it is snow-covered and whether, if soil, it is wet or dry. In his treatment of the soil moisture Deardorff (1978) uses ω, the volume of water per unit volume of soil, with ω_0 the surface value of ω and $\bar{\omega}$ the vertically averaged value of ω over a layer of depth h below which the moisture flux is negligible. Invoking results from Benoit (1976) and

Sellers (1965), Deardorff (1978) has proposed $C\delta$ as used in (51) as a function of W_0 and $\bar{W}$. His tentative suggestion derived empirically from single-column model tests is

$$C\delta = r C_0 \delta_0 + (1-r) \bar{C} \bar{\delta} \qquad (55)$$

where
$$r = 0.30 + 0.05 \frac{W_0}{\bar{W}} \qquad 0.3 < r \leq 1$$

$$C_0 = 0.27 + W_0 \qquad \text{cal cm}^{-3} \text{K}^{-1}$$

$$\bar{C} = 0.27 + \bar{W} \qquad \text{cal cm}^{-3} \text{K}^{-1}$$

and
$$\delta_0 = \left(\frac{\lambda_0 P}{C_0 \pi}\right)^{1/2}$$

$$\bar{\delta} = \left(\frac{\bar{\lambda} P}{\bar{C} \pi}\right)^{1/2}$$

where
$$\lambda_0 = 0.001 + 0.004 \, W_0^{1/3} \qquad \text{cal cm}^{-1} \text{s}^{-1} \text{K}$$

$$\bar{\lambda} = 0.001 + 0.004 \, \bar{W}^{1/3} \qquad \text{cal cm}^{-1} \text{s}^{-1} \text{K}.$$

Note that $C\delta$ from (55) is determined by coupling empirically the surface soil properties and the bulk soil properties. It is not clear from Washington et al. (1979) to what extent relationships of this form have been incorporated already into the NCAR third-generation model (see further comments at the end of Section 5). Note also that great care is needed to ensure the conservation of heat in formulations that allow C to vary as described above.

Finally in this section, the term H_I representing the conduction of heat through sea-ice from unfrozen water below is included in all the most recent versions of the models of Table 1, generally in the form, following Holloway and Manabe (1971),

$$H_I = \frac{\lambda_I (T_w - T_0)}{d_I} \qquad (56)$$

where λ_I is the thermal conductivity of ice, taken as 5×10^{-3} cal cm^{-1} s^{-1} K^{-1} (2.09 watt m^{-1} K^{-1}), d_I is the thickness of the ice in the specified ice-covered areas, assumed to be 2m in some models and 3m in others, and T_w is the temperature of the underlying water, taken as 271.2 K in some models and 273.2 K in others. It is common practice at such sea-ice points to prevent T_0 rising above the prescribed value of T_w. The scant attention given here to the modelling of the distribution, formation and effects of sea-ice reflects the working group's decision that these important topics be considered as outside the confines of land-surface processes in the present context. The interactions of the atmosphere, ice and the oceans in the polar regions have recently been comprehensively reviewed in the context of global climate by Baker et al. (1980).

The latent heat flux terms in (36), viz. Q and Q_f, are more appropriately considered within the context of the surface hydrology which is the subject of the next section.

5. SURFACE HYDROLOGY AND THE SOIL WATER BUDGET

In virtually all of the primary atmospheric general circulation models soil moisture and snow depth are now treated as variables determined by budget equations for water, snow and heat. Interactive surface hydrology was first introduced by Manabe (1969) and most of the formulations in the current general circulation models follow his approach. Where the change of soil moisture content is monitored in a single surface layer of depth h it is altered by rainfall, evaporation (condensation) snowmelt and runoff according to a water mass balance equation of the form

$$\frac{\partial m_w}{\partial t} = P_r - E + M_s - Y_o - Y_h \tag{57}$$

where m_w is the mass of liquid water per unit lateral area in the layer considered, often expressed in the equivalent alternative forms,

$$m_w = \rho_w d_w = \rho_w d_{w,max} \mu = \rho_w h w_{max} \mu \tag{58}$$

where ρ_w is the density of water, $d_w = m_w/\rho_w$ is the representative depth of water in the layer, $d_{w,max}$ is the maximum allowable value of d_w, representing the field capacity of the soil as a depth of water, $\mu = d_w/d_{w,max}$ is a normalized soil moistur content and $w_{max} = d_{w,max}/h$ is the effective porosity of the soil. (Recall Deardorff's (1978) use of w rather than μ, where w is the volume of water per unit volume of soil.) P_r is the intensity of surface rainfall, E is the surface evaporation rate, M_s is the rate of snowmelt and/or icemelt, Y_o is the intensity of surface runoff and Y_h is the intensity of percolation out of the surface layer to lower layers. Apart from an implied surface runoff term Y_o all other horizontal fluxes of soil water are neglected. It is common practice to use a single simple expression, Y, to represent the total 'runoff', $Y_o + Y_h$, from the surface layer.

A corresponding budget equation is used for snow on the 'surface', viz.,

$$\frac{\partial m_s}{\partial t} = P_s - E_s - M_s \tag{59}$$

where m_s is the mass of snow per unit area on the 'surface'; often represented as a snowdepth $d_s = m_s/\rho_s$, where ρ_s is the density of the snow, or more commonly as an equivalent depth of water $d_{sw} = m_s/\rho_w = d_s \rho_s/\rho_w \simeq 0.1 d_s$ (no allowance is made at present for the varying density of a snow pack), P_s is the intensity of snowfall and E_s is the rate of sublimation at the surface.

As with the incoming radiative fluxes at the surface, the precipitation rates P_r and P_s are regarded in the present context as determined externally and so the methods used to produce them in general circulation models will not be discussed. It remains to describe the evaluation of the surface evaporation (sublimation), runoff and snowmelt.

Evaporation. In the discussion of the surface turbulent fluxes in Section 3 it was noted that the surface specific humidity, required in the bulk transfer formula (10) to determine the surface evaporation rate, is not easy to predict explicitly. To overcome this problem it is standard practice to imply a value of q_o through relations with $q_{o,sat}(T_o)$, the saturation specific humidity at the surface which is readily determined as a function of surface temperature (and pressure) via the Clausius-Clapeyron relationship. Two common methods are:

(1) To specify a surface relative humidity r_o such that $q_o = r_o\, q_{o,sat}(T_o)$.

(2) To evaluate a potential evaporation rate

$$E_p = -\rho\, C_E \cdot V(z_\ell)\, (q(z_\ell) - q_{o,sat}(T_o)) \qquad (60)$$

and to specify a moisture availability function, β, (usually ranging from 0 for an arid surface to 1 for a saturated surface) such that the actual evaporation rate

$$E = \beta E_p. \qquad (61)$$

For a discussion and comparison of the two approaches see, for example, Nappo (1975).

The former method with fixed $r_o = 0.8$ over sea and 0.5 over land was used in the version of the MO 5-layer model described by Corby et al. (1977) which did not have variable soil moisture content or snowdepth. The choice of 0.8 over the sea, as opposed to 1, makes an empirical allowance for the fact that z_ℓ in the 5-layer model is a long way from the surface. In the RAND model of Gates and Schlesinger (1977) r_o was equated to the variable normalized soil moisture content, μ, so that potential evaporation could occur only when the soil moisture content had reached field capacity. The second method is by far the more commonly adopted. Fixed values of β can still be found as, for example, in the ANMRC model (McAvaney et al., 1979) where $\beta = 0.25$ over land and snow and 1 over ocean and sea-ice, but the most common method now employed is to express β as a function of the variable soil moisture content and the hydrological state of the surface. Based on the suggestions of Budyko (1956), Manabe (1969) proposed the following linear relationship between β and the soil moisture content,

$$\beta = \begin{cases} \dfrac{\mu}{\mu_p} & 0 \leq \mu \leq \mu_p \\ 1 & \mu > \mu_p \end{cases} \qquad (62)$$

where $\mu_p = dw,p/dw,max$ gives the value dw,p of the soil moisture content heralding the onset of evaporation at the potential rate expressed as a fraction, μ_p, of the field capacity. Over ocean, snow-covered and ice-covered surfaces the surface evaporation (sublimation) rate is the potential rate given by (60). The potential rate is also often adopted over a land surface when $(q(z_\ell) - q_{o,sat}(T_o)) > 0$ on the assumption that dew deposition will establish very quickly or maintain a moist surface. Otherwise, i.e. for upward surface water vapour flux over land, equations (60)-(62) are used with μ_p specified.

In present models using this formulation no allowance is made for the spatial variability of soil types or the effects of different vegetative coverings and the already oversimplified parameterization of E is reduced further by assuming single global values for the soil field capacity and for the threshold soil moisture content for potential evaporation and hence for μ_p. For a surface soil layer 1m deep Manabe (1969) originally selected 15 cm as his field capacity and took $\mu_p = 0.75$, implying $d_{w,p} = 11.25$ cm. Note from Table 6 that values used for the field capacity range from 10-30 cm and for μ_p from 1/3-3/4.

Note from Table 6 that a slight deviation from Manabe's (1969) approach is used in the MO 11- and 5-layer models. The bulk transfer coefficients for determining the surface turbulent exchanges in the MO 11-layer model are prescribed as functions of the bulk Richardson number (equation (25)) which in turn depends on $\Delta q(z_\ell)$. In order to avoid spuriously too unstable bulk Richardson numbers and hence too large bulk transfer coefficients for sensible heat and momentum transfer at unsaturated surfaces by the incorrect use of $q_{o,sat}(T_o)$ rather than q_o in $\Delta q(z_\ell)$ in equation (25) (see full discussion in Carson and Roberts (1977)) it was decided to use

$$\Delta q(z_\ell) = \beta \left(q(z_\ell) - q_{o,sat}(T_o) \right) \tag{63}$$

which can be written

$$q_o = \beta q_{o,sat}(T_o) + (1-\beta) q(z_\ell) \tag{64}$$

where a field capacity of 15 cm is used and $d_{w,p}$ is chosen to be 5 cm based on the results of Priestley and Taylor (1972). Therefore although the potential evaporation rate is not identified explicitly in this model's formulation, equation (63) was adopted because it is compatible with the relation (61) between actual and potential evaporation rates which essentially relates the gradients of q in the surface layer, rather than the surface values q_o and $q_{o,sat}$. A similar formula to (64) with $d_{w,max} = 20$ cm and $d_{w,p} = 10$ cm is now used in the MO 5-layer model with a coefficient to allow for the fact that $q(z_\ell)$ in a model with such poor vertical resolution is not a good estimate of the near-surface value of q.

It has been drawn to the reviewer's attention by Deardorff (private communication) that equations (64) and (61) (which reduces to (64) in many of the models where C_E is defined independently of stability or the surface humidity or is incorrectly evaluated using $q_{o,sat}$ instead of q_o) are fundamentally wrong except in the two limits $\beta = 0$ and $\beta = 1$. As it stands equation (64) implies that, for general β, q_o is a function of z_ℓ or, conversely, that $q(z_\ell)$ is independent of z_ℓ. Both of these assumptions are clearly strictly incorrect.

In all schemes employing the budget equation (57) evaporation is not allowed to remove more water from the soil than is available.

<u>Runoff</u>. In all models employing the single-layer water mass balance equation (57) the total runoff $Y = Y_o + Y_h$ is specified simply and naively (see Table 6). The simplest method is used in the models at GFDL, MO, CCSAS and AES where the excess of d_w over the field capacity is implicitly assumed to be runoff water which plays no further role in the model's hydrological cycle,

$$Y = Y_o + Y_h = \begin{cases} P_r - E + M_s & \text{if } P_r + M_s > E \text{ and } \mu = 1 \\ 0 & \text{otherwise.} \end{cases} \tag{65}$$

Table 6. Summary of land-surface hydrological relationships used in atmospheric general circulation models.

MODEL	REFERENCE	HYDROLOGICAL MASS BALANCE EQUATIONS	EVAPORATION	RUNOFF	SNOWMELT
AES	Boer and McFarlane (1979)	Equations (57), (59) $d_{w,max} = 20$ cm	Equations (60)–(62) $d_{w,p} = 10$ cm $\mu_p = 0.5$	Equation (65)	Equation (70) Also freezing/ melting of soil water
ANMRC	McAvaney et al (1979)	NONE	$E = \beta E_p$ $\beta = \begin{cases} 0.25 & \text{land/snow} \\ 1 & \text{ocean/sea-ice} \end{cases}$	NONE	NONE
CCSAS	Marchuk et al (1979)	Equations (57), (59) $d_{w,max} = 15$ cm As GFDL	Equations (60)–(62) $d_{w,p} = 11.25$ cm $\mu_p = 0.75$	Equation (65)	Equation (71) As GFDL
GFDL	Manabe (1969)	Equations (57), (59) $d_{w,max} = 15$ cm	Equations (60)–(62) $d_{w,p} = 11.25$ cm $\mu_p = 0.75$	Equation (65)	Equation (71)
GLAS	Halem et al (1979)	Prognostic equations (57), (59) for water and snow masses: Evaporation, runoff and snowmelt following developments of Lin et al (1978).			
GISS	Somerville et al (1974) Stone et al (1977)	NONE	Equations (60)–(62) with β prescribed geographically.	NONE	NONE
MO	Corby et al (1977) (5-level model)	NONE	$E = -\rho C_E V(z_\ell)(q(z_\ell) - r_0 q_{0,sat}(T_0))$ where $r_0 = \begin{cases} 0.5 & \text{land} \\ 0.8 & \text{sea} \end{cases}$	NONE	NONE

Table 6 (continued)

MODEL	REFERENCE	HYDROLOGICAL MASS BALANCE EQUATIONS	EVAPORATION	RUNOFF	SNOWMELT
MO	Slingo (1979) (5-level model)	Equations (57), (59) $d_{w,max} = 20\ cm$	$E = -\rho C_E V(z_\ell) \Delta q(z_\ell)$ $= \beta (q(z_\ell) - q_{v,sat}(T_0))$ β as in eqn. (62) $d_{w,p} = 10\ cm$ $\mu_p = 0.5$	Equation (65)	Equation (70)
	Carson and Roberts (1977) (11-level model)	Equations (57), (59) $d_{w,max} = 15\ cm$	$E = -\rho C_E V(z_\ell) \Delta q(z_\ell)$ $= \beta (q(z_\ell) - q_{v,sat}(T_0))$ β as in eqn. (62) $d_{w,p} = 5\ cm$ $\mu_p = 0.33$	Equation (65)	Equation (70)
NCAR	Washington and Williamson (1977)	Same as GFDL	Same as GFDL	Same as GFDL	Same as GFDL
	Washington et al (1979)	Surface hydrology in third-generation model based on proposals of Deardorff (1978).			
RAND	Gates and Schlesinger (1977) Gates et al (1971)	Equations (57), (59) $d_{w,max} = 30\ cm$	$E = -\rho C_E V(z_\ell) \cdot (q(z_\ell) - r_0 q_{v,sat}(T_0))$ where $r_0 = \mu$	Equation (66)	Equation (70)
UCLA	Arakawa (1972)	Equations (57), (59) $d_{w,max} = 10\ cm$	Equations (60)-(62) $d_{w,p} = 5\ cm$ $\mu_p = 0.5$	Equations (68), (69)	Equation (70) Also freezing/melting of soil water.

More complicated but nonetheless highly empirical formulations have been used for example at RAND, GISS and UCLA. In the RAND model of Gates and Schlesinger (1977) there is runoff for $\mu < 1$ but an abrupt change in the runoff rate at $\mu = 1$,

$$Y = \begin{cases} P_r - E & \text{if } P_r > E \text{ and } \mu = 1 \\ \frac{\mu}{2}(P_r - E) & \text{if } P_r > E \text{ and } \mu < 1 \\ 0 & \text{if } P_r \leq E. \end{cases} \qquad (66)$$

In a GISS specification, recorded in the report of the Working Group on Land-surface Processes meeting held in Dublin, 1978, and attributed to a private communication from J. Shukla,

$$Y = P_r (0.1 + 0.9 \, \mu^{3/4}) \qquad (67)$$

Note that evaporation does not appear in the formula, 10% of any rain runs off when the soil is dry ($\mu = 0$) and all rain is runoff when field capacity is reached ($\mu = 1$). Another complicated looking formulation relating total runoff to the rainfall rate and the soil moisture content is that due to Arakawa (1972) who used

$$Y = (P_r^3 + D_w^3)^{1/3} - D_w \qquad (68)$$

where D_w, the 'water deficiency' in the soil layer is defined by

$$D_w = (1 - \mu) \, \rho_w \, d_{w,\max} \qquad (69)$$

and $d_{w,\max} = 10$ cm. Again the runoff is independent of the evaporation rate and all rain is run off when $\mu = 1$. (Note that D_w as defined has dimensions of mass per unit area which strictly are not compatible with the dimensions of Y and P_r).

Snowmelt. In most of the present atmospheric general circulation models which treat the surface snow mass as a prognostic variable as in (59) the rate of snowmelt M_s is evaluated simply and crudely in conjunction with the surface energy balance equation (36). In general when snow is lying T_o is not allowed to rise above 273 K and the snowdepth accumulates without limit or decreases according to the net value of $(P_s - E_s)$. If however snow is lying and the solution of the heat balance equation (36) excluding the term Q_f produces an interim surface temperature value $T_o' > 273$ K then sufficient snow (if available) is allowed to melt to keep the surface temperature at 273 K. It is usually assumed that the snowpack has no moisture holding capacity; all melted snow is added directly to the soil moisture content (see (57)) with a corresponding reduction in the snow mass (see (59)).

In models with specified non-zero surface thermal capacity used to relate G_o to the rate of change of T_o as in (41) the heat to melt the snow and reduce T_o to 273 K is effectively extracted from that surface layer such that

$$Q_f = L_f M_s = -L_f \rho_w \frac{\delta d_{sw}}{\delta t} = \frac{C^*(T_o' - 273)}{\delta t} \qquad (70)$$

where δt is the model's time-step for the physical processes and δd_{sw} is the change in the water equivalent snowdepth resulting from the melting. If an interim value $T_o' > 273$ K is diagnosed at a snow-covered surface in models such as those at GFDL and CCSAS which assume $G_o \equiv 0$, i.e. zero heat capacity, then the excess energy made available for melting snow is evaluated by

$$Q_f = L_f M_s = \begin{cases} [R_N - H - Q]_{T_0 = 273k} & \text{if } [R_N - H - Q] > 0 \\ 0 & \text{if } [R_N - H - Q] \leq 0 \end{cases} \quad (71)$$

In all cases only when the snow disappears through melting or sublimation does evaporation of moisture resume at the surface.

The basic scheme has been extended in a simple manner by Arakawa (1972) (UCLA) and Boer and McFarlane (1979) (AES) to allow for the latent heat flux associated with the freezing and melting of the soil water content. Recall from Section 2 that the surface albedo is frequently related not only to the presence of snow but also its varying depth. However, at present the thermal and hydrological properties of the snow pack are generally represented very crudely, if at all, in atmospheric general circulation models.

Recent more advanced approaches. The representation of the hydrological processes in general circulation models is currently receiving considerable attention. For example, new formulations for runoff, snowmelt and evaporation developed by Lin et al. (1978) are being introduced into the GLAS model (Halem et al., 1979) and Deardorff's (1978) proposals for parameterizing the surface temperature and moisture while allowing explicitly for vegetation have already been incorporated into the NCAR third-generation model (Washington et al., 1979). Deardorff (1978) formulates his equations for soil moisture content in terms of ω, the volume of water per unit volume of soil, and the essence of his treatment is his use of predictive equations for not only $\overline{\omega}$ (the vertically averaged value of ω over a surface layer of depth h) but also for ω_0, the surface value of ω. By carrying ω_0 and using it to determine the evaporation rate through a relation between actual and potential evaporation corresponding to equations (60)-(62) the model has the facility to cater for surface moisture changes on a short time-scale; for example, the surface may become saturated quickly during rainfall although the soil throughout depth h remains unsaturated on average. The surface moisture may then be evaporated fairly quickly.

It is not entirely clear to the reviewer to what extent these proposals have been incorporated into the NCAR model. Although Washington et al. (1979) cite Deardorff (1978) as the basis of their parameterization of the surface hydrology their description of the model runoff criterion (see their Section 4.4) is consistent with the simple single-layer approach where runoff occurs when the implied soil moisture content of the layer is in excess of a field capacity of 15 cm. This appears to be contrary to the spirit of Deardorff's (1978) proposals which imply that surface runoff can occur without the full layer being at field capacity. Deardorff's (1978) and other similar proposals are at the fore of current activity in the parameterization of land-surface processes in atmospheric general circulation models and clearly require much further development and discussion before being universally accepted and adopted. Indeed Monteith (1981), in his first Presidential address to the Royal Meteorological Society, has stated, 'because of the very complex nature of water flow through plants and of the response of stomata to a restricted supply of water, the specification of surface wetness for atmospheric models is not likely to improve rapidly and there is no immediate hope of replacing the rather arbitrary empirical expressions in current use by more fundamental relationships. I therefore see little point in constructing physically elaborate but morphologically unrealistic models of vegetation as proposed by Deardorff (1978)'.

6. CONCLUDING REMARKS

It is hoped that this paper has provided a reasonably comprehensive review of the current practices in the parameterization of the land-surface properties and processes employed in the most advanced atmospheric general circulation models being developed for climate studies. It should be evident that, although the current general circulation models are complex in structure and make increasingly large demands on the most advanced computers, in many respects their representations of the physical processes and properties at and beneath the land surface are naive, crude and simple. Although there is some evidence that changing, for example, the land surface boundary conditions can produce significant changes in the simulations of existing general circulation models a comprehensive study of the sensitivity of simulated climate to land-surface properties and processes, particularly the hydrological cycle, is lacking.

A most striking example of the generally simplistic representation of the land surface is the almost universal neglect at present of any explicit effects of vegetation on all the aspects of the surface properties and exchanges discussed above. As yet only the NCAR third-generation model (Washington et al., 1979) incorporates directly such effects following the speculative and tentative proposals of Deardorff (1978).

A stage of general development of the general circulation model now reached and the desire to extend its period of integration to climatic time-scales make now appropriate for a more active and concerted effort to improve our knowledge of the land-surface processes in order to determine to what extent each facet needs to be included in the models and ultimately to provide a physically more acceptable basis to the parameterizations.

REFERENCES

Arakawa, A., 1972: Design of the UCLA general circulation model. Numerical Simulation of Weather and Climate Technical Report No. 7, Dept. of Met., UCLA.

―――― and V. R. Lamb, 1977: Computational design of the basic dynamical processes of the UCLA general circulation model. Methods in Computational Physics, 17, 173-265, Academic Press, New York.

Arya, S. P. S., 1975: Geostrophic drag and heat transfer relations for the atmospheric boundary layer. Quart. J. R. Met. Soc., 101, 147-161.

Baker, D. J., Jr. and others (The Polar Group), 1980: Polar atmosphere-ice-ocean processes: A review of polar problems in climate research. Rev. Geophys. and Space Phys., 18, 525-543.

Baumgartner, A., H. Mayer and W. Metz., 1977: Weltweite Verteilung des Rauhigkeitsparameters zo mit Anwendung auf die Energie dissipation an der Erdoberfläche. Meteorol. Rdsch., 30, 43-48.

Benoit, R., 1976: A comprehensive parameterization of the atmospheric boundary layer for general circulation models. Ph.D. Thesis, Dept. of Met., McGill Univ., Montreal, Quebec.

Beuttner, K. J. K. and C. D. Kern, 1965: The determination of infrared emissivities of terrestrial surfaces. J. Geophys. Res., 70, 1329-1337.

Bhumralkar, C. M., 1975: Numerical experiments on the computation of ground surface temperature in an atmospheric general circulation model. J. Appl. Met., 14, 1246-1258.

Boer, G. J. and N. A. McFarlane, 1979: The AES atmospheric general circulation model. Report of the JOC Study Conference on Climate Models, GARP No. 22, Vol. I, 409-460.

Bourke, W., B. McAvaney, K. Puri and R. Thurling, 1977: Global modelling of atmospheric flow by spectral methods. Methods in Computational Physics, 17, 267-324, Academic Press, New York.

Budyko, M. I., 1956: Heat balance of the Earth's surface. Gidrometeoizdat, Leningrad, 255 pp. English translation by N. A. Stepanova, Office of Technical Services, US Dept. of Commerce, Washington, 1958.

Burridge, D. M. and J. Haseler, 1977: A model for medium-range weather forecasting - Adiabatic formulation. ECMWF Tech. Report No. 4.

Carson, D. J. and W. M. Cunnington, 1979: General circulation experiments with an 11-layer model. GARP Programme on Numerical Experimentation, Research Activities in Atmos. and Oceanic Modelling, Report No. 19, 85-89.

―――― and P. J. R. Richards, 1978: Modelling surface turbulent fluxes in stable conditions. Boundary-layer Met., 14, 67-81.

―――― and S. M. Roberts, 1977: Concerning the evaluation of the turbulent fluxes at unsaturated surfaces. Unpublished Met. Office Report, Met O 20 Tech. Note No. II/103.

Charney, J., W. J. Quirk, S-H. Chow and J. Kornfield, 1977: A comparative study of the effects of albedo change on drought in semi-arid regions. J. Atmos. Sci., 34, 1366-1385.

Charnock, H., 1955: Wind stress on a water surface. Quart J. R. Met. Soc., 81, p. 639.

Clarke, R. H., 1970: Recommended methods for the treatment of the boundary layer in numerical models. Australian Met. Mag., 18, 51-73.

Corby, G. A., A. Gilchrist and P. R. Rowntree, 1977: United Kingdom Meteorological Office five-level general circulation model. Methods in Computational Physics, 17, 67-110, Academic Press, New York.

Cressman, G. P., 1960: Improved terrain effects in barotropic forecasts. Mon. Weath. Rev., 88, 327-342.

Deardorff, J. W., 1978: Efficient prediction of ground surface temperature and moisture, with inclusion of a layer of vegetation. J. Geophys. Res., 83, 1889-1903.

Delsol, F., K. Miyakoda and R. H. Clarke, 1971: Parameterized processes in the surface boundary layer of an atmospheric circulation model. Quart. J. R. Met. Soc., 97, 181-208.

Dyer, A. J. and B. B. Hicks, 1970: Flux gradient relationships in the constant flux layer. Ibid., 96, 715-721.

Gadd, A. J. and J. F. Keers, 1970: Surface exchanges of sensible and latent heat in a 10-level model atmosphere. Ibid., 96, 297-308.

Gates, W. L. and M. E. Schlesinger, 1977: Numerical sumulation of the January and July global climate with a two-level atmospheric model. J. Atmos. Sci., 34, 36-76.

_____, E. S. Batten, A. B. Kahle and A. B. Nelson, 1971: A documentation of the Mintz-Arakawa two-level atmospheric general circulation model. R-877-ARPA, The Rand Corporation, 408 pp.

Gilchrist, A., 1979: The Meteorological Office 5-layer general circulation model. Report of the JOC Study Conference on Climate Models, GARP No. 22, Vol. I, 254-295.

Halem, M., J. Shukla, Y. Mintz, M. L. Wu, R. Godbole, G. Herman and Y. Sud, 1979: Comparisons of observed seasonal climate features with a winter and summer numerical simulation produced with the GLAS general circulation model. Ibid., 207-253.

Haugen, D. A., 1973: Workshop on Micrometeorology, D. A. Haugen (Editor), American Met. Soc., Boston.

Hicks, B. B., 1976: Wind profile relationships from the 'Wangara' Experiment. Quart. J. R. Met. Soc., 102, 535-551.

Holloway, J. L., Jr. and S. Manabe, 1971: Simulation of climate by a global general circulation model I Hydrological cycle and heat balance. Mon. Weath. Rev., 99, 335-370.

Kasahara, A. and W. M. Washington, 1971: General circulation experiments with a six-layer NCAR model, including orography, cloudiness and surface temperature calculations. J. Atmos. Sci., 28, 657-701.

Kondratyev, K. Ya., 1972: Radiation processes in the atmosphere. Second IMO Lecture, WMO No. 309, WMO.

Kung, E. C., R. A. Bryson and D. H. Lenschow, 1964: Study of a continental surface albedo on the basis of flight measurements and structure of the Earth's surface cover over North America. Mon. Weath. Rev., 92, 543-564.

Lettau, H. H. and B. Davidson, 1957: Exploring the atmosphere's first mile. Pergamon Press, London, New York and Paris, 2 Vols.

Lin, J. D., J. Alfano and P. Bock, 1978: Documentation of a ground hydrology parameterization for use in the GISS atmospheric GCM. Washington, NASA, Contract Rep. CR-158766 School of Engineering, Univ. of Connecticut, Storrs, Connecticut.

Louis, J.-F., 1979: A parametric model of vertical eddy fluxes in the atmosphere. Boundary-layer Met., 17, 187-202.

Manabe, S., 1969: Climate and the ocean circulation. I The atmospheric circulation and the hydrology of the Earth's surface. Mon. Weath. Rev., 97, 739-774.

_____ and J. L. Holloway, Jr., 1975: The seasonal variation of the hydrologic cycle as simulated by a global model of the atmosphere. J. Geophys. Res., 80, 1617-1649.

_____ and T. B. Terpstra, 1974: The effects of mountains on the general circulation of the atmosphere as identified by numerical experiments. J. Atmos. Sci., 31, 3-42.

_____, D. G. Hahn and J. L. Holloway, Jr., 1974: The seasonal variation of the tropical circulation as simulated by a global model of the atmosphere. Ibid., 31, 43-83.

_____, _____ and _____, 1979: Climate simulations with GFDL spectral models of the atmosphere: Effect of spectral truncation. Report of the JOC Study Conference on Climate Models, GARP No. 22, Vol. I, 41-94.

_____, J. L. Holloway, Jr. and M. J. Spelman, 1974: GFDL global 9-level atmospheric model. Modelling for the First GARP Global Experiment, GARP No. 14, 7-26.

Marchuk, G. I., V. P. Dymnikov, V. N. Lykosov, V. Ya. Galin, I. M. Bobyleva and V. L. Perov, 1979: Numerical simulation of the global circulation of the atmosphere. Report of the JOC Study Conference on Climate Models, GARP No. 22, Vol. I, 318-370.

McAvaney, B. J., W. Bourke and K. Puri, 1978: A global spectral model for simulation of the general circulation. J. Atmos. Sci., 35, 1557-1583.

_____, _____ and _____, 1979: A simulation of the January global circulation using a spectral model. Report of the JOC Study Conference on Climate Models, GARP No. 22, Vol. I, 296-317.

McBean, G. A., K. Bernhardt, S. Bodin, Z. Litynska, A. P. Van Ulden and
J. C. Wyngaard, 1979: The Planetary Boundary Layer, G. A. McBean (Editor),
Chairman CAS Working Group on Atmospheric Boundary-layer Problems, WMO
No. 530, Tech. Note No. 165, Geneva.

Miyakoda, K. and J. Sirutis, 1977: Comparative integrations of global models with
various parameterized processes of sub grid-scale vertical transports:
Descriptions of the parameterizations. Contr. Atmos. Phys., 50, 445-487.

Monin, A. S. and S. S. Zilitinkevich, 1967: Planetary boundary layer and large-
scale atmospheric dynamics. Report on the GARP Study Conference, 1967,
Stockholm, Appendix V, ICSU/IUGG and WMO.

Monteith, J. L., 1981: Evaporation and surface temperature. Quart. J. R. Met. Soc.,
107, 1-27.

Nappo, C. J., Jr., 1975: Parameterization of surface moisture and evaporation rate
in a planetary boundary layer model. J. Appl. Met., 14, 289-296.

Nickerson, E. C. and V. E. Smiley, 1975: Surface layer and energy budget parameterizations for mesoscale models. Ibid., 14, 297-300.

Paltridge, G. W. and C. M. R. Platt, 1976: Radiative Processes in Meteorology and
Climatology, Developments in Atmospheric Science, 5, Elsevier Sci. Pub. Co.,
Amsterdam, Oxford, New York.

Posey, J. W. and P. F. Clapp, 1964: Global distribution of normal surface albedo.
Geofisica Intl., 4, 33-48.

Preuss, H. J. and J.-F. Geleyn, 1980: Surface albedos derived from satellite data
and their impact on forecast models. Arch. Met. Geoph. Biokl., Ser. A, 29,
345-356.

Priestley, C. H. B. and R. J. Taylor, 1972: On the assessment of surface heat flux
and evaporation using large-scale parameters. Mon. Weath. Rev., 100, 81-92.

Rowntree, P. R., 1975: The representation of radiation and surface heat exchange in
a general circulation model. Unpublished Met. Office Report, Met O 20 Tech.
Note No. II/58.

Saker, N. J., 1975: An 11-layer general circulation model. Ibid., Met O 20 Tech.
Note No. II/30.

Sasamori, T., 1970: A numerical study of atmospheric and soil boundary layers.
J. Atmos. Sci., 27, 1122-1137.

Schlesinger, M. E. and W. L. Gates, 1979: Performance of the Oregon State University
two-level atmospheric general circulation model. Report of the JOC Study
Conference on Climate Models, GARP No. 22, Vol. I, 139-206.

Sellers, W. D., 1965: Physical Climatology. The Univ. of Chicago Press, Chicago
and London.

Sheppard, P. A., 1969: The atmospheric boundary layer in relation to large-scale
dynamics. The Global Circulation of the Atmosphere, G. A. Corby (Editor),
91-112, R. Met. Soc., London.

Slingo, J., 1979: A new interactive radiation scheme for the 5-level model, Unpublished Met. Office Report, Met O 20 Tech. Note No. II/135

Smith, F.B. and Carson, D.J., 1977: Some thoughts on the specification of the boundary layer relevant to numerical modelling, Boundary-layer Met., 12, 307-330

Somerville, R.C.J. et al., 1974: The GISS model of the global atmosphere, J. Atmos. Sci., 31, 84-117

Stone, P.H. Chow, S., and Quirk, W.J., 1977: The July climate and a comparison of the January and July climates simulated by the GISS general circulation model, Mon. Weath. Rev., 105, 170-194.

Tiedtke, M., Geleyn, J.-F., 1979: ECMWF model - Parameterization of sub grid-scale processes, ECMWF Tech. Report No. 10

Washington, W.M. and Williamson, D.L., 1977: A description of the NCAR global circulation models, Methods in Computational Physics, 17, 111-172 Academic Press, New York

Washington, W.M., et al., 1979: Preliminary atmospheric simulation with the third-generation NCAR general circulation model: January and July, Report of the JOC Study Conference on Climate Models, GARP No. 22, Vol. 1, 95-138

Webb, E.K., 1970: Profile relationships: the log-linear range and extension to strong stability, Quart. J.R. Met. Soc., 96, 67-90.

THE SENSITIVITY OF NUMERICALLY SIMULATED CLIMATES TO LAND SURFACE CONDITIONS *

by

Y. Mintz

Univ. of Maryland
College Park, Maryland
and
NASA GSFC Laboratory for Atmospheric Sciences
Greenbelt, Maryland

Abstract

Several experiments with atmospheric general circulation models have shown that changing the land surface boundary conditions can produce significant changes in the simulated rainfall and circulation.

(1) Walker and Rowntree (1977), using a simplified version of a three-dimensional atmospheric general circulation model, showed that over North Africa an initial soil dryness and lack of precipitation persists; but that if the soil is initially moistened it maintains itself in that state for at least several weeks by a rainfall recycling of the order of 5-10 mm day^{-1}. This suggests to them that soil moisture anomalies might contribute to the observed persistence of rainfall anomalies in the northern Sahel.

(2) Rowntree and Bolton (1978), with a comprehensive general circulation model, showed that if the amount of water initially in the soil over central Europe is changed from 0 to 5 to 15 gm cm^{-2}, it respectively doubles and quadruples the local mean rainfall (from about 1 to 2 to 5 mm day^{-1}) over the last 30 days of 50 day integrations.

(3) Manabe (1975) has made a massive global irrigation simulation experiment in which over all of the continents the soil is constantly saturated with water. Compared with the control run, in which there was no such imposed constraint on the soil water, there were large and significant increases in the rainfall, especially in the arid and semi-arid regions of the continents.

---

* The complete paper will be published shortly by NASA and can be obtained from the Global Modeling and Simulation Branch, Code 911, NASA/Goddard Space Flight Center, Greenbelt, Maryland 20771, USA.

(4) Charney et al (1977) compared two different formulations for surface evapotranspiration, in a numerical general circulation model that was otherwise the same, and found significant differences in the simulated atmospheric circulation and rainfall.

(5) With the same model Charney et al. (1977) used a fixed formulation for the evapotranspiration, but changed the prescribed surface albedo of some land regions; and again found significant differences in the simulated circulation and rainfall.

(6) Washington and Chervin (1979) used an atmospheric general circulation model to study the climatic impact of the thermal energy (90 W m^{-2}) from an anticipated United States east coast megalopolis, circa 2000 A.D., and found large and significant changes in precipitation and soil moisture in the energy release region.

(7) Chervin and Washington (1977) used the same atmospheric general circulation model, as in (6) above, to test the sensitivity of the simulated climate to the prescribed surface albedo of the prairie region of North America and the desert and steppe regions of North Africa. They found large and significant differences in the surface temperatures, rainfall rates and soil moisture.

My review paper will cover these (and perhaps additional studies) in some detail; and will include some analysis of how the large scale climate and its changes (and especially the rainfall) is coupled to the land surface properties and processes in these experiments.

REFERENCES

Walker, J. and P.R. Rowntree, 1977: The Effect of soil moisture on circulation and rainfall in a tropical model. Quart. J.R. Met. Soc., 103, pp. 29-46.

Rownteee, P.R. and J.A. Bolton, 1978: Experiments with soil moisture anomalies over Europe. The GARP Programme on Numerical Experimentation. Research Activities in Atmospheric and Oceanic Modelling. (ed., R. Asselin), GARP, Report No. 18, August 1978, p. 63. (figures by personal communication).

Manabe, S., 1975: A study of the interaction between the hydrological cycle and climate using a mathematical model of the atmosphere. <u>Proceedings of Conference on Weather and Food, Endicott House, Mass. Inst. Tech., Cambridge, Mass., 9-11 May 1975.</u> 10 pp. (and additional figure by personal communication).

Charney, J., W.J. Quirk, S-H. Chow, and J. Kornfield, 1977: A comparative study of the effects of albedo change on drought in semi-arid regions. <u>J. Atmos. Sci.</u>, 34, pp. 1366-1385.

Washington, W.M. And R.M. Chervin, 1979: Regional climatic effects of large-scale thermal pollution: simulation studies with the NCAR general circulation model. <u>J. Appl. Met.</u>, 18, pp. 3-16

Chervin, R.M. and W.M. Washington. 1977: Sensitivity of the NCAR general circulation model to changes in surface albedo: implications for desertification. <u>Book of Abstracts, International Conference on the Meteorology of Semi-arid Zones, Tel Aviv, 31 October - 4 November 1977,</u> p. 29 (Figures by personal communication).

SESSION II

THE MICROPHYSICAL PROCESSES OF MOMENTUM, HEAT AND WATER TRANSFERS

ACROSS AND NEAR THE SURFACE OF THE LAND

VERTICAL FLUX OF MOISTURE AND HEAT AT A BARE SOIL SURFACE

by

Wilfried Brutsaert
Cornell University
Ithaca, NY 14853, USA

This paper presents a brief introduction to some of the physical concepts that are used to describe heat and mass transfer at bare soil surfaces. In arid regions large portions of the land surface can be considered bare. Even in more humid regions where under natural conditions the surface is covered with vegetation, land maintained under agriculture is often bare as plowed, fallow or stubbled fields. The bare soil case is also of some interest in the study of vegetated surfaces, since in certain canopy models it is used to specify the lower boundary condition.

Depending on the objectives of the analysis or on the availability of the data, the transfer at the atmosphere - bare soil interface can be studied either in terms of mechanisms within the soil or in terms of those in the lower atmosphere. This paper is concerned with the transfer phenomena in the top layers of the soil; the transfer mechanisms at the atmospheric side of the interface are treated elsewhere in great detail (Brutsaert, 1981). The transport phenomena in the soil profile are considered locally, that is roughly at the so-called Darcy scale [see (1)] or slightly larger. In other words, the length scales considered are somewhere intermediate between the Navier-Stokes scales for the water and air in the soil, on the one hand, and field size scales, on the other.

1. Isothermal Water Transport in Partly Saturated Soil

   a. Mathematical Formulation

   The specific flux of liquid water in a soil, $\underline{v}_s$, can be described by Darcy's (1856) law, which is written as

   $$\underline{v}_s = -k(\frac{1}{\rho_w g} \nabla p_w + \nabla z) \tag{1}$$

   where k is the hydraulic conductivity, (a symmetrical second order tensor for an anisotropic soil but a scalar for an isotropic soil), ρ_w the density of the water, g the acceleration of gravity, p_w the pressure in the water and z the vertical coordinate. Originally, this law was proposed for porous materials saturated with a single fluid, but meanwhile (Buckingham, 1907; Richards, 1931) it has been found that (1) is also applicable to flow of a fluid which only partly fills the pores of the soil. In the case of a partly saturated soil, the hydraulic or "capillary" conductivity and the water - or "capillary" pressure, are both functions of the volumetric soil water content θ.

   The derivation of a general equation to describe the flow in a porous material is not a trivial matter, on account of such factors as the compressibility of the material and of the fluids, the shrinking and swelling of clays and the presence of solutes in the water. However, if it can be assumed that the soil and the water have a constant density, and that the soil is isotropic, the combination of the equation of continuity with (1) yields Richards's (1931) equation. For strictly vertical flow, this can be written as

   $$\frac{\partial \theta}{\partial t} = \frac{\partial}{\partial z} [k \frac{\partial (p_w/\rho_w g)}{\partial z}] + \frac{\partial k}{\partial z} \tag{2}$$

   b. Soil-Water Characteristics

   <u>Water Content and Pressure.</u> As the water content of a soil is reduced, the soil water pressure generally becomes smaller than that in the atmospheric air which displaces the water in the pores. Examples of such relationships, which are also referred to as moisture characteristic curves are shown in Figure 1.

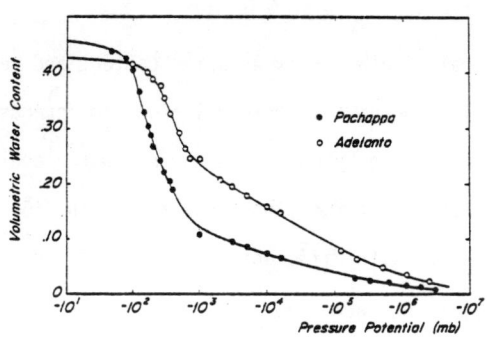

Figure 1.   Desorption moisture curves for Adelanto loam and Pachappa loam (from Jackson et al. 1965).

Numerous empirical functions have been proposed to describe this relationship. Among the better known is that of Brooks and Corey (1964),

$$S_e = 1 \qquad \text{for } -p_w < -p_b$$

$$S_e = \left(\frac{p_b}{p_w}\right)^b \qquad \text{for } -p_w \geq -p_b \qquad (3)$$

where b and p_b are constants which are characteristic for a given soil; the later is also referred to as bubbling pressure. The variable S_e is a normalized water content, referred to as the effective saturation, and defined by

$$S_e = \frac{\theta - \theta_r}{\theta_0 - \theta_r} \qquad (4)$$

where θ_0 is the water content at saturation, that is the porosity, and θ_r is another constant; θ_r is sometimes called the residual water content, because it is a water content at which k is quite small and negligible as compared to what it is at θ_0.

Another example of an empirical function is (Brutsaert, 1966)

$$S_e = \frac{a}{a + (-p_w)^b} \qquad (5)$$

where a and b are constants for a given soil. The advantage of (5) is that it yields a smooth transition from $S_e = 1$ (or $\theta = \theta_0$) down to $S_e = 0$ (or $\theta = \theta_r$).

Hysteresis. The relationship between water content and capillary pressure exhibits hysteresis. This means that this relationship depends on the sequence of events of wetting and drying by which the current water content of the soil is attained. It also means that functions such as (3) and (5) are applicable only for a drying cycle or (with different values of the constants) for a wetting cycle, but not for situations involving alternate wetting and drying. In Figures 2 and 3 some examples are presented of hysteresis for different soils. The bounding curves of the hysteresis regions shown in these figures are called wetting and drying boundary curves; any point inside the hysteresis region can be reached by scanning curves; the scanning curves starting from the drying and wetting boundary curves, and terminating at the extremes can be called primary wetting and drying scanning curves, respectively. It is obvious by now that there is an infinity of possible scanning curves in the hysteresis region. To describe these quantitatively, some type of interpolation scheme must be devised.

The phenomenon of hysteresis was discussed already in physical terms by Haines (1930). One of the first attempts to treat capillary hysteresis quantitatively was published by Poulovassilis (1962), who made use of the concept of the independent domain, proposed by Néel (1942, 1943) and Everett (1954; 1955). Briefly, this concept involves the assumpation that each element (or "pore") of the total pore space is specified completely by the (negative) pressure range over which it is emptied, and by that over which it is filled. Implicit in this is that any such element is either full or empty, with a transition sometimes referred to as a Haines-jump. The independent domain model was found to compare favorably with experimental data by Poulovassilis (1962) and Talsma (1970) (see Figure 2) but not so favorably by Topp (1969; 1971) (see Figure 3). In theoretical studies, the independent domain model has been simplified by various similarity assumptions by Philip (1964), Mualem (1973), Parlange (1976) and others. Further extensions and attempts to obtain a simpler formulation (requiring fewer data) and more accurate results have been presented by Mualem (1974, 1977), Mualem and Miller (1979), and Mualem and Morel-Seytoux (1978). Examples of the application of this model to the problem of intermittent infiltration with redistribution of soil water are the numerical studies of Ibrahim and Brutsaert (1968) and Staple (1970, 1976).

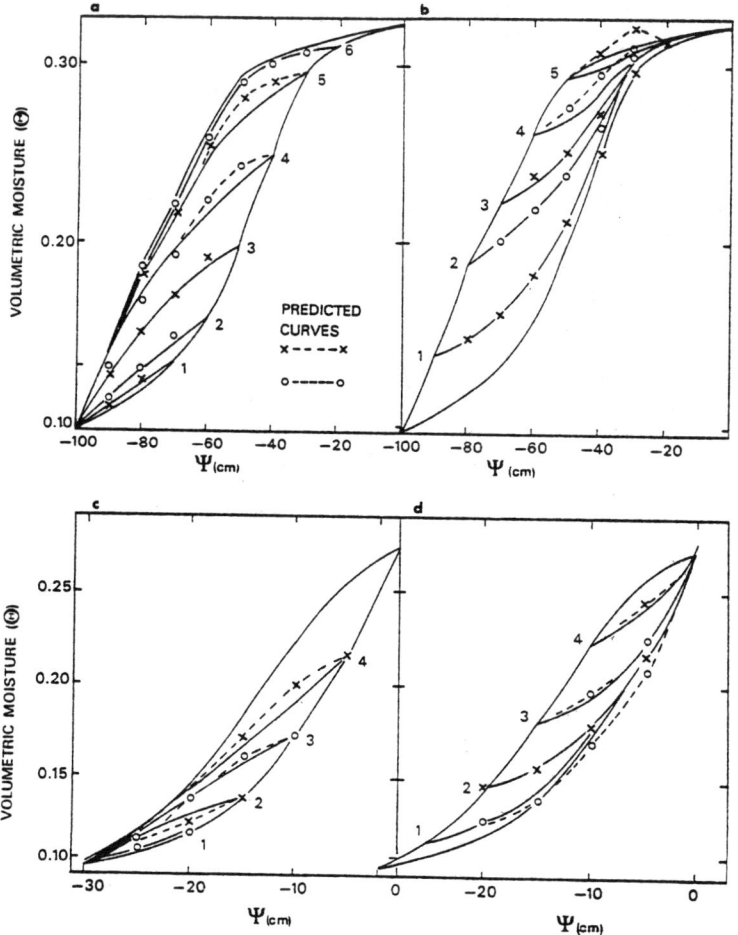

Figure 2. Boundary and primary scanning curves for (a) Adelaide dune sand, draining; (b) Adelaide dune sand, wetting; (c) Molonglo sand, draining; and (d) Molonglo sand, wetting. The solid lines are best fit through the data, while the dashed lines are calculated by means of the independent domain model (from Talsma, 1970). [Here Ψ denotes $p_w/\rho_w g$].

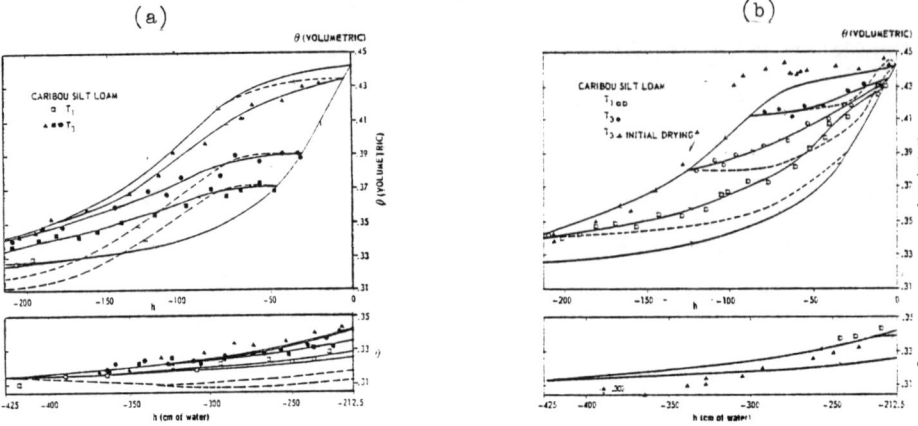

Figure 3.   Same as Figure 2 for Caribou silt loam (from Topp, 1971). [Here h denotes $p_w/\rho_w\ g$].

The experimental determination of hysteresis curves is very difficult in the laboratory and close to impossible in the field. Therefore, although the independent domain model and its simpler versions are perhaps rather crude approximations, they should be quite useful in practical simulation of soil water flow problems. Certainly, the error resulting from this approximation will be much smaller than the usually large errors resulting from sampling, inhomogeneity of the soil, anisotropy, and uncertainty in porosity and capillary conductivity.

Capillary Conductivity and Water Content. As the water content of the soil is decreased, also the capillary conductivity decreases. Reasons for this are that fewer pores are available for flow, and that the flow paths, avoiding the empty (i.e. air-filled) pores, become more tortuous. Because the larger pores are emptied first, the initial decrease in conductivity for a certain decrease in water content is larger than that later on at lower water contents. All this is illustrated in Figure 4 which shows the relative permeabilities for water and air in two soils. In the present context only that of water is of interest, since it is usually assumed that the air movement takes place under negligibly small pressure gradients.

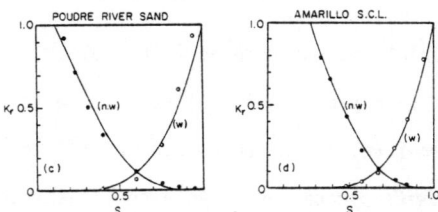

Figure 4. Relative permeabilities (k/k_o) for wetting (w. - e.g. water) and non-wetting (n.w. - e.g. air) phases as a function of the degree of saturation $S = \theta/\theta_o$, in a sandy soil and in a silty clay loam (from Brooks and Corey, 1966).

The capillary conductivity as a function of water content θ can be determined by different methods. For the purpose of simulating flow problems in the field, it is clearly preferable to determine $k(\theta)$ for the undisturbed soil profile. Various experimental determinations, consisting in general of the inverse application of the finite difference form of (2) with (1), in the absence of precipitation and by preventing evaporation at the surface, have been carried out by Ogata and Richards (1957), Nielsen et al. (1964, 1973) Davidson et al. (1969) and Baker et al. (1974). However, measurements of soil water content and water pressure, at several levels in the profile and over extended periods of time, are not easy and they require many precautions. Thus any such field method is probably only feasible under favorable conditions; field methods are usually hard, if not impossible to apply when one of the following conditions is present: a water table close to the surface, frequent and large precipitation, non-negligible or unknown net lateral inflows, a large vertical drainage rate at the lower end of the profile, and large variability in the soil properties. In the absence of field data, $k(\theta)$ can also be determined by various laboratory methods (e.g. Nielsen and Biggar, 1961) or calculations (e.g. Brutsaert, 1967; Klute, 1972; Gardner, 1974). But even in the laboratory, the experimental determination of $k(\theta)$ is exceedingly difficult. On the other hand, although several of the computation methods are probably reliable enough for routine applications in simulation, they also still require measurements, such as the conductivity at saturation $k_o = k(\theta_o)$, and the moisture characteristic $\theta(p_w)$, which are by no means trivial.

In many flow calculations encountered in hydrology, the accuracy of k at high water contents is fairly critical, but at lower water contents inaccuracies can be tolerated. Therefore, it has been found useful to represent $k(\theta)$ by relatively simple parametric equations. As also illustrated in Figure 4, the following has been used widely, viz.

$$k = k_o S_e^n \quad (6)$$

where n is a constant; clearly (6) requires the determination of four parameters, namely k_o, θ_o, θ_r and n. Interestingly, (6) has been derived on the basis of some widely different theoretical models; Averjanov (Polubarinova-Kochina, 1952) proposed (6) with n = 3.5, and Irmay (1954) with n = 3; Brooks and Corey (1966) derived n = 3 + 2/b, where b is the same as in (3); with the same b but using a more realistic theoretical model, Brutsaert (1967) obtained n = 2 + 2/b; more recently Mualem (1978) obtained the empirical relationship n = 0.015 w + 3, in which $w = \rho_w g \int_{\theta_o}^{\theta_r} p_w \, d\theta$. Inspection of past determinations of n shows that it may be as low as 1 and as high as 20, but that typical values lie around 4 to 5; n appears to be small for materials with a narrow pore size distribution and larger for wider pore size distributions. As mentioned however, and as also observed in numerical simulations of infiltration (Ibrahim and Brutsaert, 1968) the exact value of n is not very critical for the accuracy of the calculated results.

Since θ is a function of $(-p_w)$, it is also possible to express k as a function of p_w. Gardner (1958) has proposed an empirical function, which can be fitted to data for many different soils, viz.

$$k = \frac{a}{b + (-p_w/\rho_w g)^n} \quad (7)$$

where a, b and n are constants; note that (a/b) is equal to k_o, the hydraulic conductivity at saturation; b is the value of $(-p_w/\rho_w g)^n$ for $k = k_o/2$, and the range of n lies between about 2 for clayey soils and 4 or more for sandy soils.

It can be seen that (7) is of the right general shape to fit to experimental data such as shown in Figure 5.

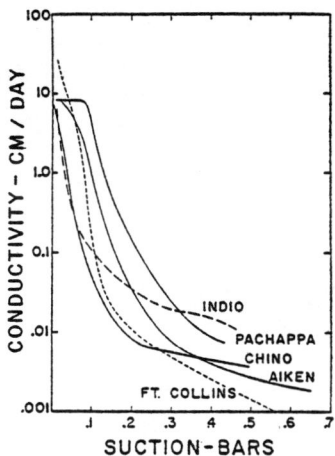

Figure 5. Capillary conductivity plotted as a function of negative pressure $-p_w$ for the drainage (desorption) cycle for different soils; Pachappa sandy loam, Indio sandy loam, Fort Collins loam, Aiken clay loam, Chino clay (from Gardner and Miklich, 1962).

There is no evidence in the literature that $k = k(\theta)$ exhibits any significant hysteresis (e.g. Jackson et al. 1965; Topp, 1971). However, since there is considerable hysteresis in $\theta = \theta(p_w)$, also $k = k(p_w)$ is quite hysteretic. An example of this is shown in Figure 6.

Soil-Water Diffusivity. In the solution of certain flow problems it has been found convenient to write (2) as a diffusion equation

$$\frac{\partial \theta}{\partial t} = \frac{\partial}{\partial z}\left(D_w \frac{\partial \theta}{\partial z}\right) + \frac{\partial k}{\partial z} \qquad (8)$$

by defining (e.g. Klute, 1952) the soil water diffusivity $D_w(\theta)$ by

$$D_w = k \frac{d(p_w/\rho_w g)}{d\theta} \qquad (9)$$

From the physical point of view (8) does not contain any more information than (2), and in simulation on digital computer there may be no significant advantage in this diffusion formulation. Actually, when part of the flow domain is

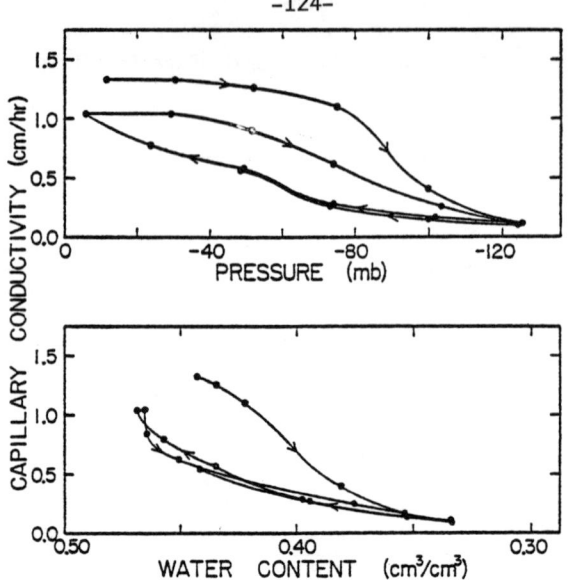

Figure 6.   Capillary conductivity curves for Columbia silt loam (from Nielsen and Biggar, 1961).

saturated,(8) with (9) may cause some difficulties which are avoidable by (2). Nevertheless, it is of some interest to consider briefly empirical equations that have been used to describe $D_w(\theta)$. A diffusivity function which has been useful in soil water problems is of the following exponential form

$$D_w = D_{wi} \exp[\beta(\theta - \theta_i)/(\theta_o - \theta_i)] \qquad (10)$$

where β is a constant, D_{wi} is the diffusivity at some reference moisture content θ_i, and θ_o is the water content of the soil at satiation. Gardner and Mayhugh (1958) used (10) in the numerical solution of the problem of sorption, or horizontal infiltration (see below), and they presented theirs and Wagner's (1952) results for different values of β. Reichardt et al. (1972) made use of (10) to scale experimental data on horizontal infiltration obtained for eight different air-dry soils, so that they could represent the results by a single regression equation involving dimensionless variables. Miller and Bresler (1977) who reconsidered the analysis of Reichardt et al. (1972) have found that the exponential diffusivity can be related to the rate of advance of the wetting front during horizontal infiltration, by a simple regression equation as follows

$$D_w = \alpha \phi_f^2 \exp[\beta(\theta - \theta_d)/(\theta_o - \theta_d)] \tag{11}$$

where α and β are constants and ϕ_f is the Boltzmann variable at the wetting front. The latter is the similarity variable arising in the solution of the sorption problem (see below) which is defined by

$$\phi_f = x_f t^{-1/2} \tag{12}$$

in which t is the time and x_f is the distance to the wetting front from $x = 0$, where the water content is maintained at $\theta = \theta_o$, and where water is allowed to infiltrate horizontally into a soil of an initially air-dry uniform water content, $\theta_i = \theta_d$. The analysis of Miller and Bresler (1977) showed that to the extent that the soil samples of Reichardt et al. (1972) cover a sufficiently wide range of soil types encountered in the field, α and β may be universal constants, for which the following values were suggested

$$\alpha = 10^{-3} \qquad \beta = 8 \tag{13}$$

approximately. It was subsequently shown by Brutsaert (1979) that (11) can also be obtained theoretically on the basis of the analytical solution of the sorption problem, and that in this relationship the constants α and β are not independent but one is a function of the other. It was then shown that the diffusivity can also be expressed in terms of the infiltrated volume of water, namely (Brutsaert, 1979)

$$D_w = \gamma[A_o/(\theta_o - \theta_d)]^2 \exp[\beta(\theta - \theta_d)/(\theta_o - \theta_d)] \tag{14}$$

where γ is a parameter which is a function of β, and where A_o is the sorptivity defined by

$$F = A_o t^{1/2} \tag{15}$$

where F is the infiltrated water volume during sorption. For the value of $\beta = 8$, obtained from the data of Reichardt et al. (1972), it was found that

$$\gamma = 1.4 \ 10^{-3} \tag{16}$$

approximately. It is now understood that the parameters α, β and γ are not quite universal, but that they vary slightly, depending on the soil and its bulk density; nevertheless, (13) and (16) may be useful as close estimates when no other information is available.

A second diffusivity equation which has been used to parameterize soil properties for hydrological purposes (e.g. Brutsaert, 1968; Verma and Brutsaert, 1970), results from the combination of (5) with (6). Thus by virtue of (9) it can be written as

$$D_w = k_o \ \alpha \ S_e^{\beta} \ (1 - S_e)^{\gamma} \tag{17}$$

where $\alpha = a^{1/b}$, $\beta = (n - b^{-1} - 1)$, and $\gamma = (b^{-1} - 1)$ are constants in which the parameters a, b and n are those of (5) and (6). Note that α, which is a length scale, can also be used (Verma and Brutsaert, 1971) as a similarity variable to quantify the importance of the partly saturated domain in ground water flow problems.

c. Some Examples of Simple Applications.

This section does not contain an exhaustive review of all past work dealing with vertical water transport in the soil. Rather, a few problems are briefly discussed as illustrations of the approaches and of the results that have been obtained so far.

<u>Sorption: Horizontal Potential Infiltration.</u> The problem of sorption, as applied to horizontal infiltration of water into a partly but uniformly saturated soil when the movement of the displaced air is unimpeded, has received considerable attention in the hydrological literature. Although this type of one-dimensional flow may not be very common in nature, the solution of this problem is of

practical importance. First, it is often a very good approximation of the initial stages of vertical infiltration of water ponded on the soil surface, while the effects of gravity are negligible. Second, it is frequently an essential part of, or a building block for, solutions for the later stages, obtained by perturbation, linearization, or other techniques.

The problem may be formulated by Richards's (1931) equation (2) without the second term on the right, or in the form of a diffusion equation, (8)

$$\frac{\partial \theta}{\partial t} = \frac{\partial}{\partial x}\left(D_w \frac{\partial \theta}{\partial x}\right) \tag{18}$$

and by the following boundary conditions

$$\begin{array}{lll} \theta = \theta_i & x > 0 & t = 0 \\ \theta = \theta_o & x = 0 & t \geq 0 \end{array} \tag{19}$$

where θ_i is the initial water content and θ_o that at the surface $x = 0$. By the application of Boltzmann's transformation

$$\phi = x\, t^{-1/2} \tag{20}$$

(18) can be simplified (Matano, 1933) to an ordinary differential equation

$$\frac{d}{d\phi}\left(D_w \frac{d\theta}{d\phi}\right) + \frac{\phi}{2}\frac{d\theta}{d\phi} = 0 \tag{21}$$

with the conditions

$$\begin{array}{ll} \theta = \theta_i & \phi \to \infty \\ \theta = \theta_o & \phi = 0 \end{array} \tag{22}$$

In the hydrological literature many numerical solutions of (21) with (22) have been published starting with those of Klute (1952) and Philip (1955). Although such solutions are quite accurate, the regional or basinwide application of (21) with (22) is a cumbersome task that far exceeds the capabilities of hydrological catchments in use today. For this reason it is imperative that the problem be parameterized to allow solutions that satisfy the dual requirement of physical consistency and computational simplicity. In the past few years a number of simple analytical solutions has been formulated, which may satisfy this requirement. Although these solutions are approximate, their accuracy is reasonable, and they involve smaller errors than most experimental studies. Moreover, they are concise, so that unlike most numerical, iterative, or perturbation series solutions, they are easy to apply in practical calculations. In a review of these solutions (Brutsaert, 1976) it has been shown that several of them can be combined in the general form

$$\phi = (2/\int_0^1 D_w S_n^a \, dS_n)^{1/2} \int_{S_n}^1 D_w(y) \, y^b \, dy \tag{23}$$

where S_n is the normalized water content

$$S_n = \frac{\theta - \theta_i}{\theta_o - \theta_i} \tag{24}$$

and where a and b are constants which assume different values depending on the method of solution. For example, in the quasi-steady state solution (Macey, 1959; Parlange, 1971), a = 1 and b = 0; in a second approximation of the quasi-steady state solution (Parlange, 1973) a = 0 and b = -1; in the sharp front solution (Brutsaert, 1974) a = b = -1; finally (Brutsaert, 1976) in a first weighting solution a = -b = 1/2, and in a second weighting solution a = b = - 3/2. The error involved in all these solutions for ϕ is at most of the order of 3% or 4%; however, the first weighting solution with a = -b = 1/2 is accurate within a few thousandths.

The parameter ϕ_f used in (11) and defined in (12) can be obtained by simply putting $S_n = 0$ as the lower limit of integration in (23). The cumulative infiltration is given in general by

$$F = \int_{\theta_i}^{\theta_o} x \, d\theta \tag{25}$$

Hence, by virtue of (20), the sorptivity defined in (15) is

$$A_o = \int_{\theta_i}^{\theta_o} \phi \, d\theta \tag{26}$$

With the solution (23) this produces

$$A_o = (\theta_o - \theta_i)(2/\int_0^1 D_w \, S_n^a \, dS_n)^{1/2} \int_0^1 D_w \, S_n^{b+1} \, dS_n \tag{27}$$

or, in its most accurate form, simply

$$A_o = (\theta_o - \theta_i)(2 \int_0^1 D_w \, S_n^{1/2} \, dS_n)^{1/2} \tag{28}$$

Infiltration Capacity or Potential Infiltration. Vertical infiltration into a partly saturated or dry soil can be described by (8). When the flow takes place from a thin layer of water ponded at the surface, into a soil with a uniform initial water content θ_i, the boundary conditions are (19), but with x replaced by z. Note, however, that when as is usual z is taken pointing down, the second term in (2) or (8) should be preceded by a minus instead of by a plus sign.

For this problem numerous methods of solution on digital computer are available. Again however, for applications at the scale of a catchment and of a region, it is necessary to describe the phenomenon by some suitable parameterization. Gardner (1967) and Philip (1969) have reviewed the development of the various underlying concepts, and Talsma and Parlange (1972) have compared some of the concise infiltration equations with experimental data.

A method, which is of some interest here, is the time expansion by Philip (1957a); this solution of (8), which is in fact equivalent to a perturbation series of the solution of (18), can be written as follows

$$z = \phi\, t^{1/2} + \chi\, t + \psi\, t^{3/2} + \omega\, t^2 + \ldots \tag{29}$$

where the functions $\phi = \phi(\theta)$ [see (20)], $\chi = \chi(\theta)$, $\psi = \psi(\theta)$, $\omega = \omega(\theta)$, etc. are each governed by a separate [one of them is (21)] ordinary differential equation; for each of these equations, Philip (1957a) presented a numerical method of solution. Although such time expansion may have certain drawbacks (e.g. Parlange, 1975), it has been reported to produce good agreement with experimental data in the laboratory by Davidson et al. (1963) (see Figure 7). The functions ϕ, χ and ψ that were used in this calculation are shown in Figure 8. Clearly, the main disadvantage of the series solution (29) is that eventually, i.e., for large values of t, it fails to behave properly. This can readily be seen by considering the rate of infiltration

$$f = dF/dt \tag{30}$$

where F is given by [cf. (25)]

$$F = \int_{\theta_i}^{\theta_0} z\, d\theta + k_i\, t \tag{31}$$

where k_i is the capillary conductivity at $\theta = \theta_i$, which is presumably negligibly small in most cases. Substitution of (29) in (30) produces

$$f = \tfrac{1}{2} A_0\, t^{-1/2} + (A_1 + k_i) + \tfrac{3}{2} A_2\, t^{1/2} + 2 A_3\, t + \ldots \tag{32}$$

where A_0 is the sorptivity defined in (26), and

$$A_1 = \int_{\theta_i}^{\theta_0} \chi\, d\theta \tag{33}$$

$$A_2 = \int_{\theta_i}^{\theta_0} \psi\, d\theta \tag{34}$$

$$A_3 = \int_{\theta_i}^{\theta_o} \omega \, d\theta \tag{35}$$

Common experience and the assymptotic solution of (8) (e.g. Philip, 1969) indicate that for $t \to \infty$, f should approach a constant value equal to k_o. Thus (29) and (32), which fail to converge for large values of t, can only be applied for small and intermediate values of time.

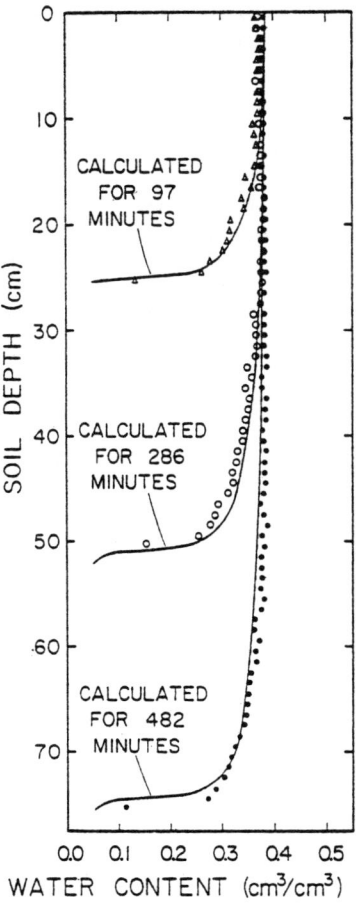

Figure 7. Calculated [by means of Philip's (1957) numerical method] and measured soil water profiles for air-dry Hesperia soil allowed to wet at $\theta_o = 0.385$ cm^3/cm^3 (from Davidson et al. 1963).

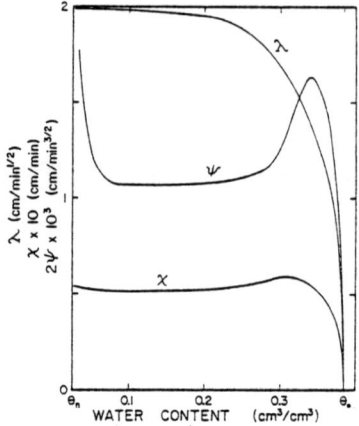

Figure 8. Calculated values of φ (shown here as λ) χ and ψ defined in (29) for Hesperia sandy loam, and obtained by means of Philip's (1957) numerical method of solution (from Davidson et al. 1963).

Nevertheless, even though the series expansion itself diverges, for some cases it can be shown (Brutsaert, 1977) to lead to a formulation for the infiltration rate which is well behaved both for small and for large t. This formulation can be obtained by making use of an approximate - but quite accurate - method of solving the differential equations for φ, χ, ψ and ω. By using the diffusivity (17) for the two extreme cases of very large b and small b (which in (5) represent soils with narrow and wide pore size distributions, respectively) one finds that the infiltration rate (32) can be approximated quite closely by

$$f = k_0 + \frac{1}{2} A_0 t^{-1/2}(1 - 2y + 3y^2 - 4y^3...) \qquad (36)$$

where $y = k_0 t^{1/2} \beta_0/A_0$ and A_0 is the sorptivity; β_0 is a constant which depends on the pore-size distribution of the soil which for most field soils is of the order of

$$\beta_0 = 2/3 \qquad (37)$$

The point of interest in (36) is that for y < 1 it can be expressed in closed form as a two-parameter "algebraic" infiltration equation, viz.

$$f = k_o + \frac{1}{2} A_o t^{-1/2}[1 + \beta_o(\frac{k_o t^{1/2}}{A_o})]^{-2} \tag{38}$$

which does not diverge for large t but tends to the proper limit $f = k_o$; also for small t, (38) approaches the proper limit, viz. $f = \frac{1}{2} A_o t^{-1/2}$ as required by (15). This correct behavior at low and high values of t is also an indication that (38) is relatively insensitive to the exact value of β_o. The convergence criterion for (36), suggests that the "small to intermediate time period" for which (29) is valid is of the order of

$$t < (1.5 \, A_o/k_o)^2 \tag{39}$$

The formulation of the infiltration problem reviewed in this section is for a rather simple and idealized case. In field situations there may be such important problems as the effect of air movement, areal non-uniformity and stratification of soil properties, the effect of crusting at the soil surface, non-uniform initial water content after redistribution during a period of drainage and evaporation, and others. Some of these are briefly touched upon in what follows.

Under certain conditions the movement of air can greatly reduce the movement of infiltrating water. For example, it is not unusual that, when there is a shallow impermeable layer or a shallow water table below a relatively flat surface ponded with water, bubbling of air can be observed; this is evidence of a counterflow of air, which undoubtedly (e.g. Linden and Dixon, 1975) reduces the water intake rate. Several mathematical formulations have been developed to describe infiltration as a flow problem of two immiscible fluids. Examples of these can be found in the studies by McWhorter (1971) Phuc and Morel-Seytoux (1972) and Sonu and Morel-Seytoux (1976). When the objective is a rigorous physical description of the flow at the Darcy scale, infiltration should be considered as a two-phase flow phenomenon. However, when the objective is the derivation of a parametric equation to describe the phenomenon at the field scale, the one-phase flow assumption, with Richards's equation (2), is likely to be quite adequate. This is especially true when the soil profile is deep, without a shallow water table, or when, as is often the case in the field, there is some surface connected macroporosity (not accounted for by the soil moisture characteristic) as a result of shrinkage cracks, worm holes or root channels. Parlange and Hill (1979) have studied the air effect by comparing solutions in which the air movement is considered with that resulting from Richards's equation. For the case where the column is sealed at the bottom, the difference was found to be quite large; however, their results showed a difference of only 2% in water intake for the case where the air can move ahead of the wetting front without an appreciable pressure buildup. In experiments dealing with soils, a difference of 2% is very difficult to detect.

The spatial variability of soil properties has been studied in the field (e.g. Nielsen et al. 1973; Rogowski, 1972; Jeppson et al. 1975; Peck et al. 1977, Warrick et al. 1977), but it is still very difficult to use this type of information to determine infiltration over a larger area. The effects of stratification of soil properties and of crusts at the surface have also received considerable attention (e.g. Miller and Gardner, 1962; Philip, 1967; Bouwer, 1969; Hillel and Gardner, 1970; Farrell and Larson, 1972; Ahuja and Swartzendruber, 1973; Reeves and Miller, 1975; Bruce et al. 1976; White et al. 1977; Whisler et al. 1979); but a detailed review would lead beyond the scope of this paper.

Rain Infiltration. The rainfall rate only rarely exceeds the initial infiltration capacity of the soil. Thus, for a certain initial period at least, all the rainfall infiltrates into the soil. During this initial period, the surface water content gradually increases and the absorptive capacity of the soil decreases until it equals the rainfall rate. From that time onward, ponding takes place and the excess precipitation is evacuated by surface runoff. During this initial period, the boundary conditions are of the flux-type rather than of the concentration type as (19). If $t = t_p$ denotes the inception of ponding, these are

$$\theta = \theta_i \qquad z > 0 \qquad t = 0$$

$$r = -k \frac{\partial p_w}{\partial z} + k \qquad z = 0 \qquad 0 < t \leq t_p \qquad (40)$$

or

$$r = -D_w \frac{\partial \theta}{\partial z} + k$$

where $r = r(t)$ is the rainfall rate. Several methods of solution have been presented for (2) [or (8)] subject to conditions (40) (e.g. Rubin, 1966; Youngs, 1972; Smith, 1972; White et al. 1979). An important part of any solution, for practical purposes, is the determination of the time to ponding t_p and the water content distribution at that time; this distribution must then be used as initial condition together with the second of (19) to solve (8) after ponding has started. This is not an easy problem. A number of simpler approaches have provided ways of parameterizing the time to ponding and the subsequent potential infiltration rate. These have consisted of extensions of the approach of Green and Ampt (e.g. Mein and Larson, 1973; Swartzendruber, 1974; Swartzendruber and Hillel, 1975; Chu, 1978); empirical equations derived from the numerical solution of (8) (e.g. Smith, 1972; Smith and Chery, 1973); and equations derived by the analytical solution of (8) on the basis of the quasi-steady state or other approximations (e.g. Parlange, 1972; Smith and Parlange, 1978).

Most of the parameterizations that have been proposed involve the concept of "time condensation" or some assumption similar to it. Briefly, the underlying assumption is that the potential infiltration rate at any given time depends only on the previous cumulative infiltration volume, regardless of the previous rainfall history. The concept of time condensation was introduced by hydrologists in the forties (e.g. Sherman, 1943; Holtan, 1945) in the context of partitioning the rainfall on a watershed into runoff and infiltration.

The validity of the time-condensation assumption can be assessed from the analysis of intermittent potential infiltration events by Ibrahim and Brutsaert (1968); this study consisted of the numerical solution of (2) for different initial and boundary conditions representing alternating potential infiltration and drainage (or redistribution) cycles. The hysteresis in the water content - pressure relationship was taken into account by means of the concept of independent domain. An example of the results for the cumulative infiltration during the reinfiltration phase (i.e. after the drainge phase) is shown in Figure 9. Analysis of these curves shows that the cumulative infiltration after a certain reinfiltration time τ_{rec} can be obtained by merely translating the initial cumulative infiltration curve horizontally to the right over a constant distance, say τ_t. This means that the process takes place as if initial infiltration starts at $\tau = \tau_t$, instead of at $\tau = 0$, and then proceeds without interruption. The values of τ_{rec} and also τ_t appear to depend on the drainage starting time, the drainage period, τ_d, and the initial saturation. Now, if the time-condensation assumption were strictly valid, one would have $\tau_{rec} = 0$ and $\tau_t = \tau_d$.

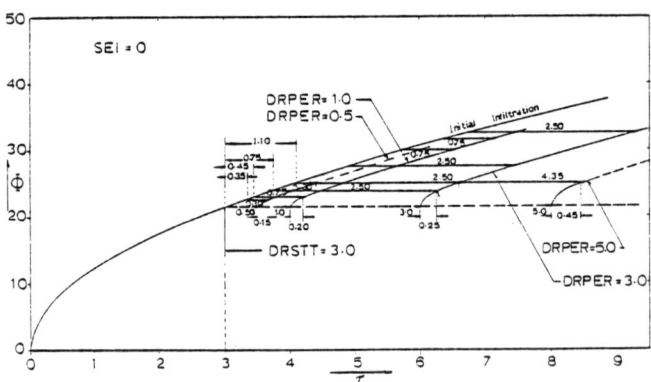

Figure 9. Illustration of cumulative infiltration-time relations for different cases of intermittent potential infiltration. DRSTT is the drainage starting time, DRPER is the drainage period, SEI is the initial effective saturation S_e, $\Phi = \int_0^\infty [S_e(\tau) - SEI]d\zeta$ is the cumulative infiltration in which $\zeta = z/\Delta z$, $\tau = t\, k_0/[(\theta_0 - \theta_r)\Delta z]$ and $\Delta z = 2$ cm (from Ibrahim and Brutsaert, 1968).

As can be seen in Figure 9, this is not quite the case; for example, for DRSTT = 3.0, and τ_d = 3, one has τ_{rec} = 0.25 and τ_t = 2.5. However, it is clear that, since Φ is roughly proportional to $\tau^{1/2}$, the error involved in the time condensation approach would be small, except for small times of reinfiltration, which are of the order of τ_{rec}. Later, similar results were also published by Reeves and Miller (1975); they compared the results obtained by means of the time condensation method (see Figure 10) with those obtained by means of the numerical solution of (8), with the inclusion of hysteresis for redistribution. These results confirmed the finding of Ibrahim and Brutsaert (1968) that the time condensation method under-estimates the infiltration; in some extreme cases they reported the error

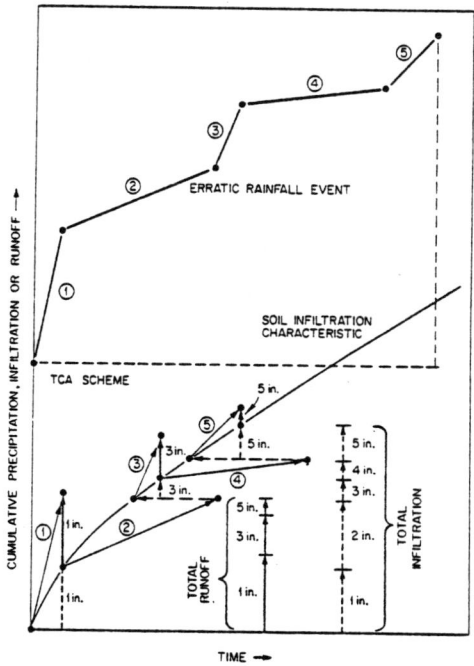

Figure 10. Illustration of the application of the time condensation concept for erratic rainfall (from Reeves and Miller, 1975).

to be as large as 15 to 20%. Nevertheless, in their appraisal they felt that the time-condensation approximation may be adequate and useful for application in watershed modeling.

Effiltration: Evaporation from Bare Soil. The water which evaporates at a bare soil surface is transported to the surface through the underlying layers of the soil profile. As will be seen in Section 3, the exact formulation of this transport is rather complicated, since the water transport takes place both in the liquid and vapor phase, involving not only pressure gradients and gravity, but often also temperature gradients with a soil heat flux, and salt concentration gradients. Nevertheless, it has been found, that in many situations of hydrological interest the main features of the evaporation at the soil surface can be obtained on the basis of the isothermal flow equation (1). Two such situations are of interest here, namely evaporation in the presence of a water table, and unsteady evaporation from a soil profile without water table.

(i) Steady Evaporation from a Water Table. The water flows from the water table through the soil profile to the soil surface, where it is taken away by evaporation. For a vertical coordinate system pointing upward with $z = 0$ at the water table, where $p_w = 0$, (1) yields, since $E = v_{sz}$

$$z = -\frac{1}{\rho_w g} \int_0^{x=p_w} \frac{dx}{[1 + E/k(x)]} \tag{41}$$

This can be readily integrated for a uniform soil profile, provided the capillary conductivity $k(p_w)$ is known as a function of the soil water pressure. Gardner (1958) presented a solution of (41) with k given by (7) for values of $n = 1, 3/2, 2, 3$ and 4.

Equation (41) produces the vertical pressure distribution of the soil water for any given rate of evaporation E. For relatively low values of E or for a soil profile with a water table at a shallow depth below the surface, the value of

$(-p_w)$, that is the soil water suction at the soil surface, is relatively small, and the soil surface is close to saturated. Hence in such a case, the rate of evaporation is governed by the prevailing atmospheric conditions, and not by the ability of the soil profile to transmit water. However, as the drying power of the air, or the depth of the water table, is increased, so that also the suction $(-p_w)$ at the soil surface increases, the rate, at which water moves upward and evaporates, increases. But eventually a limit is approached beyond which E cannot increase; in the limit E is totally controlled by the ability of the profile to transmit water, regardless of the drying power of the air, that is the potential evaporation. For most practical purposes, it is probably sufficiently accurate, to assume that the actual evaporation at any time is the lesser of the potential evaporation and of the limiting evaporation E_{lim}.

A satisfactory approximation of this limiting value E_{lim} can be obtained by assuming that the soil surface at $z = d_w$ is nearly dry or at field capacity, so that $(-p_w) \to \infty$ and $k_w \to 0$. Integration of (41) with (7) produces then in general (e.g. Cisler, 1969) the following relationship between the limiting rate of evaporation and the depth of the water table.

$$d_w = \frac{\pi}{n \sin(\pi/n)} \left(\frac{a}{a + b E_{lim}}\right) \left(\frac{a + b E_{lim}}{E_{lim}}\right)^{1/n} \qquad (42)$$

Since in many cases $a \gg b E_{lim}$, this is to a good approximation

$$E_{lim} = a \left[\frac{\pi}{n \sin(\pi/n)}\right]^n d_w^{-n} \qquad (43)$$

The model on which (43) is based is clearly an oversimplification. Especially near the soil surface water vapor transport may be important, so that the limiting evaporation rate is probably larger than that predicted. However, Gardner (1958) has estimated that the increase is not likely to exceed 20%. Similarly, experimental results presented by Gardner and Fireman (1958) support the adequacy of the isothermal flow model for steady evaporation in the presence of a water table. Equation (41) was used by Willis (1960) to study the steady state flow from a water table in the case of a soil profile consisting of two layers of different texture. He concluded that for many practical purposes the presence of inhomogeneities may have little effect on E when d_w is relatively large; the effect of stratification was pronounced for a system with the coarse-textured soil overlying the fine-textured soil, but not for the reversed condition.

(ii) Unsteady Drying of Soil Profile Without Water Table. A high water table at a constant depth is not a common occurrence; more often than not the water which evaporates from the soil surface is supplied by a release from storage in the soil profile. To facilitate the solution of this problem, it is convenient to consider two stages in this drying process of the soil profile.

In the first stage, which prevails as long as the soil is still sufficiently moist, the evaporation rate is primarily controlled by the atmospheric conditions. The duration of the first stage depends on the rate of evaporation and the ability of the soil profile to supply this rate. The rate of evaporation during this stage can thus probably best be calculated on the basis of measurements in the atmosphere.

As the soil near the surface dries out, the water supply to the surface eventually falls below that required by the atmospheric conditions. In this second or falling-rate stage, the rate of evaporation is limited by the conditions and properties of the soil profile. The transition from the first to the second stage may be quite abrupt at a given point on the surface, but on a field-wide scale, it is usually more gradual.

In the second stage of drying, water moves also through the profile by diffusion of water vapor. And especially after the soil has become quite dry the water transport in the profile is sensitive to the temperature gradients in the soil. However, when the profile has become dry, the rate of evaporation is usually so small that it is of little significance hydrologically. Thus, at least initially in the falling rate stage, the water moves primarily as a liquid. Although the matter is more complicated (e.g. Philip, 1957b; Cary, 1967), just like for the steady case, the available evidence shows that some of the more important features of the falling-rate stage of drying can be obtained by means of the simple isothermal flow model.

A simplified problem formulation, whose solution still has considerable practical relevance, can be obtained by considering the second stage of drying as a problem of desorption; this formulation, which was first used by Gardner (1959), involves the following assumptions. First, it is assumed that the effect of gravity is negligible, so that (18) is the governing equation, instead of (8). Second, the boundary conditions are taken as

$$\begin{aligned} \theta &= \theta_i & z &\geq 0 & t &= 0 \\ \theta &= \theta_o & z &= 0 & t &> 0 \end{aligned} \qquad (44)$$

where, as before, θ_i is the initial water content of the soil but θ_o is now the water content at the presumably dry soil surface. In the first of (44) it is assumed that the initial water content is uniform, and in the second that the water content at the surface is always very low. These conditions are equivalent with the assumption that the energy-limiting drying rate, i.e. the potential evaporation is so large, that the duration of the first stage of drying is negligibly short.

To date, no general exact solution has been obtained for (18) with (44), but only approximate solutions or for certain types of diffusivity functions. Gardner (1959) made use of two solutions. One was the linearized solution obtained by means of a weighted-mean diffusivity calculated by Crank's method. The other solution, which was presented graphically was obtained by iteration for the exponential-type diffusivity (11). In the present context, however, the most interesting feature of any solution of (18) with (44), regardless of the solution method and regardless of the assumed diffusivity function $D_w(\theta)$, is that, just like in the sorption problem, the total water volume lost from the soil profile is proportional to the square root of time; this can be seen from the Boltzmann transform (20), which is used to reduce (18) to the ordinary differential equation (21). Thus the rate of evaporation is given by

$$E = \frac{1}{2} De\, t^{-1/2} \qquad (45)$$

where De, which is a constant for a given soil and given values of θ_i and θ_o, is commonly referred to as desorptivity, by analogy with the sorptivity defined in (15).

Good agreement was obtained by Gardner (1959) between (45) and the evaporation rate from a 100 cm-long, initially uniformly moist column of clay soil, subjected in the laboratory to a large potential evaporation rate of about 4 cm/day. Similar field data on evaporation from a bare sand surface are shown in Figure 11.

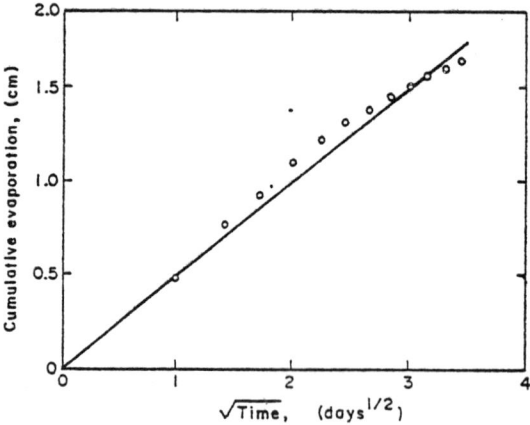

Figure 11. Cumulative evaporation from a bare sand surface as a function of the square root of time, obtained in the field by means of a lysimeter. The straight line represents (45) with a desorptivity De = 0.496 cm/(day)$^{1/2}$ (after Black et al. 1969).

These data, which were obtained by Black et al. (1969) by means of a weighable lysimeter in Wisconsin, suggest a value of the desorptivity around $De = 0.496$ cm/(day)$^{1/2}$. Black et al. (1969) compared this result with the linearized solution for the desorptivity, viz. $De = 2(\theta_i - \theta_o)(\bar{D}_w/\pi)^{1/2}$ in which $\bar{D}_w$ is the weighted-mean diffusivity. Applying Crank's method, viz. $\bar{D}_w = [1.85/(\theta_i - \theta_o)^{1.85}] \int_{\theta_o}^{\theta_i} (\theta_i - \theta)^{0.85} D_w(\theta) d\theta$, they estimated $\bar{D}_w$ from soil samples to be 10 cm^2/day; thus with $(\theta_i - \theta_o) = 0.12$ the linearized solution produces $De = 0.43$ cm/(day)$^{1/2}$ which is about 13% lower. In light of the natural variability of the soil, and also of the likely errors resulting not only from the limitations of the problem formulation but also from its linearization, the agreement may be termed good. Black et al. (1969) suspected that after a rainfall the evaporation would eventually depart from the $t^{-1/2}$ relationship, because of the finite depth of wetting. Still, they were able to model an entire summer of evaporation from the lysimeter by applying (45) after each rainfall event. Evidently, in their experiment the duration of the first stage of drying was sufficiently short that it could be neglected. This may have been due to the fact that the potential evaporation was always larger than the relatively low actual evaporation rates from the sand. Under more moderate drying conditions or for soils of finer texture, the first stage of drying should be included in the analysis.

In (44) the initial water content is assumed to be uniform. But the water content distribution at the beginning of the second stage of drying is rarely uniform, so it is clear that the subsequent evaporation rate must depend on this initial θ-distribution; this in turn depends on the rate of evaporation during the first stage and on its duration. Nonetheless, Gardner and Hillel (1962) have observed in the laboratory that this effect is of relatively short duration, so that soon after the end of the first stage the rate of evaporation becomes independent of the initial drying rate, and that it depends only on the water content of the soil. They suggested that the same drying function, namely the solution (45) for high potential evapotranspiration, should yield a good representation of the cumulative evaporation for any potential evapotranspiration, by a proper translation of the time variable. In other words, the evaporation rate, after it drops below the first stage value, can be assumed to follow the same decrease with time as (45); as a first approximation, the value of t to be used at the beginning of the second stage of drying can be taken as $t = (\Sigma\, E_{p1}/De)^2$ where $\Sigma\, E_{p1}$ is the value of the cumulative evaporation at the end of the first stage. Although the connection has not been made in the literature, it is clear that this procedure is in fact the "time condensation" discussed in the section on rain infiltration.

The problem formulation resulting in (45) and its experimental verification refer to idealized situations. In most practical field cases, such factors as the stratification of the soil profile, deep percolation or downward seepage, uncertainty concerning the end of the first stage of drying, and others tend to add complications. Despite its shortcomings and possible theoretical objections, however, under certain conditions (45) can serve as a simple parametric relationship to predict the evaporation rate from a bare soil in the second stage of drying. In practice, De is probably best determined from a field experiment during one or two drying episodes when E can be determined independently. When this is impossible, De may be estimated by the solution of (18) with (44).

2. Heat Transfer in the Soil

    a. Formulation as a Conduction Phenomenon

Heat transfer in the soil takes place primarily by conduction, so that, even though convection and radiation also play a role, the most important features can be described by casting it in the form of a conduction phenomenon. Thus the specific flux of heat is given by Fourier's law. For the vertical flux, which is of main concern here, this can be written as

$$q_H = - K_T \frac{\partial T}{\partial z} \qquad (46)$$

where K_T is called the thermal conductivity. Because the heat transfer under a temperature gradient does not involve only conduction but other mechanisms as well (primarily vapor movement with distillation) K_T is also referred to as the apparent thermal conductivity. Substitution of (46) in the equation of conservation of heat produces

$$C_s \frac{\partial T}{\partial t} = \frac{\partial}{\partial z}\left(K_T \frac{\partial T}{\partial z}\right) \tag{47}$$

where C_s is the volumetric heat capacity of the soil. In certain situations, when K_T can be assumed to be independent of z, (47) can be simplified to

$$\frac{\partial T}{\partial t} = D_T \frac{\partial^2 T}{\partial z^2} \tag{48}$$

where $D_T = (K_T/C_s) = (K_T/\rho_s c_s)$ is the thermal diffusivity, and c_s the specific heat and ρ_s the density of the soil.

b. Soil Thermal Properties

Specific Heat. A knowledge of the volume fractions of mineral soil, θ_m, organic matter θ_c, water θ, and air θ_a, allows the determination of the volumetric heat capacity $C_s = \rho_s c_s$ as follows

$$C_s = \rho_m \theta_m c_m + \rho_c \theta_c c_c + \rho_w \theta c_w + \rho_a \theta_a c_a \tag{49}$$

where the c terms are the specific heats and the ρ terms the densities as indicated by the subscripts. Using the values of these properties of the soil components, as compiled by De Vries (1963), one obtains the volumetric heat capacity in ($J\ m^{-3}\ K^{-1}$) from

$$C_s = (1.94\ \theta_m + 2.50\ \theta_c + 4.19\ \theta)10^6 \tag{50}$$

Thermal Conductivity. Not only are field soils rarely homogeneous, but it is usually also quite difficult to reproduce their structure in small samples required for laboratory measurements. For hydrological purposes, the soil thermal properties are preferably measured in situ. Field methods to determine the thermal conductivity, involving special probes have been proposed in the literature (e.g. De Vries and Peck, 1958 a,b; Janse and Borel, 1965; Fritton et al. 1974). However, these are not without difficulties; errors may result from air entrapment (Nagpal and Boersma, 1973) and from imperfect probe-soil contact (Hadas, 1974).

Since the main role of K_T is to serve as parameter in (46) or (47), it can also be determined by inverse calculations from known values of Q_H and T. For example, such a method has been used by Kimball and Jackson (1975) as follows; for the times of the day when a zero thermal gradient exists somewhere in the profile, it is possible to calculate the soil heat flux Q_H at any level by means of the calorimetric method [see (55)]; K_T at that level is then computed by dividing Q_H by the local temperature gradient.

In addition, methods have been proposed to calculate K_T. Kersten (1949) has found empirically that for a given soil K_T can be represented by a linear function of the logarithm of the water content (as percent of dry weight). De Vries (1963) has developed a physical model for heat conduction, by making use of earlier ideas by Burger (1915). The basic assumption is that the soil is a suspension of soil particles and small air pockets in water. The effect of the vapor transport, involving distillation due to the temperature gradient, is included by adapting the formulation of Krischer and Rohnalter (1940). De Vries (1963) showed that values of K_T calculated with his theoretical model were in satisfactory (usually better than ten percent) agreement with experimental data. The method was tested with field data on a loam soil by Kimball et al. (1976 a,b); but they were able to obtain fair agreement between measured and computed soil heat fluxes, only by modifying an air shape factor curve and by ignoring heat transfer due to water vapor movement; thus they concluded that heat transfer by pure conduction was far more important than any other mechanism. Similarly, the laboratory experiments of Moench and Evans (1970) showed that the apparent conductivity is usually only a few percent (around 5) different from the real thermal conductivity.

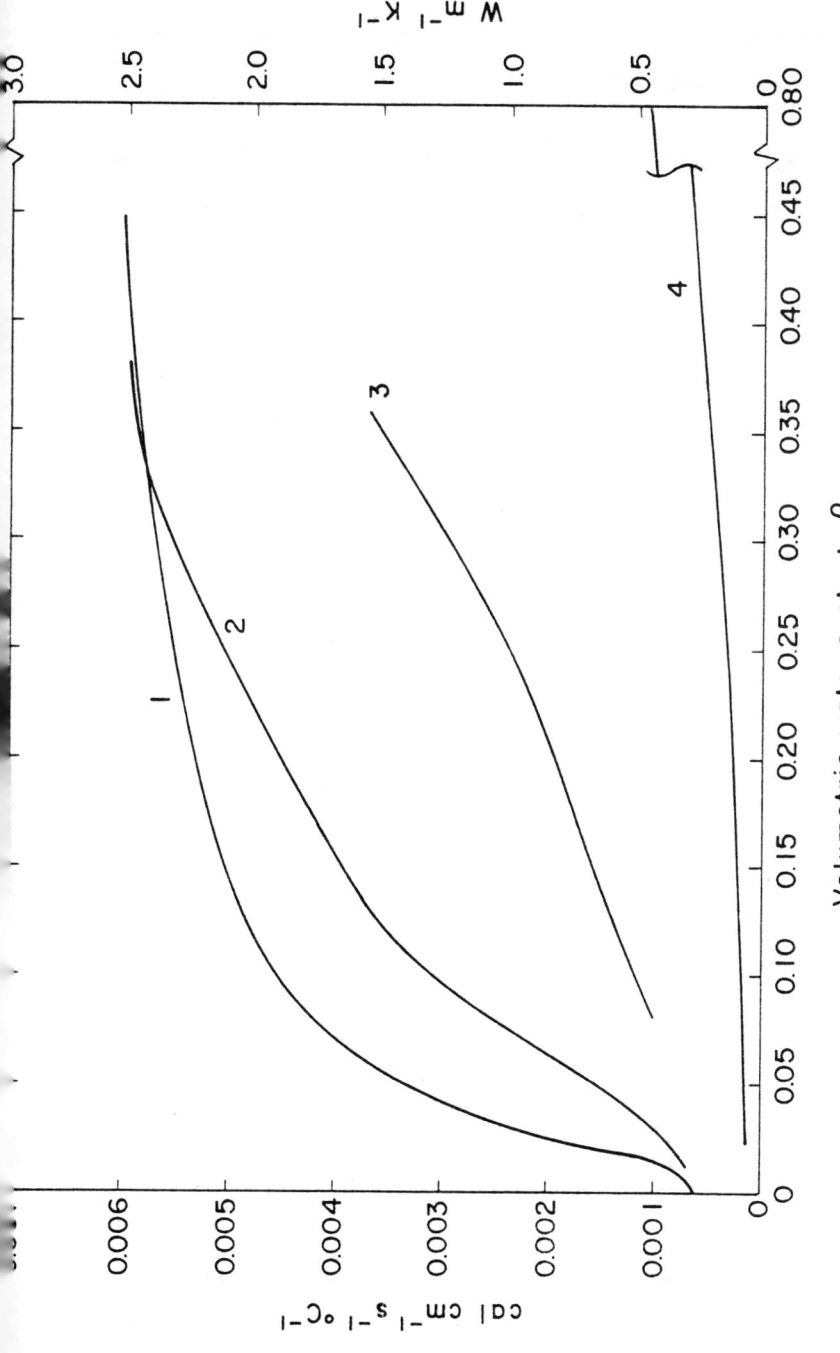

Figure 12. Thermal conductivities K_T of four types of soils as functions of water content. Curve 1 represents laboratory results on a quartz sand (porosity 0.43) obtained by De Vries (1963); curve 2 represents laboratory data on a sand loam (porosity 0.38) of Moench and Evans (1970); curve 3 is the median line through field data on a loaom (mean porosity 0.41) from Kimball et al. (1976a); curve 4 represents laboratory data on a peat soil (porosity around 0.8) from Kersten (1949).

To illustrate the order of magnitude in Figure 12, experimental curves are shown of the thermal conductivity as function of water content, for a quartz sand, a sandy loam, a loam and a peat soil.

Thermal Diffusivity. With C_s and K_T known as functions of depth and water content, (47) can be solved to calculate the soil temperature and heat flux profiles. However, since the soil heat flux is usually considerably smaller than the other terms of the surface energy budget, extreme precision in the calculation of the soil heat flux may sometimes not be necessary. Hence for certain problems involving the energy budget or the propagation of temperature waves into the ground, it may be adequate to make use of solutions of (48) with a constant diffusivity D_T. Since both K_T and C_s, and thus also D_T may be quite variable, it is not obvious what this constant value of D_T should be to obtain optimal results.

Methods proposed in the past to determine D_T consist mainly of fitting observed temperature measurements to the appropriate solution of (48) with constant diffusivity. For example, use can be made of the solution for a sinusoidal temperature wave at the surface (e.g. Chudnovskii, 1962; Van Wijk and De Vries, 1963; Jackson and Kirkham, 1958). This solution of (48) can be written as

$$T = T_{sa} + a_0 \exp[-(\omega/2\ D_T)^{1/2} z] \sin[\omega t - (\frac{\omega}{2\ D_T})^{1/2} z + C] \tag{51}$$

where T_{sa} is the average soil surface temperature, a_0 the amplitude of the sine wave at $z = 0$, $\omega = (2\pi/p)$ the radial frequency, p the period in the same units as t, and C a phase constant. This solution shows that the maxima of the temperature wave occur at later and later times with increasing depths. The maximum of the sinusoidal wave occurs when the argument of the sine-function equals $\pi/2$; thus, if the maximum of the wave for depth z_1 occurs at t_1 and for z_2 at t_2, one has from (51), that $\omega(t_2 - t_1) - (\omega/2\ D_T)^{1/2} (z_2 - z_1) = 0$ from which one obtains the diffusivity

$$D_t = \frac{p(z_2 - z_1)^2}{4\pi (t_2 - t_1)^2} \tag{52}$$

The diffusivity can be calculated from observations of when the maxima of a sinusoidal temperature wave, applied at the surface, penetrate into the soil profile at different depths. A second feature from which the diffusivity can be determined is the amplitude of the wave at depth z; from (51) this is

$$a_{oz} = a_o \exp[-(\pi/D_T\, p)^{1/2} z] \tag{53}$$

Equations (52) and (53) are simple in principle, and they have been used extensively during the past one hundred years (e.g. Chudnovskii, 1962) in methods to determine soil properties. However, in the field these methods are of limited applicability, because the surface temperature wave is rarely sinusoidal.

Some methods have been proposed based on solutions of (48) for more general (i.e. non-sinusoidal) temperature variations at the surface. For example, Van Wijk (1963, p. 157) suggested a method when temperature records are available at two depths, z_1 and z_2. The method can be summarized as follows. The Laplace transform of the solution of (48) for a constant initial temperature T_i is

$$L\{T - T_i\} = \text{const.} \exp[-z(s/D_T)^{1/2}] \tag{54}$$

where s is the Laplace Transform variable. The temperature records are Laplace-transformed by multiplying the record by $\exp(-st)$, and integrating for any suitabily chosen value of s. The diffusivity D_T is obtained by taking the ratio of the two integrals which equals $\exp[-(z_1 - z_2)(s/D_T)^{1/2}]$. The value of s is chosen so that it gives a proper weight to the important portions of the record; if s is large, only the record for small t contributes to the record. In his example, with a record of one hour at 1 cm depth, Van Wijk (1963) used $s = 0.001$. This method, applied with field data, has been found (Feddes, 1971) to be quite sensitive to the vertical uniformity of the initial temperature T_i; but this sensitivity may be an indication of the limitations inherent in (48), namely the assumed constancy of K_T and C_s.

A method, based on the Green's function solution of (48) for arbitrary temperature input at the surface, has been proposed by Laikhtman and Chudnovskii (1962, p. 59). Also this method requires only simple numerical integrations which can readily be executed in practical calculations.

c. Some Methods to Determine Surface Soil Heat Flux

The choice of a method, to determine the heat flux at the soil surface, $G = Q_H(0,t)$, is governed by the available data and by the required accuracy. For a known temperature variation at the surface, and with known $K_T = K_T(z,\theta)$ and $C_s = C_s(z,\theta)$, it is in principle possible to calculate $T = T(z,t)$ and $Q_H = Q_H(z,t)$ from (46) and (47). However, since $\theta = \theta(z,t)$, this type of calculation also involves a solution of the water flow problem as formulated in Section 1. In most cases, the available information and boundary conditions for the combined θ and T formulation are incomplete and simpler methods must be used to solve the heat flux problems.

Measurement with Heat Flux Plate. This device usually consists of a thin plate or sheet of insulating material which is placed in the soil normal to the direction of the heat flux; the temperature difference, which is measured across the plate, is a direct measure of the heat conduction through the plate, so that it can be related to the heat flux in the surrounding soil by a proper calibration. Although heat flux plates are simple to use, their construction calibration and installation require great care (e.g. Deacon, 1950; Portman, 1958; Philip, 1961; Fuchs and Tanner, 1968; Idso, 1972). The thermal properties of the plate material are likely to be different from those of the soil which vary with moisture content. If this difference is large or if the plate is placed too close to the surface, the soil heat flux pattern may be distorted considerably.

In addition, proximity to the soil surface may cause sampling difficulties resulting from surface heterogeneity. Therefore, soil heat flux plates should probably be placed at least 5 to 10 cm below the soil surface. If the soil thermal properties vary over too wide a range, the calibration may be invalid or it may have to be adjusted for moisture content. Problems may also arise from poor contact between the plate and the soil, and from possible interference of the plate with the soil water movement.

Soil Calorimetry. The soil heat flux can be determined from changes in soil heat storage. The procedure is based on the integral of (47) with (46), viz.

$$Q_{H1} - Q_{H2} = \int_{z_1}^{z_2} C_s(z) \frac{\partial T}{\partial t} dz \tag{55}$$

where Q_{H1} and Q_{H2} are the heat flux densities at levels z_1 and z_2, respectively. Thus if z_1 refers to the soil surface and z_2 to some lower level where Q_{H2} is known, the surface heat flux $G = Q_{H1}$ during a certain time interval may be calculated by numerical integration of (55) for measured soil temperature and moisture content profiles at the beginning and at the end of the time interval. The volumetric heat capacity can be calculated from (49). If the depth z_2 is sufficiently large, the heat flux Q_{H2} can be assumed to be negligible. If the depth z_2 is not sufficiently large to allow this assumption, Q_{H2} can be determined by one of the following methods.

Temperature Gradient Method. The soil heat flux at a given depth can, in principle, be calculated by means of (46) from measurements of the soil temperature gradient, provided the thermal conductivity of the soil is known. However, the gradient obtained from differences involves often large errors, and as seen above, it is not easy to determine K_T. Therefore, this method is not suitable for direct calculation of the heat flux at the surface, but only to determine the heat flux at some larger depth, when its accuracy may not be critical.

This can then serve as Q_{H2} in (55) to determine the surface heat flux G by means of the calorimetric method.

Combination Method. This method, suggested by C. B. Tanner in Wisconsin, is based on the combination of heat flux plate measurements with calorimetric measurements by means of (55). The heat flux plate is placed at a depth of 5 to 10 cm below the surface where it determines the flux Q_{H2}. The integral in (55) is then determined from successive temperature profiles measured above the level of the heat flux plate. This combination eliminates some of the undesirable features of both the heat flux plate and of the calorimetric method. Since the plate is installed at greater depth, there is less interference with the heat and moisture flow pattern near the surface. It also removes some of the uncertainties of the calorimetric method which, especially for computations over periods shorter than a day (e.g. Hanks and Tanner, 1972), may sometimes require accurate temperatures to depths of 1 m or more.

Null-Alignment Method. Also this method, described by Kimball and Jackson (1975), is essentially calorimetric, but null points in the temperature gradient are used to provide known zero heat fluxes at known depths in the profile. In Figure 13, several examples are shown of profiles with a well-defined zero temperature gradient, which were measured shortly after sunrise and after sundown. For the times of day when a zero-gradient is observed at a level z_2 in the T-profile, the soil heat flux can be determined at any level (down to say, 30 cm) in the soil by means of (55) both above and below z_2 where $Q_{H2} = 0$. This information allows then the determination of the thermal conductivity $Q_{H1}/(dT/dz)$ at some reference depth, e.g. 20 cm, where the moisture content does not vary much so that K_T remains approximately constant through the day. Hence the soil heat flux for those times of day when no zero-gradient exists, can be determined by means of (55) with Q_{H2} as the heat flux at the reference depth calculated from the observed temperature gradient.

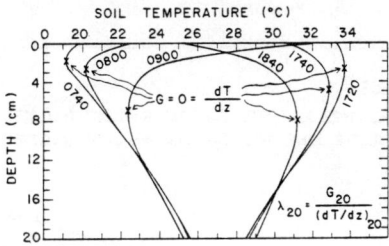

Figure 13.   Soil temperature profiles obtained at 0740, 0800, 0900, 1720, 1740 and 1840 on 17 May 1973 in Avondale loam in Arizona. The heat flux can be assumed to be zero at the level of the zero temperature gradient (from Kimball and Jackson, 1975).

Empirical Relationships. When necessary measurements are not available, the surface soil heat flux may be estimated on the basis of empirical relationships. The simplest assumption here is that the surface flux G is proportional to some other term in the energy budget equation. An obvious choice is the sensible heat flux into the air H, or

$$G = c_H H \tag{56}$$

where c_H is a constant; for bare soil Kasahara and Washington (1971) have taken $c_H = 1/3$, which was suggested by Sasamori's (1970) numerical simulation of field data of Lettau and Davidson (1957). The soil heat flux can also be assumed to be proportional to net radiation, or

$$G = c_R R_n \tag{57}$$

where, again, c_R is an empirical constant. For bare soil, Fuchs and Hadas (1972) found that on average $c_R = 0.3$, approximately; as shown in Figure 14, (57) appeared to be best satisfied for moist soil, and it displayed some hysteresis for the dry soil. Nickerson and Smiley (1975) concluded from the data of Lettau and Davidson (1957) that c_R is 0.19 for $R_n > 0$ during the day time and 0.32 for $R_n < 0$.

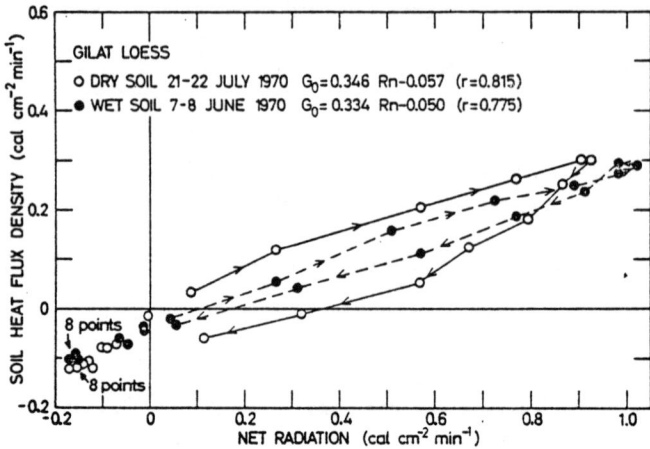

Figure 14. Relationship between soil heat flux at the surface and net radiation on a bare loess soil (from Fuchs and Hadas, 1972).

For a stand of maize, Perrier (1975) estimated that $c_R = 0.2$, when R_n is taken at the soil surface. The data of Isdo et al. (1975) showed some variation of c_R with moisture content; but inspection shows that combining all their data one would obtain a c_R value of approximately 0.4. In the light of these different studies it seems, that (57) with $c_R = 0.3$ may be used as a good compromise value for bare soils, or when R_n is taken at the soil surface. It goes without saying, that both (56) and (57) are oversimplifications since G is related not to one but to all the terms of the surface energy budget equation; therefore, such simple relationships must be calibrated anew for each given problem, and the values given for the constants c_H and c_R are accurate only for specific conditions. Nevertheless, in certain practical applications, (56) and (57) can be quite useful.

Analytical Solutions for Simplified Cases. Some analytical solutions are available of (48) for certain boundary conditions, which albeit idealized, may illustrate or give an order of magnitude of certain features of the surface heat flux phenomenon. Perhaps the simplest example among these is the sine-function solution (51) for a harmonic temperature variation at the surface; the heat flux in the soil is $-K_T \, dT/dz$, or

$$Q_H = (K_T C_S \omega)^{1/2} a_0 \exp[-(\frac{\omega}{2 D_T})^{1/2} z] \sin[\omega t - (\frac{\omega}{2 D_T})^{1/2} z + \frac{\pi}{4} + C] \quad (58)$$

or at the surface

$$G = a_0 (K_T C_S \omega)^{1/2} \sin(\omega t + \frac{\pi}{4} + C) \quad (59)$$

When an average thermal conductivity, an average specific heat and the soil surface temperature amplitude a_0 are known, (59) can be readily applied for a diurnal or for an annual cycle. While this solution is derived for idealized conditions, it can be used to obtain rough estimates of G and of the amount of heat taken up or released from storage. Equation (58) gives a rough estimate of the penetration of the temperature wave by considering the term $(\omega/2 \, D_T)^{1/2}$. For example, 95% of the wave is damped at a depth $z = 3(2 \, D_T/\omega)^{1/2}$. Further details on this and some other solutions of (48) for averaged soil thermal properties can be found in the works of Van Wijk and De Vries (1963), Van Wijk (1963), and De Vries (1975).

3. Simultaneous Water and Heat Transport in the Soil

In Sections 1 and 2, water transport and heat transport have been considered separately without any interaction. In many applications, this approach is quite satisfactory. It is clear, however, that the flows of water and heat actually take place simultaneously, and that in a rigorous analysis they should be treated as interdependent phenomena. For example, an input of heat at some point in a partly saturated soil can be accompanied by local vaporization, which in turn sets up a specific humidity gradient, and thus a water vapor transport in the air filled pore space. In addition, changes in temperature may result in changes in surface tension which affect the pressure p_w for a given water content, and thus the liquid transport. Conversely, water transport, such as infiltration of water warmer or colder than the soil, affects the soil temperature profile.

On account of the complexity of the various phenomena and their interaction within the soil, at present there is apparently no theory available which is generally accepted. Nevertheless, in recent years, many problems have been studied within the mechanistic framework developed by Philip and De Vries (1957; De Vries, 1958). According to this model, the total specific flux of soil water can be written as [cf. (1)]

$$\underline{v}_{st} = -(D_w + D_v)\nabla \theta - (D_{Tw} + D_{Tv})\nabla T - k \nabla z \tag{60}$$

where $\underline{v}_{st} = (\underline{v}_{sv} + \underline{v}_s)$ is the sum of vapor and liquid flux (expressed as volume flow of liquid per unit area and unit time), D_w the isothermal liquid diffusivity [see (9)], D_v the isothermal vapor diffusivity, D_{Tw} the thermal liquid diffusivity, D_{Tv} the thermal vapor diffusivity, and z is pointing upward. The heat flux can be written as [cf. (46)]

$$Q_H = -K_T \nabla T - L_e \rho_w D_v \nabla \theta + c_{pv}(T - T_o) \rho_w \underline{v}_{sv} + c_w (T - T_o) \rho_w \underline{v}_s \tag{61}$$

where K_T is the apparent thermal conductivity, L_e is the latent heat of vaporization, c_{pv} the specific heat of water vapor at constant pressure, $\underline{v}_{sv}$ the flux of water vapor expressed as volume flow of liquid, and T_o an arbitrary reference temperature.

The term with the isothermal vapor diffusivity D_v in (60) represents the vapor flux under a relative humidity gradient resulting from the water content gradient. This term can be simply calculated. The third term on the right of (60) with D_{Tw} is the flux of liquid water under a temperature gradient by virtue of the temperature dependence of the surface tension at the air-water interfaces in the pores, and therefore, the temperature dependence of the water pressure, p_w. The fourth term on the right of (60) with D_{Tv} represents the vapor transport under a temperature gradient. The main contribution in the formulation of Philip and De Vries (1957) was the assumption that this vapor transport is a repeated series-parallel process of flow through regions of vapor and liquid. This process involves condensation at the upstream end of liquid islands and evaporation at the downstream end, which produce changes in the curvature of the menisci at each end, resulting in a virtually zero resistance for liquid flow through the island; in addition, it was found that temperature gradients across air-filled pores may be of the order of twice the mean temperature gradient in the bulk soil as a whole. The combination of these two effects yielded values of D_{Tv}, varying between about three times that predicted by "simple" theory for very dry soils and about eight times that predicted for wet soils. The last term of (60) obviously represents the effect of gravity.

The first term on the right of (61) is the same as given in (46). Thus K_T also includes the effect of latent heat by thermally induced vapor transport, for which a theory was proposed by De Vries (1963) following the work of Krischer and Rohnalter (1940). The second term on the right of (61) is the transport of latent heat by vapor transport induced by the relative humidity gradient. The third and fourth terms are the transport of sensible heat by vapor and liquid transport, respectively; both terms were neglected in the formulation of Philip and De Vries (1957) and in most cases the sensible heat transport by vapor is probably negigible (see De Vries 1958). From the above discussion, it is clear that the diffusivities in (60) and (61) are functions of θ, or of T, or of both. For further details, the reader is referred to the papers by Philip and De Vries (1957) and De Vries (1958; 1963; 1975).

Results obtainable by means of (60) have been compared by Jackson et al. (1974) with experimental data on soil temperature, soil-water content determined by sampling, and evaporation obtained with weighable lysimeters. A conclusion of that study was that the theory of Philip and De Vries yielded better agreement with the measured values at intermediate water contents; but the isothermal theory of (1) gave better predictions of the measured values at high and very low water contents. Similar soil temperature and water content data obtained at the same experimental site in Arizona were used by Kimball et al. (1976a) to test (61) (without the last two terms). They found that the measurements could be predicted only when an air shape factor in the model of De Vries (1958; 1963) for K_T was modified and when the heat transfer due to water vapor transport (viz as incorporated in K_T and in D_v) was ignored. Thus they inferred that the conductive heat transfer mechanism was most important, and that the thermal (incorporated in K_T) and isothermal (involving D_v) vapor fluxes cancelled during the day and were negligibly small during the night. In other words, it was concluded that the theory overestimated vapor transport. This led to the recommendation that at present the theory should still be calibrated for any particular soil, before it can be applied to obtain reliable predictions of the heat flux. Hadas (1977 a,b) tested the predictive capabilities of the De Vries (1963) and Philip and De Vries (1957) models for heat transfer by water vapor by laboratory and field experiments. The model was found to predict this transfer quite satisfactorily under steady conditions, but it was found to underestimate under unsteady state conditions. Hadas suggested that it may be necessary to add a third equation to the formulation to account for air movement by pressure and density gradients; more attention should also be devoted to the convective processes involved. A study by Farrell et al (1966) showed that the vapor flux in the near-surface soil may be intensified by turbulence due to wind.

Beside the mechanistic approach to the description of interactive heat and water transport, as exemplified in the formulation of Philip and De Vries, in the literature there is a second major approach to this problem, which is based on the concepts of irreversible thermodynamics. An exploration of the theoretical and practical compatibility of the two types of approaches can be found in a paper by Raats (1975).

A complete theory, for simultaneous water and heat transport in the soil, is at present evidently still not available. Thus, more research is requried to arrive at a better understanding of all the physical aspects of the problem. Certainly, on account of the many interactive phenomena involved, it will always be quite difficult to design suitable field or laboratory experiments to validate any theory. Moreover, even presently known formulations, such as given by (60) and (61), appear perhaps too complicated to be useful in practical computations. Nevertheless, recent improvements in computational facilities make the complexity of such formulations less and less of a hindrance. Results of calculations have been presented by Slegel et al. (1975) and Rosema (1975).

Acknowledgements. This research has been supported and financed, in part, by the Office of Water Research and Technology, USDI, under the Matching Grant Program.

REFERENCES

Ahuja, L. R. and Swartzendruber, D. (1973), Horizontal soil-water intake through a thin zone of reduced permeability, Jour. Hydrol. $\underline{19}$, 71-89.

Baker, F. G., Veneman, P. L. M. and Bouma, J. (1974), Limitations of the instantaneous profile method for field measurement of unsaturated hydraulic conductivity, Soil Sci. Soc. Amer. Proc., $\underline{38}$, 885-888.

Black, T. A., Gardner, W. R. and Thurtell, G. W. (1969), The prediction of evaporation, drainage, and soil water storage for a bare soil, Soil Sci. Soc. Amer. Proc., $\underline{33}$, 655-660.

Bouwer, H. (1969), Infiltration of water into non-uniform soil, Irrig. Drain. Div., Proc. ASCE, $\underline{95}$, 451-462.

Brooks, R. H. and Corey, A. T. (1966), Properties of porous media affecting fluid flow, Jour. Irrig. Drain. Div., Proc. ASCE, $\underline{92}$, (IR2), 62-88.

Bruce, R. R., Thomas, A. W. and Whisler, F. D. (1976), Prediction of infiltration into layered field soils in relation to profile characteristics, Trans. Am. Soc. Agric. Eng., $\underline{19}$, 693-698, 703.

Brutsaert, W. (1966), Probability laws for pore-size distributions, Soil Sci., $\underline{101}$, 85-92.

Brutsaert, W. (1967), Some methods for calculating unsaturated permeability, Trans. Amer. Soc. Agr. Engrs. $\underline{10}$, 400-404.

Brutsaert, W. (1968), A solution for vertical infiltration into a dry porous medium, Water Resour. Res., $\underline{4}$, 1031-1038.

Brutsaert, W. (1974), More on an approximate solution for nonlinear diffusion, Water Resour. Res., $\underline{10}$, 1251-1252.

Brutsaert, W. (1976), The concise formulation of diffusive sorption of water in a dry soil, Water Resour. Res., $\underline{12}$, 1118-1124.

Brutsaert, W. (1977), Vertical infiltration in dry soil, Water Resour. Res., $\underline{13}$, 363-368.

Brutsaert, W. (1979), Universal constants for scaling the exponential soil water diffusivity?, Water Resour. Res., $\underline{15}$, 481-483.

Brutsaert, W. (1981), Evaporation into the atmosphere: Theory, history and applications, D. Reidel Publ. Co., Dordrecht and Boston, in press.

Buckingham, E. (1907), Studies on the movement of soil moisture, Bureau of Soils, Bull. No. 38, U.S. Dept. Agr., Washington, 61 pp.

Burger, H. C. (1915), Das Leitvermögen verdünnter mischkristallfreier Lösungen, Phys. Zeits., $\underline{20}$, 73-76.

Cary, J. W. (1967), The drying of soil: thermal regimes and ambient pressure, Agr. Met., $\underline{4}$, 357-365.

Chu, S. T. (1978), Infiltration during an unsteady rain, Water Resour. Res., $\underline{14}$, 461-466.

Chudnovskii, A. F. (1962), Heat transfer in the soil, Translated from Russian, Israel Program for Scientific Translations, Jerusalem, 164 pp.

Cisler, J. (1969), The solution for maximum velocity of isothermal steady flow of water upward from water table to soil surface, Soil Sci., $\underline{108}$, 148.

Darcy, H. (1856), Les fontaines publiques de la ville de Dijon, Victor Dalmont, Paris.

Davidson, J. M., Nielsen, D. R., Biggar, J. W. (1963), The measurement and description of water flow through Columbia silt loam and Hesperia sandy loam, Hilgardia, $\underline{34}$, (15), 601-617.

Davidson, J. M., Stone, L. R., Nielsen, D. R. and LaRue, M. E. (1969), Field measurement and use of soil-water properties, Water Resour. Res., $\underline{5}$, 1312-1321.

Deacon, E. L. (1950), The measurement and recording of the heat flux into the soil, Quart. Jour. Roy Meteor. Soc., $\underline{76}$, 479-483.

De Vries, D. A. (1958), Simultaneous transfer of heat and moisture in porous media, Trans. Amer. Geophys. Un., $\underline{39}$, 909-916.

De Vries, D. A. (1963), Thermal properties of soils, pp. 210-235, Ch. 7 in Van Wijk, W. R. (Ed.), "Physics of plant environment", North-Holland Pub. Co., Amsterdam, 382 pp.

De Vries, D. A. (1975), Heat transfer in soils, pp. 5-28, in D. A. De Vries and N.H. Afgan (Ed.), "Heat and Mass Transfer in the Biosphere", John Wiley & Sons, New York, 594 pp.

De Vries, D. A. and Peck, A. J. (1958, a,b), On the cylindrical probe method of measuring thermal conductivity with special reference to soils, Austral. Jour. Phys., $\underline{11}$, 225-271, 409-423.

Everett, D. H. (1954; 1955), A general approach to hysteresis, Trans. Faraday Soc., $\underline{50}$, 1077-1096; $\underline{51}$, 1551-1557.

Farrell, D. A., Greacen, E. L. and Gurr, C. G. (1966), Vapor transfer in soil due to air turbulence, Soil Sci., $\underline{102}$, 305-313.

Farrell, D. A. and Larson, W. E. (1972), Dynamics of the soil-water system during a rainstorm, Soil Sci., $\underline{113}$, 88-95.

Feddes, R. A. (1971), Water, heat and crop growth, Meded. Landbouwhogeschool Wageningen, (Comm. Agric. Univ.) $\underline{71\text{-}12}$, 1-184.

Fritton, D. D., Busscher, W. J. and Alpert, J. E. (1974), An inexpensive but durable thermal conductivity probe for field use, Soil Sci. Soc. Amer. Proc., $\underline{38}$, 854-855.

Fuchs, M. and Hadas A. (1972), The heat flux density in a non-homogeneous bare loessial soil, Boundary-Layer Met. $\underline{3}$, 191-200.

Fuchs, M. and Tanner, C. B. (1968), Calibration and field test of soil heat flux plates, Soil Sc. Soc. Am. Proc., $\underline{32}$, 326-328.

Gardner, W. R. (1958), Some steady-state solutions of the unsaturated moisture flow equation with application to evaporation from a water table, Soil Sci., $\underline{85}$, 228-232.

Gardner, W. R. (1959), Solutions of the flow equation for the drying of soils and other porous media, Soil Sci. Soc. Am. Proc., $\underline{23}$, 183-187.

Gardner, W. R. (1967), Development of modern infiltration theory and application in hydrology, Trans. Am. Soc. Agric. Eng., $\underline{10}$, 379-381.

Gardner, W. R. (1974), The permeability problem, Soil Sci., $\underline{117}$, 243-249.

Gardner, W. R. and Fireman, M. (1958), Laboratory studies of evaporation from soil columns in the presence of a water table, Soil Sci., $\underline{85}$, 244-249.

Gardner, W. R. and Hillel, D. I. (1962), The relation of external evaporative conditions to the drying of soils, Jour. Geophys. Res., $\underline{67}$, 4319-4325.

Gardner, W. R. and Mayhugh, M. S. (1958), Solutions and tests of the diffusion equation for the movement of water in soil, Soil Sci. Soc. Amer. Proc., $\underline{22}$, 197-201.

Gardner, W. R. and Miklich, F. J. (1962), Unsaturated conductivity and diffusivity measurements by a constant flux method, Soil Sci., $\underline{93}$, 271-274.

Hadas, A. (1974), Problems involved in measuring the soil thermal conductivity and diffusivity in a moist soil, Agric. Met., $\underline{13}$, 105-113.

Hadas, A. (1977a), Evaluation of theoretically predicted thermal conductivities of soils under field and laboratory conditions, Soil Sci. Soc. Am. Jour., $\underline{41}$, 460-466.

Hadas, A. (1977b), Heat transfer in dry aggregated soil: I. Heat conduction, Soil Sci. Soc. Am. Jour., $\underline{41}$, 1055-1059.

Haines, W. B. (1930), Studies in the physical properties of soils 5, Jour. Agric. Sci. $\underline{20}$, 97-116.

Hanks, R. J. and Tanner, C. B. (1972), Calorimetric and flux meter measurements of soil heat flow, Soil Sci. Soc. Amer. Proc., $\underline{36}$, 537-538.

Hillel, D. and Gardner, W. R. (1970), Transient infiltration into crust-topped profiles, Soil Sci., $\underline{109}$, 69-76.

Holtan, H. N. (1945), Time condensation in hydrograph analysis, Trans. Am. Geophys. Un., $\underline{26}$, 407-413.

Ibrahim, H. A. and Brutsaert, W. (1968), Intermittent infiltration into soils with hysteresis, Jour. Hydraul. Div., Proc. ASCE, $\underline{94}$, 113-137.

Idso, S. B. (1972), Calibration of soil heat-flux plates by a radiation technique, Agric. Met., 10, 467-471.

Idso, S. B., Aase, J. K. and Jackson, R. D. (1975), Net radiation - soil heat flux relations as influenced by soil water content variations, Boundary-Layer Meteor., 9, 113-122.

Irmay, S. (1954), On the hydraulic conductivity of unsaturated soils, Trans. Am. Geophys. Un., 35, 463-467.

Jackson, R. D. and Kirkham, D. (1958), Method of measurement of the real thermal diffusivity of moist soil, Soil Sci. Soc. Am. Proc., 22, 479-482.

Jackson, R. D., Reginato, R. J. and Van Bavel, C. H. M. (1965), Comparison of measured and calculated hydraulic conductivities of unsaturated soils, Water Resour. Res., 1, 375-380.

Jackson, R. D., Reginato, R. J., Kimball, B. A. and Nakayama, F. S. (1974), Diurnal soil-water evaporation: comparison of measured and calculated soil-water fluxes, Soil Sci. Soc. Amer. Proc. 38, 861-866.

Janse, A. R. P. and Borel, G. (1965), Measurement of thermal conductivity in situ in mixed materials, e.g. soils, Netherl. Jour. Agr. Sci. 13, 57-62.

Jeppson, R. W., Rawls, W. J., Hamon, W. R. and Schreiber, D. L. (1975), Use of axisymmetric infiltration model and field data to determine hydraulic properties of soils, Water Resour. Res., 11, 127-138.

Kasahara, A. and Washington, W. M. (1971), General circulation experiments with a six-layer NCAR model, including orography, cloudiness and surface temperature calculation, Jour. Atmos. Sci., 28, 657-701.

Kersten, M. S. (1949), Thermal properties of soils, Bull. Univ. Minnesota Inst. of Techn., Engin. Exper. Stat., Bull. 28, 227 pp.

Kimball, B. A. and Jackson, R. D. (1975), Soil heat flux determination: a null-alignment method, Agric. Meteor., 15, 1-9.

Kimball, B. A., Jackson, R. D., Reginato, R. J., Nakayama, F. S. and Idso, S. B. (1976a), Comparision of field-measured and calculated soil-heat fluxes, Soil Sci. Soc. Amer. Jour., 40, 18-25.

Kimball, B. A., Jackson, R. D., Nakayama, F. S., Idso, S. B. and Reginato, R. J. (1976b), Soil-heat flux determination: temperature gradient method with computed thermal conductivities, Soil Sci. Soc. Amer. Jour., 40, 25-28.

Klute, A. (1952), A numerical method for solving the flow equation for water in unsaturated materials, Soil Sci., 73, 105-116.

Klute, A. (1972), The determination of the hydraulic conductivity and diffusivity of unsaturated soils, Soil Sci., 113, 264-276.

Krischer, O. and Rohnalter, H. (1940), Wärmeleitung und Dampfdiffusion in feuchten Gütern, VDI Forschungsheft 402.

Lettau, H. and Davidson, B. (1957), Exploring the atmosphere's first mile. Vols. 1-2, Pergamon Press, N.Y.

Linden, D. R. and Dixon, R. M. (1975), Water table position as affected by soil air pressure, Water Resour. Res., 11, 139-143.

Macey, R. I. (1959), A quasi-steady-state approximation method for diffusion problems: I. concentration dependent diffusion coefficients, Bull. of Math. Biophys. 21, 19-32.

Matano, C. (1933), On the relation between the diffusion-coefficients and the concentrations of solid metals (the nickel-copper system), Japanese Jour. of Physics, 8 (3), 109-113.

McWhorter, D. B. (1971), Infiltration affected by flow of air, Hydrology Paper No. 49, Colorado State Univ., 43 pp.

Mein, R. G. and Larson, C. L. (1973), Modeling infiltration during a steady rain, Water Resour. Res., 9, 384-394.

Miller, D. E. and Gardner, W. H. (1962), Water infiltration into stratified soil, Soil Sci. Soc. Am. Proc., 26, 115-119.

Miller, R. D. and Bresler, E. (1977), A quick method for estimating soil water diffusivity functions, Soil Sci. Soc. Amer. Proc., 41, 1021-1022.

Moench, A. F. and Evans, D. D. (1970), Thermal conductivity and diffusivity of soil using a cylindrical heat source, Soil Sci. Soc. Amer. Proc., 34, 377-381.

Mualem, Y. (1973), Modified approach to capillary hysteresis based on a similarity hypothesis, Water Resour. Res., 9, 1324-1331.

Mualem, Y. (1974), A conceptual model of hysteresis, Water Resour. Res., 10, 514-520.

Mualem, Y. (1977), Extension of the similarity hypothesis used for modeling the soil water characteristics, Water Resour. Res., 13, 773-780.

Mualem, Y. (1978), Hydraulic conductivity of unsaturated porous media: generalized macroscopic approach, Water Resour. Res., 14, 325-334.

Mualem, Y. and Miller, E. E. (1979), A hysteresis model based on an explicit domain-dependence function, Soil Sci. Soc. Am. J., 43, 1067-1073.

Mualem, Y. and Morel-Seytoux, H. J. (1978), Analysis of a capillary hysteresis model based on a one-variable distribution function, Water Resour. Res., 14, 605-610.

Nagpal, N. K. and Boersma, L. (1973), Air entrapment as a possible source of error in the use of a cylindrical heat probe, Soil Sci. Soc. Amer. Proc., 37, 828-832.

Néel, L. (1942, 1943), Théories des lois d'aimantation de Lord Rayleigh, Cah. Phys., 12, 1-20; 13, 19-30.

Nickerson, E. C. and Smiley, V. E. (1975), Surface layer and energy budget parameterizations for mesoscale models, Jour. Appl. Meteor., 14, 297-300.

Nielsen, D. R. and Biggar, J. W. (1961), Measuring capillary conductivity, Soil Sci., 92, 192-193.

Nielsen, D. R., Biggar, J. W. and Erh, K. T. (1973), Spatial variability of field-measured soil-water properties, Hilgardia, 42, 215-259.

Nielsen, D. R., Davidson, J. M., Biggar, J. W. and Miller, R. J. (1964), Water movement through Panoche clay loam soil, Hilgardia, 35, 491-506.

Ogata, G. and Richards, L. A. (1957), Water content changes following irrigation of bare-field soil that is protected from evaporation, Soil Sci. Soc. Am. Proc., 21, 355-356.

Parlange, J.-Y. (1971), Theory of water movements in soils: I. One-dimensional absorption, Soil Sci., 111, 134-137

Parlange, J.-Y. (1972), Theory of water movement in soils: 8. One-dimensional infiltration with constant flux at the surface, 114, 1-4.

Parlange, J.-Y. (1973), Horizontal infiltration of water in soils: a theoretical interpretation of recent experiments, Soil Sci. Soc. Amer. Proc., 37, 329-330.

Parlange, J.-Y. (1975), Convergence and validity of time expansion solutions: a comparison of exact and approximate solutions, Soil Sci. Soc. Amer. Proc., 39, 3-6.

Parlange, J.-Y. (1976), Capillary hysteresis and the relationship between drying and wetting curves, Water Resour. Res., 12, 224-228.

Parlange, J.-Y. and Hill, D. E. (1979), Air and water movement in porous media - Compressibility effects, Soil Sci., 127, 257-263.

Peck, A. J., Luxmore, R. J. and Stolzy, J. L. (1977), Effects of spatial variability of soil hydraulic properties in water budget modeling, Water Resour. Res., 13, 348-354.

Perrier, A. (1975), Assimilation nette, utilisation de l'eau et microclimat d'un champ de maïs, Ann. agron., 26, 139-157.

Philip, J. R. (1955), Numerical solution of equations of the diffusion type with diffusivity concentration-dependent, Trans. Faraday Soc., 51, 885-892.

Philip, J. R. (1957a), Numerical solution of equations of the diffusion type with diffusivity concentration-dependent, 2, Austral. Jour. Phys., 10, 29-42.

Philip, J. R. (1957b), Evaporation, and moisture and heat fields in the soil, Jour. Meteor., 14, 354-366.

Philip, J. R. (1961), The theory of heat flux meters, Jour. Geophys. Res., 66, 571-579.

Philip, J. R. (1964), Similarity hypothesis for capillary hysteresis in porous materials, Jour. Geophys. Res., 69, 1553-1562.

Philip, J. R. (1967), Sorption and infiltration in heterogeneous media, Austral. Jour. Soil Res., 5, 1-10.

Philip, J. R. (1969), Theory of infiltration, Advances in Hydroscience, 5, 215-296.

Philip, J. R. and De Vries, D. A. (1957), Moisture movement in porous materials under temperature gradients, Trans. Amer. Geophys. Un., 38, 222-232.

Phuc, L. V. and Morel-Seytoux, H. J. (1972), Effect of soil air movement and compressibility on infiltration rates, Soil Sci. Soc. Amer. Proc., 36, 237-241.

Polubarinova-Kochina, P. Ya. (1952), Theory of ground water movement, Translated from the Russian by J. M. R. De Wiest, (1962), Princeton Univ. Press, 613 pp.

Portman, D. J. (1958), Conductivity and length relationships in heat-flow transducer performance, Trans. Am. Geophys. Un., 39, 1089-1094.

Poulovassilis, A. (1962), Hysteresis of pore water, an application of the concept of independent domains, Soil Sci., 93, 405-412.

Raats, P. A. C. (1975), Transformations of fluxes and forces describing the simultaneous transport of water and heat in unsaturated porous media, Water Resour. Res., 11, 938-942.

Reeves, M. and Miller, E. E. (1975), Estimating infiltration for erratic rainfall, Water Resour. Res., 11, 102-110.

Reichardt, K., Nielsen, D. R. and Biggar, J. W. (1972), Scaling of horizontal infiltration into homogeneous soils, Soil Sci. Soc. Amer. Proc., 36, 241-245.

Richards, L. A. (1931), Capillary conduction of liquids through porous mediums, Physics, 1, 318-333.

Rogowski, A. S. (1972), Watershed physics: soil variability criteria, Water Resour. Res., 8, 1015-1023.

Rosema, A. (1975), Simulation of the thermal behavior of bare soils for remote sensing purposes, pp. 109-123 in D. A. De Vries and N. H. Afgan (Eds.) "Heat and Mass Transfer in the Biosphere", John Wiley & Sons, N.Y., 594 pp.

Rubin, J. (1966), Theory of rainfall uptake by soils initially drier than their field capacity and its applications, Water Resour. Res., 2, 739-749.

Sasamori, T. (1970), A numerical study of atmospheric and soil boundary layers, Jour. Atmos. Sci., 27, 1122-1137.

Sherman, L. K. (1943), Comparison of F-curves derived by the methods of Sharp and Holtan and of Sherman and Mayer, Trans. Am. Geophys. Un., 24, 465-467.

Slegel, D., Davis, L. and Boersma, L. (1975), Simultaneous heat and mass transfer in soils with subsurface heated porous pipes, pp. 87-96 in D. A. De Vries and N. H. Afgan (Eds.), "Heat and Mass Transfer in the Biosphere", John Wiley & Sons, N.Y., 594, pp.

Smith, R. E. (1972), The infiltration envelope: results from a theoretical infiltrometer, Jour. Hydrol., 17, 1-21.

Smith, R. E. and Chery, D. L. (1973), Rainfall excess model from soil water flow theory, Jour. Hydraul. Div., 99, 1337-1351.

Smith, R. E. and Parlange, J.-Y. (1978), A parameter-efficient hydrologic infiltration model, Water Resour. Res., 14, 533-538.

Sonu, J. and Morel-Seytoux, H. J. (1976), Water and air movement in a bounded deep homogeneous soil, Jour. Hydrol., 29, 23-42.

Staple, W. J. (1970), Predicting moisture distribution in rewetted soils, Soil Sci. Soc. Amer. Proc., 34, 387-392.

Staple, W. J. (1976), Prediction of evaporation from columns of soil during alternate periods of wetting and drying, Soil Sci. Soc. Amer. Jour., 40, 756-761.

Swartzendruber, D. (1974), Infiltration of constant-flux rainfall into soil as analyzed by the approach of Green and Ampt, Soil Sci., 117, 272-281.

Swartzendruber, D. and Hillel, D. (1975), Infiltration and runoff for small field plots under constant intensity rainfall, Water Resour. Res., 11, 445-451.

Talsma, T. (1970), Hysteresis in two sands and the independent domain model, Water Resour. Res., 6, 964-970.

Talsma, T. and Parlange, J.-Y. (1972), One-dimensional vertical infiltration, Austral. Jour. Soil Res., 10, 143-150.

Topp, G. C. (1969), Soil-water hysteresis measured in a sandy loam and compared with the hysteretic domain model, Soil Sci. Soc. Amer. Proc., 33, 645-651.

Topp, G. C. (1971), Soil water hysteresis in silt loam and clay loam soils, Water Resour. Res., 7, 914-920.

Van Wijk, W. R. (1963), General temperature variations in a homogeneous soil, pp. 144-170, Ch. 5 in Van Wijk, W. R. (Ed.) "Physics of Plant Environment", North Holland Pub. Co., Amsterdam, 382 pp.

Van Wijk, W. R. and De Vries, D. A. (1963), Periodic temperature variations in a homogeneous soil, pp. 102-143, Ch. 4 in Van Wijk, W. R. (Ed.), "Physics of Plant Environment", North Holland Pub. Co., Amsterdam, 382 pp.

Verma, R. D. and Brutsaert, W. (1970), Unconfined aquifer seepage by capillary flow theory, Proc. Amer. Soc. Civ. Eng., Jour. Hydr. Div., 96, (HY6), 1331-1344.

Verma, R. D. and Brutsaert, W. (1971), Similitude criteria for flow from unconfined aquifers, Proc. Amer. Soc. Civ. Eng., Jour. Hydr. Div., 97, (HY9), 1493-1509.

Wagner, C. (1952), On the solution to diffusion problems involving concentration-dependent diffusion coefficients, Jour. of Metals, Trans. AIME, 4, 91-96.

Warrick, A. W., Mullen, G. J. and Nielsen, D. R. (1977), Scaling field-measured soil hydraulic properties using a similar media concept, Water Resour. Res., 13, 355-362.

Whisler, F. D., Curtis, A. A., Niknam, A. and Römkens, M. J. M. (1979), Modeling infiltration as affected by soil crusting, Proc. 3d. Intern. Hydrology Symposium, Fort Collins, Colorado (June, 1977).

White, I., Colombera, P. M. and Philip, J. R. (1977), Experimental studies of wetting front instability induced by gradual change of pressure gradient and by heterogeneous porous media, Soil Sci. Soc. Amer. Jour., 41, 483-489.

White, I., Smiles, D. E. and Perroux, K. M. (1979), Absorption of water by soil: the constant flux boundary condition, Soil Sci. Soc. Amer. Jour., 43, 659-664.

Willis, W. O. (1960), Evaporation from layered soils in the presence of a water table, Soil Sci. Soc. Amer. Proc., 24, 239-242.

Youngs, E. G. (1972), An approximate evaluation of surface concentration during one-dimensional diffusion with surface flux a function of time for concentration-dependent diffusivity, Jour. Phys. D: Appl. Phys., 5, 1592-1595.

THE VERTICAL FLUXES OF HEAT AND MOISTURE AT A VEGETATED LAND SURFACE

Leo J. Fritschen

College of Forest Resources
University of Washington
Seattle, WA 98195
U.S.A.

ABSTRACT

A model which computes evapotranspiration is presented to illustrate the complex interaction of the soil, plant, and the atmosphere to the partitioning of the available energy to sensible and latent heat fluxes. Methods used to obtain evapotranspiration values are presented and discussed with respect to various plant systems. Selected ET data are presented for arid rangeland, dryland and irrigated crops, and forest. Conclusions on relative evapotranspiration from various plant systems are presented for different climatic conditions.

1. INTRODUCTION

The modeling of sensible and latent heat fluxes from vegetated surfaces is more complex than for bare soil, water and snow surfaces. In order of decreasing complexity, I would rank the surfaces as vegetated > snow > soil > water. The increased complexity arises from (1) the fact that both plant and soil or snow surfaces are present, (2) multiple plant layers may exist, (3) plant canopies change with age and season, (4) plant root depth increases with developing canopies, (5) plant stomates control the evaporation rate via complicated relationships, (6) plant reflection and structure change with stress, and (7) plants have a feedback mechanism which alters their development and partitioning of photosynthate in relation to continued stress.

The partitioning of the available energy into sensible and latent heat may be modulated by vegetation. The ratio of sensible to latent heat flux (Bowen ratio) varies from -1 to +2 or more for various vegetated surfaces. Evaporation rates can exceed those of water surfaces at one extreme and can approach those of dry soil at the other extreme. Plants, by virtue of their root systems, tend to extend the rapid evaporation stage of a wetted soil and thus ameliorate the climate.

Vegetation systems vary widely. Using the United States as an example, of its 5.6×10^9 ha about one-fifth is classed as cropland, more than one-fourth is used primarily for grazing, nearly one-third is forested, and the final one-fifth comprises a variety of nonagricultural uses (Frey 1979). The proportion of land in crops, pasture, forestry, and other uses varies greatly over the country. Physical conditions control the land use choices, permitting a choice between crops, pasture, and forestry in some regions. In other regions, large areas are only suitable for grazing or other extensive uses. Croplands account for 21 percent of the land nationally but it is concentrated in the central United States. Croplands occupy 15 percent of the Northeast and about 10 percent of the mountain and Pacific regions where forests predominate.

Some vegetated systems are evergreen, others are deciduous. Most cropland systems are annuals. The phenology of the vegetation will alter the partitioning of energy during the year. For example, the wheat growing area (Central United States) is a green, actively transpiring surface during the spring. Wheat matures and is harvested during June in Texas. These surfaces are then converted to bare dry soil by plowing. Maturation and harvest moves northward during the summer, reaching

Canada in late September. Thus, the dynamic nature of croplands adds additional complexity to the computation of a Bowen ratio for a large area.

This paper (1) reviews a soil-plant-atmospheric model to further illustrate the complexity of the problem, (2) reviews the methods which have been used to obtain evapotranspiration data, (3) presents selected data from various vegetated systems, natural and cultivated, dryland and irrigated, and (4) generalizes relative water use and Bowen ratio for various systems.

2. A SOIL-PLANT-ATMOSPHERIC CONTINUUM MODEL

Many models have been prepared to simulate the variable and complex nature of the soil-plant-atmospheric continuum. These models usually divide the soil-root zone or the canopy-atmosphere zone into many layers. Some of the layered canopy models assume that the soil is uniformly and continuously wet (Ehleringer and Miller 1975). Others assume that the soil dries out uniformly with depth (Denmead and Millar 1979). Some models utilize a simple relation for stomatal control of transpiration (Nimah and Hanks 1973) or a complex functional relation of several variables (Waring and Running 1976, Saxton et al. 1974). Most models utilize some form of an atmospheric evaporative demand, the Penman combination type equation or some well-watered reference crop (e.g., alfalfa or grass). Other models are presented by Benecke (1976) and Gash et al. (1978). The model selected for illustration purposes is a single canopy multilayered soil model called soil-plant-air-water (SPAW) model (Saxton et al. 1974). Vertical flows of energy and water are assumed. These flows are shown in Fig. 1 where the solid lines represent energy and water flow.

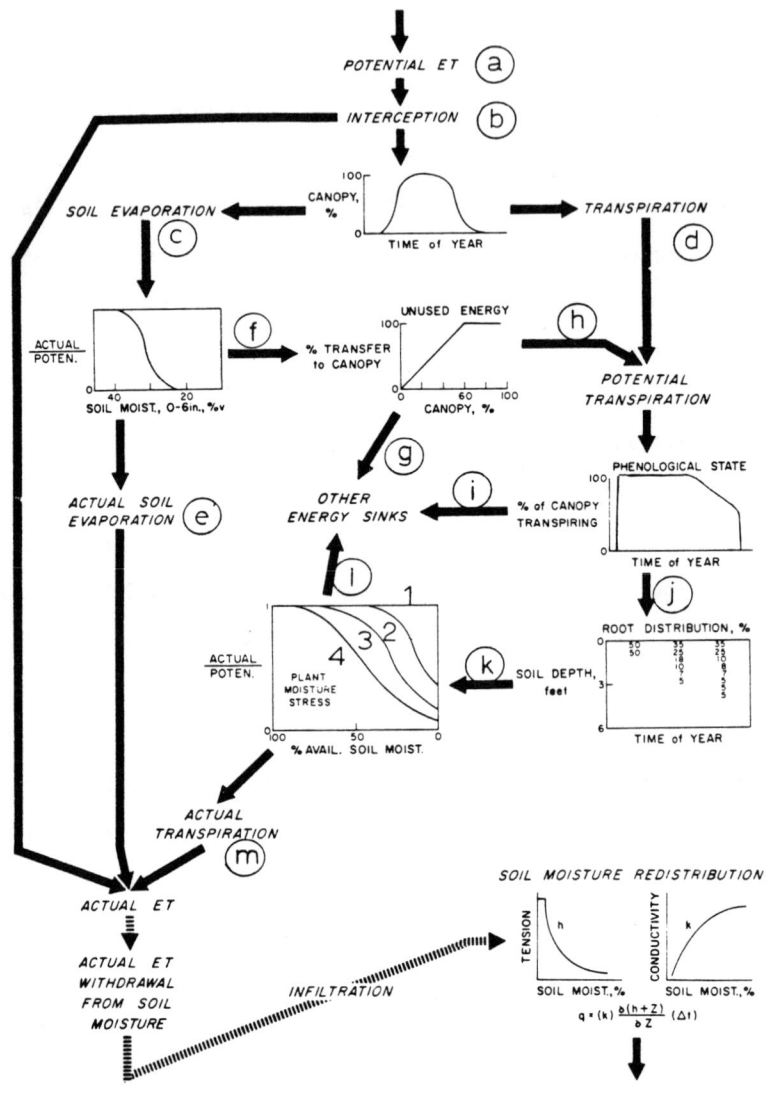

Figure 1 - Schematic representation of the soil-plant-air-water (SPAW) model (Saxton, 1974).

2.1 Potential Evaporation

Many techniques have been used to compute a potential evaporation, PE,[1] (Fig. 1,A) such as pan evaporation (Saxton et al. 1974), reference crops (Jensen et al. 1971), and the Penman type equation (Penman 1948). Penman's 1948 equation is expressed as

$$PE = (\Delta Q^* + \gamma (e_s - e) f(U_2))/(\Delta + \gamma) \tag{1}$$

where: Δ, the slope of the saturation vapor pressure curve; Q^*, net radiation; γ, psychrometric constant; e_s, saturation vapor pressure; and e, vapor pressure of the air. The function of wind is given by

$$f(U_2) = 0.35 (1 + 9.8 \times 10^{-3} U_2) \tag{2}$$

where U_2 is the daily wind run at 2 m in miles. Van Bavel's (1966) modification of (1) for specific surfaces including soil heat flux, G, and surface roughness, z_o is

$$PE = -\frac{\Delta(Q^* - G) + LK (e_s - e) U_a/(\ln (z_a/z_o))^2}{\Delta + \gamma} \tag{3}$$

$$K = \gamma \rho \varepsilon \, k^2/p \tag{4}$$

where: L, is the latent heat of vaporization; Za, the measurement height, ρ, the air density; ε, the ratio of molecular weight of wet to dry air; k, von Karman's constant; and p, atmospheric pressure. Although (1) lacks the surface specific values of G and z_o of (3) it can be applied universally for modeling purposes. The surface specific values can be accounted for with a vegetation coefficient. Doorenbos and Pruitt (1975) recommended

$$f(U_2) = 0.27 (1 + U_2/100) \tag{5}$$

---

[1] A list of symbols and definitions is included in Appendix 1.

for (2) where U_2 is expressed in km/day. This form increases the contribution of the second term of (1) under windy conditions.

2.2 Interception Loss

Water intercepted by the plant canopy, e.g. precipitation or dew, is assumed to evaporate at the potential rate (Fig. 1,B). This evaporation reduces the demand for soil water. The amount of water intercepted depends upon the canopy, history of previous storms, storm duration and intensity. Since the amount of interception evaporation is relatively small for crop surfaces compared to evapotranspiration, a fixed value of storage is used; e.g., 2.5 mm. Under some climatic conditions, the interception loss from forest can be significant and may need to be treated in more detail (Benecke 1976, Fritschen et al. 1977, Fritschen and Doraiswamy 1973).

2.3 Soil Evaporation

The remaining energy (A - B) is direct soil evaporation or transpiration by the percentage of the vegetative canopy covering the ground. The percent shading is rather constant for evergreen plants but changes drastically for deciduous trees and annual crops. Examples of canopy shading are given in Fig. 2.

Energy reaching a wet bare soil (Fig. 1,C) is assumed to evaporate water at the potential rate. Evaporation from the soil is reduced as the soil surface dries. The rates of evaporation reduction are dependent upon the soil type. Evaporation in this stage can be computed by

$$E = \frac{D_w \theta_v \pi^2}{4 Z_w} \tag{6}$$

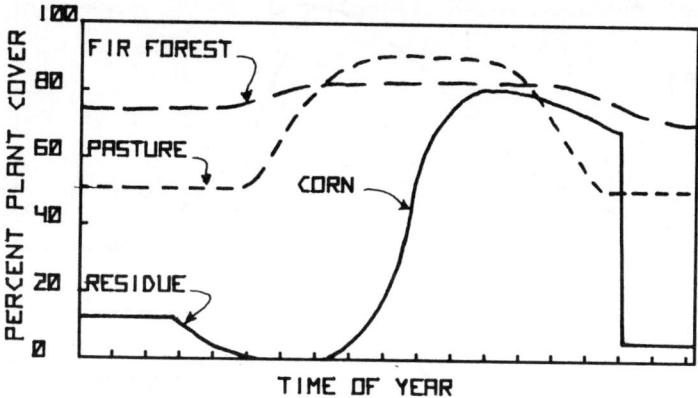

Figure 2 - Examples of canopy cover curves.

where: D_w is the soil water diffusivity; θ_v, the volumetric water content; and Z_w, the depth to the wetting front. Reduction of evaporation is faster on sandy soils than on clayey soils.

As the soil dries, an increasingly larger portion of the potential energy is unused for soil evaporation (Fig. 1,F). Some of this energy is transferred back to the canopy for transpirational use (H). The remainder is assumed to be used to heat the soil and the air (G). The portion transferred to the canopy varied with plant cover, c, according to

$$H = 1.67 c, \quad c \leq 60\% \qquad (7)$$

Mulch or crop residue present on the soil surface impedes evaporation and is not considered here.

2.4 Potential Transpiration

Energy available for potential transpiration is the sum (D + H). The amount of this energy that can be used in transpiration is dependent upon the canopy's physiological development, which is an indication of its ability to transpire. Green leaf area index is a good indicator of an annual crop's ability to transpire (Fig. 3).

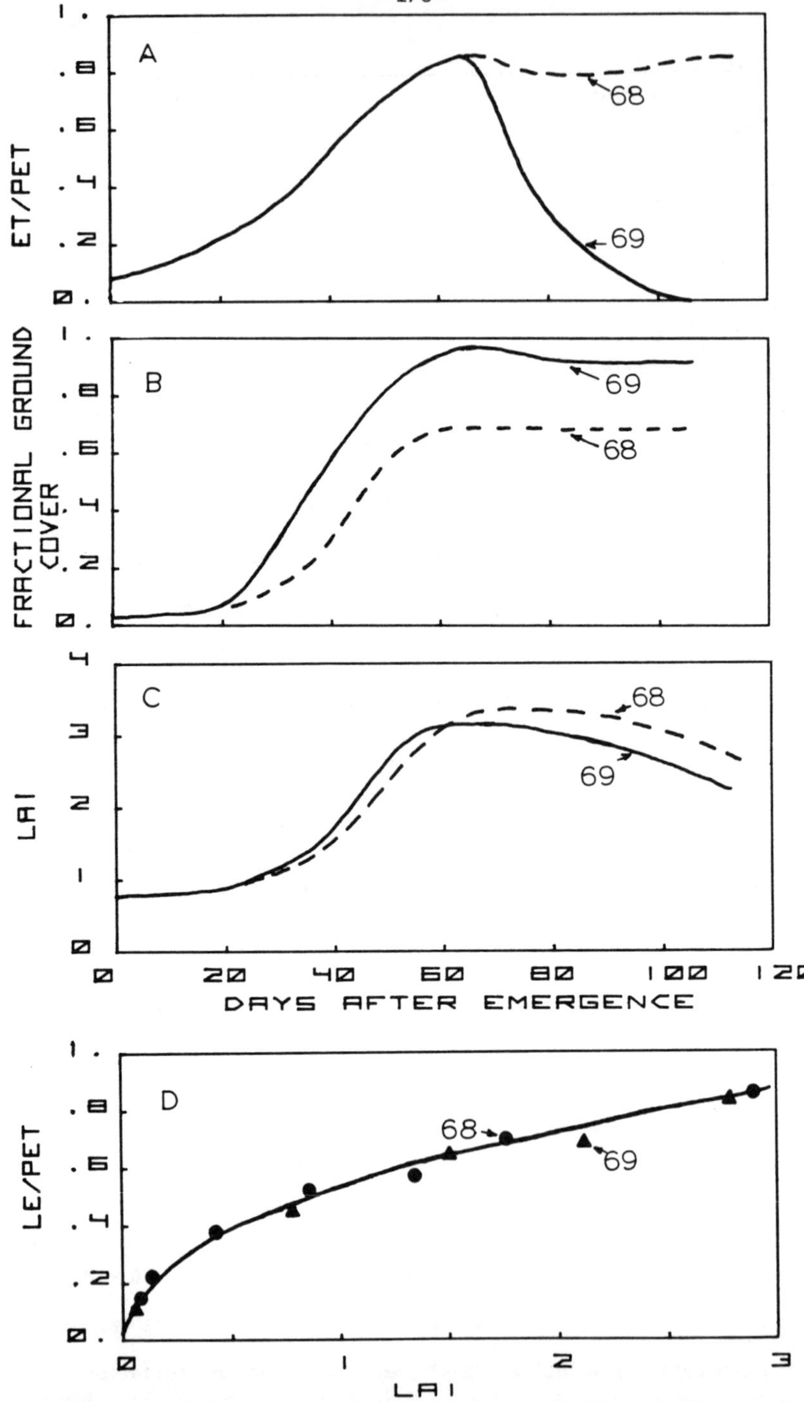

Figure 3 - Data for non-irrigated grain sorghum near Temple, Texas, in 1968 and 1969. (A) The ratio of ET/PET, (B) Fractional ground cover, and (C) LAI versus days after emergence. (D) The ratio LE/PET versus LAI. Rainfall during June and July was 213 mm in 1968 and 27 mm in 1979 (Ritchie 1971, Ritchie and Burnett 1971).

Evergreen trees carry their needles for as long as eight years. Stomates in older needles become plugged with waxes and other materials. As a result, needles 3 years and older do not transpire as rapidly as younger needles. Therefore, the green leaf area index has to be modified to be used as an indicator of transpirability.

Active root distribution by layers and time of year are considered next. A value of K (Fig. 1) is assigned to each layer which is the product of J multiplied by the percentage of active roots in each layer of soil. The values of K for each layer are multiplied by the plant moisture stress values obtained by entering the graph with available soil moisture in the layer and the total daily PE demand of the atmosphere (A). The resulting values for each root layer are summed to obtain the actual transpiration (M). The percent available soil moisture in each layer resulted from the previous day's calculation.

The curves (1, 2, 3, and 4) in the plant stress figure indicate the crop's ability to transpire relative to a potential rate given the available soil moisture and the atmospheric demand. Curves of this type are available for some agricultural crops but are not available for other plants. A similar set of curves need to be developed for other major plant species. The vertical axis would be canopy conductance, the horizontal axis would be soil water potential and the family of curves would represent values of atmospheric saturation deficit.

2.5 Actual Evapotranspiration

The amount of energy used in actual evapotranspiration is the sum of energies used for interception loss (B), soil evaporation (E), and transpiration (M).

2.6 Soil Moisture

The soil moisture status for each layer is computed daily. The amount of soil evaporation is subtracted from the available soil moisture in the first 15-cm layer, and transpiration amounts are substracted from the appropriate layers. Daily infiltration (e.g., precipitation interception runoff) expressed as depth equivalent is added to the soil by layers. The infiltration is considered to fill the 15-cm layer to 90 percent saturation. Excess infiltration would overflow into deeper layers.

Redistribution of soil moisture in 15-cm layer was computed using a one-dimensional Darcy type equation for unsaturated flow. Knowledge of the soil moisture characteristic and conductivity curves are required for each layer of soil. Steady state flow rates were used for 4-hr increments. Soil moisture contents, tensions, and conductivities were adjusted for the next 4-hr calculation.

2.7 Model Evaluation

The SPAW model has been expanded to include crop water stress effects upon growth and yield. The model was tested on corn and soybeans in Missouri and Iowa. The results indicate that the model can predict evapotranspiration, soil moisture, and yield satisfactorily (Sudar et al. 1979).

3. METHODS OF ASSESSING WATER USE BY PLANTS

Four general methods are used to determine water use by vegetated surfaces. They are: (1) hydrologic water balance; (2) soil moisture sampling; (3) lysimetry; and (4) vaporflow methods. These methods will be discussed briefly to illustrate the applicability to various vegetated surfaces.

3.1 Hydrologic Water Balance

Evapotranspiration, ET, is considered as a residual of the hydrologic balance equation

$$ET = P + I + \Delta S + D + R \qquad (8)$$

subject to the availability of energy and plant control. The symbols are defined as depths of water for precipitation, P; irrigation, I; change in soil water storage, ΔS; percolation below the root zone, D; and surface runoff, R. Evaporation from a watershed may be determined by measuring precipitation and runoff. Care is necessary in the selection of the gauging station to minimize underground drainage. However, the magnitude of underground drainage is always the big unanswered question. If one starts and ends the water budget period when the soil profile is fully charged, then ΔS can be ignored. This procedure is questionable for periods of less than 1 year.

The hydrologic water balance application, while physically sound, is strictly a long-term method; and even when applied on a yearly basis, the error may be 10 percent or more because of precipitation sampling, stream gauging, and underground drainage errors.

3.2 Soil Moisture Sampling

To obtain ET for shorter periods, ΔS must be evaluated. Two main methods are available. They are soil moisture sampling and lysimetry.

In uniform soils, the soil moisture sampling techniques are very time consuming and seldom give a high overall accuracy. Nevertheless, they are important in some areas, especially where a permanently dry subsoil exists, such as in dryland agriculture or arid range lands. Caution must be exercised in applying these techniques to other areas.

Normally soil moisture is sampled in layers. The average moisture content of the layer is represented at the midpoint. After a time (e.g., one day or week), the soil moisture is resampled. The difference is assumed to be evaporation, but may include deep percolation. Deep percolation has been known to persist 7 to 10 days after heavy irrigation or rain (van Bavel 1968). Therefore, soil moisture depletion data should not be collected during the first week following such an event unless the hydraulic gradient has been determined to be zero.

Due to soil variability and plant cover, many soil moisture samples need to be taken to obtain representative values. For example, 46, 11.5, and 3 soil samples were required to show a difference of 0.5, 1.0, and 2.0 percent water on an oven dry basis between any two sets of samples due to the variability of clay content only.

The two most widely used soil moisture sampling methods are gravimetric and neutron scattering. Gravimetric is the simplest and most reliable, but requires a lot of work and bulk density information. The neutron method is volumetric, covers the full range of soil water content, and may be used frequently at the same site. The error involved in soil moisture sampling is approximately equivalent to one day's evapotranspiration.

3.3 Lysimetry

The best method to measure ΔS is with a properly designed, well-exposed, and correctly operated weighing lysimeter. Examples are found at Cocshocton, Ohio; Davis, California; Tempe, Arizona; and Aspendale, Australia. Accuracies of 0.02 mm of water change have been reported by the Tempe group (van Bavel and Meyers 1962). This is equivalent to 5-min evaporation on a good day. Their initial cost is justified by the reliable data available for short periods of time.

A good lysimeter installation should have adequate fetch treated similar to the lysimeter, preferably a monolith soil and a drainage system at the bottom of the soil column. Lysimeters are easy to use with agricultural crops and range. However, they are difficult to use in forest applications because of the size of installation required (Fritschen et al. 1973).

3.4 Vaporflow Methods

Vaporflow methods have been developed from physical theory to evaluate the magnitude of the fluxes of energy and mass and have several advantages: (1) they are non-destructive; (2) they can be used continuously or can be used for sampling at several sites; (3) they can be employed with a great deal of mobility; and (4) they do not require assumptions concerning the wetness of the surface or the status of soil water.

Meteorological techniques also have several disadvantages. The most serious in forestry is the assumption that the fluxes are vertical; i.e., no horizontal gradients. This assumption may not always be met over forests. Vertical gradients will exist in most ecosystems under forced convection conditions. However, in many locations, free convection prevails more often than does the forced convection regime that is required for application of many meteorological exchange models.

The energy budget equation is a statement of the conservation of energy; i.e., the sum of all energy fluxes must be zero for any system. The major fluxes in the soil-plant-atmosphere system at the earth's surface are conveniently expressed:

$$Q^* + LE + H + G + P_n = 0 \qquad (9)$$

The terms are defined as: Q^*, net radiation flux; G, flux of sensible canopy and soil heat flux density; LE, evaporative flux; H, atmospheric sensible heat flux; P_n, net energy used in photosynthesis and respiration. The convention used here is that the system interface (either surface, soil or plant canopy) is the reference plane and that all fluxes toward that plane are considered positive, whether from above or below the plane, and all fluxes away from the reference plane are negative. Lateral transfer of energy is assumed to be absent.

Net radiation flux is a readily measurable energy budget component. Instrumental developments in recent years have made the direct assessment of Q^* for plane surfaces relatively easy, so there is no need to review the governing relationships for this energy budget component.

The canopy and soil heat flux density below the reference plane, whether it be the soil surface or the top of the canopy, can be described by either the heat capacity equation (10) or the thermal conductivity relation (11) or a combination of both. The heat capacity equation is

$$G = \sum_{i=1}^{n} C_i (\Delta \overline{T}/\Delta t) \, z_i \qquad (10)$$

where the space below the reference plane is divided into n strata of Δz thickness with discrete volumetric heat capacity, C_i, and change in average temperature $\Delta \overline{T}$ over time interval Δt. In equation (10) the volumetric heat capacity of air and vegetation may be combined into a weighted mean, when used with appropriate temperature measurements. The latent energy storage associated with any time rate of change of specific humidity $\Delta \overline{q}/\Delta t$ should also be taken into account in the total storage flux. The thermal conduction relation is

$$G = k \, (\Delta T/\Delta z) \qquad (11)$$

where k is the thermal conductivity and $\Delta T/\Delta z$ the temperature gradient of the layer.

The fluxes of sensible and latent heat are most directly expressed in terms of the eddy transfer equations, so called because transfer is effected largely through the eddying motion of the air in turbulent flow:

$$H = -\rho c_p K_h (\partial \theta / \partial z) \quad (12a)$$

$$LE = -(\rho L\varepsilon/p) K_v (\partial e / \partial z) \quad (12b)$$

c_p is the specific heat at constant pressure and θ is potential temperature. In these equations the facility with which transfer may proceed across the potential gradients is represented by the eddy transfer coefficients, K_h and K_v. Since these transfer coefficients are not known, it is necessary to employ the equations in combination, or with some other relationships, resulting in either the energy balance or aerodynamic models.

3.4.1 Energy balance model

The ratio of the eddy transfer equations (12a) and (12b) is known as the Bowen ration, β. In finite difference form:

$$\beta = \frac{H}{LE} = \frac{c_p}{p} \frac{L\varepsilon}{} \frac{K_h}{K_v} \frac{\Delta \theta}{\Delta e} \quad (13)$$

When β is substituted in the energy balance equation (9), a solution for either H or LE is obtained

$$LE = \frac{-(Q^* + G)}{(1+\beta)} = H/\beta \quad (14)$$

The common assumption is that the ratio of the eddy diffusivities for heat and vapor is a constant, equal to one, over the distance, Δz, across which the potential differences are measured.

3.4.2 Aerodynamic model

The aerodynamic model is based on the eddy transfer equation for momentum

$$\tau = \rho K_m (\partial U/\partial z) \quad (15)$$

and utilizes a ratio of either (12a) or (12b) to (15) to obtain solutions for H and LE. In finite difference form, the resultant equations are:

$$H = -\rho c_p k^2 \frac{K_h}{K_m} \frac{\Delta\theta\,\Delta U}{(\Delta \ln[z-D])^2} \quad (16a)$$

and

$$LE = -\frac{\rho \varepsilon L k^2}{p} \frac{K_v}{K_m} \frac{\Delta e\,\Delta U}{(\Delta \ln[z-D])^2} \quad (16b)$$

The effect of the forest in displacing the reference plane above the ground is accounted for by scaling the instrument heights by D, an "effective" displacement height. The actual height of the reference plane is given by $d = D + z_o$, where z_o is the roughness length of the surface.

In equations (16a) and (16b) equality of the eddy diffusivities may be assumed. This is a reasonable assumption only when the temperature profile indicates neutral stability conditions. Equations (16a) and (16b) are usually corrected for atmospheric stability by multiplying by ϕ. Many expressions have been proposed for ϕ. However, most take the form of

$$\phi = (1 + a\,Ri)^b \quad (17)$$

where a and b are constants and Ri is the Richardson number. The constants in equation (17) appear to be related to height of measurement and to surface roughness. One formulation developed over grass (Paulson 1970) suggests that $a = -15$ and $b = -0.75$.

The Richardson number is a ratio of the contributions to transfer of the buoyancy due to heating in the atmosphere and that due to mechanical energy of the wind. In finite difference form it may be expressed

$$Ri = (g/\bar{\theta})\,(\Delta\theta/\Delta z)/(\Delta u/\Delta z)^2 \qquad (18)$$

where $\bar{\theta}$ is the mean potential temperature (K) of the layer.

3.4.3 Eddy correlation model

The time averaged product of the deviations from the mean of either temperature or vapor pressure and the vertical wind yields the eddy correlation solutions for H and LE, respectively.

$$H = \rho c_p \,\overline{T'w'} \qquad (19a)$$

and

$$LE = \frac{\rho \varepsilon L}{p}\,\overline{e'w'} \qquad (19b)$$

The eddy correlation model is theoretically very sound. It does not require assumptions about equality of the transfer coefficients, height determinations, nor correction for atmospheric stability. Its main limitation is the instrumentation required. However, an approximate estimate of LE may be obtained by combining (19a) with (9) and by assuming G to be nil to obtain

$$LE = -(Q^* + \rho c_p \,\overline{T'w'}) \qquad (20)$$

This approximation simplifies the instrumentation required by avoiding troublesome vapor pressure measurements, and it may prove extremely useful for forested lands.

3.4.4 Application of vaporflow methods

The basic micrometeorological methods (energy balance, aerodynamic, and eddy correlation) are each derived with certain fundamental assumptions.

Tanner (1967) has reviewed these assumptions and discussed their effect upon application of the methods in agriculture. The application of the methods require three conditions: (1) steady-state conditions, (2) surface homogeneity, and (3) boundary layer integrity.

The surface energy budget is in a steady-state when the fluxes do not change with respect to time. Although the fluxes do change continuously throughout the day, the steady-state condition need prevail only during the time required to sample the mean of the environmental variables. The actual time required depends upon the characteristics of the surface and upon the characteristics of the recorder and of the sensors. The interrelationship between these characteristics is not well defined, and it is apparent that many other, more transient effects occur in the environment. Some examples are: the effects of clouds on the radiation flux, the moving pattern of light and shadow at the floor of the forest, and the variability observed in the wind.

Busch (1973) suggests that steady-state requirements may seldom be met in nature to the degree required by aerodynamic (mean profile) methods. The same pessimistic view applies to the Bowen ratio method.

Since the surface characteristics of forests tend to enhance mixing, this may thus reduce the time during which steady-state conditions must prevail. For example, Dyer (1963) discusses the time required for a change in sensible heat flux at the surface to be propagated up through the boundary layer above a grass surface. He reported that almost a half-minute was needed to effect a 90 percent adjustment of the profiles through a distance of a half-meter. Profile adjustment should take place more rapidly above a forest because of the higher intensity of turbulence found near the rough canopy, thus reducing the period required for steady-state conditions. On the other hand, forest profile measurements are commonly made at levels as high as 10 or 20 m above the height of the zero plane displacement. While large heights will obviously require longer periods for flux changes to be propagated up through the profile, no definitive information is available. Studies are needed on the effects of forest roughness and scale on steady-state requirements.

A homogeneous surface is one with uniformly distributed sources and sinks for latent and sensible heat and momentum. Relatively speaking, cultivated crops generally fulfill the homogeneous surface assumption, as do dense, young forest plantations. However, some row crops, natural brush fields, desert vegetation, and forested areas of sparse vegetation are far from homogeneous. Furthermore, as stands mature they become inhomogeneous. The criteria for determining the age or condition when a stand becomes inhomogeneous are not set down. Tanner (1968) states that the inhomogeneity scale of forests is especially taxing and the problems of horizontal flux divergence and sampling become of real importance.

The difficulties appear to be even greater in forests, as the scale of homogeneity must be extended to account for the appreciable depth of the forest canopy. Denmead (1964) found that the sources of heat and vapor and the sink for momentum occurred at different depths within the canopy of a young pine plantation.

Differences in the source/sink levels may be even more pronounced in mature forests with deeper, more diffusely distributed canopies. If such differences exist, it would then be improper to consider that temperature and humidity readings from a given psychrometer represent energy exchanged by sensible and latent heat transfer from the same level in the canopy. The extent and the consequences of different source/sink levels are not worked out for forests, but the effects may be considerable.

Boundary layer integrity prevails when the properties of the atmosphere derive primarily from exchange with the underlying surface. This layer, when in equilibrium with the surface, will be characterized by relatively constant fluxes; e.g., within 10 or 20 percent of surface values (Lumley and Panofsky 1964). Within such a boundary layer developed over a homogeneous surface, all of the net energy transfers will be a vertical direction and horizontal gradients will not exist.

Measurements made outside of this boundary layer will not properly represent the exchanges taking place at the surface of interest regardless of whether the aerodynamic, Bowen ratio, or eddy correlation technique is used. It is common practice to consider the height of the sensors with respect to the thickness of the surface boundary layer, generally with reference to the fetch of the site. However, over forests one may easily extend the measurement heights down to levels below the boundary layer (e.g., below the constant flux zone) and into the region dominated by local effects associated with the projecting crowns of the canopy.

Crowns of different size and shape, when sunlit, may act as sources of heat and vapor with resulting buoyant plumes that create horizontal variability in the atmospheric properties. Shaded crowns, or spaces between crowns, may serve as sinks for sensible heat. Measurements within this region near the canopy are affected by inhomogeneity and horizontal transport; they may be as erroneous as measurements made above the upper limit of the surface boundary layer.

There are thus two limits on sensor heights to consider in order to ensure that measurements are made within the boundary layer of the surface of interest. The only guidance available for selecting the appropriate minimum height is Lettau's (1959) rule of thumb that sensors should be at least five times the average roughness length above the surface. With forests, the actual exchange surface is ill defined. One would presumably refer measurements to the level of the zero-plane displacement. There is also contradiction in guidelines for selecting the maximum instrument height. A number of formulations developed for the growth of boundary layer thickness over smooth surfaces or low vegetation suggest that the ratios of fetch to instrument height (again, presumably measured above the reference plane) are in the order of 100-500 to 1.

Dyer (1963) presents fetch to height requirements for 90 percent profile adjustment. The fetch/height ratio increases from 140 at 0.5-m height to 270 at 5-m height or a fetch of 1,350 m. Brooks (1961) stated that at least eight times the height of the upwind obstructions has to be added to the fetch requirement. It is unclear how this rule is applied in hilly or mountainous terrain.

If such relationships are extrapolated to forests, then the fetch requirements may reach the range of tens of kilometers. It is evident that none of the forest sites studied to date can meet such stringent fetch requirements. Since it seems likely that profile adjustment will be achieved more rapidly over a rough forest surface than over lower, smoother vegetation (Pasquill 1972), it may be that fetch requirements can be eased considerably. Again, this problem lacks a comprehensive treatment for forest conditions.

3.5 Application of Methods of Assessing Water Use by Plants

Application of the methods will be discussed by land use classifications (e.g., agriculture, rangeland, wildlands, and desert and forest). Agricultural lands can be subdivided into irrigated and non-irrigated lands. The largest past effort in assessing water use has been on irrigated crops. Soil moisture sampling, lysimetry, and the energy balance methods have been used extensively to evaluate ET of irrigated crops. The main use of the data has been to determine soil water levels needed for maximum yield and to schedule irrigation. Unfortunately, accompanying meteorological data are not always collected. Water use from non-irrigated crops has received less attention. However, a large data base does exist in various reports and publications.

Less ET information is available for range and wildlands. This data comes primarily from soil moisture sampling and lysimetry, although vaporflow methods are applicable.

As one moves into deserts and forest, data become limited because soil moisture sampling and lysimetry are not necessarily applicable. As the aerodynamic roughness increases, the difficulty of use of vaporflow methods also increases. The eddy correlation technique appears to be the only adequate method for use over tall rough forest. Cost and complexity have limited its use. Therefore, in spite of the fact that a large portion of the earth is forest covered, there is very little data on ET of forest. In addition, the topographic variations of forest lands adds to the complexity of the problem, consequently the hydrologic balance of watersheds is used to obtain ET.

4.   ENERGY BALANCE INFORMATION

Evapotranspiration, ET, from most non-irrigated surfaces may be limited by the lack of available water during the summer half of the year. Lack of energy may limit the water loss from some land surfaces during the winter months.

Solar radiation directly supplies energy needed for ET most of the time. When water is not limiting, generally east of the Mississippi River, 70 to 80 percent of the solar energy is used in evaporation. There are cases, usually irrigated areas, where the horizontal advection of energy augments the energy available for ET. These areas are referred to as "oasis" or wetted areas in desert regions. However, large-scale advection from south to north occurs in the central United States as the wheat crop is removed and solar energy is converted into sensible heat.

In the drier regions of the United States, ET is limited to the amount of precipitation with the bulk of solar energy being converted into sensible heat. These areas can be considered sensible heat source regions, they occur in the west and southwest. The vegetation cover ranges from grass and forest to desert types like sage brush, mesquite, etc.

The ET process requires tremendous amounts of energy relative to other processes. An average daily ET of 2.1 mm from a Douglas-fir forest amounts to 7.6×10^6 ℓ $ha^{-1} yr^{-1}$ which is equivalent to about 19×10^{12} J $ha^{-1} yr^{-1}$, or 2 580 barrels of No. 2 fuel oil $ha^{-1} yr^{-1}$ (Fritschen et al. 1977). The long-term average solar radiation was 55×10^{12} J $ha^{-1} yr^{-1}$, or 7 440 barrels of No. 2 oil $ha^{-1} yr^{-1}$. These values suggest that about 35 percent of the solar energy was used in ET. Assuming 10 percent albedo, the yearly Bowen ratio would be 1.6.

An ET rate of 2.1 mm day^{-1} is low compared to irrigated crops. Values as large as 15 mm day^{-1} have been reported for alfalfa. Irrigated sudangrass transpired 9.8 mm day^{-1} on a day with 15.8 MJ m^{-2} day^{-1} net radiation yielding a Bowen ratio of -0.32 (van Bavel et al. 1963).

4.1 Dryland Evapotranspiration

In the western half of the United States evaporation is generally water limited. Sims et al. (1978) reported ET from native grass at 10 sites in the western United States for a 3-year period. With the exception of Osage, Kansas (eastern edge), and Bridger, Montana (a high altitude site, 2340 m), growing season ET equalled growing season P (Fig. 4). Regression of growing season ET on growing season precipitation yielded a regression coefficient of 0.984 and a R^2 of 0.997. The relation was not as good, $R^2 = 0.83$, when annual values were used because runoff occurred when the grass was dormant.

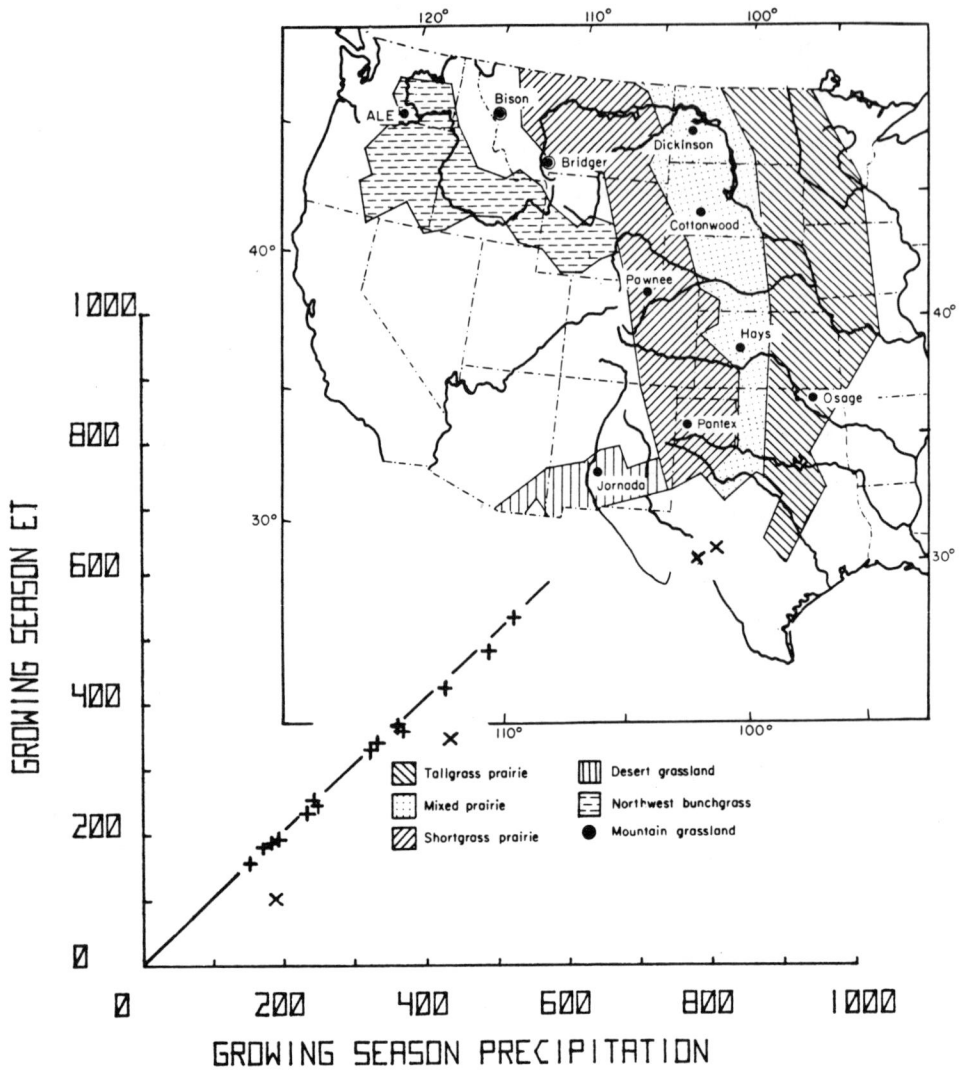

Figure 4 - The relation between growing season ET and precipitation at 10 grassland sites (Sims et al., 1978). X's are from Bridger and Osage which have been omitted from the regression.

The average monthly ET from the Pawnee site shortgrass steppe in northeastern Colorado for 1972-1975 is shown in Fig. 5 along with P, PET, and LAI. ET is water limited at this site and is approximately equal to P. The ratio ET/PET ranged from 0.25 to 0.30 during the growing season. Sixty-seven percent of the precipitation events were ≤ 4 mm and contributed only 17 percent of the total precipitation. Since PET was greater than 5 mm per day, the water added by the small precipitation events would be evaporated within a day or so.

Under dryland conditions ET from crops is also water limited. Hanks et al. (1968) reported ET from crops which exceeded precipitation during the year with the excess water being derived from previous fallow storage (Table 1). Maximum daily ET rates varied with stage of development, time of year, and precipitation. For 7-day periods they were: native grass, 8.2 mm/day in 1965; sorghum, 8.9 mm/day in 1965; oats, 11.5 mm/day in 1965; sudangrass 7.4 mm/day in 1967. Evaporation from bare soil was 81 percent of precipitation for the 3-year period. The Bowen ratios for native grass and oats were 0.33 and 0.05 in 1966 and 0.4 for winter wheat in 1967.

Near Bozeman, Montana (annual precipitation of 160 mm), dryland winter wheat used 221, 272, and 315 mm of water when fertilized with 0, 67, and 268 kg/ha of nitrogen (Brown 1971). Koehler (1960) reported that wheat grown in eastern Washington used 292 mm of water. This value increased to 310 mm when the wheat was fertilized with 179 kg/ha of nitrogen. The increased fertility promoted root development.

At Temple, Texas, where the annual precipitation is 861 mm, ET from grain sorghum was 497 mm in 1968, and 312 mm in 1969. The peak ET rate was 6.4 mm/day in early June. ET from cotton was 510 mm in 1968 and only 290 mm in 1969. The peak ET rate was 6.7 mm/day in mid-July. The primary difference for the 68-69 ET was that 213 mm of rain fell in June and July, 1968, while only 27 mm fell during the same period of 1969 (Ritchie 1971, Ritchie and Burnett 1971).

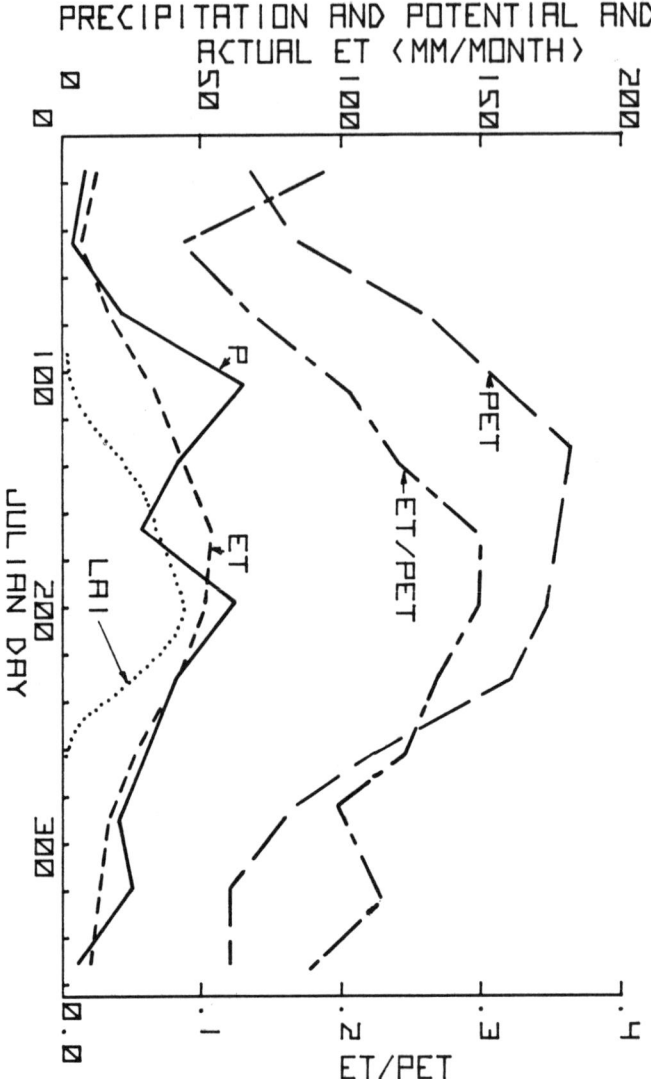

Figure 5 - Average monthly PET, ET, P, and ET/PET for 1972-1975 at the Pawnee Site in Northeastern Colorado (Parton et al., 1981). LAI for 1970-1971 at the same site (Knight 1973).

Table 1. Evapotranspiration of various surfaces under dryland conditions at Akron, Colorado, from Hanks et al. 1968.

Surface	1965	1966	1967
	---- mm ----		
Precipitation	570	350	470
Fallow	410	250	410
Native grass	510	390	510
Sorghum	510	450	
Oats	610	540	
Sudangrass	660		
Wheat		570	530
Millet			560

4.2 Evapotranspiration from Irrigated Crops

Under arid conditions, supplemental addition of water generally increases ET to the extent that the Bowen ratio becomes negative. However, under similar meteorological conditions, the Bowen ratio over water and wet bare soil is positive. This difference is attributed to the difference in aerodynamic roughness or exposure of the evaporating surface to their air mass. Energy balance data over shallow water, wet and drying bare soils is shown in Table 2. The Bowen ratio over water average 0.22. As the wet bare soil surface dried, the Bowen ratio increased from near 0 to 0.30 in three days' time.

Energy balance data for alfalfa, wheat, and cotton are shown in Figs. 6 and 7. The Bowen ratio was negative for irrigated alfalfa during the period of measurement. With a canopy height of 0.5 to 1 m, alfalfa was able to extract about 10 $MJ/(m^2\ day)$ additional energy from the air mass to support the large evaporative flux.

Table 2. Energy balance (MJ m^{-2}day^{-1}) for shallow water, wet and drying bare soil at Tempe, Arizona, 1961 (van Bavel et al. 1963).

Surface	Date	Q*	G	LE	H	β
Shallow water	25 Apr.	17.6*	0.3	-14.9	-3.0	0.20
	26	17.6	-1.0	-13.3	-3.1	0.24
Wet bare soil	28	15.8	0.5	-14.6	-1.7	0.11
	29	17.0	0.4	-17.4	-0.1	0.01
Drying bare soil	30	15.8	-0.9	-15.0	0.1	-0.01
	1 May	13.9	-1.0	-11.3	-1.6	0.14
	2	13.8	-1.4	-9.5	-2.8	0.30

*Divide by 2.44 to obtain mm/day at 25°C.

Plant development alters the distribution of energy. The Bowen ratio was positive over developing cotton for the first 40 days. After canopy development the Bowen ratio became negative approaching a -0.4 (Fig. 7). Similar results were found for wheat, barley, and grain sorghum.

Ritchie and Burnett (1971) showed that the ET/PET ratio increased with LAI for well-watered cotton and grain sorghum which also confirms a decreasing Bowen ratio with canopy development (Fig. 3,D).

Consumptive water use patterns for selected crops are shown in Fig. 8. Seasonal totals were: 1740 mm for alfalfa, 1060 mm for cotton, 1450 mm for sunflower, 522 mm for potatoes, and 1019 mm for Bermuda grass lawn. Additional data for well-watered common crops in the United States and Canada are given in Table 3. Daily ET rates during the summer growing season tend to be larger than 12 month rates. For example the yearly ET rates for grass were 3.3 mm/day and for fruit trees was 3.0 mm/day, while the seasonal rates for corn were 4.4, wheat 3.4, and sugar beets 4.9 mm/day. These values will be compared with forest ET rates in section 5.3.

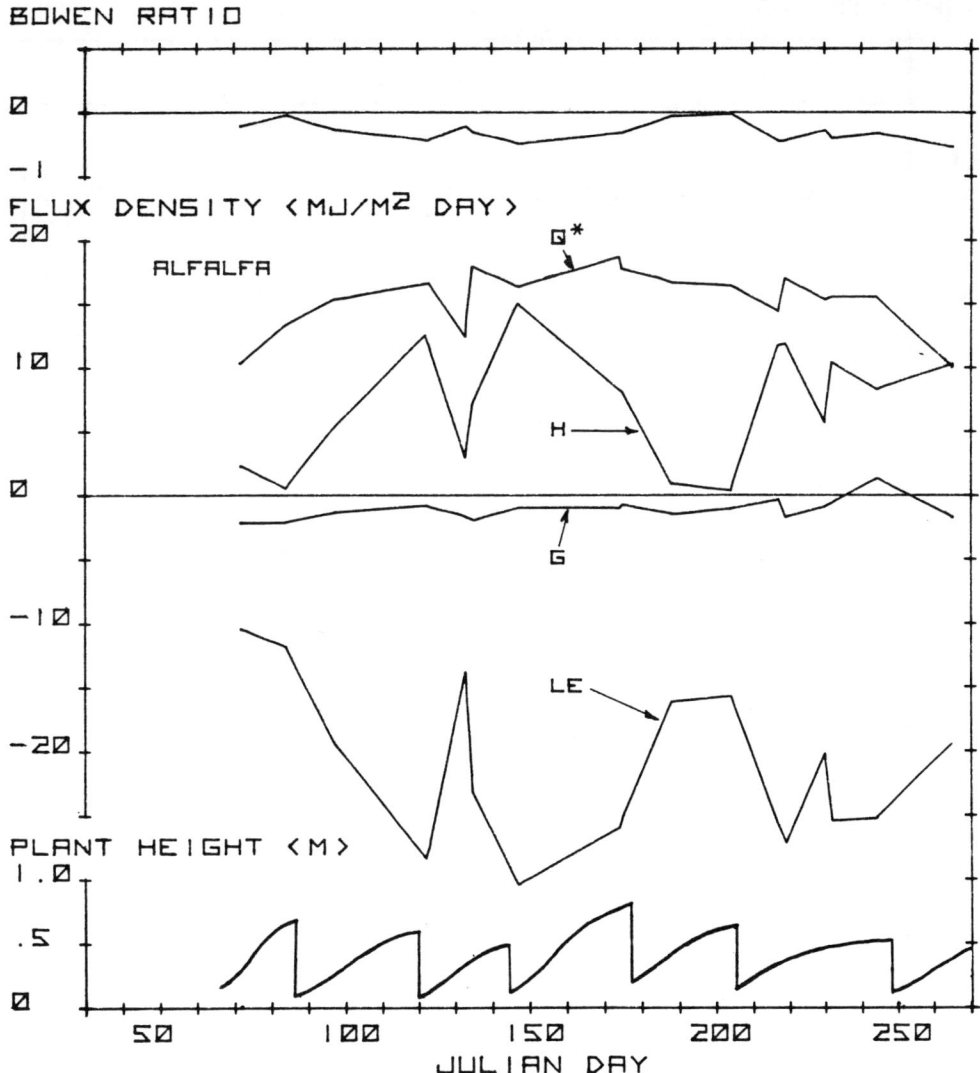

Figure 6 - Energy balance data for irrigated alfalfa at Tempe, Arizona, 1965 (Fritschen 1966).

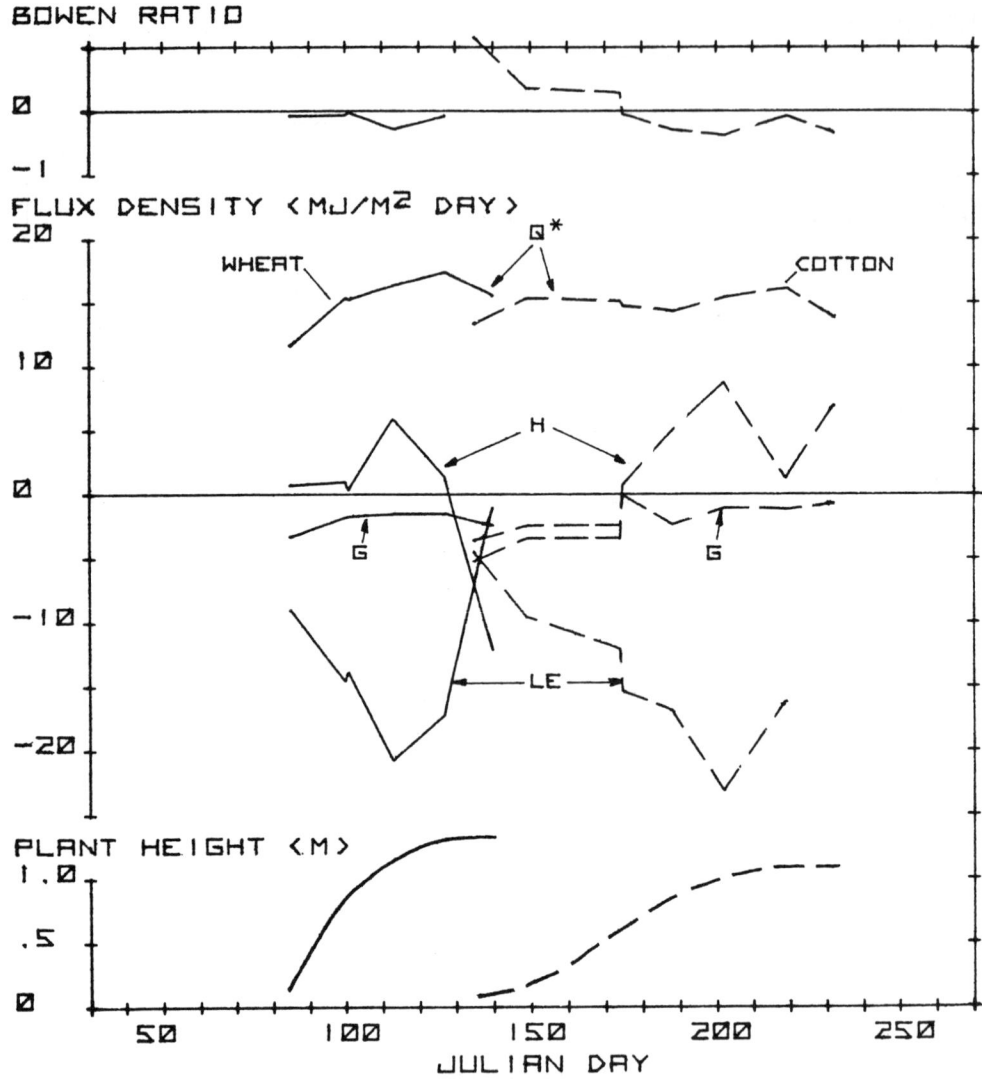

Figure 7 - Energy balance data for irrigated wheat and cotton at Tempe, Arizona, 1965 (Fritschen 1966).

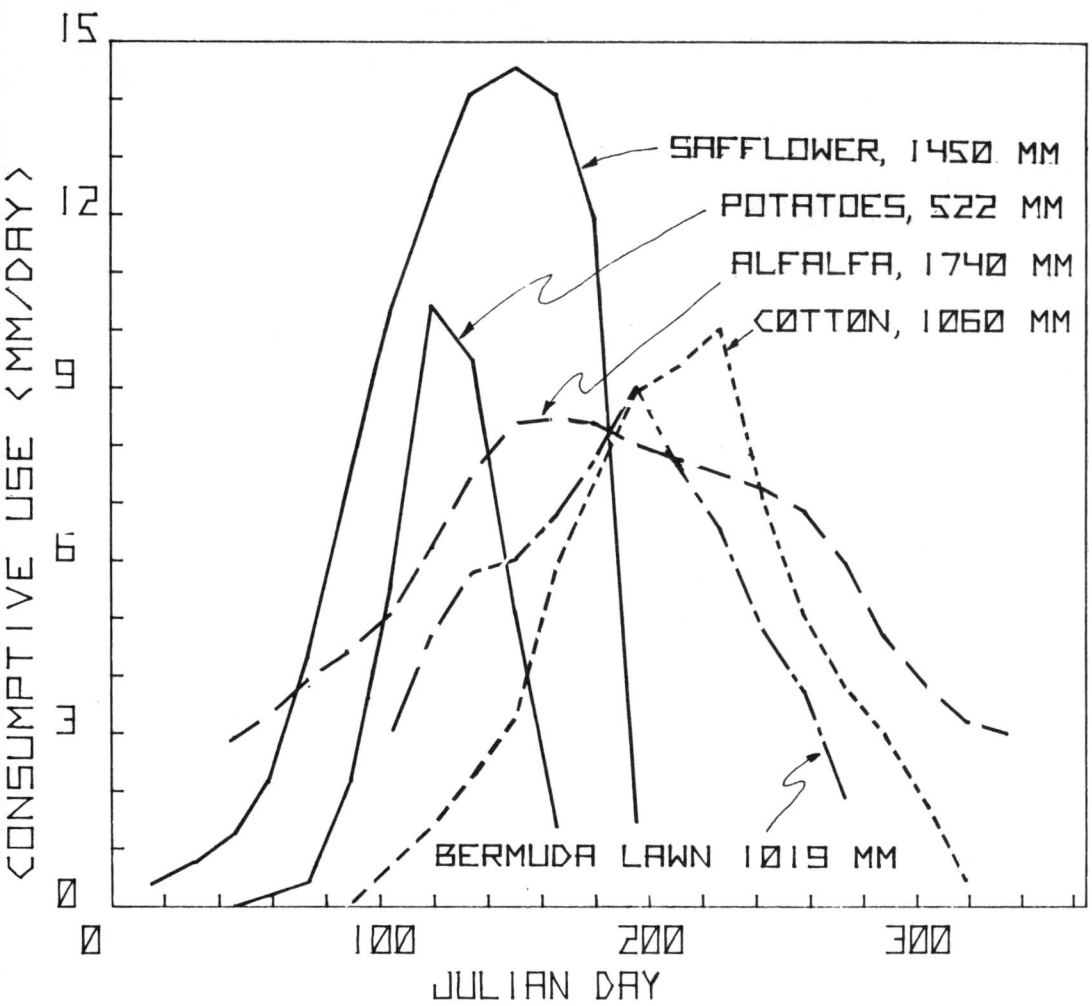

Figure 8 - Consumptive water use for alfalfa (1963), cotton (1962), safflower (1963), potatoes (1963), and Bermuda grass lawn (1963). Personal communication with Leonard Erie, U.S. Water Conservation Laboratory, Phoenix, Arizona.

Table 3. Selected examples of observed seasonal evapotranspiration for well-watered, common crops in the United States and Canada (Jensen 1973).

Crops	Location	Periods and years measurements		Evapotranspiration (mm)	(mm/day)	Reference
FORAGE CROPS						
Alfalfa	Kimberley	---	(1957-58)	594		18
Alfalfa	Swift Current, Sask.	---	(1961-64)	615		23
Alfalfa	S. Alberta, Can.	---	(1950-61)	648		26
Alfalfa	Upham, N.D.	143 days	(1954-56)	594	4.2	3
Alfalfa	Mitchell, Nebr.	---	(1966-68)	747		4
Clover, ladino	Prosser, Wash.	23 May-28 Oct.	(1955)	859	4.6	21
Alfalfa	Kimberley, Ida.	1 May-30 Sept.	(1969-71)	916	6.0	31
Alfalfa	Reno, Nev.	124 days	(1959-61)	1013	8.2	28**
Alfalfa	Arvin, Calif.	12 months	(1961)	1275	3.5	2
Alfalfa	Mesa and Tempe, Ariz.	12 months (1946, 50, 62-63)		1887	5.2	9
Alf.-Reed Can. Br.	Swift Current, Sask.	---	(1961-64)	605		23
Alf.-Cr. Wht. Grass	Swift Current Sask.	---	(1961-64)	594		23
Alf.-Br. Grass	Swift Current, Sask.	---	(1961-64)	610		23
Alf-Reed Canary	Swift Current, Sask.	---	(1961-64)	643		23
Alf.-Inter. Wht Grass	Swift Current Sask.	---	(1961-64)	660		23
Alf. Timothy	Swift Current, Sask.	---	(1961-64)	691		23
Grass	Kimberley, B.C.	---	(1957-59)	579		18
Grass pasture	S. Alberta, Can.	---	(1950-61)	599		26
Grass	Davis, California (Sacramento Valley)	12 months	(1959-71)	1316	3.6	22
Grass	Arvin, California (San Joaquin Valley)	12 months	(1960-65)	1308	3.6	2a
Grass	Thornton, California (Delta)	12 months	(1964-68)	1196	3.3	2a
Grass	Soledad, California (Salinas Valley)	12 months	(1963-70)	1232	3.4	2a
Grass	Guadalupe, California (Coastal)	12 months	(1963-67)	1006	2.8	2a

GRAIN AND FIELD CROPS

Crop	Location	Period (Years)			
Barley	S. Alberta, Can.	---	(1950-61)	409		26
Barley	Powell, Wyo.	18 May-16 Aug.	(1956-57)	386	4.3	1
Barley	Mesa, Ariz.	16 Dec.-15 May	(1952-53,56)	643	2.6	9
Barley	Davis, Calif.	1 Nov.-31 May	(1969-70)	384	1.8	22
Beans	Powell, Wyo.	28 May-3 Sept.	(1957)	396	5.9	1
Beans	Redfield, S. Dak.	105 days	(1952-53)	417	4.0	7
Beans	Davis, Calif.	21 June-24 Sept.	(1968)	404	4.3	22
Corn	S. Alberta, Can.	---	(1950-61)	373		26
Corn	Upham, N. Dak.	107 days	(1953-56)	445	4.2	3
Corn	Redfield, S. Dak.	---	(1951-53)	422		7

RAIN AND FIELD CROPS

Crop	Location	Period (Years)			
Corn	Powell, Wyo.	30 May-6 Sept.	(1958-60)	414	4.2	1
Corn	Coshocton, Ohio	23 May-25 Sept.	(1961)	470	3.8	5
Corn	Hot Springs, S. Dak.	124 days	(1955)	536	4.3	11
Corn	Bushland, Tex.	7 May-8 Sept.	(1970)	617	5.0	25
Corn	Davis, Calif.	15 May-20 Sept.	(1970-71)	640	5.0	22
Oats	S. Alberta, Can.	---	(1950-61)	409		26
Potatoes	Upham, N. Dak.	112 days	(1953-56)	467	4.2	3
Potatoes	Mandan, N. Dak.	128 days	(1953)	455	3.6	3
Potatoes	S. Alberta, Can.	---	(1950-61)	505		26
Potatoes	Phoenix, Ariz.	15 Feb.-15 June	(1959-63)	617	5.1	9
Rice	Davis, Calif.	1 May-30 Sept.	(1968-69)	920	6.1	20
Sorghum	Garden City, Kans.	---	(1957-58)	551		10
Sorghum	Bushland, Tex.	15 June-20 Oct.	(1956-59)	549	4.3	17
Sorghum	Mesa, Ariz.	1 July-31 Oct.	(1955-58, 60)	645	3.6	9
Wheat	Redfield, S. Dak.	---	(1953)	414		7
Wheat, hard	S. Alberta, Can.	---	(1950-61)	462		26
Wheat, soft	S. Alberta, Can.	---	(1950-61)	493		26
Wheat	Mesa, Ariz.	1 Jan.-31 May	(1959-60)	582	3.9	9
Wheat, Mexican	Mesa, Ariz.	15 Nov.-15 May	(1969-70)	655	3.6	6
Wheat, winter	Bushland, Tex.	10 Oct.-25 June	(1956-58)	719	2.7	16

Table 3 (continued).

Crops	Location	Periods and years measurements	Evapotranspiration (mm)	(mm/day)	Reference
		SUGAR CROPS			
Sugarbeet	S. Alberta, Can.	--- (1950-61)	546		26
Sugarbeet	Huntley, Mont.	20 Apr.-27 Sept.(1953)	572	5.2	19
Sugarbeet	Redfield, S. Dak.	--- (1951-53)	610		7
Sugarbeet	Kimberley, Ida.	15 Apr.-17 Oct.(1965-67)	617	3.4	15
Sugarbeet	Davis, Calif.	25 Mar.-20 Sept.(1966)	851	4.6	22
Sugarbeet	Garden City, Kans.	10 Apr.-1 Nov.(1959-60)	927	5.8	12
Sugarbeet	Bushland, Tex.	28 Mar.-18 Oct.(1964,66)	991	6.2	24
Sugarbeet	Mesa, Ariz.	1 Oct.-17 July(1965-66)	1054	4.1	8
		OIL CROPS			
Castorbean	Mesa, Ariz.	15 Apr.-15 Nov.(1958)	1128	7.5	9
Safflower	Mesa, Ariz.	1 Jan.-15 July(1958-60) and (1963-64)	1153	5.9*	9
Safflower	Kimberley, Idaho	1 Apr.-30 Sept.(1966)	635	3.5	14
Soybean	Redfield, S. Dak.	--- (1962-53)	399		7
Soybean	Mesa, Ariz.	16 June-31 Oct.(1944)	564	4.1	9
		FIBER CROPS			
Cotton	Arvin, Calif.	12 months (1961)	912	2.5	2
Cotton	Mesa and Tempe, Ariz.	1 Apr.-15 Nov.(1954-62)	1046	7.6	9
Flax	S. Alberta, Can.	--- (1950-61)	386		26
Flax	Redfield, S. Dak.	105 days (1952-53)	381	3.5	7
Flax	Mesa, Ariz.	1 Jan.-10 June(1943-44)	795	4.4	9
		VEGETABLE CROPS			
Broccoli	Mesa, Ariz.	1 Sept.-14 Feb. (1960-62)	500	2.5*	9
Cabbage, early	Mesa, Ariz.	1 Sept.-31 Jan. (1960-62)	437	2.0	9
Cabbage, late	Mesa, Ariz.	1 Sept.-15 Mar. (1960-62)	622	3.7	9
Cantaloupe	Mesa, Ariz.	1 Apr.-15 July (1959-62)	485	4.6	9

Crop	Location	Date			
Carrots	Mesa, Ariz.	16 Sept.-31 Mar. (1960-62)	422	2.5	9
Cauliflower	Mesa, Ariz.	16 Sept.-31 Jan. (1960-62)	472	1.9*	9
Corn, sweet	S. Alberta, Can.	--- (1950-61)	386		26
Corn, sweet	Mesa, Ariz.	16 Mar.-15 June(1959, 61-62)	498	5.5	9
Lettuce	Mesa, Ariz.	16 Sept.-31 Dec. (1960-62)	216	2.0	9
Onion, dry	Mesa, Ariz.	1 Nov.-15 May (1961-62, 64)	592	3.0*	9
onion, green	Mesa, Ariz.	16 Sept.-31 Jan. (1960-62)	445	3.2	9
Peas, green	S. Alberta, Can.	--- (1950-61)	340		26
Tomato	S. Alberta, Can.	--- (1950-61)	366		26
Tomato	Davis, Calif.	30 Apr.-24 Sept. (1969)	681	3.1	22
DECIDUOUS FRUIT					
Apples	B.C., Can.	--- (1962,64)	531		23
Apples (grass cover)	Wenatchee Wash.	1 Apr.-31 Oct. (1955,57-59)	1059	7.0	13
Plums	Arvin, Calif.	12 months (1962-63)	1072	2.9	2
EVERGREEN FRUIT					
Grapefruit	Phoenix, Ariz.	12 months (1931-34)	1217	3.3	9
Oranges	Phoenix, Ariz.	12 months (1931-34)	993	2.7	9
LAWNS AND ORNAMENTALS					
Bermuda	Raleigh, N.C.	30 May-22 Sept. (1958)	450	3.9	30
Bermuda	Reno, Nev.	112 days (1965-67)	509	4.5	29
Bermuda	Mesa and Tempe Ariz.	16 Apr.-15 Oct. (1959-60,63-64)	1105	6.0	9
Bermuda and St. Augustine	Fort Lauderdale, Fla.	12 months(5 yr. avg.)	1087	3.0	27
Turf	Reno, Nev.	112 days (1965-67)	554	4.9	29

* May include some drainage. ** May be high because of limited fetch.

Table 3 (continued).

(1) Burman et al. (1962)
(2,2a) California, State of (1967, 74)
(3) Carlson et al. (1961)
(4) Daigger et al. (1970)
(5) Dreibelbis and Amerman (1964)
(6) Erie et al. (1973)
(7) Erie and Dimick (1954)
(8) Erie and French (1968)
(9) Erie et al. (1965)
(10) Grimes and Musick (1959)
(11) Haise (1958)
(12) Herron et al. (1964)
(13) Jensen, M.C. et al. (1962)
(14) Jensen, M.E. (unpublished data)
(15) Jensen and Erie (1971)
(16) Jensen and Sletten (1965a)
(17) Jensen and Sletten (1965b)
(18) Krogman and Lutwick (1961)
(19) Larson and Johnston (1965)
(20) Lourence and Pruitt (1971)
(21) Pruitt (1960)
(22) Pruitt et al. (1972)
(23) Ripley (1966)
(24) Schneider and Mathers (1969)
(25) Shipley et al. (1971)
(26) Sonmor (1963)
(27) Stewart and Mills (1967)
(28) Tovey (1970)
(29) Tovey et al. (1969)
(30) van Bavel and Harris (1962)
(31) Wright, J. L. (1972) Unpub. data

At locations where winds are strong, ET is enhanced by advected energy. Blad and Rosenberg (1974) reported average ET from irrigated alfalfa in western Nebraska of 9 mm/day during June, July, and August with maximum values of 12 mm/day. ET from irrigated pasture average 6 mm/day during the same period with maximum values of 9 mm/day.

4.3 Evapotranspiration from Forest

Relative to agriculture, minimal daily or seasonal energy balance data are available for the different species and climates occupied by forests. This is due in part to the difficulty of measurement of big plants in rough terrain and also to the relative abundance of timber in the past. Generally speaking, the transpirational rates from forests are less than from well-watered agricultural crops with comparable LAI. Greater stomatal resistance of plants selected for survival over the centuries accounts for this difference. Leaf resistance for cultivated plants range from 100 to 500 s/m, while resistances from large deciduous tree leaves range from 300 to 1000 s/m and stomatal resistance for conifer needles range from 1000 to 3000 s/m (Nobel 1974).

Rutter (1968) compiled precipitation and evaporation data from forest throughout the world (Table 4). These data are largely from watersheds and may include some percolation losses. Even though the data are classified according to soil water deficit, the average annual ET ranged from 1.1 to 4.3 mm/day, the larger values being associated with larger rainfall. On an annual basis, most of the values in Table 4 are 1 to 2 mm/day less than well-watered grass (Table 3) and are considerably below alfalfa rates. The average daily rates for the growing season also appear to be 1 to 2 mm/day less than the common crops listed in Table 3.

Table 4. Precipitation and evaporation for forest according to soil water deficit classes (Rutter 1968).

Location and vegetation	Precipitation evaporation					
	annual mm	growing season mm/day	annual mm	mm/day	growing season mm/day	Reference
NEGLIGIBLE SOIL WATER DEFICIT						
Northern Taiga, USSR mainly Pinus sylvestris and Picea abies	525	50%	286	0.8		1
Middle taiga, USSR mainly Picea abies and Betula	600	50%	329	0.9		1
Yorkshire, England Picea sitchensis	1350	50%	800	2.2		2
Harg Mountains, West Germany Picea abies	1250	50%	579	1.6		3
Carpathian Mountains, Czechoslovakia mixed conifer and deciduous	1650	950	861	2.4		4
Northern Japan 40% conifer, 60% deciduous	2617	50%	542	1.5		5
Ota, Japan mixed conifer and deciduous	1600	60%	658	1.8		6
Shackham Brook, New York, U.S. 57% conifer, 27% deciduous, 16% pasture and crops	1050	500	504	1.4		7
Mt. Koskiusko, New South Wales, Australia Eucalyptus niphophia	2500	50%			3.8	8
Coweeta, North Carolina, U.S. mixed hardwoods	1730	50%	875	2.4		9
Coweeta, North Carolina, U.S. mixed hardwoods	2000	50%	700	1.9		10
SMALL SOIL WATER DEFICIT						
Southern taiga, USSR mainly Picea abies and Betula	600	50%	412	1.1		11
Bregentved, Denmark Picea abies	680	50%	470	1.3		12

Table 4 (continued)

Location and vegetation	Precipitation evaporation						Reference
	annual mm	growing season mm/day	annual mm	mm/day	growing season mm/day		

		SMALL SOIL WATER DEFICIT					
Bregentved, Denmark Picea abies,	520	60%	450	1.2			
Fagus sylvatica, Quercus and other deciduous			400	1.1			
Scania, Sweden Picea abies (wet site)	793	50%	883	2.4	4.8		13
Castricum, Holland Pinus nigra	840	50%	655	1.8			14
Berkshire, England Pinus sylvestris	686	378	655	1.8			15
Lower Michigan, U.S. Pinus resinos and Quercus		550			2.8		16
Fernow, West Virginia, U.S. mixed deciduous	1490	50%	865	2.4			17
Okayama, Japan Pinus densiflora	1153	66%	847	2.3			18
Gunnison, Colorado, U.S. Populus tremuloids and Picea engelmannii	530				4.2 3.3		19
Fraser, Colorado, U.S. Pinus contoria	620		360	1.0	1.5		20
Wagon Wheel Gap, Colorado, U.S. Pseudotsuga taxifolia and Populus tremuloids	530	50%	383	1.0			21
Kericho, Kenya evergreen rain forest	1950		1570	4.3			22
Aberdare Mountains, Kenya bamboo forest	2160		1150	3.2			23

Table 4 (continued)

Location and vegetation	Precipitation		evaporation			Reference
	annual mm	growing season mm/day	annual mm	mm/day	growing season mm/day	
MODERATE SOIL WATER DEFICIT						
Moscow region, USSR Pinus sylvestris and Picea abies	480-575	60%	437	1.2		24
Forest steppe, Russia deciduous forest	510-525	50%	405	1.1		24
Steppe zone, Eastern USSR Quercus, Fraxinus, Betula and other deciduous	375-457	50%	424	1.2		24
Lebanon State Forest, New Jersey, U.S. Pinus echinata and Quercus	1200	50%			3.0	25
Farmington, Utah, U.S. Populus tremuloides	1340	100-250	567	1.6	3.9	26
Vinton, Ohio, U.S. Pinus echinata, Quercus and shrubs	1000	30%			3.5 3.5 3.2	27
Union, South Carolina, U.S. Pinus taeda	1000	50%			3.7	28
Crossett, Arkansas, U.S. Quercus, all ages Pinus echinata + Pinus taeda	1200	460			4.2 4.5	29
SEVERE SOIL WATER DEFICIT						
North Fork, California, U.S. mixed chaparral species	1150	slight	485	1.3		30
San Gabriel Mountains, California, U.S. Rhus and Cercocarpus montana	780	slight	595	1.6		31
California, U.S. Bass Lake Pinus ponderosa	1260	slight	580	1.6		32
North Fork mixed chaparral	1050	slight	415	1.1		

Table 4 (continued)

Location and vegetation	Precipitation		evaporation			Reference
	annual mm	growing season mm/day	annual mm	mm/day	growing season mm/day	
		SEVERE SOIL WATER DEFICIT				
San Dimas						
mixed chaparral	1150	slight	560	1.5		
Pinus coulteri	1230	slight	637	1.7		33
Adenostoma fasciculatum			648	1.8		
Ceanothus crassifolius			599	1.6		
Quercus dumosa			630	1.7		
Carmel Mountains, Israel						
Maqui shrubs	650	100	509	1.4	1.9	34
		EXTREME SOIL WATER DEFICIT				
San Dimas, California, U.S.						
Quercus dumosa and other chaparral	600	slight	535	1.5	1.9	35
San Dimas, California, U.S.						
Pinus coultrei			392	1.1		36
Adenostoma fasciculatum	525	slight	430	1.2		36
Ceanothus crassifolius			455	1.2		
Quercus dumosa			428	1.2		

(1) Molchanov (1960)
(2) Law (1958)
(3) Delfs et al. (1958)
(4) Valek (1959)
(5) Nakano (1967)
(6) Hirata (1929)
(7) Schneider and Ayer (1961)
(8) Costin et al. (1964)
(9) Hoover (1944)
(10) Johnson and Kovner (1956)
(11) Molchanov (1960)
(12) Holstener-Jorgensen (1959)
(13) Stålfelt (1963)
(14) Deij (1954)
(15) Rutter (1964)
(16) Urie (1959)
(17) Reinhart and Eschner (1962)
(18) Nakano (1967)
(19) Brown and Thompson (1965)
(20) Wilm and Dunford (1948)
(21) Bates and Henry (1928)
(22) Pereira et al. (1962a)
(23) Pereira et al. (1962b)
(24) Molchanov (1960)
(25) Lull and Axley (1958)
(26) Croft and Monninger (1953)
(27) Marston (1962)
(28) Metz and Douglass (1959)
(29) Moyle and Zahner (1954)
(30) Rowe (1948)
(31) Hoyt and Troxell (1934)
(32) Rowe and Colman (1951)
(33) Patric (1961)
(34) Shachoria et al. (1967)
(35) Rowe and Reimann (1961)
(36) Patric (1961)

Fritschen et al. (1977) reported 3 years ET data from 28-m Douglas-fir in a weighing lysimeter. The results from 1972 are shown in Fig. 9.

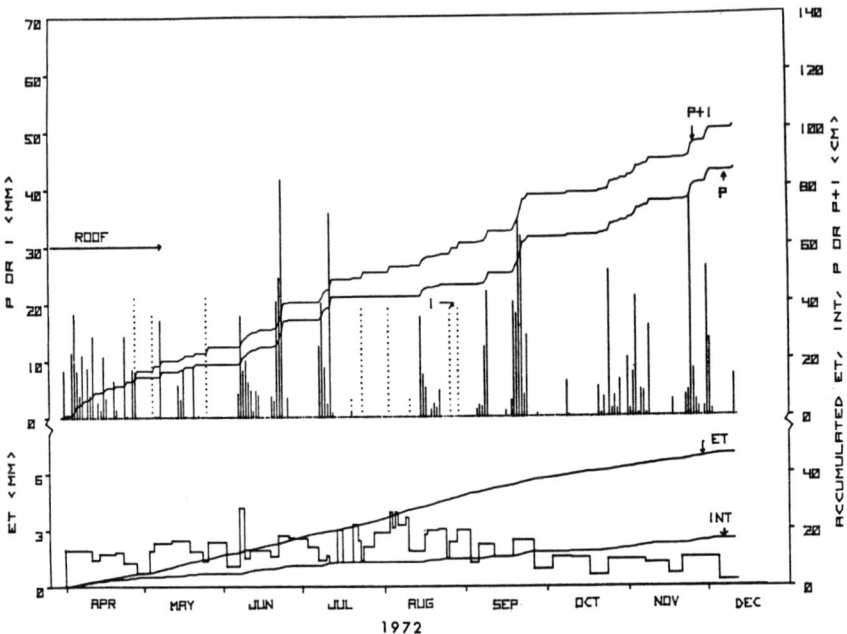

Figure 9 - Average daily and accumulated values of evapotranspiration (ET), precipitation (vertical lines), irrigation (dotted lines), their accumulation (P+l), and interception (INT) for the lysimeter tree on the Cedar River watershed near Seattle, Washington, 1972.

Monthly rates ranged from a low of 0.3 mm/day during December to 2.8 mm/day in August. The daily average for the 3 years was 2.1 mm which is in agreement with the values in Table 4. Diurnal energy balance data from a 7.8 m Douglas-fir forest shows a positive Bowen ratio throughout the day (Fig. 10).

Evaporation rates from deciduous trees throughout the year reflect the development of foliage similar to annual crops. The evaporation rates of beech increased from 0.28 mm/day in the spring to 3.93 mm/day during the summer (Table 5.) Smaller increase in evaporation rates were observed from spruce (0.48 to 3.13 mm/day). The increasing evaporation from spruce is indicative of increasing PET while the increasing rates from beech also include foliage development. Although larger peak rates were obtained from beech (5.07 mm/day) than from spruce (3.36 mm/day), the January 1-October 31 rates were larger for spruce (1.64 mm/day) than for beech (1.40 mm/day). Swank and Douglas (1974) reported that the conversion of a mature deciduous watershed cover to white pine reduced the total water yield by 20 percent and the largest reductions occurred during the dormant and early growing seasons.

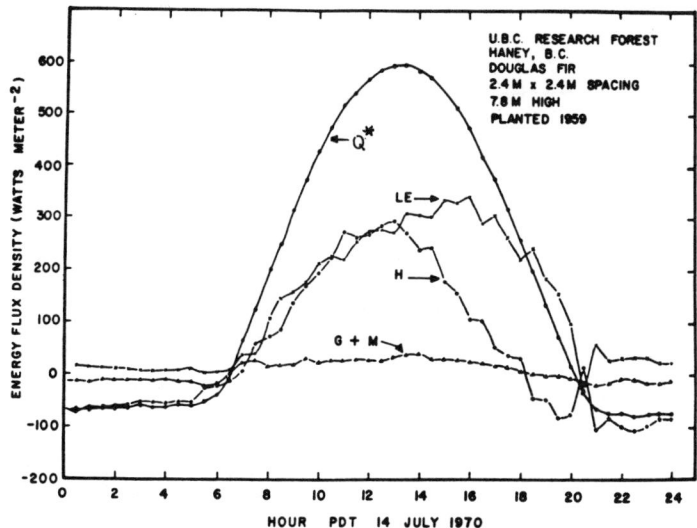

Figure 10 - Diurnal energy budget of a young Douglas-fir plantation (Black and McNaughton, 1971). LE, H and G + M are plotted negatively relative to the other figures.

Table 5. Total evaporation rates from beech and spruce for different time periods in 1969 at Solling, Germany (Benecke 1976).

Time period	Total evaporation (mm/day)	
	beech	spruce
January 1-October 31	1.40	1.64
January 1-March 31	0.28	0.48
April 1-April 30	0.33	0.63
May 1-May 31	1.17	3.28
May 31-June 30	3.11	2.16
July 5-August 5	3.92	3.13
July 15-August 1	5.07	3.36

5. SUMMARY AND CONCLUSIONS

A model to compute evaporation from vegetated surfaces was presented and discussed to illustrate the complexity of the problem of combining soil, plant and atmospheric conditions. Additional parameterization is required to apply the model to other plant species and varying topography.

Methods of obtaining evapotranspiration data were presented and their applicability to various plant systems was discussed.

Energy balance data including evapotranspiration and Bowen ratio information for natural and cultivated surfaces in arid, humid, and irrigated conditions were presented. From these data several general conclusions can be drawn:

(1) The leaf resistance of cultivated crops is less than that of native grasses, shrubs, and trees. Therefore ET from cultivated crops can be expected to be larger than ET from native grasses, shrubs, and trees with the same LAI and water not limiting. Phreatophytes (e.g., salt cedar, etc.) will be an exception to this general rule.

(2) Due to precipitation patterns and low temperatures, ET is usually energy limited during the winter months. During the summer, ET is usually water limited.

(3) Under humid conditions (Eastern United States), ET from cultivated crops is generally energy limited. The Bowen ratio from crops will be near zero and will be slightly positive from forests. When short drought periods occur, shallow root crops and grasses will show stress before the deeper rooted trees and shrubs will.

On a yearly basis ET from coniferous forest will be larger than from deciduous forest. Because of larger aerodynamic roughness, ET from forests would be larger than from shrubs and native grasses.

(4) Under arid conditions, ET is water limited. Consequently, the Bowen ratio would fluctuate from zero to large positive values depending upon the availability of water. Shallow-rooted plants would show stress before deeper rooted plants. The yearly ET from coniferous plants would be greater than from deciduous plants because of greater interception loss and transpiration during the winter months when other plants are dormant. Less annual runoff would occur from coniferous type watersheds than from deciduous type watersheds.

(5) Under irrigated conditions, ET may be larger than net radiation with additional energy being derived from the air mass. Bowen ratios vary with crop development. Positive Bowen ratios exist during early stages of plant development. After canopy closure, Bowen ratios become negative and may approach -1.

REFERENCE

Bates, C. G. and A. J. Henry. 1928. Forest and stream-flow experiment at Wagon Wheel Gap, Colorado. Monthly Weather Rev. Suppl. 30, 79 pp.

Benecke, P. 1976. Soil water relations and water exchange of forest ecosystems. In: Water and Plant Life, Ed. by O. L. Lange, L. Kappen, and E. D. Schulze. Springer-Verlag, New York. 536 pp.

Black, T. A. and K. G. McNaughton. 1971. Psychrometric apparatus for Bowen ratio determination over forest. Bound. Layer Meteorol. 2:246-254.

Blad, B. L. and N. J. Rosenberg. 1974. Evapotranspiration by sub-irrigated alfalfa and pasture in East Central Great Plains. Agron. J. 66:248-252.

Brooks, F. A. 1961. Need for measuring horizontal gradients in determining vertical eddy transfer of heat and moisture. J. Meteorol. 18:589-596.

Brown, H. E. and J. R. Thompson. 1965. Summer water use by aspen, spruce and grassland in western Colorado. J. Forestry 63, 756.

Brown, P. L. 1971. Water use and woil water depletion by dryland winter wheat as affected by nitrogen fertilization. Agron. J. 63:43-46.

Burman, R. D. and J. R. Partridge. 1962. Evapotranspiration of water by small grain, corn and beans in northwestern Wyoming. Wyo. Agr. Expt. Sta. Mimeo. Circ. 174, 14 pp.

Busch, N. E. 1973. The surface boundary layer (Part 1). Bound. Layer Meteorol. 4:213-240.

California, State of. 1967. Vegetative water use. Dept. Water Resources Bull. No. 113-2, 82 pp.

California, State of. 1974. Vegetative water use. Dept. Water Resources Bull. No. 113-3, (in press).

Carlson, C. W., D. L. Grunes, L. O. Fine et al. 1961. Soil, water, and crop management on newly irrigated lands in the Dakotas. U.S. Dept. Agr. Prod. Res. Rept. 53, 34 pp.

Costin, A. B. D. J. Wimbush, and R. N. Cromer. 1964. Studies in catchment hydrology in the Australian alps. 5. Soil moisture characteristics and evapotranspiration. Australia, CSIRO, Div. Plant Ind. Tech. Paper 20, 20 pp.

Croft, A. R. and L. V. Monninger. 1953. Evapotranspiration and other water losses on some aspen forest types in relation to water available for stream flow. Trans. Am. Geophys. Union 34, 563.

Daigger, L. A., L. S. Axthelm, and C. L. Asburn. 1970. Consumptive use of water by alfalfa in western Nebraska. Agron. J. 62:507-508.

Deij, L. J. L. 1954. The lysimeter station at Castricum (Holland). Intern. Assoc. Sci. Hydrol. Gen. Assembly Rome 3, 203.

Delfs, J., W. Friedrich, H. Kiesekamp, and A. Wagenhoff. 1958. Der Einfluss des Waldes und des Kahlschlages auf den Abflussvorgang, den Wasserhaushalt und den Bodenabtrag. Mitt. Niedersächsischen Landesforstsverwaltung 3, 223 pp.

Denmead, O. T. 1964. Evaporation sources and apparent diffusivities in a forest canopy. J. Appl. Meteorol. 3:383-389.

Denmead, O. T. and B. D. Millar. 1979. Water transport in wheat plants in the field. Agron. J. 68:297-303.

Doorenbos, J. and W. O. Pruitt. 1975. Crop water requirements. In Irrigation and Drainage 24. FAO, Rome. 178 p.

Dreibelbis, F. R. and C. R. Amerman. 1964. Land use, soil type, and practice effects on the water budget. J. Geophys. Res. 69(16): 3387-3393.

Dyer, A. J. 1963. The adjustment of profiles and eddy fluxes. Quart. J. Roy. Meteorol. Soc. 89:276-280.

Ehleringer, J. R. and P. C. Miller. 1975. A simulation model of plant water relations and production in alpine tundra, Colorado. Oecologia 19:177-193.

Erie, L. J., D. A. Bucks, and O. F. French. 1973. Consumptive use and irrigation management for high-yielding wheats in central Arizona. Progressive Agr., 25(2):14-15.

Erie, L. J. and N. A. Dimich. 1954. Soil moisture depletion by irrigated crops grown in South Dakota. So. Dak. Agr. Expt. Sta. Circ. 104, 15 pp.

Erie, L. J. and O. F. French. 1968. Water management of fall-planted sugarbeets in the Salt River Valley of Arizona. Trans. Am. Soc. Agr. Engr. 11:792-795.

Erie, L. J., O. F. French, and K. Harris. 1965. Consumptive use of water by crops in Arizona. Ariz. Agr. Expt. Sta. Tech. Bull. 169, 44 pp.

Frey, H. T. 1979. Major uses of land in the United States: 1974. AER-440. U.S. Dept. Agr., Econ. Res. Serv.

Fritschen, L. J. 1966. Evapotranspiration rates of field crops determined by the Bowen ratio method. Agron. J. 58:339-342.

Fritschen, L. J., L. Cox, and R. Kinerson. 1973. a 28-m Douglas-fir in a weighing lysimeter. For. Sci. 4:256-261.

Fritschen, L. J. and P. Doraiswamy. 1973. Dew: An addition to the hydrologic balance of Douglas-fir. Water Resources Res. 9:891-894.

Fritschen, L. J., J. Hsia, and P. Doraiswamy. 1977. Evapotranspiration of a Douglas-fir determined with a weighing lysimeter. Water Resources Res. 13:145-148.

Gash, J. H. C., C. R. Lloyd, and J. B. Stewart. 1978. A model of forest transpiration and interception, using data from an automatic weather station. In: Comparison of Forest Waters and Energy Exchange Models. Ed. by Swen Halldin. Proceedings from an IUFRO Workshop. Uppsala (Sweden), September 24-30, 1978. International Society for Ecological Modelling. Copenhagen, Denmark. 258 pp.

Grimes, D. W. and J. T. Musick. 1959. How plant spacing, fertility and irrigation affect grain sorghum production in southwestern Kansas. Kans. State Agr. Expt. Sta. Bull. 414, 17 pp.

Haise, J. R. 1958. Irrigation. In: Agronomic Trends and Problems in the Great Plains, Adv. in Agron., Vol. X. Academic Press, New York, pp. 47-56.

Hanks, R. J., H. R. Gardner, and R. L. Florian. 1968. Evapotranspiration-climate relations for several crops in the Central Great Plains. Agron. J. 60:538-542.

Herron, G. M., D. W. Grimes, and R. E. Finkner. 1964. Effect of plant spacing and fertilizer on yield, purity, chemical constituents, and evapotranspiration of sugarbeets in Kansas. I. Yield of roots, purity, percent sucrose and evapotranspiration. J. Am. Soc. Sugarbeet Technol. 12(8):686-698.

Hirata, T. 1929. Contributions to the problem of the relation between forest and water in Japan. Imp. Forestry Expt. Sta. (Meguro, Tokyo) 41 pp.

Holstener-Jorgensen, H. 1959. A contribution to elucidation of evapotranspiration of forest stands on clayey soils with a high water-table. Publ. Intern. Assoc. Sci. Hydrol. 48, 286.

Hoover, M. D. 1944. Effect of removal of forest vegetation upon water yields. Trans. Am. Geophys. Union 25, 969.

Hoyt, W. D. and H. C. Troxell. 1934. Forests and stream-flow. Trans. Am. Soc. Civil Engrs. 99, 1.

Jensen, M. C., E. S. Degman, and J. E. Middleton. 1962. Apple orchard irrigation. Wash. Agr. Expt. Sta. Circ. 402, 11 pp.

Jensen, M. E. and L. J. Erie. 1971. Irrigation and water management. In: Advances in Sugarbeet Production, R. T. Johnson et al. (ed.), Iowa State Univ. Press, Ames, pp. 189-222.

Jensen, M. E., J. L. Wright and B. J. Pratt. 1971. Estimating soil moisture depletion from climate, crop, and soil data. Transactions ASAE 14:954-959.

Jensen, M. E. and W. H. Sletten. 1965a. Evapotranspiration and soil moisture-fertilizer interrelations with irrigated winter wheat in the Southern High Plains. U.S. Dept. Agr. Conserv. Res. Rept. No. 4, 26 pp.

Jensen, M. E. and W. H. Sletten. 1965b. Evapotranspiration and soil moisture-fertilizer interrelations with irrigated grain sorghum in the Southern Great Plains. U.S. Dept. Agr. Conserv. Res. Rept. No. 5, 27 pp.

Jensen, M. E. 1973. Consumptive use of water and irrigation water requirements. Am. Soc. Civil Engr. New York, p. 215.

Johnson, E. A. and J. L. Kovner. 1956. Effect on stream-flow of cutting a forest understory. For. Sci. 2, 82.

Knight, Dennis H. 1973. Leaf area dynamics of shortgrass prairie. Ecology 891-896.

Koehler, F. E. 1960. Nitrogen uptake and moisture use by wheat. Proc. 11th Annual Pacific Northwest Fertilizer Conf. Salt Lake City, Utah, pp. 141-146.

Krogman, K. K. and L. E. Lutwick. 1961. Consumptive use of water by forage crops in Upper Kootenay River Valley. Can. J. Soil Sci. 41:1-4.

Larson, W. E. and W. B. Johnston. 1965. The effect of soil moisture level on the yield, consumptive use of water, and root development by sugar beets. Soil Sci. Soc. Am. Proc. 19:275-279.

Law, F. 1958. Measurement of rainfall, interception and evaporation losses in a plantation of Sitka spruce trees. Intern. Assoc. Sci. Hydrol. Gen. Assembly Toronto 2, 397.

Lettau, H. 1959. A review of research problems in micrometeorology. Dept. Meteorol. Univ. Wisconsin (Final contract report DA-36-39-SC-80063).

Lourence, F. J. and W. O. Pruitt. 1971. Energy balance and water use of rice grown in the Central Valley of California. Agron. J. 63:827-832.

Lull, H. W. and J. H. Axley. 1958. Forest soil-moisture relations in the Coastal Plain sands of Southern New Jersey. For. Sci. 4, 2.

Lumley, J. L. and H. A. Panofsky. 1964. The structure of atmospheric turbulence. Vol. 12, Interscience Monographs and Texts and in Physics and Astronomy. Wiley, New York. 239 pp.

Marston, R. B. 1962. Influence of vegetation cover on soil moisture in south-eastern Ohio. Soil Sci. Soc. Am. Proc. 26, 605.

Metz, L. J. and J. E. Douglass. 1959. Soil moisture depletion under several Piedmont cover types. U.S. Dept. Agr. Tech. Bull. 1207, 23 pp.

Molchanov, A. A. 1960. The Hydrological Role of Forests. (Translation by A. Gourevitch, 1963), 407 pp. Israel Program Sci. Transl., Jerusalem.

Moyle, R. C. and R. Zahner. 1954. Soil moisture as affected by stand conditions. U.S. Dept. Forest. Serv. Southern Forest Expt. Sta. Occasional Paper 137, 14 pp.

Nakano, H. 1967. Effects of changes of forest conditions on water yield, peak flow and direct run-off of small watersheds in Japan. In: Forest Hydrology (W. E. Sopper and H. W. Ludd, eds.), p. 551. Pergamon, Oxford.

Nimah, M. N. and R. J. Hanks. 1973. Model for estimating soil water, plant, and atmospheric interrelations, 1, Description and sensitivity. Soil Sci. Soc. Amer. Proc. 34:522-527.

Nobel, Park S. 1974. Introduction to Biophysical Plant Physiology. Freeman and Co. San Francisco, Calif., pp. 488.

Parton, W. J., W. J. Lauenroth, and F. M. Smith. 1981. Water loss from a shortgrass steppe. Agric. Meteor. In press.

Pasquill, F. 1972. Some aspects of boundary layer description. Quar. J. Roy. Meteorol. Soc. 98:469-494.

Patric, J. H. 1961. The San Dimas large lysimeters. J. Soil Water Conserv. 16, 13.

Paulson, C. A. 1970. The mathematical representation of wind speed and temperature profiles in the unstable atmospheric surface layer. J. Appl. Meteorol. 9:857-861.

Penman, H. L. 1948. Natural evaporation from open water, bare soil, and grass. Proc. Royal Soc. Ser. A, No. 1032, 193 pp. 120-145.

Pereira, H. C., M. Dagg, and P. H. Hosegood. 1962a. The development of tea estates in tall rain forest: The water balance of both treated and control valleys. E. African Agr. Forestry J. 27, 36.

Pereira, H. C., M. Dagg, and P. H. Hosegood. 1962b. The water balance of bamboo thicket and of newly planted pines. E. African Agr. Forestry J. 27, 95.

Pruitt, W. O. 1960. Relation of consumptive use of water to climate. Trans. Am. Soc. Agr. Engr. 3(1):9-13, 17.

Pruitt, W. O., F. J. Lourence, and S. von Oettingen. 1972. Water use by crops as affected by climate and plant factors. Calif. Agr., Oct., pp. 10-14.

Reinhart, K. G. and A. R. Eschner. 1962. Effect on stream-flow of four different forest practices in the Allegheny Mountains. J. Geophys. Res. 67, 2433.

Ripley, P. O. 1966. The use of water by crops. Intern. Comm. on irrg. and Drain., Special Session, Jan. 1966, SSR 2, S9-S21.

Ritchie, J. T. 1971. Dryland evaporative flux in a subhumid climate: I. Micrometeorological influences. Agron. J. 63:51-55.

Ritchie, J. T. and F. Burnett. 1971. Dryland evaporative flux in a subhumid climate: II. Plant influences. Agron. J. 63:56-62.

Rowe, P. B. 1948. Influence of woodland chaparral on water and soil in Central California. Calif. Dept. Nat. Resources, Div. Forestry, 70 pp.

Rowe, P. B. and E. A. Coleman. 1951. Disposition of rainfall in two mountain areas of California. U.S. Dept. Agr. Tech. Bull. 1048.

Rowe, P. B. and L. F. Reimann. 1961. Water use by brush, grass and grassforb vegetation. J. Forestry 59, 175.

Rutter, A. J. 1964. Studies in the water relations of *Pinus sylvestris* in plantation conditions. 4. Direct observation of the rates of transpiration, evaporation of intercepted water, and evaporation from the soil surface. J. Appl. Eco. 3, 393.

Rutter, A. J. 1968. Water consumption by forests. In: Water deficits and plant growth. VI. Ed. by T. T. Kozlowski. Academic Press, New York. 333 pp.

Saxton, K. E., H. P. Johnson and R. H. Shaw. 1974. Modeling evapotranspiration and soil moisture. Transactions Am. Soc. Agric. Eng. 17:673-677.

Schneider, W. J. and G. R. Ayer. 1961. Effect of reforestation on stream-flow in Central New York. U.S. Geol. Surv. Water Supply Papers 1602.

Schneider, A. D. and A. C. Mathers. 1969. Water use by irrigated sugarbeets in the Texas High Plains. Texas Agr. Expt. Sta. MP-935, 7 pp.

Shachori, A., D. Rosenzweig, and A. Poljakoff-Mayber. 1967. Effect of Mediterranean vegetation on the moisture regime. In: Forest Hydrology (W. E. Sopper and H. W. Lull, eds.), p. 291. Pergamon, Oxford.

Shipley, J. L., P. W. Unger, and C. Regier. 1971. Consumptive water use, harvestable dry matter production, and nitrogen uptake by irrigated corn, Northern Plains of Texas. Tex. Agr. Expt. Sta. PR-2898, 14 pp.

Sims, P. L., J. S. Singh, and W. E. Lauenroth. 1978. The structure and function of ten western North American grasslands. I. Abiotic and vegetational characteristics. J. Ecology 66:251-285.

Sonmor, L. G. 1963. Seasonal consumptive use of water by crops grown in Southern Alberta and its relationship to evaporation. Can. J. Soil Sci. 43:287-297.

Stålfelt, M. G. 1963. On the distribution of the precipitation in a spruce stand. In: The water relations of plants (A. J. Rutter and F. H. Whitehead, eds.), p. 116. Blackwell, Oxford.

Stewart, E. H. and W. C. Miller. 1967. Effect of depth to water table and plant density on evapotranspiration rate in Southern Florida. Trans. Am. Soc. Agr. Engr. 10:746-747.

Sudar, R. A., K. E. Saxton, and R. G. Spomer. 1979. A predictive model of water stress in corn and soybeans. Paper No. 70-2004 presented at the summer meeting of ASAE & CSAE, Winnipeg, Canada, June 24-27.

Swank, W. T. and J. E. Douglas. 1974. Streamflow greatly reduced by converting deciduous hardwood stands to pine. Science 185:857-859.

Tanner, C. B. 1968. Evaporation of water from plants in soils. In: Kozlowski, T. T. (ed.) Water Deficits and Plant Growth. Vol. I, Chapter 4. Academic Press, New York. 390 pp.

Tanner, C. B. 1967. Measurement of evapotranspiration. In: R. M. Hogen, H. R. Haise, and T. W. Edminister (eds.), Irrigation of Agriculture of Lands. Am. Soc. Agron. Monogr. 11:534-574.

Tovey, R. 1970. Alfalfa water table investigations. J. Irrig. and Drain. Div., Am. Soc. Civ. Engr. 95(IR4):525-535.

Tovey, R., J. S. Spencer, and D. C. Muckel. 1969. Turfgrass Evapotranspiration. Agron. J. 61:863-867.

Urie, D. H. 1959. Pattern of soil moisture depletion varies between Red pine and oak stands in Michigan. U.S. Dept. Agr. Forest Serv. Lake States Forest Expt. Sta. Tech. Note 564, 1 pp.

Valek, Z. D. 1959. Beitrag zur hydrologischen und hydrotechnischen Vervendbarkeit der Holzarten. Publ. Intern. Assoc. Sci. Hydrology 48, 322.

van Bavel, C. H. M. and D. G. Harris. 1962. Evapotranspiration rates from Bermudagrass and corn at Raleigh, North Carolina. Agron. J. 54:319-322.

van Bavel, C. H. M., L. J. Fritschen, and R. J. Reginato. 1963. Surface energy balance in arid lands agriculture. U.S. Dept. Agr. Production Res, Rept. No. 76, pp. 46.

van Bavel, C. H. M., L. J. Fritschen and W. E. Reeves. 1963. Transpiration by sudangrass as an extremely controlled process. Science 141:269-270.

van Bavel, C. H. M. 1966. Potential evaporation: The combination concept and its experimental verification. Water Resources Res. 12:455-467.

van Bavel, C. H. M. and L. E. Meyers. 1962. An automatic weighing lysimeter. Agric. Engineering 43:580-583, 586-588.

van Bavel, C. H. M., G. B. Stirk, and K. J. Brust. 1968. Hydraulic properties of a clay loam soil and the field measurement of water uptake by roots. Soil Sci. Soc. Am. Proc. 32:310-317.

Waring, R. H. and S. W. Running. 1976. Water uptake, storage, and transpiration by conifers: A physiological model. In water and plant life, problems and modern approaches. O. L. Lange, E. D. Schulze and L. Kappen (eds), Springer, New York.

Wilm, H. G. and E. G. Dunford. 1948. Effect of timber cutting on water available for stream-flow. U.S. Dept. Agr. Tech. Bull. 968, 43 pp.

APPENDIX 1

List of Symbols

Symbol	Unit	Definition
a		constant, coefficient, ratio
b		constant
c	%	plant cover
c_p	$J(kg\ K)^{-1}$	specific heat at constant pressure
C	$J(m^3\ K)^{-1}$	heat capacity
d	m	$D + z_o$
D	m	data displacement height
D	m/m^2	percolation below root zone
D_w	m^2/s	soil water diffusivity
E	m/m^2	soil evaporation
ET	m/m^2	evapotranspiration
e	Pa	vapor pressure
e_s	Pa	saturation vapor pressure
G	$W\ m^2$	soil heat flux density
H	%	percentage of potential evaporation that is unused in soil evaporation and is transferred back to the canopy
H	$W\ m^{-2}$	sensible heat flux of density
I	m/m^2	irrigation
k	$W\ m^{-1}s^{-1}$	thermal conductivity
k		von Karman's constant
K_h	m^2s^{-1}	eddy diffusivity for heat
K_v	m^2s^{-1}	eddy diffusivity for water vapor
K_m	m^2s^{-1}	eddy diffusivity for momentum
L	$J\ kg^{-1}$	latent heat of vaporization (2.50 X 10^6 J/kg at 0°C)

List of Symbols (Continued)

Symbol	Unit	Definition
M	W/m^2	canopy storage
P	m/m^2	precipitation
p	Pa	pressure
PE	W/m^2	potential evaporation
PET	W/m^2	potential evapotranspiration
P_n	W/m^2	photosynthesis and respiration energy
q	$kg\ kg^{-1}$	specific humidity
Q^*	$W\ m^{-2}$	net radiation flux density
R	m/m^2	surface runoff
Ri		Richardson number
ΔS	m/m^2	change in soil water storage
t	s or min	time
T	°C or K	temperature
U	$m\ s^{-1}$	wind speed
w	$m\ s^{-1}$	vertical wind speed
Z_w	m	depth to wetting front
z	m	depth or height
z_o	m	roughness length
β		Bowen ratio
γ	$Pa\ °C^{-1}$	psychrometric constant
Δ	$Pa\ °C^{-1}$	slope of saturation vapor pressure curve
ϵ		ratio of mole weight of water vapor to dry air (0.622)
θ	°K	temperature
θ_v		volumetric water content
ρ	$kg\ m^{-3}$	density
τ	kg/ms^2	momentum flux density
ϕ		stability factor

VERTICAL FLUX OF HEAT AND MOISTURE IN SNOW AND ICE

Michael Kuhn,
Institut für Meteorologie und Geophysik, Universität Innsbruck

1.   INTRODUCTION

In the course of the year, $77 \cdot 10^6$ km^2 or about 50 per cent of the Earth's land surface are covered by snow or ice, $75 \cdot 10^{-6}$ km^2 in the Northern Hemisphere winter, $18 \cdot 10^{-6}$ km^2 in summer (for details see Table 1 in Kotliakov, this volume). This cover constitutes a highly variable boundary condition for the global radiation balance and for the general atmospheric circulation.

It is the intent of this paper to describe those features of the vertical fluxes of heat and water vapour at the land surface that are particular to snow and ice, and to propose formulations of these processes that can be used in models of the global atmospheric circulation with a gridpoint spacing of several hundred kilometers.

Of all snow covered land surfaces the antarctic inland ice has the least horizontal variability and is thus easy to model on a large scale. It is a place where data from point measurements can be extrapolated over hundreds of kilometers without serious difficulties. The physiography of the antarctic ice sheet has been described by Mellor (1961), its climate by Rusin (1961), Schwerdtfeger (1970). Meteorological studies of the antarctic atmosphere were the subject of two volumes of the Antarctic Research Series (Rubin 1966, Businger 1977) and a wealth of information can be found in the Antarctic Bibliography (Thuronyi 1979). The antarctic boundary layer was first investigated by Liljequist (1957) and again in the IGY (e.g. Hoinkes 1964, Dalrymple et al., 1966, Lettau et al., 1967).

Greenland's ice cover (Bader, 1961) is less homogeneous and reaches into warmer, moister regions than Antarctica. Its climate is discussed, among others, by Putnins (1970) and a number of micrometeorological and glaciological investigations are published in Meddelelser om Grønland (see, for instance, Ambach 1963). Apart from the two major ice sheets, there are a number of minor ice caps in the Arctic (see Vowinckel and Orvig 1970 for a survey of the climate of the North Polar Basin). The largest extrapolar ice masses, the Patagonian ice fields cover an area of the order of 10^4 km^2. Smaller mountain glaciers affect the climate only on a local scale.

Although the polar ice sheets contain about 99 per cent of the Earth's fresh water ice by mass, the seasonal snow cover with its large areal extent and its high spatial and temporal variability is equally or more important to atmospheric circulation.

Although very detailed investigations and models of the heat and mass balance of the seasonal snow pack have been made in recent years (Schlatter 1972, de Quervain 1973, Kraus 1973, Anderson 1976, Colbeck 1979, Male 1980) their application to large areas is still problematic. Here again the polar and subpolar land surfaces like tundra and boreal forest are easier to handle in a model than mountain areas or the regions at the transient snow line. With these difficulties in mind one can dispense with the demands on accuracy customary in micrometeorology and emphasise further studies of extrapolation and representative averaging methods. It seems indeed that establishing useful gridpoint-averages for a broken snow cover is the central problem and it has to be admitted that the following formulation of heat and mass transfer in snow is helping only in the choice of proper parameters but does not give the solution of large-scale extrapolation.

2. A SUBDIVISION OF THE CRYOSPHERE

The cryosphere may be considered that part of the planet Earth where temperature is at or below $0°C$. It is customary to count snow, ice and permafrost to the cryosphere but when heat and vapour fluxes are concerned the freezing temperature may not be such a useful criterion as for instance the water content of the soil or other physical properties. Permafrost is defined irrespective of the presence of water (Ives 1974) and even though the soil contains water that freezes the ice in the soil matrix will depend on the physical state of the soil and the frozen ground will behave differently from a pure ice or snow surface. Frozen ground will therefore not be further considered in this paper.

The distinction between snow and ice is usually made on the basis of density, the transition occurs at 0.8 to 0.85 $Mg\ m^{-3}$ when the pore space becomes discontinuous. With this definition of snow, maximum snow depth is reached in the polar ice sheets at about 50m. Extreme values in the seasonal snow pack are considerably less. With annual precipitation at the windward coasts of New Zealand and Chile reaching values in excess of 5 $Mg\ m^{-2}$, maximum snow depth in the adjacent mountains can be expected in the range of 10 to 15m.

Benson distinguished four diagenetic facies on the Greenland ice sheet (1960, quoted in Bader 1961):

(1) The dry-snow facies - is characteristic of the whole area above the dry-snow line. Melting and percolating are negligible; the snow remains permanently dry, and the average density of the upper 5 metres is less than 0.375 g/cm^3. Snow of the dry facies covers 30% of the ice sheet area.

(2) The percolation facies - extends from the saturation line to the dry snow line. Summer melting is insufficient to soak all of the past year's accumulation. Percolation in pipe-like vertical channels can, however, penetrate to and past the previous summer layers and spread to form ice lenses and layers in older snow. The average density of the upper 5 metres varies from 0.43 to 0.39 g/cm^3. The percolation facies covers some 45% of the ice sheet area.

(3) The soaked facies - extends from the firn line to the saturation line. Summer melting produces sufficient water to soak at least the whole snow accumulation of the past year. Thus all the snow of the soaked facies has been wet at least once. Average density of the upper 5 meters is greater than 0.50 g/cm^3. The soaked facies belt covers about 10% of the ice sheet area.

(4) The ablation facies - extends from the edge of the ice to the firn line. Above the firn line, there is a permanent snow cover, i.e., the snow does not all melt away. Bare ice characterizes the area of the ablation facies, which covers no more than 15% of the ice sheet area.

The zone of superimposed ice is not included since it does not behave much different from bare glacier ice. Since dry snow has a higher albedo than soaked or crusted snow, the dry snow line and the firn line are two significant transitions on the ice sheets. From Figure 1 in Benson 1967 the following approximations can be made for the altitude of the dry snow line, DSA, as function of latitude, φ:

Greenland: $\quad DSA\ (m) = 500 + 133\ (90 - \varphi)$
Antarctica: $\quad DSA\ (m) = 133\ (87 - \varphi); \quad \varphi \leq 87°$
$\qquad\qquad\qquad\ = 0 \qquad\qquad\qquad\ ; \varphi > 87°$

3. BASIC PROCESSES

3.1 Special conditions at snow and ice surfaces

The transfer of energy and water vapour through the atmospheric boundary layer has been the subject of a sufficiently large number of papers to do without a repetition in the present article. Carson (1981, this volume) has summarized present techniques of incorporating boundary layer processes into global circulation models, Brutsaert (1981, this volume) has dealt with the peculiarities of bare soil surfaces, and Anderson (1976) has discussed the energy budget of a snow pack. This chapter emphasises the special conditions of snow and ice layers:

(1) $T_o \leq 0°C$. The surface temperature cannot exceed the melting temperature of ice.

(2) $E = E_p$. Evaporation and sublimation takes place at the potential rate.

(3) The albedo is generally high.

(4) The medium is permeable to air and water and transparent to visible radiation.

(5) The snow pack is a good thermal insulator.

(6) It has a high storage capacity for heat and water.

(7) The roughness of the surface is extremely low.

(8) Generally, the atmospheric surface layer over ice and snow is stably stratified.

3.2 The surface energy budget

This section closely follows an earlier paper (Kuhn 1979) where numerical examples for alpine conditions may be found.

A layer of snow or ice may gain or lose heat

- from the absorption or emission of radiation at the rate R $(W\ m^{-2})$;

- from condensation or evaporation (sublimation) L;

- from turbulent transfer of sensible heat H;

- by conduction through its boundary G;

- from freezing of liquid water (rain, meltwater or surface runoff) at a rate W.

The sign of these fluxes is such that heat gains are positive, irrespective of their direction.

Their sum ΣQ is used to change the snow temperature T_s or, once $T_s = 0$ and $\Sigma Q > 0$, to melt the snow:

$$R + L + H + G + W = \Sigma Q \qquad (3.1)$$

The individual fluxes may be expressed as follows:

$$R = SW\downarrow(1-a) + LW\downarrow + LW\uparrow \qquad (3.2)$$

The radiation budget is composed of the shortwave part (SW↓ is global irradiance, a is albedo) and a longwave part where

$$LW\uparrow = -\varepsilon_o \sigma T_o^4 \qquad (3.3)$$

With ε_o the surface emissivity. For the downward longwave component a form similar to (3.3) is proposed

$$LW\downarrow = \varepsilon_{eff} \sigma T_a^4 \qquad (3.4)$$

Where ε_{eff} is an effective emissivity close to unity for clouds and about 0.7 for clear sky under alpine conditions. T_a is then either cloud base temperature or air temperature in the surface layer. The values of ε_{eff} need to be calibrated with T_a at the desired level Z. When choosing this level one has to bear in mind that over a melting ice surface and at saturation vapour pressure, 50 per cent of LW↓ originate in the lowest 20m, 75 per cent in the lowest 200m as can be checked with an Elsasser diagram. This means that over melting ice equation (3.4) is a useful approximation even in the presence of clouds.

The turbulent flux of latent heat over snow has been discussed extensively by Kuz'min (1972) who proposes

$$L = \mathcal{L} E \; f(U_a) \; (e_a - e_o) \qquad (3.5)$$

where $\mathcal{L}$ is the latent heat of evaporation (sublimation), E is the evaporation in mm water equivalent, e_o and e_a are the water vapour pressures in mb at the surface and at a specified level and U_a is wind speed at the same level. For the dimensions used above, Anderson (1976) quotes values of the constants a and b in

$$f(U_a) = a + b \cdot U_a \qquad (3.6)$$

from four authors where b ranges from 19 to 42 · 10^{-4} (mm · mb^{-1} · km^{-1}) for comparable reference levels. The divergence of these constants gives an approximate picture of the accuracy that can be expected from present day evaporation formulae even under conditions of potential evaporation. For the turbulent transfer of sensible heat, there are two options. First, the use of a bulk transfer coefficient that can be derived from a logarithmic wind profile.

$$H = C_H \cdot U_a \cdot (T_a - T_o) \qquad (3.7)$$

Since in stable stratification the diffusivity of momentum, heat and vapor are identical.

$$C_H = C_M = \frac{k^2}{(\ln z/z_o)^2} \quad (Wm^{-2}m^{-1}s \; deg^{-1}) \qquad (3.8)$$

where z_o is the roughness parameter and k the Karman constant.

The second choice is the use of a transfer coefficient independent of wind speed

$$H = \alpha_H (T_a - T_o) \qquad (3.9)$$

with α_H in $Wm^{-2} deg^{-1}$. That α_H has a low standard deviation as was found for melting alpine glaciers (Kuhn 1979) reflects the small scatter of wind velocities in the kabatic flow regime over a glacier. In this sense the use of a constant transfer coefficient is better justified for longer periods, e.g., the entire ablation season.

One of the typical features of air flow in low-level inversions is the appearance of a velocity maximum near the ground. Values of U_a in the lowest dekameter are most likely to be influenced by this maximum and the logarithmic layer is reduced to the lowest meter. The use of equation (3.8) is thus very restricted over ice and snow surfaces, especially when they are inclined. For this reason Kuhn et al. (1977) used Lettau's geostrophic drag coefficient

$$C_o = (T_o/\rho_a)^{1/2} / V_{go} = U_*/V_{go} \qquad (3.10)$$

where T_o is the surface stress (Pa m^{-2}), ρ_a air density (kg m^{-3}), V_{go} the surface geostrophic wind and U_* the friction velocity.

V_{go} is furnished by a general circulation model. If the surface stress is determined independently equation (3.10) is more suitable for modelling than (3.8) since it does not need the assumption of a logarithmic wind profile. The use of "typical" values of U_* may be a helpful method but it must be stressed that U_* is not a material constant of the surface but of the turbulent boundary layer (it is the square root of the covariance of horizontal and vertical motion).

The conduction of heat through the snow pack, G, is important only for seasonal snow and short periods, in melting, isothermal snow or ice it is zero and its annual sum also vanishes in the absence of longer temperature trends. The transfer of heat within the snow is complicated by ventilation, radiation and vapour diffusion as will be discussed in the next section.

The amount of heat delivered by the solid or liquid precipitation of temperature T_p is

$$W_p = P \cdot c_p (T_p - T_s) \qquad (3.11)$$

with P precipitation rate (kg $m^{-2}s^{-1}$), c_p specific heat of precipitation (J kg^{-1} deg^{-1}). In the case of freezing

$$W_F = w \cdot \mathcal{L}_f \qquad (3.12)$$

where w is the rate of freezing (kg $m^{-2}s^{-1}$) and $\mathcal{L}_f$ is the heat of fusion (J kg^{-1}).

3.3 Energy and vapour transfer within the snow

On the scale of a global atmospheric circulation model horizontal conduction is negligible. Heat is transferred vertically by molecular conduction G_c, sublimation and diffusion of water vapour G_v and transfer of radiation G_R.

$$G_c = k_i \frac{\partial T_s}{\partial z} \tag{3.13}$$

where k_i is the molecular conductivity of ice and z is positive in the downward direction.

The bulk temperature gradients $\partial T_s/\partial z$ are not evenly distributed through ice grains and pore space (de Quervain 1973). This means that molecular diffusion takes place at locally higher gradients. In addition infrared radiation transfers energy across the pore spaces at a rate determined by their temperature gradients. Considering these additional transfers the use of an effective thermal conductivity in conjunction with the bulk temperature gradient is appropriate.

$$G_e = k_e \, \partial T/\partial z \tag{3.14}$$

where k_e is higher by a factor of up to 2 (Weller and Schwerdtfeger 1970, 1977) depending on density and geometry of pore space. The difference $k_e - k_i$ becomes negligible below 10m depth in polar snow.

From the divergence of G_e the rate of temperature change can be determined

$$\partial T_s/\partial t = (\rho_s \cdot c_i)^{-1} \, \partial/\partial z \, (k_e \partial T/\partial z). \tag{3.15}$$

The mass flux by sublimation and diffusion in the temperature gradient metamorphism carries heat from higher to lower temperature in the gradient of saturation vapour pressures $e_i(T)$

$$G_v = m \cdot \mathcal{L}_s \tag{3.16}$$

where m is the rate of diffusion of mass through unit area (kg m^{-2}s^{-2}) and $\mathcal{L}_s$ is the latent heat of sublimation. Following de Quervain (1973) and Male (1980)

$$m = -(fD_e/R\overline{T}) \, \partial e_i/\partial z \tag{3.17}$$

where f is a shape factor which is approximately unity at low snow density, D_e is the effective diffusion coefficient of water vapour in the pore space and R is the gas-constant. Expanding

$$\partial e_i/\partial z = (\partial e_i/\partial T)(\partial T/\partial z) \tag{3.18}$$

and using an approximation

$$e_i(T) = a \exp(b(T - 273)) \tag{3.19}$$

equation (3.17) now becomes

$$m = -(fDe/R\overline{T}) \cdot a \cdot b \exp(b(T - 273)) \, \partial T/\partial z \tag{3.20}$$

which now relates G_v (equation 3.16) to the temperature gradient. The rate of temperature change can be determined from the vertical derivatives of $G_e + G_v$

$$\rho_s c_i \, \partial T_s/\partial T = \partial/\partial z \, (G_e + G_v). \tag{3.21}$$

3.4 Density changes due to metamorphism and to compaction

The vertical divergence of the mass flux m causes a change in bulk density.

$$\partial \rho_s/\partial t = -\partial m/\partial z \qquad (3.22)$$

and with equation (3.20) this becomes

$$\partial \rho_s/\partial t = (a \cdot b \cdot f \cdot D_e/R\bar{T})(\partial^2 T/\partial z^2 + b(\partial T/\partial z)^2) \exp(b(T-273)) \qquad (3.23)$$

which relates densification to temperature gradients.

The effect of overburden can be described with the concept of a compactive viscosity (Bader 1960 quoted in Male 1980, Mellor 1975, 1977).

$$\eta_c = \sigma_v/\varepsilon_v \quad (\text{Pa s}) \qquad (3.24)$$

with σ_v the overburden pressure (Pa) and ε_v the vertical strain rate (s^{-1})

$$\sigma_v = \int_0^z \rho_s g\, dz \qquad (3.25)$$
$$= A\, g$$

where A is the accumulation (kg m^{-2}) and g is the acceleration of gravity.

The vertical strain rate is equal to the rate of relative densification

$$\varepsilon_v = -\partial \ln \rho/\partial t \qquad (3.26)$$

knowing η_c and net accumulation the densification can be computed for arbitrary depths. Mellor (1977, p.28) spells out a general warning that "more research is needed ... meanwhile, it is as well to treat with caution the compactive viscosity values that have been derived from application of continuous creep theory to low-density seasonal snow covers".

Nonetheless this concept should be kept in snow cover modelling where density is one of the most useful parameters.

4. RELEVANT PHYSICAL PROPERTIES OF SNOW AND ICE

4.1 Albedo

The albedo of land surfaces will be discussed in detail in Kondratyev's survey (see Appendix). Table 1 summarizes albedo values applicable to yearly totals of radiation at several Antarctic stations (Kuhn et al., 1977).

In general albedo is inversely proportional to the square root of grain size (Bohren and Barkstrom 1974). As grain size increases rapidly with the onset of melting (de Quervain 1973) the albedo decreases as soon as the snow surface is wet, the most rapid change occuring in the near infrared. The albedo of glacier ice is generally in the range of 0.25 to 0.50 and can reach values as low as 0.15 when contaminated with till.

Table 1: Albedo at selected Antarctic stations. Values are applicable to yearly totals of global radiation. (Kuhn et al., 1977).

Station	Latitude	Elevation m	Albedo
Coastal Stations			
Oazis	66°18'		0.251
Mirnyy	68°33'		0.846
Port Martin	66°49'		0.580
Mawson	67°34'		0.700
Norway	70°30'		0.849
Novolazarevskaya	70°46'		0.260
Maudheim	71°03'		0.844
Little America V	78°11'		0.862
Inland Stations			
Charcot	69°22'	2400	0.833
Pionerskaya	69°44'	2740	0.807
Vostok	78°27'	3488	0.846
Plateau	79°15'	3625	0.840
South Pole	90°00'	2800	0.840

4.2 Long wave emissivity

The value of ε_o can be assumed to be unity for all practical purposes. Values decreasing to 0.87 at temperatures of -6°C were found by Dunkle and Gier (quoted in Mellor 1977) but have not been reported since.

The effective emissivity of the clear sky was discussed above.

4.3 The roughness length

This is a constant of integration of the logarithmic wind profile which does not have a definite relation to physical roughness. Holmgren (1971) has presented group means of z_o which range from 10^{-4} to 10^{-3} at polar stations. Minimum values are associated with wind speeds around 10 m s^{-1}, while at lower wind speeds the surface is not streamlined; erosion may be the cause for the increase of z_o at higher wind speeds.

Ambach (1963) shows an impressive example of roughness change: when the snow cover in the ablation zone of Greenland melts to expose bare ice z_o increases from about $5 \cdot 10^{-4}$ to $6 \cdot 10^{-3}$.

Kuhn (1979) summarizes roughness lengths measured at melting alpine glaciers where values range from 1 to 7 · 10^{-3} m.

4.4 The friction velocity

In a study of the extremely stable surface layer of the East Antarctic Plateau Kuhn et al. (1977) found values of U_* increasing from class means of 0.07 m s^{-1} at high stability to 0.18 m s^{-1} at moderate stability.

The values found by Ambach (1963) in the ablation zone of the Greenland ice sheet range from 0.1 to 0.8 m s^{-1} with a modus in the group of 0.3 to 0.4 m s^{-1}.

Alpine glaciers (Kuhn 1979) experience friction velocities similar to the Grönland data, i.e. in the range from 0.1 to 0.5 m s^{-1}.

4.5 Bulk transfer coefficients for heat

As mentioned above, there is general agreement that with stable stratification transfer coefficients of heat and momentum are identical. With the help of equation (3.8) transfer coefficients can be computed for any roughness length z_o.

The bulk coefficient α_H in equation (3.9) was determined by various methods in the Alps (Kuhn 1979) and in the Arctic (Braithwaite 1977). Both studies agree broadly, the alpine value being 1.7 MJ m^{-2}d^{-1}K^{-1} ± 14%.

The geostrophic drag coefficient according to equation (3.10) was determined for central antarctic conditions (Kuhn et al. 1977) ranging from 0.007 in extreme to 0.015 in moderate stability. The scatter of C_o is less than that of U_* so that the former seems to be a more suitable parameter for describing the stable surface layer.

4.6 Specific and latent heats

The latent heat of fusion (Dorsey 1940 quoted in Male 1980)

$$\mathcal{L}_f = 333.66 \text{ kJ/kg}$$

and for sublimation (Keeman et al., 1969 quoted in Male 1980)

$$\mathcal{L}_s = 2834.8 \text{ kJ/kg at } 0°C$$
$$2838.9 \text{ at } -40°C$$

(See also Mellor 1977 for the temperature dependence). The specific heat of pure ice was given by Dickinson and Osborne (1915, quoted in Male 1980) as a function of temperature

$$C_i = 2.117 + 0.0078 \, t \quad (kJ \, kg^{-1} \, deg^{-1}) \qquad (4.1)$$

where t is temperature in degrees Celsius.

4.7 Thermal conductivity

Values of k_i are summarized in Weller and Schwerdtfeger (1977) and reproduced in Table 2.

Table 2: Thermal Conductivity and Diffusivity of Snow and Ice as a Function of Temperature and Density (Weller and Schwerdtfeger 1977)

Temperature, °C	Density, g cm^{-3}						
	0.3	0.4	0.5	0.6	0.7	0.8	0.917
Thermal Conductivity							
0	0.59	0.99	0.58	2.55	3.64	4.56	5.35
-60	0.82	1.38	2.20	3.54	5.06	6.35	7.44
Thermal Diffusivity							
0	0.0040	0.0050	0.0064	0.0086	0.0105	0.0115	0.0118
-60	0.0069	0.0086	0.0110	0.0148	0.0180	0.0198	0.0203

Thermal conductivity is in units of (cal cm^{-1} deg^{-1} s^{-1}) x 10^3; Thermal diffusivity is in square centimeters per second.

The effective conductivity of snow has been determined as function of density by a number of authors. Summaries are given by Schwerdtfeger (1963), Yen (1965), Mellor (1977) and Anderson (1976). It seems that Abels' formula (established in 1894) still gives reasonable figures.

$$K_e = 0.007 \, \rho^2 \quad (\text{cal cm}^{-1} \text{s}^{-1} \text{deg}^{-1}) \qquad (4.2)$$
$$= 2.94 \, \rho^2 \quad (\text{J m}^{-1} \text{s}^{-1} \text{deg}^{-1}).$$

4.8 Saturation pressure and diffusivity of water vapour

The coefficients of equation (3.19) are

a = 6.11 (mb)
b = 0.0857 (°C^{-1})

Anderson (1976) proposes

$$e_{si} = 3.5558 \cdot 10^{10} \exp\left(-\frac{6141.9}{T}\right) \qquad (4.3)$$

The coefficient of molecular diffusion of water vapour in air is given by de Quervain (1973) as $D_o = 0.22$ cm^{2}s^{-1}. The effective value D_e is given as 0.65 cm^{2}s^{-1} by Yen (1965). Mellor (1977) mentions similar values determined by other authors.

4.9 Strain rate and compactive viscosity

Mellor (1977) gives values for the strain rate of $3 \cdot 10^{-6}$ s^{-1} for a density of 0.1 Mg m^{-3}. This means a compaction of 26 per cent per day. Maximum initial values are as high as 10^{-4} s^{-1}, later ε_v slows down to 10^{-8} s^{-1}.

A compilation of compactive viscosity values by Mellor (1975) can be expressed in two approximate formulae

Polar snow: $\quad \log_{10} \eta_c = 9 + 13 \rho_s \quad$ (4.4)

Seasonal snow: $\quad \log_{10} \eta_c = 7 + 10 \rho_s \quad$ (4.5)

when η_c is in (Pa s) and ρ_s in (Mg m^{-3})

5. RECOMMENDATIONS FOR THE CHOICE OF PARAMETERS

Of the basic processes described in Chapter 3 the formulation of the energy budget is of direct interest to the modeller. Supposing that global radiation, air temperature, vapour pressure and wind speed are furnished for the planetary boundary layer by the model, snow cover extent or mean albedo for a grid point must be specified independently. Considering the large uncertainties in these data the actual transfer mechanisms of energy and water vapour can be formulated in a relative crude manner.

With bulk transfer coefficients for heat and vapour specified, the surface temperature becomes a variable of prime importance. From the scatter of transfer coefficients found in the literature, heat and vapour fluxes modelled over snow and ice should not be considered too accurate at subfreezing temperatures. However when melting sets in, these values come surprisingly close to the real values of H and L since T_o and e_i are known almost exactly.

For short period modelling, the storage of energy and water in snow and ice may be considerable. The heat conduction in the snow pack is determined with an effective thermal conductivity which is parameterized by bulk snow density. The snow density finally can be assessed from the overburden of accumulation with the help of a compactive viscosity concept, and from densification in temperature gradient metamorphism.

Vapour transfer within the snow is thus important only to the thermal regime, and potential evaporation persists at the surface regardless of vapour fluxes within.

Recommendations for further research are:

(1) Continued study of transfer coefficients and albedo in point experiments.

(2) Development of representative averaging methods to overcome the problem of non-linear effects of partial snow cover.

No doubt, this advice is easy to give but hard to carry out.

REFERENCES

Ambach, W., 1963: Untersuchungen zum Energieumsatz in der Ablationszone des Grönländischen Inlandeises Meddelelser om Grønland, 4, 174, 311 pp.

Anderson, E. A., 1976: A point energy and mass balance model of a snow cover. NOAA Technical Report NWS 19. Silver Spring, Md. 150 pp.

Bader, H., 1961: The Greenland Ice Sheet. Cold Regions Science and Engineering, I-B2, 18 pp. CRREL, Hannover NH.

Benson, C. S., 1967: Polar Regions Snow Cover. Physics of Snow and Ice, Vol. 1, part 2, pp. 1039-1063. University of Hokaido, Sapporo.

Bohren, C. and R. B. Barkstrom, 1974: Theory of the optical properties of snow. J. Geoph. Res., 79(30): 4527-4535.

Braithwaite, R. J., 1977: Air temperature and glacier ablation - a parametric approach. Thesis, McGill University Montreal. 146 pp.

Brutsaert, W., 1981: Vertical Flux of Moisture and Heat at a Bare Soil Surface. This volume.

Businger, J., editor, 1977: Meteorological Studies at Plateau Station, Antarctica. Antarctic Research Series, Vol. 25. American Geophysical Union.

Carson, D. J., 1981: Current Parameterizations of Land-Surface Processes in Atmospheric General Circulation Models. This volume.

Colbeck, S. and M. Ray, 1979: Modeling of Snow Cover Runoff, 432 pp. Cold Regions Research and Engineering Laboratory, Hanover.

Dalrymple, P., H. Lettau and S. Wollaston, 1966: South Pole Micrometeorology Program: Data Analysis. Studies in Antarctic Meteorology, ed. M. Rubin. Am. Geophys. Union, Antarctic Res. Ser. 9:13-58.

De Quervain, M., 1973: Snow structure, heat, and mass flux through snow. Proceedings of the Symposium on the Role of Snow and Ice in Hydrology (Banff, 1972): pp. 203-226.

Hoinkes, H., 1964: Glacial Meteorology. Research in Geophysics, ed. H. Odishaw, Vol. 2, pp. 391-424. M.I.T. Press, Cambridge, Mass.

Holmgren, B., 1971: Climate and Energy Exchange on a Sub-Polar Ice Cap in Summer. Arctic Institute of North America Devon Island Expedition 1961-1963. Part A-F Meteorologiska Institutionen, Uppsala Universitet. Meddelande Nr. 107-112, Uppsala 1971. Part A: Physical Climatology, 83 pp. B: Wind- and Temperature Field in the low layer on the Top Plateau of the Ice Cap, 43 pp. C: On the Katabatic Winds over the North-West Slope of the Ice Cap. Variation of Surface Roughness. 43 pp.

Ives, J. D., 1974: Permafrost. In J. D. Ives and R. G. Barry (eds.) Arctic and Alpine Environments, 159-194. Methuen, London.

Kondratyev, K., 1981: Albedos and emissivities. This volume.

Kotliakov, V. M. and A. N. Krenke, 1981: The Data on Snow Cover and Glaciers for the Global Climatic Models. This volume.

Kraus, H., 1973: Energy exchange at air-ice interface. Proceedings of the Symposium on the Role of Snow and Ice in Hydrology (Banff, 1972): pp. 128-151.

Kuhn, M., 1979: On the computation of heat transfer coefficients from energy-balance gradients on a glacier. Journal of Glaciology, 22(87): 263-272.

_____, L. S. Kundla and L. A. Stroschein, 1977: The radiation budget at Plateau Station, Antarctica 1966-1967. Antarctic Research Series, 25: 41-73, edited by J. A. Businger. AGU, Washington D.C.

_____, H. H. Lettau and A. J. Riordan, 1977: Stability related wind spiraling in the lowest 32m. Antarctic Research Series, 25:93-111, edited by J. A. Businger. AGU, Washington D.C.

Kuz'min, P. P., 1972: Melting of Snow Cover (Original 1961) 290 pp. Israel Program for Scientific Translations, Jerusalem.

Lettau, H. H., S. H. Wollaston and P. C. Dalrymple, 1967: Little America V Micrometeorology Program. Data and Analysis. US Army Materiel Command, Techn. Rep. 67-46-Es. (Contribution No. 97, Inst. of Polar Studies, Ohio State University, Columbus, Ohio).

Liljequist, G. H., 1957: Energy Exchange of an Antarctic Snow-Field Norwegian-British-Swedish Antarctic Expedition, 1949-52, Scientific Results Vol. II, Part 1D. Narsk Polarinstitut, Oslo.

Male, D. H., 1980: The seasonal snowcover. In Dynamics of Snow and Ice Masses, edited by S. C. Colbeck: pp. 305-395.

Mellor, M., 1961: The Antarctic Ice Sheet. Cold Regions Science and Engineering, I-B1, 50 pp. CRREL, Hannover NH.

_____, 1975: A review of basic snow mechanics. Proceedings of the Symposium on Snow Mechanics (Grindelwald, 1974): pp. 251-291.

_____, 1977: Engineering properties of snow. Journal of Glaciology, 19(81): 15-66.

Paterson, W. S. B., 1969: The physics of glaciers. Pergamon Press, Oxford, 250 pp.

Putnins, P., 1970: The Climate of Greenland. World Survey of Climatology, 14:3-128. Elsevier Publishing Company, Amsterdam.

Rubin, M., editor, 1966: Studies in Antarctic Meteorology. Antarctic Research Series, Vol. 9, 231 pp. American Geophysical Union.

Rusin, N. P., 1961: Meteorological and radiational regime of Antarctica. Gidromet. Izdat., Leningrad. English translation, Israel Program for Translation, Jerusalem, 1964, 355 pp.

Schlatter, T. W., 1972: The local surface energy balance and subsurface temperature regime in Antarctica. Journal of Applied Meteorology, 11(7): 1048-1062.

Schwerdtfeger, P., 1963: Theoretical deviation of the thermal conductivity and diffusivity of snow. International Association of Scientific Hydrology, Publ. No. 61: 75-81.

Schwerdtfeger, W., 1970: The Climate of Antarctica. World Survey of Climatology, Vol. 14, 253-355. Elsevier Publishing Company, Amsterdam.

Thuronyi, G., editor, 1979: Antarctic Bibliography, Vol. 10, 498 pp. Library of Congress, Washington.

Vowinckel, E. and S. Orvig, 1970: The Climate of the North Polar Basin. World Survey of Climatology, 14:129-252. Elsevier Publishing Company, Amsterdam.

Weller, G. and P. Schwerdtfeger, 1970: Thermal properties and heat transfer processes of the snow of the Central Antarctic Exploration. IASH Publ. No. 86.

_____ and _____, 1977: Thermal Properties and Heat Transfer Process of Low-Temperature Snow. Antarctic Research Series, 25(3): 27-34. American Geophysical Union, Washington D.C.

Yen, J.-C., 1965: Effective Thermal Conductivity and Water Vapour Diffusivity of Naturally Compacted Snow. Journal of Geophysical Research, 70(8):1821-1825.

SESSION III

MESOSCALE PARAMETERIZATIONS OF THE TRANSFER PROCESSES

PARAMETRIZATION OF HYDROLOGIC PROCESSES

J.C.I. Dooge

Department of Civil Engineering,
University College Dublin

1. THE HYDROLOGIC VIEWPOINT

1.1. Approaches to flow problems

The hydrologist is concerned with the occurrence and movement of water on under and over the surface of the earth and with the continuous circulation of the constant volume of water in the hydrosphere. Figure 1 shows in highly lumped form a block diagram of this hydrologic cycle. The boxes in the diagram represent forms of moisture storage and the directed arrows the transformation of moisture from one form of storage to another. These directed arrows in the diagram represent the hydrologic processes whose nature is discussed in texts on physical hydrology (e.g. Eagleson 1970).

Hydrologists are not the only scientists interested in the description of water movement. As in the case of any other phenomenon, the conceptual or mathematical models used to describe the various aspects of a particular general phenomenon depends upon the objective of the study and the disciplinary background of the investigator. The variations in approach are often strongly influenced by the scale of the particular phenomenon which is under examination and this is true for various types of description for water movement (Dooge, 1979). The theoretical physicist uses the mathematic models of quantum mechanics to explore the movement in water particles on a subatomic scale. The physical chemist studies the behaviour of water on a molecular scale and may use either the mathematical models of statistical mechanics or else simple deterministic models (e.g. conceptual models for the diffusion of oxygen across an air-water interface).

The expert in fluid mechanics usually ignores the fine structure of matter studied by the physicist and chemist and, defining his fluid as a continuum, goes on to derive such basic equations as the Navier-Stokes equations. In many practical situations, turbulence is present and the Navier-Stokes equations become not untrue but inapplicable on the scale of the description of the characteristics of the mean motion. Accordingly, the viscous effects are parametrized by means of a mixing length or of

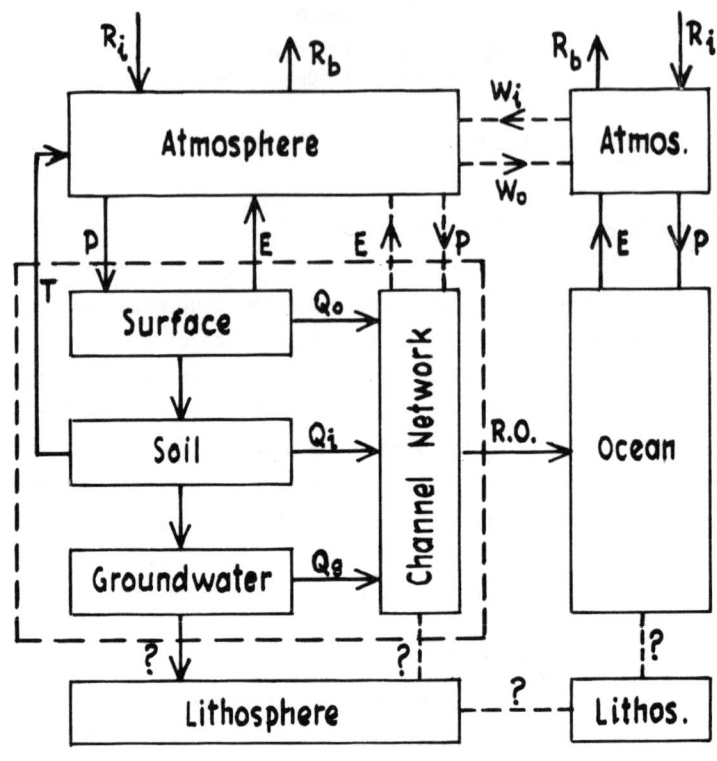

BLOCK DIAGRAM OF HYDROLOGIC CYCLE

Figure 1

R_i - incoming radiation
R_b - back radiation
W - transport of precipitable water
P - precipitation
E - evaporation
T - transpiration

F - infiltration
R - groundwater recharge
Q_o - overland flow
Q_i - interflow
Q_g - groundwater runoff
R.O. - total runoff

some other turbulence model in order to study the phenomenon. The resulting equations frequently cannot be solved without resort to numerical methods so that the scale of solution is that of the finite difference network used. The hydrologist is concerned with still larger scales of description and the climate modeller with even larger scales again. At any given level of description, the finer scale process may either be ignored or may be represented by their statistical effect on the larger scale description. In the latter case a deterministic description at the macro scale involves, either implicitly or explicitly, a statistical description of those processes at the macro-scale which affect in some way the behaviour at the scale of interest. The process of parametrization at the macro-scale of the micro-scale processes may be based either on the nature of these fine scale processes or else determined empirically at the macro-scale.

As the scale of the phenomenon under study increases, the interconnection between the components of the system become more and more complex. Under these conditions the solution of the mathematical model describing the system becomes more difficult to obtain and the computer requirements give rise to various logistic and financial problems. In addition, serious difficulties arise in relation to the observational input to the model. When these difficulties are encountered in hydrology (and in many other disciplines), resort is had to the study of operation of the overall system without reference to the detailed structure of the components or the details of the physical processes involved. Such an external description of the system depends on measurement observations and is essentially an *a posteriori* rather than an *a priori* approach as used in the finer scale descriptions. Such external descriptions may use distributed models in which variations in both space and time are taken into account or lumped models in which only variations in time are examined.

1.2. Internal descriptions of hydrologic processes

Most internal descriptions of hydrologic processes are based on the equations of fluid mechanics. Thus overland flow over the surface of the ground can be described by equations based on the conservation of mass and on the conservation of linear momentum. The continuity equation for the one-dimensional flow over a plane surface under the influence of rainfall is given by:

$$\frac{\partial q}{\partial x} + \frac{\partial y}{\partial t} = i(x,t) - f(x,t) \qquad (1)$$

where $q(x,t)$ is the rate of overland flow per unit width, $y(x,t)$ is the depth, $i(x,t)$ is the precipitation and $f(x,t)$ is the infiltration through the surface. The conservation of linear momentum in the direction of flow for such overland flow is usually written as:

$$\frac{\partial y}{\partial x} + \frac{u}{g}\frac{\partial u}{\partial x} + \frac{1}{g}\frac{\partial u}{\partial t} = S_o - S_f - \frac{u(i-f)}{gy} \qquad (2a)$$

where S_o represents the slope of the plane in the direction of flow (a downward slope being taken conventionally as positive), S_f represents the resistance term expressed as a friction slope, and the third term of the right hand side represents the momentum drag induced by the necessity to give the net lateral inflow the same momentum as the main flow. Even at this simple level of description, difficulty arises because of lack of adequate empirical data for the parametrization in the term S_f representing the frictional resistance of flow over a vegetated slope. The dynamic equation (2a) can be expressed in terms of the same variables as the linear continuity equation (1a) by expressing velocity (u) in terms of discharge per unit width (q) and depth of flow (y) and obtaining:

$$(gy^3 - q^2)\frac{\partial y}{\partial x} + 2qy\frac{\partial q}{\partial x} + y^2\frac{\partial q}{\partial t} = gy^3(S_o - S_f) \qquad (2b)$$

This dynamic equation is clearly highly non-linear.

For the case of flow in a channel network without lateral inflow the equation of continuity is given by:

$$\frac{\partial Q}{\partial x} + \frac{\partial A}{\partial t} = 0 \qquad (3)$$

where $Q(x,t)$ is the discharge and $A(x,t)$ is the area of flow.

The equation of linear momentum is usually written as:

$$\frac{\partial y}{\partial x} + \frac{u}{g}\frac{\partial u}{\partial x} + \frac{1}{g}\frac{\partial u}{\partial t} = (S_o - S_f) \qquad (4a)$$

where the variables have the same meaning as in equation (2a). As before this dynamic equation can be transformed to:

$$g\bar{y}(1-F^2)\frac{\partial A}{\partial x} + \frac{2Q}{A}\cdot\frac{\partial Q}{\partial x} + \frac{\partial Q}{\partial t} = gA(S_o - S_f) \qquad (4b)$$

in which $\bar{y}$ is the hydraulic mean depth defined by:

$$\bar{y}(x,t) = \frac{A}{T} \qquad (5a)$$

$A(x,t)$ is the area of flow and $T(x,t)$ is the surface width and is given by:

$$T(x,t) = \frac{dA}{dy} \qquad (5b)$$

which is a function of the shape of the section and the depth of flow $y(x,t)$. F in equation (4b) is the Froude number defined by:

$$F(x,t) = \frac{Q^2}{g}\cdot\frac{T}{A^3} \qquad (5c)$$

In equations (2b) and (4b) the friction slope S_f must be expressed in terms of the dependent variables by some friction formula usually Chezy or Manning or Logarithmic (Henderson 1966). Such formulae are essentially parametrizations on the scale of the mean flow described by equations (2b) and (4b) of the conversion of the energy of the mean flow to energy of turbulence and ultimately to heat through viscous dissipation, the details of which are governed by equations appropriate to a fine scale of description.

In the unsaturated soil above the water table, the equation of continuity for one-dimensional vertical flow can be written as

$$\frac{\partial w}{\partial z} + \frac{\partial c}{\partial t} = 0 \qquad (6)$$

where z is elevation above a fixed datum, w(z,t) is the face velocity measured vertically upward and c(z,t) is the moisture content on a volume basis. Since the movement is very gradual, the acceleration terms in the Navier-Stokes equation for laminar flow are neglected and the remaining terms integrated firstly across each individual pore and then across the whole horizontal section of the soil column. This results (Bear 1972) in a parametrized equation in the form of the Darcy equation for unsaturated flow:

$$w(z,t) = -K(c)\frac{\partial \phi}{\partial z} \qquad (7)$$

where K(c) is the unsaturated hydraulic conductivity of the soil in the vertical direction and ϕ (z,t) is the moisture potential. In hydrological analysis the only elements of the total potential considered are the pressure head (h) and the gravitational head (z). Combination of equations (6) and (7) gives a single equation for the unsteady vertical movement of water in an unsaturated soil as:

$$\frac{\partial c}{\partial t} = \frac{\partial}{\partial z}\left[K(c)\frac{\partial h}{\partial z}\right] + \frac{\partial K}{\partial z} \qquad (8)$$

which is commonly known as the Richard's equation. Equation (8) involves two dependent variables and hence must be supplemented by a second equation relating the pressure head (h) and the moisture content (c). Since the soil moisture plays a key role in the interface between atmospheric and hydrologic models, the solution of these equations will be referred to later. Since both equations are non-linear a closed form solution is not possible even for simple boundary conditions.

In the case of flow in the saturated zone below the water table, the continuity equation for one dimensional horizontal flow through a homogeneous soil is given by:

$$\frac{\partial q}{\partial x} + f\frac{\partial h}{\partial t} = r(x,t) \qquad (9)$$

where q(x,t) is the horizontal discharge per unit width, h(x,t) is the elevation of the water table, r(x,t) is rate of recharge at the surface of the water table and f is a parameter known as the storativity which is equal to the amount of water released for a soil column when the watertable changes by a unit amount in a vertical direction. As for the case of flow in an unsaturated soil, the equation of motion is parameterized to the Darcy equation:

$$q(x,t) = -K \cdot \frac{\partial h}{\partial x} \qquad (10)$$

where K is the saturated hydraulic conductivity in the horizontal direction. Combination of equations (9) and (10) gives

$$\frac{\partial h}{\partial t} = K\frac{\partial}{\partial x}\left[h \cdot \frac{\partial h}{\partial x}\right] + r(x,t) \qquad (11)$$

which is known as the Boussinesq equation.

If the equations of continuum mechanics described in the last few paragraphs are to be applied to an actual catchment area with spatially variable properties rather than to idealised textbook examples, then it is necessary to divide the catchment surface into modules over which the parameters can be assumed constant and to which the equations can be applied and solved and to define the topology of the interconnections of the various modules. For the highly simplified case of urban hydrology where infiltration is ignored, such a modular arrangement will involve only a single layer to represent the two dimensional impervious surface. For the case of a natural catchment, the modular representation is further complicated by the fact that it must be represented on three levels i.e. the surface, the unsaturated zone and the saturated zone. The solution of a large set of non-linear equations of this type would be a formidable task for a large computer even if the input data were readily available.

1.3. External descriptions of hydrologic systems

In the macro-scale or overall systems approach the customary mathematical formulation based on the laws of physics and thermodynamics is abandoned. Instead, the system is observed and an attempt made to infer the nature of the operation of the system from the input and output data. Such an external description can be formulated in a number of ways. Thus the relationship between input and output can be expressed in the very general form:

$$\underline{y}(t) = H\left[\underline{u}(t)\right] \qquad (12)$$

where $\underline{y}(t)$ is the output from the system, $\underline{u}(t)$ is the input to the system and H is an unknown operator which transforms the input to the output.

Alternatively one can take the viewpoint that the system will be represented by some set of differential equations as described in the last section but whose particular form is unknown due to the complexity of the physics involved and the complexity of the structure of the system. Thus one could consider that there will be a set of differential equations linking the input ($\underline{u}$) and its derivatives and the output ($\underline{y}$) and its derivatives this can be expressed as:

$$\underline{F}\left[\underline{x}, t, \underline{y}, D\underline{y}, D^2\underline{y} \ldots D^n\underline{y}\right] = \underline{G}\left[\underline{x}, t, \underline{u}, D\underline{u} \ldots D^m\underline{u}\right] \qquad (13)$$

where $\underline{F}$ and $\underline{G}$ are vectors of unknown functions, D is the differential operator, $\underline{x}$ is the vector of space co-ordinates relative to some fixed reference frame, and t is the elapsed time.

A third type of external description on a macro-scale depends on the concept of the state of the system ($\underline{X}$) which may be defined as a vector the elements of which are sufficient at any time to describe the condition of the system completely so that with a knowledge of the state at a given instant and the subsequent input it is possible to predict the changes in the system. The general state space transition equation can be written as:

$$\frac{d\underline{X}}{dt} = f\left[\underline{X}, \underline{u}\right] \qquad (14)$$

where $\underline{X}$ is the state of the system and $\underline{u}$ is the input. In order to predict the output it is necessary to have a further equation linking the output ($\underline{y}$) at any instant with the state of the system ($\underline{X}$) and the input ($\underline{u}$) at the same instant. This output equation can be expressed as:

$$\underline{y} = g\left[\underline{X}, \underline{u}\right] \qquad (15)$$

In the case of internal descriptions, the mathematical formulation can always be expressed in state-space form so that the function f in Equation (14) and the function g in Equation (15) will be given a priori. In external descriptions these functions are determined from the input and output data.

For systems encountered in hydrology, the various external descriptions given by equations (12), (13) and (14) are all perfectly general and hence equivalent to one another. They cover stochastic inputs and outputs as well as deterministic inputs and outputs. The transform formulation of equation (12) is given in terms of autocorrelation and cross-correlation functions for stochastic inputs. As in the case of the internal descriptions based on methematical physics, the first attempts at solving a systems problem on the macro-scale are based on a simplified analysis. In the development of the classical approach, preliminary analyses are made using linear ordinary differential equations with constant coefficients before proceeding to the solution of large sets of non-linear partial differential equations. Similarly, in the case of external description, attention is first focused on the analysis of linear, time-variant systems which are dynamic but in which the systems are lumped i.e. the inputs and outputs are not functions of the spatial co-ordinates. These two sets of simplifying assumptions do not merely represent similarities in approach but are identically equivalent.

For lumped, linear, time-invariant systems, the transformation formulation of equation (12) becomes:

$$y(t) = h(t) * u(t) = \int_{-\infty}^{\infty} h(\tau) \cdot u(t-\tau) d\tau \qquad (16)$$

where $u(t)$ and $y(t)$ represents the lumped input and the lumped output respectively and $h(t)$ represents the impulse response of the system i.e. the output that occurs when the input takes the form of the Dirac delta function. The impulse response $h(t)$ contains a complete mathematical description of the effect of the complex structure of the particular system under study and of the linearised version of the complex physics involved. If this impulse response can be determined on the basis of past records of input and output, then the output due to any other input can be obtained by convoluting the impulse response with the new output as indicated by equation (16). In the studies of many types of the system, the analysis is carried out in the Laplace transform domain in which the relationship for lumped, linear, time-invariant systems takes the form:

$$Y(s) = H(s) \cdot U(s) \qquad (17)$$

where $U(s)$, $H(s)$, and $Y(s)$ are the Laplace transforms of the input, the impulse response and the output respectively. $H(s)$ is often referred to as the system function.

As already mentioned the differential equation formulation for the case of the linear, lumped, time-invariant system takes the form of a linear ordinary differential equation with constant co-efficients which may be written as:

$$P_n(D) \left[y(t) \right] = Q_m(D) \left[u(t) \right] \qquad (18)$$

where $P_n(D)$ is an n-degree polynomial function of the differential operator acting on the lumped output (y) and $Q_m(D)$ is an m-degree polynomial operator acting on the lumped input (u).

The general state space formulation of equation (14) reduces for the case of a lumped, linear time-invariant system to:

$$\frac{d\underline{X}}{dt} = \underline{A}\,\underline{X} + \underline{b}\underline{u}$$
$$\underline{y} = \underline{c}^T \underline{X} + d\underline{u} \tag{19}$$

where the elements of the matrix $\underline{A}$ and of the vectors $\underline{b}$ and $\underline{c}$ are constant and the scalar d is only non-zero where part of the input bypasses the system in order to appear directly in the output.

It can be readily shown that equations (18) and (19) are identically equivalent to one another and to the special case of equation (16) where the system function of equation (17) is a rational function i.e. the ratio of two polynomials. While a rational function approximation to the system function $H(s)$ can always be found to any prescribed degree of accuracy, it may require in some cases that the polynomials are of high order thus involving the determination of a large number of parameters from data that are subject to error. In such cases a more reliable and parsimonious representation may be obtained by using a system function consisting of the product of a rational function and a negative exponential. This corresponds to an impulse response in the time domain containing a pure lag and the corresponding generalisation of equations (18) and (19) from diffential equations to differential-difference equations.

2. HYDROLOGIC MODELS

2.1. Types of hydrologic models

It is only to be expected that many types of models are used in both theoretical and in applied hydrology. No classification of models in hydrology and water resources has been developed which is generally accepted. The number of factors involved is shown by the following classification (Dooge 1978) proposed in connection with the discussion of a number of computer based models in hydrology and water resources.

1. Extent of model (point, element, catchment, region)
2. Objective of modelling (understanding, forecasting, control, design)
3. Hydrologic processes and water uses involved
4. Nature of data (meteorological, hydrological, geotechnical etc.
5. Class of model (e.g. physical equations, conceptual, black box)
6. Type of model (linear or non-linear, time variant or time invariant, deterministic or stochastic)
7. Algarithms used in solution.

It will be noted that a hierarchical element is introduced into the classification system in items (5) and (6) but that only two levels of hierarchy are involved. This was thought adequate for the purpose of that general report but might need to be considerable expanded for other purposes. The basis of any hierarchical classification of models would of necessity be arbitrary.

The nature of any hydrologic model will depend to a large extent on whether it is intended to describe (a) conditions at a point e.g. the action of a soil column in reducing potential evapotranspiration to a lesser value of actual evapotranspiration, (b) the movement of water in a catchment element such as a hill slope or a channel reach, or (c) the action of a whole catchment in transforming precipitation to runoff. The structure and detail will also depend on whether the model is to be used in order (a) to gain understanding of the hydrologic process involved or (b) for the forecasting of a hydrological event in real time or (c) for the prediction of the probability of occurrence of a hydrologic event of given magnitude in the future. Depending on the nature of the data available for verification and calibration of the model, the inputs and outputs may be treated as (a) deterministic or (b) stochastic (with time dependence between the values of the variables) or (c) probabilistic (with the events being treated as independent). The model may be also formulated with the output variables in (a) continuous or (b) quantized or (c) discrete form.

Of great importance in any discussion of types of hydrological models, are the differences between (a) models based on continuum mechanics, (b) simple conceptual models based on simple arrangements of elementary conceptual elements such as linear storage elements, and (c) black box models which depend on input-output data not only for the calibration of parameters but for the determination of their essential structure itself. It has been found in practice that the models based on continuum mechanics (i.e. internal descriptions) are too complex to allow for the spatially variable nature of hydrologic systems and they have been simplified to such an extent that they become in effect simple conceptual models with unknown parameters. Similarily in external descriptions or black box analysis, the problem of determining the nature of the system operation on the basis of data subject to measurement errors has led to the introduction of constraints on optimisation procedure which lead ultimately to the replacement of purely black box analysis by the implicit use of a conceptual model. Thus the most successful hydrologic models have a simplified structure which includes some basic principles of hydrologic physics but not the detail of models based on continuum mechanics and have a limited number of parameters which have to be determined on the basis of field measurements.

2.2. Water balances in hydrology

A water balance, whether of a continent or of a component element such as the soil moisture in a defined area, can be attempted on a number of different time scales. The longer the time scale involved, the less significant become the terms due to changes in storage and the easier it is to accomplish a balance. Individual water balance studies can be classified in terms of time scales (minutes, hours, days, weeks, years and long-term) and space scales (varying from a point through local conditions to a catchment element, a complete catchment, a region, a continent to a global scale).

The space scales and time scales of interest in hydrology are shown on figure 2. In urban hydrology and highway and airfield drainage the time scales of interest are best expressed in minutes since on any longer time scale the features of runoff due to the intense rainfalls which are of interest in drainage design will not be apparent. The storm runoff from a small natural catchment can be described on a scale of hours and the runoff from larger natural catchments on a scale of days. The large variations in initial high rate infiltration are measured on a scale of minutes, some variations will occur on a scale of hours. Changes in groundwater storage and outflow are discussed on a scale of months or years.

Hydrologic analyses are also made on varying space scales. Measurements and estimates of precipitation, infiltration and evaporation usually refer to a single point or a very small area. Experimental plots give runoff data on a scale between 10 metres and 100 metres. Catchment elements may be on a scale from 1 kilometre to 10 kilometres. Catchments on a scale larger than this constitute a region and are analysed by division into subordinate catchments.

From the point of view of the interface between hydrologic studies and General Circulation Models the relevant time and space scales are given by Table 6.4 on page 85 of GARP Publication Series No. 16 on "The Physical Basis of Climate and Climate Modelling" as about 10-15 days and 100-500 km.

Also of interest are studies on longer time scales and larger space scales which would provide for a given region the long term value which could be considered as the base values about which the variables of interest were perturbations. Examples of such large scale long term water balances are contained in such publications as the proceedings of the Reading Symposium on World Water Balance (IAHS 1970), the monograph on "World Water Balance and Water Resources of the Earth" by the USSR Committee for the International Hydrological Decade (1974, 1978) and the monograph by Baumgartner and Reichel (1975) on "The World Water Balance".

The coupling between the atmosphere and the land surface is the main theme of this Study Conference and hence is referred to in the various papers from different points of view. In the present paper attention is concentrated on the coupling through the surface of the water balance of the elements of precipitation and evaporation on the one hand and infiltration and run-off on the other. Taking any area of land surface and the associated soil profile under that surface area, the rate of change of moisture storage must equal the excess of the rate of moisture inflow over the rate of excess of moisture outflow. For the time scales of interest in climate modelling the changes in surface storage can be neglected compared with the changes in subsurface storage. For any isolated defined area the lumped continuity equation can therefore be given by:

$$\frac{\partial}{\partial t}(\underline{c}) = P - E - Y \qquad (20)$$

TIME AND SPACE SCALES

	Minutes	Hours	Days	Months	Years	Long term
Point						
Local						
Element						
Catchment			✓	✓		
Region			✓	✓		
Continent						
Global						

Figure 2

where $\underline{c}(t)$ is a vector representing the state of the soil moisture (both saturated and unsaturated) in the soil profile below the area under study, $P(t)$ is the rate of precipitation on the area, $E(t)$ is the rate of evapotranspiration from the area, and $Y(t)$ the rate of runoff from the area. The latter can for convenience of analysis be divided into direct storm runoff $Y_s(t)$ and baseflow or groundwater runoff $Y_g(t)$:

$$Y(t) = Y_s(t) + Y_g(t) \qquad (21)$$

The division during storms of the available precipitation at the surface $P(t)$ into effective precipitation $P_e(t)$, which is converted to direct storm runoff $Y_s(t)$, and the remainder, which is available for subsequent evapotranspiration $E(t)$ and subsequent runoff as baseflow $Y_g(t)$, depends on the rate of infiltration at the surface $F(t)$. Whereas the average response time for direct storm runoff is on a scale of hours or days, the response time for base flow is usually on a scale of months or years.

The left hand side of Equation (20) gives the rate of change of soil moisture $(\underline{c})$ and on the right hand side of the equation both the evaporation (E) and the runoff (Y) depends on the amount of soil moisture. From this it is clear that the hydrologic water balance is critically dependent on the soil moisture. In Equation (20), the soil moisture has been expressed as a vector $(\underline{c})$ rather than an average value $(\bar{c})$ in order to emphasise that a high degree of information may be involved. In climate modelling we are therefore concerned with the soil moisture accounting equation:

$$\frac{\partial}{\partial t}(\underline{c}) = F - E - Y_g \qquad (22)$$

The evapotranspiration $E(t)$ is coupled to the atmosphere and the goundwater run off Y_g is added to the surface runoff Y_s to obtain the total runoff.

The surface infiltration into the soil (F) is a function of the precipitation and of the infiltration capacity or potential infiltration (F_p) which is itself a function of the state of the soil moisture ($\underline{c}$):

$$F(t) = F\left[P, F_p(\underline{c})\right] \quad (23)$$

If the rate of precipitation is less than the rate of infiltration capacity then all of the precipitation will infiltrate through the surface and there will be no surface runoff from this point of the land surface; i.e.

$$\text{if} \quad P < F_p : F(t) = P(t) \quad (24a)$$

$$Y_s(t) = 0 \quad (24b)$$

On the other hand if the rate of precipitation exceeds the infiltration capacity in any area then the infiltration through the surface of the soil will be limited to the infiltration capacity and the remainder of the precipitation will appear as direct storm runoff.

$$\text{if} \quad P > F_p : F(t) = F_p \quad (25a)$$

$$Y_s(t) = P(t) - F_p(t) \quad (25b)$$

On the scale of an experimental plot or a small catchment element it is reasonable to assume that the potential infiltration is uniform over the surface area but on a catchment or regional scale this assumption is clearly open to question.

For a long period, catchment analysis was dominated by the threshold concept of infiltration (Horton 1933) which assumed that in times of heavy rainfall the rate of precipitation would exceed the potential infiltration over a fixed proportion of the catchment area and hence produce direct storm runoff. More recently the view has been gaining ground that direct storm runoff more frequently results from the rising of the water table towards the surface and hence that the area of direct surface runoff varies throughout a storm (Dunne 1978). Since the amount of infiltration available to replenish soil moisture storage depends on the amount of direct storm runoff, the question of variability over the catchment area is an important one.

A second critical question in the analysis of catchment response and of the coupling with hydrologic and atmospheric models is the variation in time of the actual evaporation (E) which in general depends on (1) the potential evaporation E_p, which is an atmospheric factor; (2) the state of the soil moisture including the position of the water table and the storage in the unsaturated and saturated zones; and (c) the state of the vegetal cover ($\underline{v}$) which includes both the leaf area and its stomatal resistance and the root system and its power to remove water from the unsaturated zone. The formulation of a suitable form of this general relationship:

$$E = E\left[E_p, \underline{c}, \underline{v}\right] \qquad (26)$$

is a problem of vital importance and largely an unsolved one. If the moisture content of the upper layers of the soil approaches saturation then the amount of evaporation will be entirely controlled by the micro-climate and will approach the potential value. Many expressions are available for the estimation of this potential evaporation (Penman 1963). On the other hand if the moisture content of the unsaturated zone is low and the water table is well below the surface, the amount of water evaporated will be largely controlled by the soil system and may approach a limiting rate of evaporation which is completely controlled by the soil (Gardner 1958). The limiting rate of evaporation will depend on the depth of the water table below the surface and hence will vary with time and in particular will decrease during long dry periods. On the time scale of climate modelling, the assumption that the rate of evapotranspiration continues at the potential rate until the limiting rate is reached is probably sufficiently accurate at the present stage of modelling land surface processes. It would however be desirable to link the limiting rate of evaporation to soil characteristics and the position of the water table.

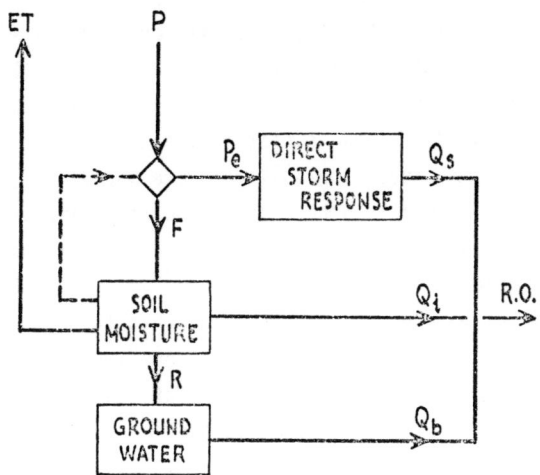

SIMPLIFIED CATCHMENT MODEL

Figure 3

2.3. The key role of soil moisture accounting

The key role of soil moisture accounting in hydrologic analysis is clear from the discussion in the previous section. Unfortunately this aspect of hydrologic analysis is also the area in which knowledge and techniques are least satisfactory. Practically all models of catchment response have the general structure shown in Figure 3 in which the catchment system is divided into three sub-systems:

(1) a sub-system responsible for the rapid response of the catchment outflow to excess precipitation (Pe) in the form of a direct storm response, (Q_s); (2) a sub-system involving the soil moisture in the unsaturated zone with its effect on infiltration (F) and on actual evapotranspiration and (ET); and (3) a sub-system modelling the response of the groundwater system in converting groundwater recharge (R) to groundwater outflow or base flow (Q_b); Figure 3 is a lumped representation and hence applied only to a catchment or portion of a catchment in which the key parameters and the inputs do not vary spatially.

Because of the concern of hydrologists with problems of flood hydrology, the sub-system of direct storm response has been the subject of more extensive study than either of the other two sub-systems. The next most intensively studied area is that of the groundwater sub-system and the least studied that involving soil moisture and its effects. In the latter area, hydrologists have been satisfied with simple empirical relationships and have not incorporated the findings of the soil physics into soundly based models of soil moisture behaviour. The position described above is partly explained by the fact that the catchment sub-system involving the soil moisture creates more serious difficulties than either of the other sub-systems. A number of these difficulties arise from the fact that there is feedback in this part of the system since the state of the soil influences the rate of infiltration into it. Another difficulty arises from the fact that there is no appreciable recharge to groundwater until the initial soil moisture deficit has been satisfied by infiltration. This produces in the soil moisture sub-system a threshold effect i.e. a concentrated non-linearity. Consequently the linear methods of analysis of direct storm runoff and of groundwater outflow which underlie many of the techniques used in applied hydrology have no counterpart in the case of the analysis of soil moisture variations, infiltration and evapotranspiration. Even though most conceptual models of catchment response provide for soil moisture accounting, this is almost always the weakest part of the model. It is unfortunate that the part of the catchment response which is of the greatest significance in the coupling of hydrologic models and general circulation models should be that in which the hydrology is weakest. However, this difficulty must be faced, and if possible, overcome.

The ground surface which represents the interface between the atmosphere and the soil may vary spatially in a number of respects which are critical from the point of view of the exchange of moisture and of energy between the atmospheric system and the soil system. Some of these major differences are reflected in the fact that at the present Conference a number of review papers are required to deal with the microphysical processes at the surface of the ground and that a number of papers are also required to discuss different aspects of global data in relation to land surfaces.

It has been mentioned above that hydrological parameterization has not been as well developed in regard to the soil system as it has been in relation to surface runoff and to ground water conditions. Such developments as there have been have concentrated on the problems of infiltration into and exfiltration from a bare soil surface and there has been relatively little work devoted to vegetated or snow covered surfaces. Even the restricted problem of soil moisture movement in the absence of vegetation or snow cover has not been successfully parameterized by hydrologists and certainly has not been parameterized on any scale approaching that of general circulation models. It is hoped that as a result of some of the other papers to be presented at the Study Conference and as a result of the subsequent discussions, there may be indications of the direction in which progress might be made to remedy this severe defect.

2.4. Estimation of soil moisture changes

Before attempting to parametrize soil moisture changes on a grid scale appropriate to general circulation models, it is necessary to refer briefly to the results that have been obtained in hydrology in regard to local changes of soil moisture. The hydrologist is primiarily interested in the rate of infiltration into the soil, the rates of evaporation from bare soil and transpiration from vegetation, and the variation in the total amount of moisture stored in the soil. Unlike the soil physicist or agronomist, the hydrologist is not greatly concerned with the actual distribution of moisture in the soil profile except in so far as this affects the rate of infiltration or the rate of evapotranspiration.

The basic principles of flow through porous media are dealt with in many articles and books. Standard reviews of the application of the principles of soil physics to the infiltration problem and the evaporation problem are contained in the works by Bear, Zaslavsky and Irmay (1968), Childs (1969), Philip (1969), and in the paper prepared for this Study Conference by Brutsaert (1980). The combination of the continuity equation with Darcy's Law (1856) gives the Richard's equation (1931) which for vertical flow can be written as:

$$\frac{\partial c}{\partial t} = \frac{\partial}{\partial z}\left[K\frac{\partial}{\partial z}\left(\frac{p}{\rho g}\right)\right] + \frac{\partial K}{\partial z} \qquad (27)$$

where $c(z,t)$ is the moisture content on a volume basis at any given level in the soil profile, K is the hydraulic conductivity which if hysteresis is neglected can be taken as a function only of moisture content and be written as $K(c)$, $p(c)$ is the soil moisture pressure which will be negative in the unsaturated zone, p is the density of the soil water, and g is the acceleration due to gravity. The above equation is non-linear due to the variation of hydraulic conductivity (K) with soil moisture content (c). The moisture characteristic curve for the individual soil which relates soil water pressure (p) to moisture content(c) will also be non-linear. Accordingly, the solution of the basic equation is difficult even for simple boundary conditions.

In fact the boundary conditions are quite complex. They may be specified either in terms of moisture content (c) or of pressure (p), the two variables being interchangable through the moisture holding characteristics of the particular soil. If the water table is used as a lower boundary, then the boundary condition will take the form of a Dirichlet condition prescribing the moisture content as the saturation value or water pressure as zero. For shallow water-tables or for longer time scales this may involve the use of a moving boundary condition. Alternatively, the lower boundary condition could be taken in the form of Neumann condition for the moisture flux on the basis of a known ground water outflow and known or assumed changes in ground water storage. In most first approaches to the problem the boundary condition at the surface is given as a Neumann boundary condition prescribing the moisture flux either as equal to a known amount of precipitation or a known (or estimated) rate of actual evaporation or else as a Dirichlet boundary condition prescribing a given moisture content or a given soil water pressure at the surface.

For a realistic approximation to field conditions, the exact boundary condition at the soil surface can not be assigned priori but only after determining which of the two types of boundary condition governs at any particular time (Rubin 1966, Feddes et al 1978). If the rate of precipitation is less than the potential infiltration rate for a soil column, then the moisture flux at the surface will be given by the precipitation rate and the moisture content at the surface will be determined as part of the solution of the problem. As the moisture content of the surface layers approaches the saturated value, the rate of potential infiltration will decrease and may become less than the rate of precipitation. If this happens, the governing boundary condition becomes the saturation at the surface and the rate of infiltration is obtained as part of the solution of the problem. An analogous situation exists in the case of evaporation from a soil surface. If the rate of potential evaporation is less than the maximum rate at which moisture can be transmitted through the soil profile to the surface, then evaporation will continue at the potential rate i.e. the evaporation will be atmosphere-controlled. If, on the other hand, the potential rate of evaporation is greater than the limiting rate at which moisture can be transmitted to the surface, then evaporation will occur at the latter rate and the evaporation process will be soil-controlled.

The problem of ponded infiltration has been solved analytically but only by a severe simplification of the equations and the boundary conditions. Even though the assumptions are unrealistic, the various analytical solutions give certain insights into the nature of the solution for the more general problem and of the sensitivity of that solution to certain factors. During the early stages of infiltration the contribution of gravity to the total hydraulic potential is considerably less than the contribution of the capillary potential. Under such circumstances, equation can be simplified by the omission of the second term on the right hand side of the equation. On the basis of the resulting diffusion-type equation, the form of the variation of infiltration with time can be found if the following simplifying assumptions are made:

(1) the soil profile is semi-infinite, i.e. the water table is at infinite depth below the soil surface;

(2) the initial condition is one of a constant initial moisture content (c_o) which is equivalent to assuming that there is a constant initial rate of infiltration equal to the hydraulic conductivity corresponding to this initial moisture content;

(3) the moisture content at the surface changes instantly from the initial value to the saturation value

For these assumptions it can be shown using the Boltzmann Transformation that the total volume of infiltration (F) is given by:

$$F(t) = A \cdot t^{\frac{1}{2}} + B \cdot t \qquad (28)$$

where A (usually termed the sorptivity) is a function of the moisture characteristics of the particular soil as well as of the initial moisture content (c_o) and (B) is a parameter that depends on the hydraulic conductivity. The family of simplified solutions represented by equation (28) are in fact, self-similar solutions of the simplified form of equation (27). Thus, for any particular assumption relating the soil moisture suction to the moisture content, the developing moisture profiles in the soil column will be similar to one another. If the gravity form is included a solution may be found in the form of a series of which equation (28) gives the first two terms (Brutsaert 1980)

A comparison of the value of the sorptivity in equation (28) corresponding to a given initial moisture content for two widely differing shapes of moisture profile during infiltration indicates that the sorptivity is not very sensitive to the shape of profile and therefore is not sensitive to the particular simplified model used. The difference in the numerical value of the sorptivity due to the assumption of different simplified models would be completely swamped by differences in the hydraulic conductivity (K) and the soil moisture suction corresponding to the initial moisture content (c_o). These two parameters could easily vary by one or more orders of magnitude over an area well below the grid scale of a general circulation model.

Assumption (1) above, that the ground water table is at an infinite distance below the surface, and assumption (2), that the initial moisture content is constant, are particularly unrealistic except under arid conditions. For the condition of more variable soil water storage, serious errors could arise from these assumptions. In connection with the contributing area concept of catchment runoff referred to in section 2.2. above (Dunne 1978), the time taken for a whole soil profile to become saturated and for surface runoff to be initiated is of particular interest. At present, some hydrologic models combine the contradictory assumption of a water table at the finite distance below the surface and constant initial moisture content from the surface to the water table. Preliminary analysis indicates that the more realistic but still highly simplified assumption of a ground water table at finite depth and a linear decrease of moisture content from saturation at the water table predicts a time to saturation of the whole profile which is only one-third the value which is given by assuming a constant initial moisture content.

During long dry periods, the condition of a steady capillary rise through the unsaturated soil from the water table to the surface may be established. If the variation of the soil moisture pressure with water content and of the hydraulic conductivity with water content are known, the moisture profile for this condition can be established by solving the steady state form of equation 27 in order to obtain the elevation at which a particular moisture content will occur. For simplified values of the soil moisture relationships analytical solutions can be obtained. When the soil in the neighbourhood of the surface becomes extremely dry the hydraulic conductivity becomes very small and the movement of moisture to the surface and its subsequent evaporation is entirely controlled by conditions within the soil profile. Gardner (1958) introduced the concept of the limiting rate of evaporation which would occur under the circumstances and which will depend on the depth to the water table and the moisture holding characteristics of the soil.

The solution for the transient case of the drying of the soil by evaporation is naturally more complex. As in the case of the estimation of potential infiltration, the problem can be solved analytically only by the use of highly simplifying assumptions. The assumption of similarity of profiles will predict that evaporation also varies inversely with the square root of the elapsed time and in the case of assumed initial constant moisture content (c_o) will be proportional to the moisture content. Again, the predicted evaporation would not be expected to vary greatly even for widely differing assumptions in regard to the nature of the similar profiles.

Just as the infiltration case can be solved for a semi infinite profile with constant initial moisture content by assuming that the surface of the ground is rapidly saturated, so the case of exfiltration or evaporation from the soil surface can be solved for the same initial condition by assuming that the surface of the ground rapidly dries out to zero. A second approach is to assume that the evaporation continues at the potential rate as long as a gradient of moisture potential sufficient to maintain this potential rate of evaporation can be established. The period of climate-controlled evaporation at the potential rate can be assumed to continue until the moisture content of the surface reduces to zero. The latter approach corresponds to the Budyko type assumption of a moisture availability function which is used in most general circulation models (Carson 1980). There is scope here for the use of some of the highly simplified models used in infiltration and exfiltration studies to relate the value of the threshold soil moisture content below which the evaporation rate is less than the potential to the soil properties for which information is readily available. In this way the value of this critical moisture content as used in general circulation models might be made less arbitrary.

3. EXTENSION FROM LOCAL VALUES TO GCM GRID SCALE.

3.1 Spatial variability on a local scale

The discussion in the preceding sections has been concerned with the elements of the soil moisture balance on time scales and space scales which are several orders of magnitude below those appropriate to general circulation models. The conclusion was reached that errors in the prediction of soil moisture balance (and hence of the coupling with the general circulation of the atmosphere) arising from the choice of model for soil moisture accounting are likely to be far less than those arising from the variation of soil parameters throughout an area. There is no generally accepted procedure in hydrology for dealing with the key question of how to parameterize the hydrologic phenomena involved on the scale of the grid of the general circulation model. Only recently have hydrologists sought to deal with the question of the parameterization on a field and catchment scale and to seek linkages between the parameters of the internal descriptions of hydrologic phenomena at local scale and the parameters of the conceptual models used on a catchment scale. Even the small amount of work which has been done has been largely devoted to free surface flow and groundwater flow rather than to soil moisture phenomena. The larger regional scale studies have been confined to overall long-term water balances and the development of empirical relationships. Accordingly, it may be said that the parameterization of hydrologic processes to the grid scale of general circulation models is a problem that has not been tackled, let alone solved.

In this section of the paper, some work will be reviewed of the parameterization of soil and moisture phenomena on a field scale and a catchment scale which might help to indicate the direction in which studies might be undertaken. Field studies of the spatial variation of soil factors likely to influence water movement have only been undertaken in recent years. Typical of such studies are those by Nielsen, Biggar and Erh (1973) and by Sharma, Gander and Hunt (1980). Nielsen and his colleagues studied the variability of soil suction and hydraulic conductivity at twenty points in a 150 hectare field in California. Whereas the full field was fairly uniform in soil classification, the steady infiltration rate varied from 5 to 457 mm/day. The hydraulic conductivity ranged between about 1 mm/day and 1,000 mm/day. The authors concluded from their observations that variations in the water content were normally distributed with depth and with horizontal distance, while the values of hydraulic conductivity and soil water diffusivity were log-normally distributed.

Sharma and his colleagues (1980) studied conditions at twenty six sites in a 9.6 hectare field in Oklahoma. Instead of studying the soil moisture characteristics and the hydraulic conductivity and then attempting to predict the parameters affecting infiltration, they studied directly the variations of the parameters in an acceptable infiltration equation. They chose as their model for infiltration the simplified relationship discussed in section 2.4 above:

$$F(t) = A \cdot t^{\frac{1}{2}} + B \cdot t \qquad (28)$$

where $F(t)$ is the cumulative infiltration, t is the elapsed time since the start of infiltration, A is a parameter usually designated as sorptivity, B is a parameter representing the final steady rate of infiltration. We have seen in Section 2.4 this equation represents a family of models of infiltration in which the shape of the wetting front may take any of a number of widely different shapes provided that the soil moisture profiles are similar at all times. This model has also been shown to give a very good approximation to more general solutions (Philip 1957). If infiltration measurements in the field are available, the values of the parameters A and B can be determined from this data.

Sharma and his colleagues studied the variation of the parameters A and B directly rather than studying the variation of the soil suction, hydraulic conductivity and other soil properties on which these parameters depend. They found that over the 9.6 hectare field the value of the sorptivity A among the twenty six sites varied by an order of magnitude and that the parameter B varied by two orders of magnitude. There appeared to be no definite pattern to this variation and it was not possible to identify the effect of the soil type or the topography on the variations. They confirmed the reasonableness of the model given by equation (28) by using the derived parameter values at each of the individual sites to write equation (28) in dimensionless form in terms of these parameters as:

$$\frac{B}{A^2} \cdot F(t) = \left(\frac{B^2}{A^2} \cdot t\right)^{\frac{1}{2}} + \frac{B^2}{A^2} \cdot t \qquad (29)$$

and by plotting the dimensionless infiltration which appears on the left-hand side of equation (29) against the dimensionless time which appears in the two terms on the right-hand side of the equation. When this was done, the large scatter obtained by plotting F(t) against t for all twenty six sites was reduced to an insignificant scatter around a single curve with the form of equation (29).

3.2. Soil similarity and scaling factors

The analysis of the spatial variability of soil parameters can be aided by the use of the concept of soil similarity and scaling introduced by Miller and Miller (1956). This approach is based on a concept of soils of equal porosity and geometrically similar internal geometry. The internal geometry of the two soils is assumed to differ only in a characteristic length scale (λ). It then follows from dimensional analysis that the soil suction of geometrically similar soils at equal degrees of saturation will be given by:

$$\lambda_i S_1 \left(\frac{c}{c_{sat}}\right) = \lambda_i S_2 \left(\frac{c}{c_{sat}}\right) \qquad (30a)$$

and a soil of characteristic length scale λ can be standardised through:

$$\lambda_i S_i(c) = \bar{\lambda} \cdot \bar{S}(c) \qquad (30b)$$

We can define a scaling factor (α) in terms of the standard length scale by

$$\alpha_i = \frac{\lambda_i}{\bar{\lambda}} \qquad (31a)$$

and hence write the standardised soil suction for the given degree of saturation as:

$$\alpha_i \cdot S_i \left(\frac{c}{c_{sat}}\right) = \bar{S}\left(\frac{c}{c_{sat}}\right) \quad (31b)$$

On the basis of the hydraulic conductivity being a parameterization on a bulk scale of Stokes equation for microscopic flow within the capillary pore system, the scale relation for hydraulic conductivity can be written as:

$$K_i \left(\frac{c}{c_{sat}}\right) = \alpha_i^2 \cdot \bar{K}\left(\frac{c}{c_{sat}}\right) \quad (32)$$

The validity of this scaling for geometrically similar porous media has been experimentally verified in the laboratory. (Klute and Wilkinson 1958, Wilkinson and Klute 1959).

Philip(1967) recognised that, in general, the spatial variation of $S_i(^c/csat)$ and of $K_i^c/csat$ are not independent. He suggested as a practical simplication that it might be useful to assume that the internal geometry of the soil preserved similarity. The spatial variation of the properties $S_i (^c/csat)$ and $K_i (^c/csat)$ can then be represented purely through the spatial variation of α_i.

Scaling analysis be applied directly to the parameters A and B in equation (28). The relation for the sorptivity A is:

$$A_i = \alpha_i^{\frac{1}{2}} \cdot \bar{A} \quad (33)$$

and the relationship for the parameter B is:

$$B_i = \alpha_i^2 \cdot \bar{B} \quad (34)$$

The effect of the geometrical scale of the pore structure of the soil on the process of infiltration is seen by substituting from equations

(33) and (34) to get an alternative dimensionless form of equation (29)

$$\alpha_i F_i = \bar{A}(\alpha_i^3 t)^{\frac{1}{2}} + \bar{B}(\alpha_i^3 t) \qquad (35)$$

Even though this analysis has reduced the question of spatial variability to that of the variability of a single factor (i.e. the scale ratio), the problem of determining the average behaviour over an area for such spatial variability of a scale-heterogeneous soil system is not an easy one (Philip, 1967, 1980). Philip has shown that the problem does become simple for the special case of:

$$\bar{S} = -\exp(-\beta c) \qquad (36a)$$

as the variation between the standardised soil suction ($\bar{S}$) and the moisture content (c) and :

$$\bar{K} = \frac{K}{\bar{S}^2} \qquad (36b)$$

for the relationship between the standardised hydraulic conductivity and the standardised soil suction. Philip however points out that without this assumption many of the benefits of the classical analysis for homogeneous soil are lost and that without the assumption of geometrical similarity in regard to the porosity and pore size distribution, the problem becomes very difficult (Philip 1967, 1980).

Warrick, Mullen and Nielsen (1977) applied the concept of scaling based on micro-scale similarity to the data of Nielsen, Bigger and Erh (1973) referred to in Section 3.1 above. They found that, by the use of scaling, the sum of the squares of the deviations of predicted soil suction for the twenty field locations could be reduced to 15% of its original value and that the scatter in the predicted unsaturated hydraulic conductivity could be reduced to 14% of its original value. The values of the scaling factors were determined by an iterative optimisation technique. The analysis was also applied to thirty six sites randomly selected within an area of 87 hectares measured by Coehlo (1974) and eight sites, three of them within 0.15km .and all within

7.2km, measured by Keisling (1977). Separate values of the scaling factor (α) were determined from the soil suction data and from the hydraulic conductivity data at each site. It was found that the values of the scaling factor determined from the soil suction data appeared to be more reliable than those determined from the hydraulic conductivity. In all cases the distribution of the scaling factors appeared to be a log-normal and the coefficients of variation were between 0.48 and 1.70.

Sharma, Gander and Hunt (1980) applied scaling analysis to their investigation of the spatial variation at 26 sites in a 9.6 hectare field referred to in Section 3.1 above. Three sets of scaling factors were determined: (1) a set of scaling factors (α_A) based the estimated values of A in equation (33) and a value of $\bar{A}$ of 6.39 mm/min$^{\frac{1}{2}}$; (2) a set of 26 scaling factors (α_B) based on the estimated values of B in equation (34) and the value of $\bar{B}$ of 0.254 mm/min; (3) a set of scaling factors (α_{opt}) based on the individual cumulative infiltration for each site and the use of equation (35) with the same values for $\bar{A}$ and $\bar{B}$. If the soils of the various sites were truly geometrically similar and if the model of equation (28) were exact then one would expect that the values of α and their distribution would be identical in the three cases. In fact the relationship between the scale factors based on A and on B was not good and α_A was far more variable than α_B. The coefficient of correlation between α_B and α_{opt} was R= 0.913 compared with R = 0.927 for α_A. The coefficient of correlation R for the arithmetic mean of the two estimates $\alpha_{\overline{AB}}$ with α_{opt} was 0.960, whereas the harmonic mean gave the very high value of R = 0.991. In each case the distribution of the scaling factor was found to be log-normal with the coefficient of variation of about 0.6.

3.3. Spatial variability and soil moisture accounting

Some studies have been made of the effect of spatial variation in soil properties on the actual elements of the water balance. Peck, Luxmoore and Stolzy (1977) simulated scale heterogeniety for a 9.7 hectare catchment in Tennessee. The mean values were determined from field observations and the spatial variability in the scaling factor taken as being normally distributed with a coefficient of variation of 0.25. The authors concluded that this type of spatial variability did not give rise to significant changes in the water balance of the catchment and that the lumping of the parameters at their mean value was justified.

Sharma and Luxmoore (1979) made similar studies based on Monte Carlo simulation but came to different conclusions. They assumed the scaling factor to be log-normally distributed with a coefficient of variation of 0.6 and used values of α between 0 and 4. They applied Monte Carlo simulation to the same 9.6 hectare catchment studied by Sharma and others (1980). Using the Monteith (1965) form of the combination equation for the estimation of evapotranspiration, they calculated the water balance changes at fifteen minute intervals during precipitation and at hourly intervals in the absence of precipitation. Monthly water balance elements were calculated for fourteen synthetic soils i.e. for fourteen different values of α. They found that deep drainage increased with increasing scale factor α, while evapotranspiration and surface runoff decreased. Once surface runoff was generated, it increased very rapidly with the decreasing value of α. The effects of spatial variability on the monthly water balance were found to be highly dependent on the soil-plant-atmosphere combinations. The effects were amplified in months with high rainfall and low evapotranspiration and when the normal distribution was used in place of the log-normal. In contrast to the results found by Peck (1977), the water balance simulated by the mean scale factor was significantly different from the integrated response predicted by either the log-normal or the normal distribution.

A recent study by Freeze (1980) also involved Monte Carlo simulation of spatial variability. The hydraulic properties of unsaturated soil were characterized through the saturated hydraulic conductivity, the porosity and a soil storage parameter. This simulation study used stochastic rainfall patterns applied to both the Horton infiltration model (Horton 1933) and a variable source area model Dunne (1978) with an assumed autocorrelation structure in the hydraulic conductivity distribution in each case. The results of the simulation were analysed from the point of view of the statistical distribution of the runoff rather than the actual infiltration. It was found that the statistical properties of the predicted runoffs generated from a stochastic sequence of some events would be greatly in error if the mean hydraulic conductivity was used. The results indicated that the distribution of hydraulic conductivity in catchment significantly affected the catchment response and should be included in any parametric representation of the catchment response.

The studies reviewed above raise the question of the amount of data required to characterise the soil water regime. Of interest in this connection are the results of a study by the Netherlands Soil Survey Institute (Bouma et al 1980) in which various amounts of field information were used as input to a sophisticated mathematical model of sub-surface storage and flow and the simulated depth to water table and the simulated evapotranspiration compared for seven levels of information input. The study was based on the application of the simulation program to an area in the east of the Netherlands 6 kilometres by 6 kilometres in extent and divided into nodal areas of 500 m by 500 m. In the first phase it was assumed that soil was uniform throughout the 36 square kilometers and the parameters required were based on measurements made elsewhere on the dominant soil type of the area under study. In phase 2 the parameters for each nodal were based on the soil type given by the soil map from compared soils. In phase 3 one boring was made in each nodal area to obtain in situ data on the texture, organic matter content and bulk density as the basis for more reliable estimation of the physico-hydrological parameters and to obtain a value for the rooting depth. In phase 4 physico-hydrological data were based on measurements in the field and in the laboratory. It was concluded in retrospect that data collection for the project would have been most cost-efficient if the phase 2 approach had been used for all nodal areas combined with selective use of the phase 3 and phase 4 approach in major mapping units.

3.4. From field scale to catchment scale

The work reviewed in the last few sections relates to the effect of spatial variability on areas of the order of 10 to 100 hectares i.e. surface areas of less than 1 sq. km. A substantial degree of parametrisation is involved in linking phenomena on that scale to the usual scale of catchment analysis i.e. areas between 100 and 1,000 sq.km. Though a number of approaches have been tried, this problem is largely an unsolved one. The two important problems at catchment

scale relevant for general circulation models are the estimation
of actual infiltration and of actual evaporation.

Classical methods of the hydrological analysis of the runoff
cycle are dominated by the infiltration concept due to Horton (1933).
This assumes that at the beginning of a storm the potential
infiltration exceeds the rainfall intensity due to the relatively
dry condition of the soil and consequently there is no accumulation
of water on the soil surface and no direct surface runoff. Later
during the storm the rate of potential infiltration will decline and
may fall below the rate of precipitation, thus generating surface
storage and surface runoff. Because of this threshold effect the
relationship of runoff to precipitation is highly non-linear.
This non-linearity means that the use of a single rate of potential
infiltration i.e. infiltration capacity for the whole catchment area
would only be justified if the soil properties and the spatial moisture
content were uniform throughout the catchment area.

Betson (1964) applied this infiltration approach to predict the
volumes of storm runoff in drainage basins between 0.01 km^2 and 100 km^2 and
compared these with field measurements.   In order to make the
predicted values correspond to the measured values he found it necessary
to apply a scaling factor between 0.05 and 0.86 which he interpreted as
meaning that the infiltration capacity (i.e. potential rate of infiltration)
was exceeded on only part of the drainage basin and that the proportion
producing runoff varied in different storms between 5% and 86% of the total
catchment area. An alternative model of direct storm runoff starts from
the assumption that most of the precipitation infiltrates the surface
but that a proportion of it may enter the drainage system relatively
quickly through lateral inflow resulting from the fact that lower layers
in the soil may have a hydraulic conductivity less than the rate of
downward percolation (Hewlett 1961). In this model, direct overland
flow would only occur in portions of the catchment **where the soil moisture**
profile became completely saturated. This alternative model
suggests that the storm runoff occurs from those parts of the
catchment immediately adjacent to the drainage network and that
the contributing area would vary throughout a storm increasing as the
soil became saturated in areas of shallow groundwater and decreasing
during the recession (Hewlett and Nutter 1970, Dunne 1978).

No matter which model is adopted to explain and estimate the generation of surface runoff, the rate of infiltration will clearly vary widely throughout the catchment area. In the conceptual models developed for the analysis of catchment response, this distribution is either ignored or handled in a relatively crude way. Kohler(1963) pointed out the conflict in modelling of soil moisture accounting between the desire to represent the soil column by a multilayer system and the need to allow for variations in infiltration capacity throughout the catchment by what he described as multi-capacity soil moisture accounting. A large number of conceptual models have been developed for the simulation of catchment behaviour (Fleming 1975, WMO 1975) but it has already been remarked that the soil moisture accounting section of these conceptual models is usually the least satisfactory part. In a number of models, the variation in infiltration capacity is taken as being a uniform distribution between zero capacity and a maximum infiltration capacity which is a parameter to be optimised in the conceptual model or to be determined from available physical data or assumed on some other basis. This assumption of a linear variation in infiltration capacity over the catchment is then combined with either the infiltration model or the variable source area model for the generation of direct storm runoff.

The second important question at the catchment scale is the parameterization of the actual evaporation. In this case the high spatial variation of vegetation as well as of soil creates particular problems. The problem is similar to that for the case of infiltration and indeed represents the problem of the rate of exfiltration (Eagleson 1978, Brutsaert 1980). Penman (1951) suggested a distinction between riparian and non-riparian conditions. Under riparian conditions, the water table would be close to the surface and would be able to supply moisture to the surface for evaporation at the potential rate. During precipitation, infiltration through the surface would be able to recharge the ground water directly. In case of non-riparian areas, the groundwater would be well below the surface and the state of the unsaturated zone would determine both the amount of infiltration and the amount of actual evapotranspiration at the surface. The partial source area concept of direct runoff generation (Dunne 1978) follows the same general line.

It is not clear whether the solution to the problem of parameterization on catchment scale of 100 km^2 will be found through successful parametrization of the processes studied in physical hydrology or through the development of reliable conceptual models and the systematic analysis of parameter values determined by optimization for the fitted of these conceptual models to catchment records. If the objective is to be achieved through either of these methods, there is a need for a thorough discussion of the problem and for the systematic planning of the research programme. The development of models for the analysis of catchment response has been bedevilled by the development of a myriad of separate models whose operation satisfies the author but whose relative performance has not been compared except in a few isolated studies (WMO 1975). The result of the WMO study and of other comparisons (Naef 1977) indicate that the complicated conceptual models now widely used in catchment hydrology cannot be shown to predict the elements of the water balance more accurately than simpler models involving fewer parameters.

3.5 From catchment scale to regional scale

Even if the problem of parameterization on a catchment scale were solved by hydrologists, there remains the problem of the final parameterization from a catchment scale of 100 km^2 to a GCM scale of between 10,000 and 100,000 km^2. The fact that the spatial distribution of soil moisture properties is significant in the parameterization from the local scale to the field scale of 1 km^2 or less as discussed in Sections 3.1 to 3.3. does not necessarily mean that the non-homogeneity of catchments within a region of 10,000 km^2 would also be significant but does indicate that the problem must be investigated. In some studies of regional water balance this problem has been adverted to in passing but it has not been systematically studied in the same way as has been described in relation to the parameterization from local scale to field scale. If adequate studies of catchment behaviour were available for some of the world's great rivers these would be compatible in scale with GCMs.

The difficulties of allowing for the high spatial variability in soil and vegetation in estimating evapotranspiration has led to the proposal of an approach for the estimation of regional evapotranspiration which depends on the micro-climate near the land surface since the atmospheric variables involved will be more uniform over a wide area than the soil and vegetable properties. This approach, which has been referred to as the advection-aridity approach, is based in practice on micro-climatological measurements and therefore it is the concepts behind the approach rather than the present methods used which would be of relevance in connection with General Circulation Models. In fact, General Circulation Models might be used to check the validity of the approach.

In the advection-aridity approach (Bouchet 1963, Morton 1965, Brutsaert and Stricker 1978), a distinction is made between the potential evaporation under potential conditions (E_{po}) and the estimated potential evaporation under non-potential conditions (E_p). In the former case, water is not limiting and consequently the actual evaporation (E) is equal to the potential evaporation under potential conditions (E_{po}) i.e.:

$$E = E_{po} \qquad (37)$$

Under non-potential conditions, less energy is used in the form of the latent heat of vaporisation and the amount of energy thus made available (q):

$$q = L(E_{po} - E) \qquad (38)$$

can be used for the transfer of sensible heat or for the creation of turbulence etc. near the surface of the ground. Such changes will increase the apparent potential evaporation (E_p) based on a combination type formulae which has been calibrated for potential evaporating conditions. Consequently, if the general circulation of the atmosphere outside the planetary boundary layer remains unaffected by the changed conditions, the apparent potential evaporation (E_p) will be greater than the potential evaporation under potential conditions (E_{po}) by an amount corresponding the extra energy made available i.e.:

$$E_p = E_{po} + \frac{q}{L} \qquad (39)$$

Combining the relationships given by equations (38) and (39) we obtain the final relationship for the actual evaporation under non-potential conditions:

$$E = 2 E_{po} - E_p \qquad (40)$$

which indicates that the actual evaporation (E) decreases as the apparent potential evapotranspiration (E_p) increases. For the application of this method it is necessary to estimate both the potential evaporation (E_p) from climatological data and the potential evaporation under potential conditions (E_{po}) for the area concerned.

There will be similar problems in regard to the estimation of groundwater outflow on the grid size scale. The approaches used will be intermediate between those used in catchment studies (Freeze and Cherry 1979) and those used in large scale water balances referred to in Section 2.2. above. On the catchment scale, the only form of parameterization which has been widely used is the use of catchment data to estimate the equivalent values for the basic parameters of transmissibility and storativity in groundwater reservoirs and to use more detailed measurements to study the variations of these parameters within a groundwater basin. There is room here for sensitivity studies which would lead the way to a soundly based parameterization.

Another problem of relevance at the grid scale of general circulation models is the hydrological characterization of various parts of the land surface. There are many ways in which one can attempt to delimit hydrological regions. Firstly, there are the classical methods used by geographers based either on the topography of the land surface or on long-term climatic averages of temperature and precipitation.

Of special relevance in this connection is the radiation index of dryness of Budyko (1977) which is defined as the ratio of

potential evaporation to precipitation.   Secondly, there are the classifications based on the hydrologic behaviour of the region which include such factors as whether the origin of the runoff is snow or glacier drainage or rain, the division of the volume of infiltration between subsequent evaporation and delayed runoff etc. A dialogue between experts in general circulation modelling for climate purposes and hydrologic experts in large scale water balance studies could prove useful in this connection. If a method of classification of the land surface could be
developed which was acceptable from a hydrological viewpoint, it would enable climate modellers to differentiate the land surface into a small number of surface types whose differing hydrological behaviour would be significant for the general circulation of the atmosphere.

4. REFERENCES

Baumgartner A. and Reichel E (1975):
The world water balance
Elsevier. Amsterdam

Bear J, Zaslavsky D, and Irmay S. (1968)
Physical principles of water percolation and seepage
UNESCO. Paris.

Bear, J., (1972)
Dynamics of Fluids in Porous Media
Elsevier, New York.

Betson, R.P. (1964)
What is watershed runoff?
Journal of Geophysical Research Vol. 69 No. 8 pp. 1541-1551

Bouma J., de Laat P.J.M., A water
R.H.C.M., van Heesen H.C., van Holst
A.F. and van de Nes Th. J. (1980)
Use of soil survey date in a model for simulating regional
soil moisture regimes
Soil Sci. Soc. Am. J. vol 44 pp 808 -814

Bouchet, R.J. (1963)
Évapotranspiration, réele et potentielle, signification climatique
General Assembly of Berkeley
IASH Publication No. 62. pp. 134-142

Brutsaert, W. and Stricker, H. (1979)
An advection-aridity approach to estimate actual regional evapotranspiration
Water Resources Research vol 15, No 2, pp 443-450

Brutsaert W. (1980)
Vertical flux of moisture and heat at a bare soil surface
This volume.

Budydo, M.I. (1956)
Teplovoi balans zennoi poverkhnosti
Gidrometeoizdat. Leningrad.
Translated by N.A. Stepanova as Heat balance of the earth's surface
U.S. Weather Bureau. Washington. 1958 MGA 8. 5-20, 13E-386

Budyko, M.I. (1977)
Global'naja Ecologija
English translation by M. Shevtsov.
Progress Publishers. Moscow, 1980

Carson D.J. (1980)
Current parametrizations of land-surface processes in atmospheric
  general circulation models
This volume.

Childs, E.C. (1969)
An introduction to the physical basis of soil water phenomena.
Wiley. New York.

Coehlo, M.A. (1974)
Spatial variability of water related soil physical parameters,
Ph.D. dissertation, 110 pp., Univ. of Ariz., Tucson.
(Available as 75-11, 061 from Xerox University Microfilms, Ann Arbor, Mich).

Dooge, J.C.I. (1978)
General report on model structure and development
International Symposium on Logistics and Benefits of using mathematical
models of hydrological and water resource system.
Pisa. October 1978.

Dooge, J.C.I. (1979)
Alternative approaches to flow problems.
General Lecture to XVIII Congress of IAHR.
Cagliari. September 1979

Dunne, T. (1978)
Field studies of hillslope flow processes, in Hillslope Hydrology,
edited by M.J. Kirkby, pp. 227-293,
Wiley-Interscience, New York, 1978.

Eagleson, P.S. (1969)
Dynamic hydrology
McGraw Hill. New York.

Eagleson, P.S. (1978)
Climate, soil and vegetation. Introduction to water balance dynamics.
Water Resources Research Vol. 14, No. 5. (October 1978) pp. 705-776

Feddes R.A., Kowalik P.J., and
Zaradny H. (1978)
Simulation of field water use
and crop yield
PU-DOC, Wageningen, Netherlands

Fleming, George (1975)
Computer Simulation Techniques in Hydrology,
Americal Elsevier Publishing Company, New York.

Freeze R.A. and Cherry (1979)
Groundwater
Prentice Hall. New Jersey.

Freeze, R.A. (1980)
A Stochastic-Conceptual Analysis of Rainfall-Runoff Processes on
a Hillslope.
Dept. of Geological Sciences, University of British Columbia,
Vancouver B.C.

Gardner, W.R. (1958)
Some steady state solutions of the unsaturated moisture flow
equations with applications to evaporation from a water table.
Soil Science Vol. 85 pp. 228-232.

G.A.R.P. (1975)
The Physical Basis of Climate and Climate Modelling.
GARP Publications Series No. 16 - WMO. Geneva

Henderson F.M. (1969)
Open channel flow
Macmillan. New York

Hewlett, J.D. and Nutter, W.L. (1970)
Varying source area of streamflow from upland basins
Symposium on interdisciplinary aspects of water management
Montana State University, Bozeman, August 1970.

Hewlett, J.D. (1961)
Soil moisture as a source of base flow from steep mountain watersheds
US Department Agric. Forest Ser., Southeastern Forest Experiment
Station, Asheville, North Carolina, Station Paper No. 132, 11 pp.

Horton, R.E. (1933)
The role of infiltration in the hydrologic cycle
Transaction of the American Geophysical Union Vol 14 (1933)
pp. 446-460.

IAHS/UNESCO

Proceedings of symposium on World Water Balance (Reading 1970)
IAHS Publications No. 92 and 93

Keisling, T.C., Davidson, J.M., Weeks, D.L. and Morrison, R.D.,(1977)
Precision with which selected soil parameters can be estimated.
Soil Sci., 124: 241-248

Klute, A., and G.E. Wilkinson. (1958)
Some tests of the similar media concept of capillary flow, 1, Reduced
capillary conductivity and moisture characteristic data,
Soil Sci. Soc. Amer. Proc., 22, 278-281.

Kohler, M.A. (1963)
Rainfall-Runoff Models
Symposium on Surface Waters. General Assembly of Berkeley. IAHS
Publication No. 63. pp 479-491.

Miller, E.E., and R.D. Miller, (1956)
Physical theory for capillary flow phenomena,
J. Appl. Phys. 27, 324-332, 1956.

Monteith, J.L. (1965)
Evaporation and environment,
in The State and Movement of Water in Living Organisms,
edited by G.E. Fogg, pp. 205-234, Academic Press, New York,1965.

Morton, F.I. (1965)
Potential evaporation and river basin evaporation
Journal of Hydraulics Divison of ASCE. Vol.91. No. HY6, pp. 67-97.

Naef F. (1977)
Ein Vergleich van mathematischen
Niederschag - Abfluss - Modellen
Mitteilung No. 26 der Versuchsanstalt fur Wasserbau, Hydrologie und
Glaziologie. Zurich

Nielsen, D.R., J.W. Biggar, and K.T. Erh,
Spatial variability of field measured soil-water properties,
Hilgardia, 42(7), 215-260, 1973.

Peck, A.J., R.J. Luxmoore, and Janice L. Stolzy, (1977)
Effects of Spatial Variability of Soil Hydraulic Properties in
Water Budget Modelling, Water Resources Research, Vol 13 No.2
pp 348-354.

Penman, H.L. (1951)
Vegetation and Hydrology,
Technical Communication No. 53, Commonwealth Bureau of Soils, Harpenden.

Philip, J.R. (1967)
Sorption and infiltration in heterogeneous media,
Aust. J. Soil Res. 5, 1-10.

Philip, J.R. (1969)
Theory of infiltration, pp. 216-296 of
Advances in hydro science Vol 5.
Edited by Ven Te Chow.
Academic Press.

Philip, J.R. (1980)
Field heterogeneity : some basic issues
Water Resources Research, Vol 16, No.2. Pages 443-448, April 1980.

Rubin J. (1966)
Theory of rainfall uptake by soils initially drier than
their field capacities and its applications
Water Resources Research
Vol. 2 No.4 pp 739 -749.

Sharma, M.L. and Luxmoore, R.J., (1979)
Soil spatial variability and its consequences on simulated
water balance.
Water Resour. Res. Vol. 15 No.6 (December 1979) pp 1567 -1573

Sharma, M.S., Gander, G.A., and C.G. Hunt.
Spatial variability of infiltration in a watershed.
Journal of Hydrology, 45 (1980) 101-122.
Elsevier Scientific Publishing Company, Amsterdam.

USSR Committee for the IHD
Mirovoi Vodnyi Balanse vodnyi resursy semli
Translated into English as
World water balance and water resources of the earth
UNESCO. Paris. 1978.

Warrick, A.W., G.J. Mullen and D.R. Nielsen (1977)
Scaling Field-Measured Soil Hydraulic Properties Using a Similar
Media Concept Water Resources Research, Vol 13 No. 2.

Wilkinson, G.E., and A. Klute,
Some tests of the similar media concept of capillary flow, 11 Flow systems
data, Soil Sci. Soc. Amer. Proc. 23, 434-437, 1959.

W.M.O. (1975)
Intercomparison of conceptual models used in operational hydrological
forecasting.
Operational Report No. 7. W.M.O. Geneva.

DYNAMIC HYDRO-THERMAL BALANCES AT MACROSCALE

Peter S. Eagleson

Massachusetts Institute of Technology

Cambridge, Massachusetts U.S.A.

ABSTRACT

Atmospheric dynamics and energetics are influenced significantly by the coupled flux of moisture and heat across the land surfaces. In rainfed systems, the principal state variable controlling both fluxes is the near-surface concentration of soil moisture whose prediction is a critical problem in specifying the atmospheric boundary conditions. Soil moisture is commonly evaluated through closure of a conservation equation in a process called soil moisture accounting or water balance, and the methods used vary with the time and space scales of interest. In humid regions where the moisture supply is large, the system is driven by the energy supply, while in arid regions where energy is plentiful, the system is driven by the moisture supply. Here we describe the land surface coupling of the heat and moisture cycles for both rainfed and snowfed systems, and their expedient decoupling in the rainfed case to allow independent analysis of the dominant moisture cycle. We then review the techniques most applicable to water and heat balance evaluation at the large scales of primary concern in atmospheric general circulation models, and we suggest some important unsolved problems of parameterization at these scales.

INTRODUCTION

Precipitation on natural land surfaces is divided among evapotranspiration, runoff and moisture storage change according to the state of the surface. For rainfall, the moisture state of the near-surface soil controls the partition, while for snowfall, the thermal state of the snowpack governs.

For rainfall, the evapotranspiration component provides, through latent heat, the primary coupling between the surface heat and moisture budgets. By regulating the partition of net radiation into latent and sensible heats, the evapotranspiration exerts a thermostatic control over the thermal state of the surface and thereby establishes the moisture budget as the dominant member of the pair. The estimation of near-surface soil moisture concentration, at scales varying from that of a single catchment to that of an entire climatic region, is thus a critical problem in parameterizing the land surface boundary condition in atmospheric models.

Consider two climates in which the same annual rainfall is delivered in significantly different fashion. In the first case, the rain occurs in a few intense storms, and because the surface soil quickly becomes (temporarily) saturated during these events, most of the precipitation becomes surface runoff and does not contribute to long-term (i.e., "climatic") soil moisture levels. In the second case, the rain occurs in many less intense storms and a much larger fraction enters the soil. All else being constant, we expect larger average soil moisture concentration and thus, larger evapotranspiration and lower surface temperature in the latter case.

It is clear, therefore, that a realistic parameterization of both soil moisture and surface temperature must reflect the dynamics of moisture flux into and out of the near-surface soil. These fluxes are excited by alternating intervals of soil moisture demand (i.e., infiltration capacity) during rainfall, and of atmospheric moisture demand (i.e., potential evapotranspiration) between rainstorms. They are regulated by the availability of moisture, and by the physical properties of the soil and the vegetation.

For snowfall, the moisture dynamics are relatively unimportant. The heat stored in the snowpack is the critical element as it controls the losses of water mass due both to evaporation and to melt.

In discussing these problems, we must specify the time and space scales of interest. When dealing with fundamental physics, we will work at the "microscale" which we define as the lateral distance over which the atmospheric inputs and the system properties may be considered as homogeneous and the system behaves one-dimensionally. On the land-surface, the scale is on the order of 1 meter. For practical application of this physics to hydrologic problems, we work at the macroscale. This may be a small catchment (1 km) or an entire climatically-homogeneous region (10^2 km). "Short" times will refer to the times necessary to define the microscale physics, i.e., the characteristic times of these one-dimensional processes. This will vary from minutes for precipitation to months for soil moisture drainage. "Long" times are those necessary to obtain a "large" sample of the phenomenon of interest. Assuming stationary processes, this time scale may be 1 year when sampling humid climate rainstorms or 10^2 years when sampling annual quantities in arid climates.

The one-dimensional, microscale dynamics of these processes are well-known, but there are two major problems associated with their adaptation for use at macroscale:

1. Parameterizations of the processes which trade off physical fidelity in exchange for economy in data requirements (and in computation) for <u>homogeneous</u> systems of large areal extent.

2. Modification of these formulations to account for heterogeneities in both the climatic inputs and the physical parameters.

Although the difficulty of these problems has prompted alternative, empirical approaches to their solution, we choose the guidance and constraint offered by established physical laws. A summary of the one-dimensional microscale dynamics of these processes follows.

ONE-DIMENSIONAL MICROSCALE DYNAMICS

A. The Hydrologic Cycle

    1. Rainfed Systems

The hydrologic cycle is characterized in the short-term and at small-scale by a mixture of discontinuous moisture fluxes which are activated and deactivated by the achievement of certain saturation moisture levels.

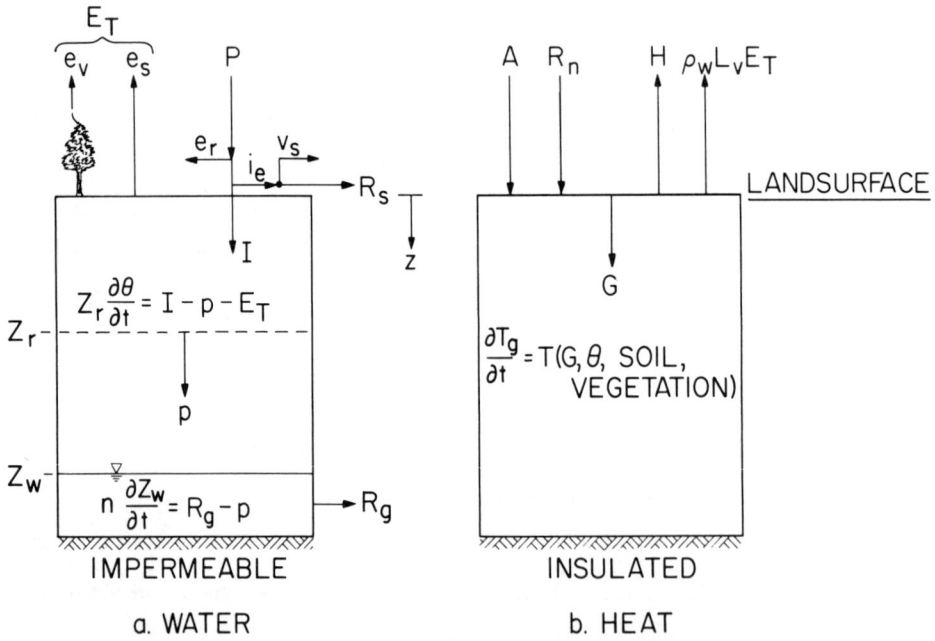

Figure 1

ONE-DIMENSIONAL FLUXES OF WATER AND HEAT AT A LANDSURFACE

Figure 1-a illustrates the instantaneous disposition of water mass at and below the landsurface in the absence of snow, ice or freezing temperatures.

Rainfall strikes the surface at rate P (cm/hr). There are initial withdrawals of water at the rate e_r = P as small depths are put into surface retention storage on the impermeable microsurfaces of both soil and vegetation, and are held there by interfacial surface "tensions." The surface retention capacity, on the order of 1 mm (perhaps greater on heavily vegetated surfaces), must be satisfied before water is available for the remainder of the process. It, therefore, constitutes a threshold or concentrated non-linearity in the system behavior. Water in surface retention is evaporated at the potential evaporation rate, E_p.

Rainfall exceeding the surface retention capacity is divided between infiltration, I, into the soil matrix and rainfall excess, i_e, according to the current moisture state, θ, of the near-surface soil.

After satisfaction of the surface retention capacity at the beginning of a rainstorm, the near-surface soil is often sufficiently dry that its capacity for infiltration exceeds the rainfall rate for a time. In this case, the actual infiltration rate equals the rainfall rate. With increasing time, however, the infiltrating water raises the near-surface soil moisture state which, in turn, modulates all soil moisture fluxes. Increasing soil moisture decreases the infiltration capacity until it equals the rainfall rate. At this time, called the ponding time, the surface soil is saturated, the generation of rainfall excess begins, and control of the infiltration process shifts to the soil. Here is another threshold. The varying relation between these rates is sketched in the upper half of Figure 2 for a random sequence of (idealized) rectangular rainfall intensity-pulses. The rainfall rate may be considered to be a potential infiltration rate.

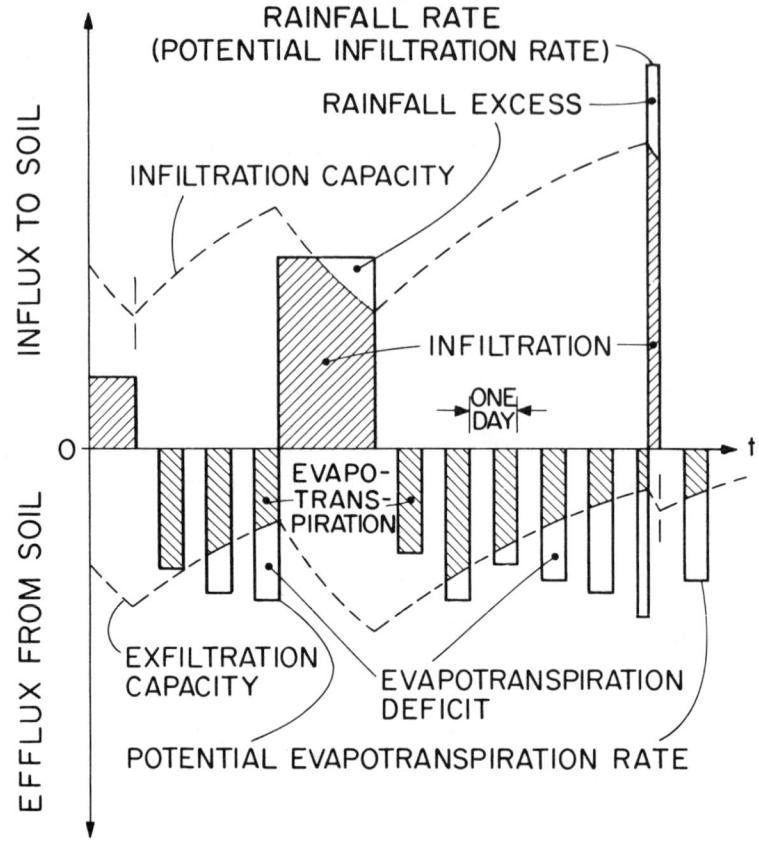

Figure 2

BARE SOIL SURFACE MOISTURE BOUNDARY CONDITIONS FOR
IDEALIZED STORM AND INTERSTORM PERIODS

(after Eagleson, 1978)

The rainfall excess must first fill the local soil moisture depressions whose capacity may be on the order of 1 cm depending on surface texture and slope. This <u>surface depression storage</u> occurs at rate v_s and constitutes another threshold in the system. Water in depression storage is either evaporated or infiltrated following cessation of rainfall.

When the depression storage is filled, the excess water moves laterally over the surface at rate R_s in a gravity-induced flow called surface runoff. This water is collected and concentrated by the stream channel network which conveys it to the ocean. Along the way, of course, there is evaporation from the water surfaces of the network, and the streamflow may be augmented by inflowing groundwater and/or diminished by further infiltration.

The increasing soil moistures which result from infiltration also increase the outflows from stored soil moisture. These outflows consist of downward percolation, p (largely gravitational), to the water table, evaporation, e_s, from the soil, and transpiration, e_v, by the plants.

At the end of a rainstorm, the near-surface soil will be relatively moist and thus the capacity of the soil to exfiltrate water to the drying surface is high. Often this early interstorm exfiltration capacity will exceed the current (vertical) atmospheric vapor transport capacity. In such a case, the latter rate, which hydrologists refer to as the potential evapotranspiration rate, E_p, governs the actual evaporation, e_s. That is, $e_s = E_p$. With continuing evapotranspiration, the near-surface soil moisture declines, thereby decreasing the exfiltration capacity until it equals the potential evapotranspiration rate. At this time, the surface is dry, an evapotranspiration deficit appears, and control of evaporation shifts to the soil. The appearance of an evapotranspiration deficit is significant from the standpoint of energetics since it marks the shifting of energy flux from latent heat to sensible heat with an accompanying rise in surface temperature. Here is still another threshold.

The vegetated fraction of the surface, M, contributes transpiration at a rate e_v that is also sensitive to the soil moisture concentration in the root zone. When the soil moisture exceeds some critical concentration, this rate will equal the maximum value e_{v_o}, which is a function of e_p and of the plant species. For soil moistures less than this critical value, the transpiration rate will be correspondingly reduced and the plant becomes stressed. To the first order, evapotranspiration from the composite surface is simply $E_T = (1 - M) e_s + M e_v$.

Alternate wetting and drying of the surface propagates a gravity-modulated soil moisture disturbance into the soil column. The effective penetration depth of this wave determines the soil moisture zone and hence the vegetation root depth. This is denoted by Z_r in Figure 1-a. It may vary from a few millimeters in fine-grained desert soils to several meters in extremely humid climates. Under average conditions, it is on the order of 1 meter.

We may now write a conservation equation for the water mass in this boundary layer. Neglecting infiltration from surface depression storage, we write

$$Z_r \frac{\partial \theta}{\partial t} = I - p - E_T \qquad (1)$$

where

θ = volumetric soil moisture content (i.e., volume of water/total volume)

and where the partial derivative is used as a reminder that the system is spatially variable.

Expanding I in terms of the soil column inputs and outputs, Eq. (1) has the alternative form

$$Z_r \frac{\partial \theta}{\partial t} = P - E_T - R_s - p - e_r - v_s \qquad (2)$$

The water percolating out the bottom of this boundary layer becomes an inflow to the saturated zone which overlies the impermeable bottom boundary of the soil column. A second conservation equation written for the saturated zone is

$$n \frac{\partial Z_w}{\partial t} = R_g - p \qquad (3)$$

where

Z_w = depth to water table

R_g = net groundwater runoff

n = porosity of the soil = volume of voids/total volume

2. Snowfed Systems

Snow falling on bare soil may melt, due to heat flow from the soil, and the resulting liquid water behaves as the rainfall just discussed.

When bare soil is cold or is insulated from the snow by leaves or compressed grass, the snow will accumulate at a density which increases with time as the growing pack compacts.

With a pack at sub-freezing temperatures, incoming rainfall will infiltrate into the snow, losing heat to the snowpack by conduction until the percolating water freezes and gives up its latent heat of fusion also. In a similar fashion, warm atmospheric temperatures may produce surface melting. This heat of fusion is then advected downward by the percolating melt water which refreezes at a lower depth. Other heat exchange processes at both the soil and atmospheric surfaces may contribute to snowpack warming (or cooling, of course).

When the whole pack has been raised to the freezing point, additional melt will be held within the pack by capillary forces until this <u>liquid-water-holding capacity</u> is filled. At this time, the pack is considered <u>ripe</u>. Further melt percolates to the soil surface where it may 1) infiltrate if the sum of the various capillary and gravity forces is downward, 2) accumulate until this condition occurs, or 3) run off across the surface.

For snowfall at rate P_s on a frozen surface (a common situation), the moisture fluxes are as sketched in Figure 3-a. Here the snowpack mass (per unit of surface area) is M_s, and the densities of the snowpack and liquid water are ρ_s and ρ_w, respectively. In this case, the mass conservation is simply

$$\frac{\partial M_s}{\partial t} = \rho_s P_s - \rho_w (E_T + R_s) \qquad (4)$$

In this case, because of the low temperatures, the sublimation rate, E_T, is often small, and here the existence of surface runoff, R_s, demands a snowpack temperature, $T_p > 0°C$.

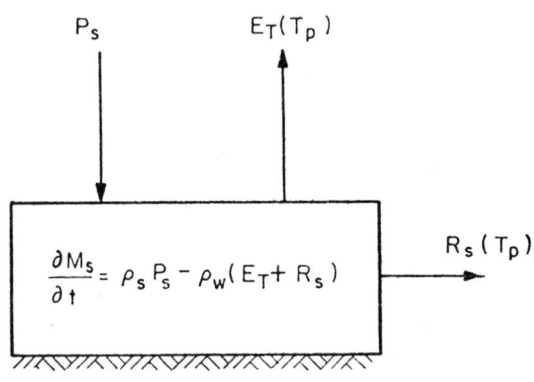

a. WATER SUBSTANCE

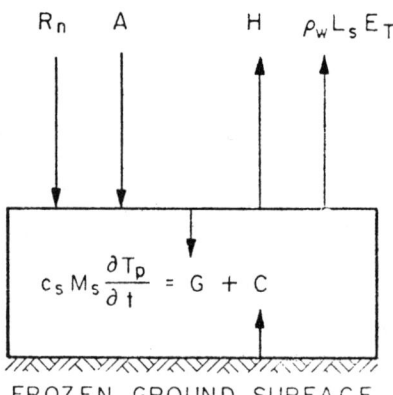

b. HEAT

Figure 3

ONE-DIMENSIONAL FLUXES OF WATER SUBSTANCE AND HEAT IN A SNOWPACK

B. The Thermal Energy Balance

The fluxes of thermal energy are shown in Fig. 1-b for the rainfed system and in Figure 3-b for the case of snow.

In these sketches, R_n is the net influx of radiant energy (shortwave and longwave), A is the net flux of advected energy by rainfall and runoff, H is the sensible heat flux, and L_v and L_s are the latent heats of vaporization and sublimation, respectively. The net surface exchange with the atmosphere determines the rate, G, at which heat is conducted downward. That is

$$G = R_n + A - H - \rho_w L E_T \tag{5}$$

This may be used in either a heat flux or a bulk energy conservation relation to express the surface temperature in the functional form

$$\frac{\partial T}{\partial t} = T(G, \theta, \text{soil}, \text{vegetation}) \tag{6}$$

Formulation of the function, T( ), will be discussed later.

For a snowpack (Figure 3-b), A must include the latent heat advected by incoming <u>rainfall</u> or by outgoing meltwater, and we must add the flux, C, of heat into the pack from the underlying ground. In this case, assuming an isothermal snowpack, energy conservation gives simply

$$G + C = \begin{cases} c_s M_s \dfrac{\partial T_p}{\partial t}, & T_p < 0°C \\ \\ L_f [R_s + m_s], & T_p = 0°C \end{cases} \tag{7}$$

where c_s and M_s are the snowpack specific heat and mass, respectively, L_f is the latent heat of fusion, and m_s is the change of storage of meltwater within the pack.

C. Coupled Heat and Moisture Exchange

The <u>primary</u> couplings of the vertical heat and moisture fluxes at a landsurface and a snowpack are illustrated by the solid arrows of Figures 4 and 5, respectively. In these figures, the "independent" atmospheric variables are the precipitation rate, P; the cloud cover, N; the atmospheric temperature, T_a; the specific humidity, q_a; the wind speed, W_s; and the atmospheric density, ρ_a. In both cases, the atmospheric vapor transport capacity (i.e., the potential rate of evapotranspiration), E_p, is given its "aerodynamic", "bulk", or "diffusion" formulation which is (Eagleson, 1970, pg. 216)

$$E_p = C_s \frac{\rho_a}{\rho_w} |W_s| [q^*(T_g) - q_a], \quad \text{cm/hr} \qquad (8)$$

where $q^*(T_g)$ is the saturated specific humidity of the air at surface temperature and q_a is the actual (i.e., ambient) specific humidity at screen height, and C_s is a coefficient related to the surface roughness and to the thermal stability of the atmosphere.

Also indicated on these figures, as dashed lines, are the primary <u>external</u> feedbacks from the landsurface fluxes to the atmospheric state variables. These are important issues at the macroscale where distributed surface injection of water vapor (i.e., evaporation) and of energy (i.e., sensible heat) into a developing boundary layer affects the downwind values of atmospheric humidity and temperature, respectively.

As was mentioned earlier, the primary coupling of the heat and moisture systems results from the dependence of the latent heat flux upon the evapotranspiration rate. A secondary tie results from the dependence of soil or snowpack heat capacity upon its moisture content. Feedback <u>from</u> the heat balance <u>to</u> the moisture balance exists through the dependence of the limiting, potential evapotranspiration rate upon surface temperature.

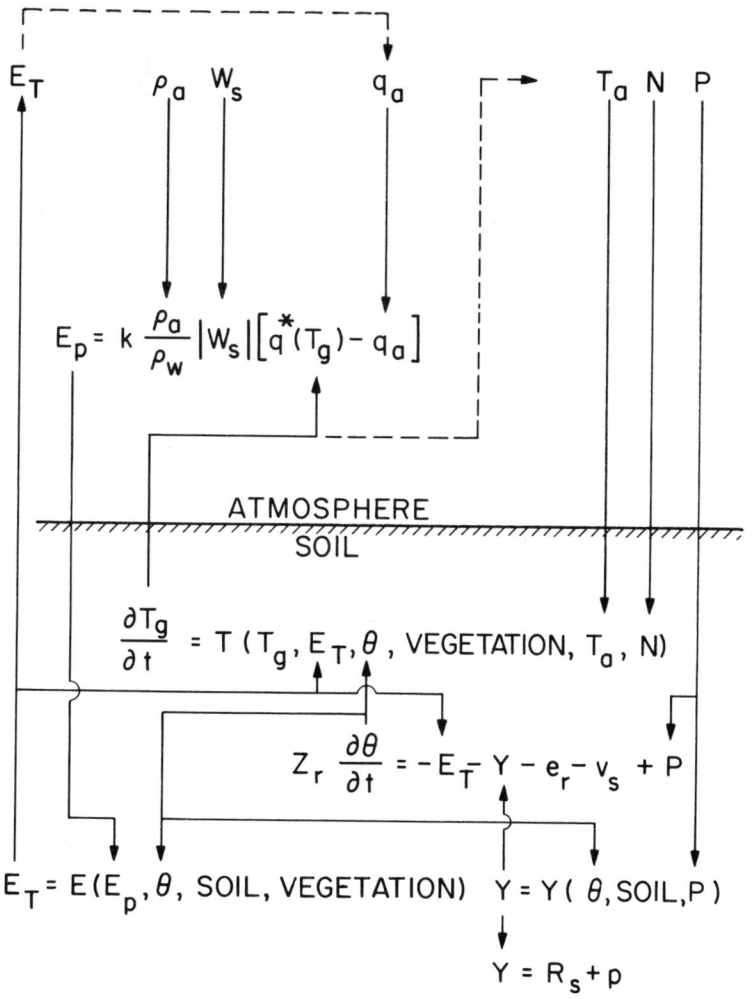

Figure 4

COUPLED HEAT AND MOISTURE EXCHANGE AT A LANDSURFACE

D. **Simplification of the Coupled Hydro-thermal System**

Economical parameterization of the systems illustrated in Figures 4 and 5 calls for a reduction in their complexity. As a guide in this, it will be helpful to consider the long-term time-averaged behavior of the one-dimensional, rainfed system. We use overbars to designate the time averages of the heat and water balance elements.

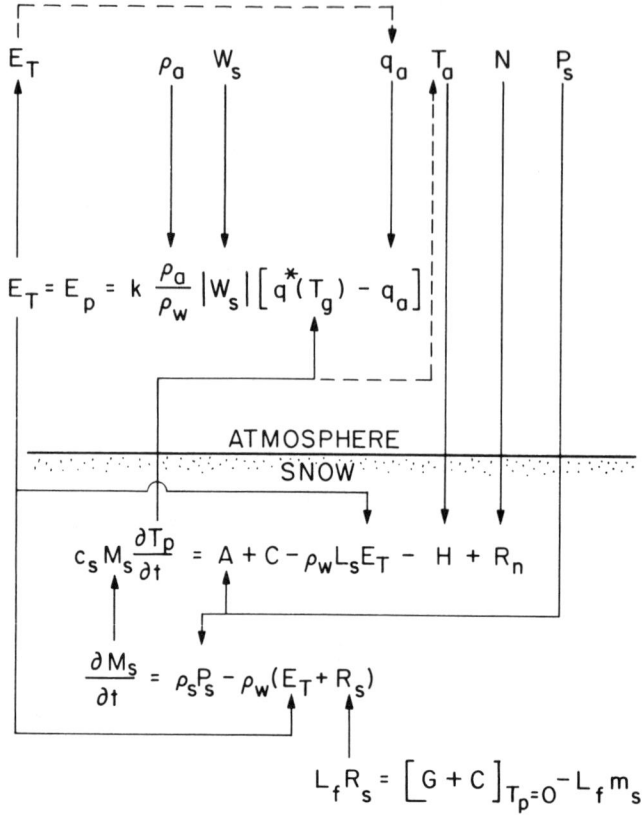

Figure 5

COUPLED HEAT AND MOISTURE EXCHANGE AT A SNOW SURFACE

Time averaging Eqs. (2) and (3) over a large number of full years causes the soil moisture and water table derivatives to vanish (for stationary systems) and assigns the surface storage terms to evaporation, their approximate ultimate fate. The resulting long-term average conservation relation is called the "average annual water balance" equation and is written

$$\overline{P}_A = \overline{E}_{T_A} + \overline{Y}_A \qquad (9)$$

where

$$\bar{Y}_A = \bar{R}_{S_A} + \bar{R}_{g_A} \qquad (10)$$

and

P_A = annual precipitation, (mm)

E_{T_A} = annual total evapotranspiration

R_{S_A} = annual surface runoff

R_{g_A} = annual groundwater runoff

Y_A = annual water yield (i.e., total runoff) = $R_{S_A} + P_A$

The general behavior of the average annual one-dimensional water balance is sketched in Figure 6 for a wide range of average annual precipitation. The overbars have been dropped here for the sake of simplicity, and the superscripts indicate normalization of all terms by the average annual potential evapotranspiration, $\bar{E}_{P_A}$. When the water supply is unlimited (large P_A^o), we expect the evapotranspiration component to approach its potential value. In this humid extreme, therefore, partition of the precipitation into the water balance components of Eq. (9) is independent of the soil moisture dynamics and is thus governed by the energetics of the system. At the other extreme, where P_A^o becomes very small, evapotranspiration will be limited by the ability of the soil and vegetation to deliver water to the surface. In this arid extreme, the terms on the right hand side of Eq. (9) will be controlled by soil moisture dynamics and will be independent of the energetics.

In the case of snowpack, the supply of moisture for sublimation is again unlimited so the heat equation controls the water balance.

These limiting behaviors provide a guide to simplification of the coupled hydro-thermal system:

To define the water balance components at anything but the climatic extremes, we must retain both the soil moisture dynamics and the energy fluxes.

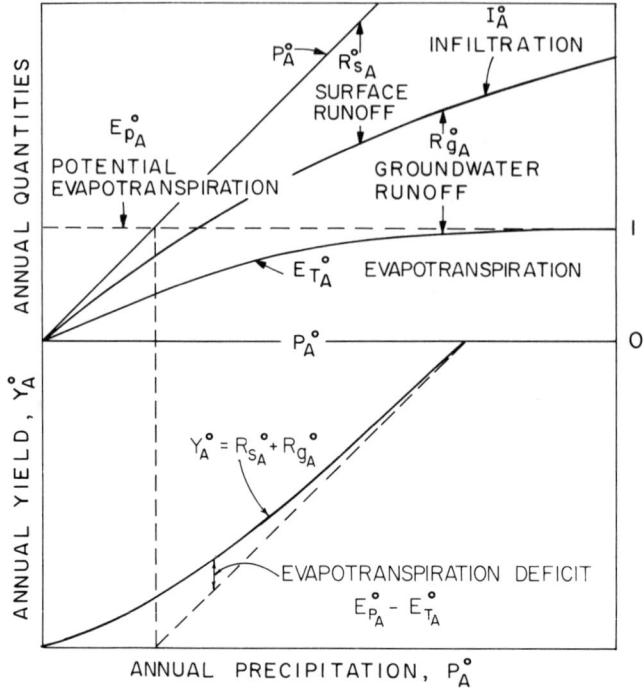

Figure 6

GENERALIZED AVERAGE ANNUAL ONE-DIMENSIONAL WATER BALANCE (Eagleson, 1978)
(all terms normalized by $\bar{E}_{P_A}$)

However, we can cut the feedback path from the energetics to the water balance in the rainfed case, by writing the (short-term) one-dimensional potential evapotranspiration in a form which is <u>largely</u> independent of the surface temperature. This was done first by Penman (1948) with later improvements by Tanner and Pelton (1960), and others. Known as the "combination method," it combines an energy balance with aerodynamic transfers of sensible heat and water vapor to obtain

$$\rho_w L_v E_p = \frac{\Delta(R_n - G) + \rho_a L_v [q^*(T_a) - q_a]\, \gamma/r_a}{\Delta + \gamma} \qquad (11)$$

where

E_p = atmospheric vapor transport capacity, cm hr^{-1}

R_n = net radiation, langleys hr^{-1}

Δ/γ = tabulated function of T_a (Van Bavel, 1966)

Δ = slope of the saturated vapor pressure-temperature curve at $T = T_a$

γ = psychometric constant (Monteith, 1973)

r_a = atmospheric diffusion resistance

$$= \frac{\ln^2\left(\frac{z_a - d}{z_o - d}\right)}{k^2 |W_s|}, \quad \text{hr cm}^{-1} \tag{12}$$

d = elevation of zero windspeed, i.e., "zero plane displacement"

k = Kármán "constant" = 0.4

z_a = elevation of atmospheric measurements (usually 2 m screen elev.)

z_o = surface roughness dimension

Equation (11) contains two primary assumptions (Van Bavel, 1966):

1. the vertical divergence of the sensible heat and water vapor fluxes between $z = d$ and $z = z_a$ is negligible, and

2. the turbulent transfer coefficients for sensible heat, momentum and water vapor are equal.

Eq. (11) has been verified in a series of careful experiments by Van Bavel (1966).

The low impedance of the evaporative heat transfer path from a wet surface makes it reasonable to neglect the residual rate of heat storage in the soil, G (Tanner and Pelton, 1960). If, in addition, the terrestrial radiative portion of R_n is expressed (approximately) in terms of T_a rather than T_g (Linacre, 1968), Eq. (11) depends primarily upon observations at screen height, and <u>for saturation of the given surface</u>, E_p becomes effectively an "independent" atmospheric variable. The hydro-thermal balance for rainfed systems reduces then to solution of the water balance equation for the soil moisture state in the fashion illustrated in Figure 7. The bypassed energy equation (Eq. 6) can then be entered with θ and E_T to evaluate T_g and the other energy flux terms if this information is desired.

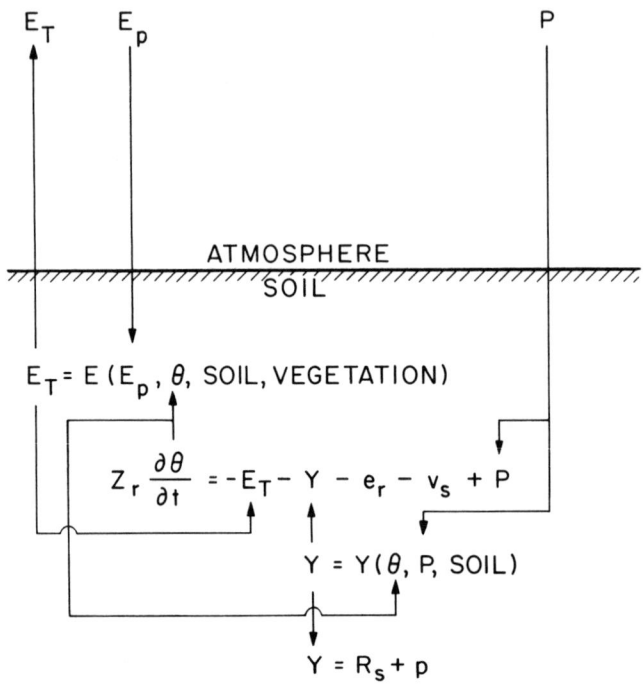

Figure 7

DECOUPLED HEAT AND MOISTURE EXCHANGE AT A LANDSURFACE

The presence of snow assures an unlimited supply of moisture for evaporation (i.e., sublimation), thus the snowfed system behaves as a humid one and is under control of the energy balance. Since we are retaining the energy equation, we must use the aerodynamic expression for $E_p \equiv E_T$. This system cannot be decoupled and remains as shown in Fig. 5.

MACROSCALE PARAMETERIZATIONS

A. The Parameterization Problem

Time variations in the atmospheric inputs of Eqs. (2), (6) and (7) are predominant at the scale of one day, due to diurnal radiation variations and to storm precipitation; and at the scale of six months, due to seasonal modulations of these quantities. The spatial scales of interest must, therefore, include not only the GCM grid scale at which the atmospheric system is defined, but also the sub-grid scale of the precipitation-related hydrologic system.

There are two primary problems related to representation of the landsurface hydro-thermal system at these scales:

1. Formulation of the elements of Eqs. (2), (6) and (7) for homogeneous systems of large areal extent, and

2. Modifications of these formulations to account for non-homogeneities in both the inputs (and thus the outputs) of the system, and in the system parameters.

The second of these problems is largely unsolved. In practice, the non-homogeneities are averaged to determine average or "effective" values. These average values are then used in the microscale dynamics to obtain a "lumped" representation of the macroscale system. While this may be adequate for the system parameters, it is unsatisfactory when applied to the input variables, precipitation in particular. The dependence of the hydro-thermal response upon local soil moisture concentration demands an input parameterization which copes with the spatial variability.

We will concentrate here on reviewing progress with the first of these problems, formulation of the hydro-thermal functions for homogeneous systems.

As represented one-dimensionally in Figs. 5 and 7, the hydro-thermal system has six basic elements which must be parameterized:

1. Atmospheric vapor transport capacity, E_p - the one-dimensional, energy-based Eq. (11) must be modified to incorporate feedback within the atmospheric boundary layer which makes local E_p a function of the extent of the upwind evaporating surface.

2. Evapotranspiration rate, E_T - a dynamic equation of soil moisture flux as governed by the soil moisture state and the physical properties of the soil and vegetation, and as limited by the atmospheric vapor transport capacity. Functionally, this is

$$\beta = \frac{E_T}{E_p} = \begin{cases} \beta(\theta, E_p, t; \text{soil vegetation}), & \beta < 1 \\ 1, & \text{otherwise} \end{cases} \quad (13)$$

in which β is commonly referred to as the "evaporation efficiency."

3. Water yield rate, y - dynamic equations of soil moisture flux as governed by the soil moisture state and the physical properties of the soil, and as limited by the rainfall rate. Functionally, this is

$$Y \equiv R_s + p = R_s + R_g - n \frac{\partial Z_w}{\partial t} = Y(\theta, P, t; \text{soil}) \quad (14)$$

4. Surface temperature, T_g - a dynamic equation for the time rate of surface temperature change due to heat storage.

5. Surface water losses, e_r and v_s - empirical relations giving the surface retention and detention as functions of the character of the surface. While the single-surface storage capacities are small, the multiple surfaces presented by forest canopies, and the depth and time distribution of storm rainfall may make these important water balance elements in certain climates.

6. Soil moisture layer thickness, Z_r - this important determinant of both thermal and hydrologic inertia of the landsurface is commonly taken to be one meter but actually it varies with both the soil properties and the statistics of the rainstorm arrivals.

These standard components have been combined by many investigators in as many ways to build complete models of the water and heat balances. The important classes of these complete models will be summarized at the end of this paper, but detailed attention will now be given to the various methods of representing these six elements.

B. Atmospheric Vapor Transport Capacity

The concept of potential evapotranspiration has been brought into this problem to provide an upper limit to the evapotranspiration which would occur from a particular surface under a given set of meteorological conditions if only one change is made, i.e., the water supply is made unlimited and the surface albedo is given the associated, "wet" value (Tanner and Fuchs, 1968). This "conditional" potential is useful for point evaporation calculations provided, in the later case, that the atmospheric observations used are suitably averaged in the direction of boundary layer growth. This is because the supply of water to a dry surface will cause progressive atmospheric temperature and humidity changes in the downwind direction (Morton, 1965). When only point (i.e., station) atmospheric observations are available but areal evaporation estimates are desired, a correction to Eq. (11) is needed which accounts for this so-called "ventilation" or "advection" effect.

Bouchet (1963) appears to have been the first to formulate this problem, followed by Morton (1965, 1971), Priestley and Taylor (1972) and Brutsaert and Stricker (1979), among others. The reader is referred to the last of these references for a summary of the developments. The basis of the approach is Penman's (1948) original equation (Eq. 11) in its two-term form as first written by Slatyer and McIlroy (1961)

$$\rho_w L_v E_p = \frac{\Delta(R_n - G)}{\Delta + \gamma} + \frac{\gamma E_a}{\Delta + \gamma} \qquad (15)$$

where E_a is called (Brutsaert and Stricker, 1979) the "drying power" of the air. From Eq. (11), it is defined as

$$E_a \equiv \frac{\rho_a L_v}{r_a} [q^*(T_a) - q_a] \qquad (16)$$

Because of the proportionality between vapor pressure deficit and wet bulb depression, this second term may also be interpreted (Monteith, 1973) as the rate at which the atmosphere supplies heat to the wet ground, in analogy with the behavior of a psychrometer. For a wet surface of infinite extent, advection of evaporated moisture and sensible heat will cause E_a to decrease in the downwind direction.

The first term of Eq. (15) is often assumed to be the minimum potential evapotranspiration from a finite area and is sometimes called the "equilibrium" value (Slatyer and McIlory, 1961). Priestley and Taylor (1972) found empirically, however, that for both ocean and saturated land surfaces this limiting value is

$$E_{p_o} = \alpha \frac{\Delta(R_n - G)}{\Delta + \gamma} \qquad (17)$$

According to an analysis of available data carried out by Brutsaert and Stricker (1979), $\alpha = 1.28$, which suggested to them that a condition of negligible advection may never be reached in nature.

This suggests that the potential evapotranspiration rate has a maximum value at the "leading edge" (i.e., $x = 0$) of a uniform region where advection is maximum, and that it declines with distance downwind as advection decreases, approaching the value given by Eq. (17) in the limit. The value of E_p at $x = 0$ will depend upon the local drying power of the atmosphere, and the rate of decay of E_p with distance will be governed by the growth of the thermal and vapor boundary layers due to continuous surface injection of moisture at rate, E_T, and of sensible heat at rate H.

McNaughton (1976-a,b), in a formulation of this boundary layer problem, writes Eq. (15)

$$E_p(x) = \frac{\Delta(\tilde{R}_n - \tilde{G})}{\Delta + \gamma} + \frac{\gamma}{\Delta + \gamma} E_a(x) \qquad (18)$$

where the overbars signify values calculated at the areal mean surface temperature and where $E_a(x)$ is again the advective enhancement term. For evaporation from a wet surface with a fully-mixed atmosphere beneath a 1000 m inversion, McNaughton (1976-b) finds $E_a(x)$ to decrease exponentially with a characteristic decay length of 10 km.

From this work, it appears that Eq. (17) should be considered to be only a quasi-equilibrium value which is "verified" within the limited accuracy of field observations due to the relative dominance of the first term when Eq. (18) is integrated over large, wet surfaces. Evidently more work is needed on this subject.

C. Evapotranspiration Rate

When the soil is unable to deliver moisture to the surface at the rate, E_p, demanded by the atmosphere, evaporation comes under soil control and takes place at a rate, E_T, that is less than E_p and is governed by the soil moisture state, as well as by the physical properties of the soil and of the vegetation. It is this dependence of the moisture flux upon the moisture state which makes the hydro-thermal balance a dynamic phenomenon. The relationship

$$E_T = E(E_p, \theta; \text{soil, vegetation}) \qquad (19)$$

is the primary element of any water balance, and the literature is full of attempts at its generalization. Because of the singular importance of this function, we will review its principal representations categorized first according to the method used for incorporating the moisture state, and second according to the time-scale of the evaporation estimate.

1. Empirical - (long-term average annual)

Since at least the beginning of this century, hydrologists concerned with river basin behavior have used observations of long-term average precipitation and streamflow to estimate basin evaporation by closure of the average water balance, Eq. (9). Investigators interested in generalized climatology have used the ratio $\overline{P}_A/\overline{R}_n$ as a measure of basin "wetness" to unify these observations in an empirical expression of Eq. (19).

Replacing $\overline{R}_n$ by its approximate equivalent, $\overline{E}_{P_A}$, (see Eq. 17) we have put these empirical relations in common format in Table 1 and Figure 8. In the former, the overbars signifying the long-term average nature of the annual values have been omitted for simplicity. Notice the similarity between these curves and the $E_{T_A}^o$ curve of Figure 6.

The ratio $\overline{E}_{T_A}/\overline{E}_{P_A}$ is again called the "evapotranspiration efficiency".

2. Divergence of Atmospheric Vapor Flux - (long-term average)

Although not related to soil moisture, another method of estimating long-term evapotranspiration by water balance closure deserves mention here. Working with the atmospheric branch of the hydrologic cycle in a vertical column, the conservation of water vapor is (Rasmussen, 1977)

$$\frac{\partial W}{\partial t} + \nabla \cdot \vec{Q} = E_T - P \qquad (20)$$

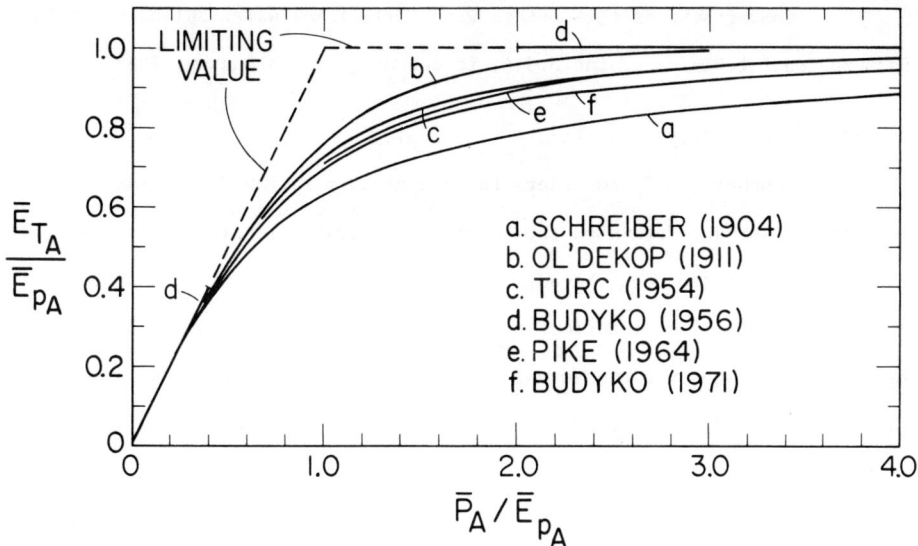

Figure 8

EMPIRICAL LONG-TERM AVERAGE ANNUAL REGIONAL
EVAPOTRANSPIRATION EFFICIENCY

where

W = total water vapor content of column (per unit area)

$\vec{Q}$ = total horizontal water vapor flux (per unit length of column)

Averaging in space over the area A_g (spatial averages are indicated by $< \quad >$), and in time (temporal averages are again indicated by overbars) gives the steady-state <u>regional</u> atmospheric water balance

$$\frac{1}{A_g} \int_c \overline{\vec{Q}} \cdot \vec{n} \, dc = <\overline{E_T - P}> \qquad (21)$$

where

$\vec{n}$ = eastward normal to boundary curve, c, of area, A_g

Atmospheric probes of <u>sufficient</u> density and frequency allow estimation of the left-hand side of Eq. (21). Associated surface precipitation observations then lead to $<\overline{E_T}>$ through water balance closure.

Comparisons of this method with terrestrial water balances for the Eastern Central Plains of the U.S.A. is encouraging (Rasmussen, 1968).

3. Advection-Aridity - (arbitrary time-scale)

Bouchet (1963) considers the evapotranspiration from a large homogeneous surface for which the leading edge transitional effects are neglected. He introduces the advective negative feedback of evaporation and sensible heat flux into regional potential evapotranspiration by reasoning that wherever the actual regional evaporation, $<E_T>$, is less than the limiting or "equilibrium" potential evapotranspiration, $<E_{P_o}>$, because of insufficient moisture supply, the sensible heat, which is equal to the difference $<E_{P_o}> - <E_T>$, becomes available for evaporation. This energy manifests itself in increased "drying power" of the air, which adds to the local potential for evapotranspiration. By this symmetry argument, we have

$$E_p - <E_{P_o}> = <E_{P_o}> - <E_T> \qquad (22)$$

As discussed by Morton (1965, 1968, 1971) and Solomon (1967), this suggests the variability sketched in Figure 9. The behavior of E_T in Figure 9 is qualitatively similar to that shown earlier in Figures 6 and 8.

If we choose a particular value, $<P_o>$, of the regional precipitation as indicated in Figure 9, we can reason the qualitative variation of E_T with distance downwind of the leading edge of a region. This is sketched in Figure 10. For large x, $E_p - E_T = 2A$ as was indicated in Figure 9. For small x, E_T will be greater due to larger E_p. More work is needed to define the scale and form of this spatial decay.

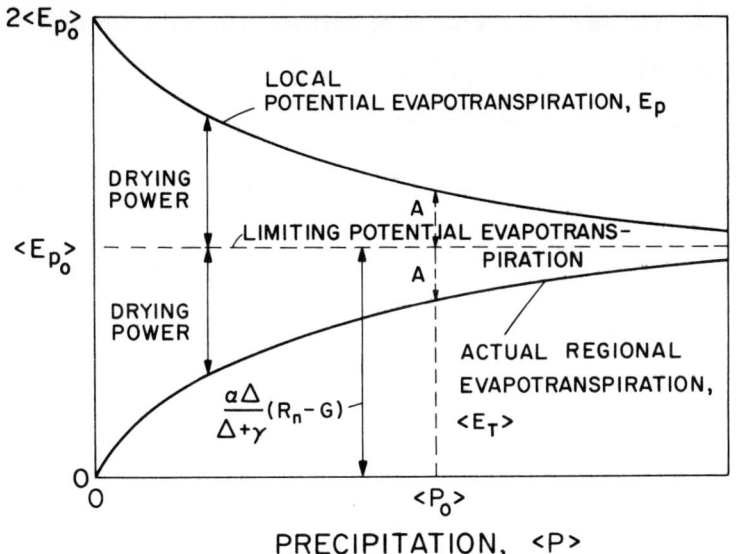

Figure 9

REGIONAL EVAPOTRANSPIRATION
(after Solomon, 1967)

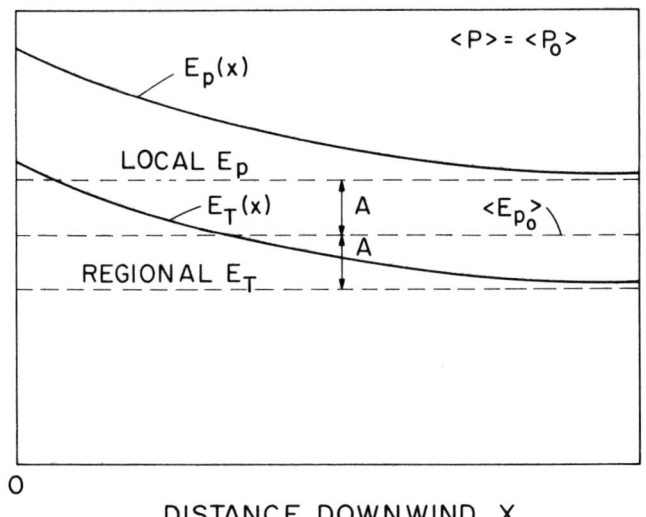

Figure 10

DOWNWIND VARIATION OF EVAPOTRANSPIRATION

We can use the above ideas to further illustrate the coupling of the regional heat and water balances. Dropping the areal average notation and referring to Eq. (2), we can write

$$P = S + E_T + Y \qquad (23)$$

where

$$S = \text{storage rate} = Z_r \frac{\partial \theta}{\partial t} + v_s + e_r$$

and

$$Y = \text{yield} = r_s + p$$

Under conditions where $\alpha\Delta/(\Delta+\gamma) \approx 1$ (i.e., $T_a \approx 30°C$ for $\alpha = 1.28$), Eq. (17) can be used to write Eq. (5) in the manner of Lettau (1969)

$$R_n - G \approx E_p = E_T + H \qquad (24)$$

Equations (23) and (24) are sketched jointly as a function of moisture availability in Figure 11. In this illustration, we see that the sensible heat flux, H, is to the energy balance what the yield, Y, is to the water balance, and the evapotranspiration function plays a common, dominant role in each balance.

Brutsaert and Stricker (1979) use Eqs. (15) and (17) in Eq. (22) to write

$$\frac{E_T}{E_p} = \frac{2\alpha(\overline{R}_n - \overline{G})}{\overline{R}_n - \overline{G} + \frac{\gamma}{\Delta} E_a} - 1 \qquad (25)$$

which they call the "advection-aridity" approach to regional evapotranspiration. Through comparisons with observations, they demonstrate that Eq. (25) is good for daily values and improves with the averaging period. They demonstrate its insensitivity to the exact value of α and to the form of the "wind function," E_a. They recommend (Penman, 1948)

$$E_a = 0.35(1 + 0.54 W_s)(e_a^* - e_a), \text{ mm d}^{-1} \quad (26)$$

where

W_s = 2 m wind speed, m sec^{-1}

e_a^*, e_a = 2 m vapor pressures, saturated and actual, mm Hg

There is need for additional work to define the x-dependence of E_a and of Δ/γ.

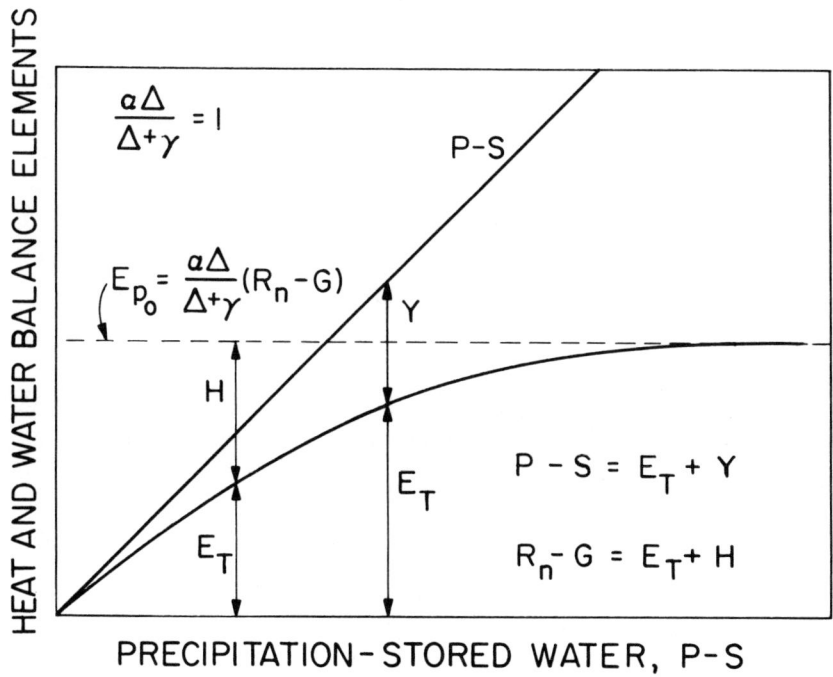

Figure 11

INTERRELATION OF REGIONAL HEAT AND WATER BALANCES

4. Soil Moisture Surrogate - (arbitrary time scale)

Slatyer and McIlroy (1961), Monteith (1965), and Barton (1979), all derive evapotranspiration equations based upon Eq. (11') which incorporate soil moisture indirectly through inclusion of a moisture-related surface parameter. These relationships are of the common form

$$\frac{E_T}{E_p} = \frac{1}{1 + \frac{\gamma}{\Delta+\gamma} B} \qquad (27)$$

where

$$B = K_a/K_s, \text{ Slatyer and McIlroy (1961)} \qquad (27\text{-a})$$

with

K_a, K_s = atmospheric and surface heat conductances,

and

$$B = r_c/r_a, \text{ Monteith (1965)} \qquad (27\text{-b})$$

with

r_a = atmospheric diffusion resistance

r_c = surface diffusion resistance

The definition of r_c (Monteith, 1965) relates the resistance, through a Fickian (i.e., linear) diffusion expression, to the evaporation rate, and to the difference between the vapor pressure at the leaf surface and the saturated value at surface temperature. The latter is assumed to represent conditions within the sub-stomatal cavities. This formulation is one of the first to suggest a means for explicit introduction of the evaporation rate of vegetation.

A third relationship is

$$B = (1 - S_s)/S_s, \text{ Barton (1979)} \qquad (27\text{-c})$$

with

S_s = saturation ratio at the evaporating surface.

From two data sets, one bare soil and one vegetated, Barton (1979) found the empirical relation

$$S_s = \begin{cases} 1.8\theta_w/(\theta_w + 30) &, \theta_w < 37.5 \\ 1 &, \text{otherwise} \end{cases} \qquad (28)$$

where

θ_w = soil moisture as a percent of dry weight

This form of evapotranspiration computation is not specifically designed for macroscale application but for such use requires estimation of areally-averaged values of the atmospheric temperature and the soil moisture surrogate, B. Furthermore, these relationships do not allow accounting for the difference between the albedos of the dry and the wet surfaces.

Tanner and Fuchs (1968) have further criticized the form of Eq. (27) since it implies a common temperature for the surface and the internal vapor source. They suggest instead

$$\frac{E_T}{E_p} = 1 + \frac{\gamma}{\Delta} - \frac{\rho c_p (e_s^* - e_a)}{\Delta r_a E_p} \qquad (29)$$

where E_p is expressed in units of energy flux density and

c_p = specific heat

e_s^* = saturation vapor pressure at surface temperature

e_a = vapor pressure at screen height

This does require knowledge of areal average surface temperature, however.

Van den Honert (1948) was apparently the first to suggest the expression of vegetal water use in the form of Ohms' law

$$E_T = \frac{\psi_s - \psi_r}{r_s} = \frac{\psi_r - \psi_\ell}{r_p} \qquad (30)$$

where r_s and r_p are the resistances (in seconds) of the soil and plant pathways, respectively, and ψ_s, ψ_r and ψ_ℓ are the water potentials in the soil matrix, at the root surface and in the leaf stomatal cavities, respectively. Cowan (1965) modified this relation to

$$E_T = \frac{\psi_s(\bar{\theta}) - \psi_r}{\alpha} Z_r K(\bar{\theta}) \qquad (31)$$

where $K(\bar{\theta})$ is the hydraulic conductivity of the soil at the average root zone moisture content, $\bar{\theta}$. The parameter α is related to root geometry and density (Cowan, 1965), while ψ_r involves the plant physiology.

Many investigators have sought general expressions for ψ_r in terms of observable plant characteristics. Prominent among these are Rijtema (1965) and Federer (1977) who related the leaf potential empirically to the canopy resistance, r_c, of Monteith (1965). This allowed Federer (1979) to combine the Monteith form of Eq. (27) with Eq. (31) to eliminate r_c and ψ_r and thus to find E_T as a function of soil moisture content.

In a one-dimensional statistical-dynamic model of the long-term average annual water balance (to be discussed more fully later), Eagleson (1978) used ecological optimality hypotheses to avoid the description of plant physiology. He hypothesized that __natural__ vegetation is in equilibrium with climate and soil at a canopy density (i.e., shaded surface fraction), M_o, which minimizes plant stress through maximization of soil moisture. He further hypothesized that such "optimum" natural systems will transpire, on the average, at their potential rate, $E_p = e_{v_o}$, where

$$e_{v_o} \equiv k_v E_{P_S} \tag{32}$$

in which E_{P_S} is the potential rate for a bare soil surface. Assuming the albedos of the vegetation and the wet bare soil to be the same, we can use Eq. (27) to write (Shuttleworth, 1979)

$$k_v = \text{"plant coefficient"} = \frac{1}{1 + \frac{\gamma}{\Delta+\gamma} B} \tag{33}$$

where $k_v \leq 1$. With these hypotheses, Eagleson (1978) finds that the average evapotranspiration efficiency, E_T/E_p, is a function primarily of the canopy density, M_o, and of k_v (which is species-related). Since the plant coefficient is believed by many to have the value unity for all species, this equilibrium hypothesis offers the possibility of estimating E_T/E_p for large natural areas solely through observation of vegetation canopy density, which is possible through remote sensing. The average evapotranspiration efficiency derived by Eagleson (1978) and modified by Tellers and Eagleson (1980) can be approximated with a polynomial in M_o as follows

$$\frac{E_T}{E_p} = \begin{cases} 0.11 + 2.22M_o - 1.87M_o^2 + 0.54M_o^3, & k_v = 1 \\ 0.11 + 1.25M_o + 0.27M_o^2 - 0.63M_o^3, & k_v = 0.7 \end{cases} \quad (33)$$

Eagleson and Tellers (1981) have compared Eqs. (33) with available observations, as shown in Figure 12. The results are encouraging.

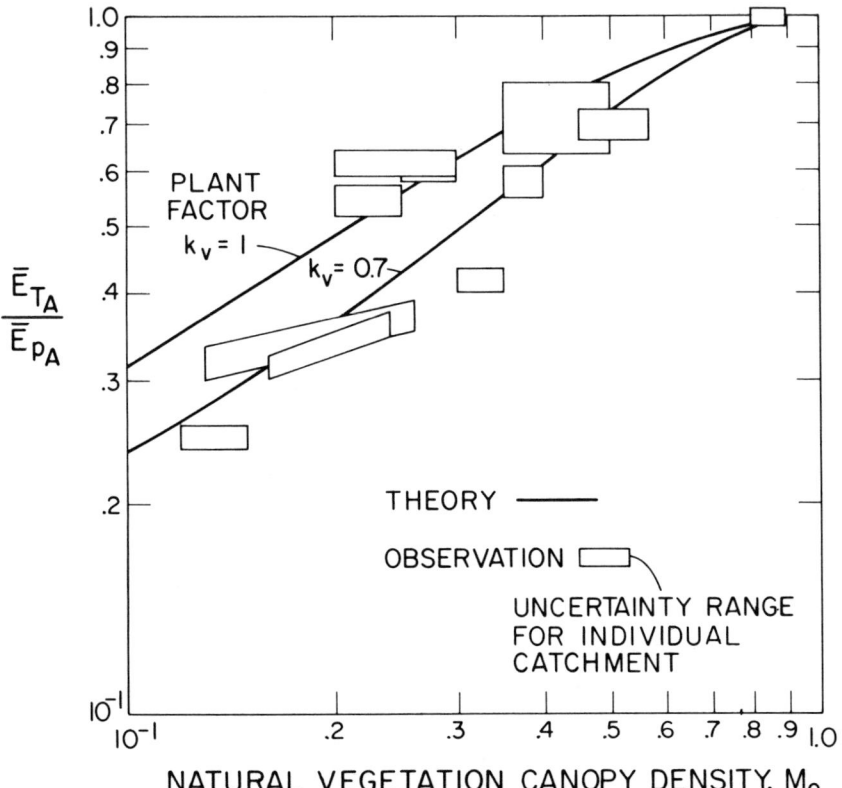

Figure 12

AVERAGE EVAPORATION EFFICIENCY FROM VEGETATION CANOPY DENSITY
(after Eagleson and Tellers, 1980)

5. Soil Moisture Accounting

   a. Short Term

The term soil moisture accounting is used to describe water balances at short time scales in which each flux on the right-hand side of Eq. (1) is expressed as a function of the soil moisture in layer Z_r. The water balance is closed in each time interval by solution of Eq. (1) for θ.

Thornthwaite and Mather (1955) appear to have been the first to carry out soil moisture accounting using a linear relationship between E_T and θ although such models are more often attributed to Budyko (1956). The Thornthwaite-Budyko model is sketched in Figure 13. It utilizes the concept of soil "field capacity", which is defined as that soil moisture, θ_{fc}, to which a saturated soil will drain by gravity. At field capacity, the upward forces of capillarity are assumed to just balance the downward force of gravity. It is assumed that

$$\frac{E_T}{E_p} = \begin{cases} 1 & , \quad \theta \geq \theta_{fc} \\ \theta/\theta_{fc} & , \quad 0 \leq \theta < \theta_{fc} \end{cases} \qquad (34)$$

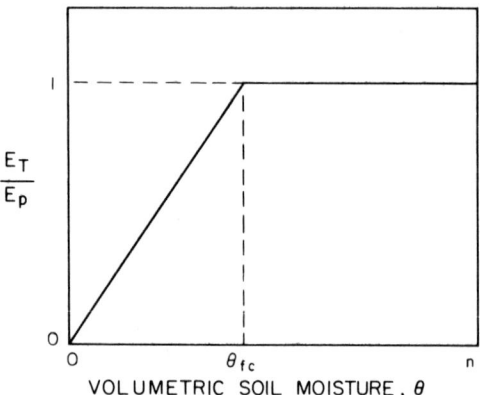

Figure 13

THORNTHWAITE-BUDYKO EVAPOTRANSPIRATION EFFICIENCY

Most of the current atmospheric general circulation models use a variant of this linearization (GARP, 1978), the major difference being an increase in the range over which evaporation occurs at the potential rate. That is

$$\frac{E_T}{E_p} = \begin{cases} 1 & \theta \geq k_\theta \, \theta_{fc} \\ \theta/k_\theta \, \theta_{fc} & 0 \leq \theta < k_\theta \, \theta_{fc} \end{cases} \qquad (35)$$

in which $k_\theta = 0.75$ (Manabe, 1969) or $k_\theta = 0.50$ (Arakawa, 1972; Gates and Schlesinger, 1977; Corby, et al., 1977).

Experimental evidence of this linear variation is presented by Novák (1980) who finds that the critical soil moisture, $k_\theta \, \theta_{fc}$, is a function of E_p.

Somerville, et al. (1976) assumed the saturated specific humidity at ground temperature to vary linearly with soil moisture and used Eq. (8) to write

$$\frac{E_T}{E_p} = \frac{(\theta/\theta_{fc}) - q_a/q^*(T_g)}{1 - q_a/q^*(T_g)} \qquad (36)$$

Other studies have shown (Philip, 1957; Hillel, 1971) that:

i. soil moisture decays essentially exponentially with time by gravity drainage to a lower limit that is governed by molecular surface forces rather than by capillarity. "Field capacity", therefore, has no physical significance and must be regarded solely as a convenient linearization parameter.

ii. the relationship between evapotranspiration and soil moisture is often highly non-linear (Lowry, 1959) depending upon the type of soil and on the relative fractions of bare soil and vegetated surface. Typical relationships are sketched in Figure 14. In this figure

   n = soil porosity = volume of pores/total volume

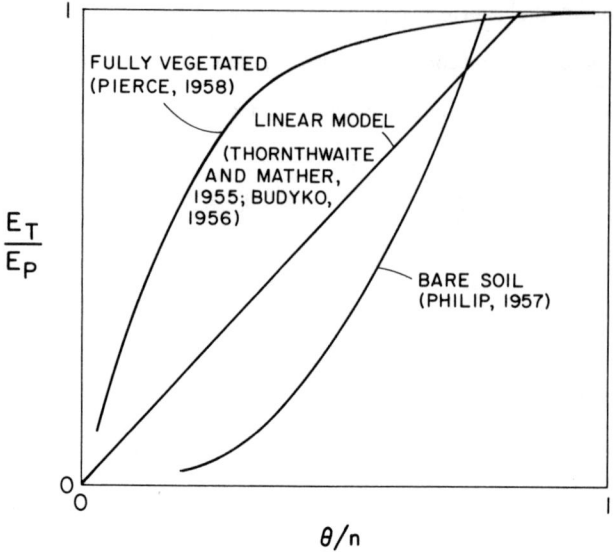

Figure 14

EVAPOTRANSPIRATION DEPENDENCE UPON SOIL MOISTURE
(after Lowry, 1959)

Other, non-linear evapotranspiration relations, based upon the work of Philip (1969) and others will be discussed in a later section dealing with complete moisture-accounting water balance models.

b. Long-term Average Annual

In a statistical-dynamic approach which will be described more fully later, Eagleson (1978) derived the long-term average annual evapotranspiration efficiency, $\overline{E}_{T_A}/\overline{E}_{P_A}$, as a function of the long-term average near-surface soil moisture, $\overline{\theta}$, and parameters of the local climate, soil and vegetation.

D. Water Yield Rate

1. Empirical - (long-term average annual)

Because of the general recognition that the way in which precipitation occurs is at least as important as the total amount in determining runoff, there are few empirical expressions for long-term yield of the type presented for long-term evapotranspiration in Table 1. Lettau (1969), however, does suggest

$$\frac{\overline{Y}_A}{\overline{P}_A} = 1 - \tanh \frac{\overline{E}_{P_A}}{\overline{P}_A} \qquad (37)$$

and others use observed precipitation and the empirical evapotranspiration expressions of Table 1 to estimate yield by closure of Eq. (9) (see Budyko, 1971, for example).

Table 1

LONG-TERM AVERAGE ANNUAL REGIONAL EVAPOTRANSPIRATION EFFICIENCY

Reference	Empirical Relation
(a) Schreiber (1904)	$\dfrac{E_{T_A}}{E_{P_A}} = \dfrac{P_A}{m_{P_A}} [1 - \exp(-E_{P_A}/P_A)]$
(b) Ol'dekop (1911)	$\dfrac{E_{T_A}}{E_{P_A}} = \tanh[P_A/E_{P_A}]$
(c) Turc (1954)	$\dfrac{E_{T_A}}{E_{P_A}} = \begin{cases} [1 + 0.9(E_{P_A}/P_A)^2]^{-1/2}, & \left(\dfrac{P_A}{E_{P_A}}\right)^2 \geq 0.1 \\ 1, & \text{otherwise} \end{cases}$
(d) Budyko (1956)	$\dfrac{E_{T_A}}{E_{P_A}} = \begin{cases} 1, & P_A/E_{P_A} \geq 2 \\ P_A/E_{P_A}, & P_A/E_{P_A} \leq 0.4 \\ \text{not specified}, & \text{otherwise} \end{cases}$
(e) Pike (1964)	$\dfrac{E_{T_A}}{E_{P_A}} = [1 + (E_{P_A}/P_A)^2]^{-1/2}$
(f) Budyko (1971)	$\dfrac{E_{T_A}}{E_{P_A}} = \left\{\dfrac{P_A}{E_{P_A}} \tanh\left(\dfrac{P_A}{E_{P_A}}\right) [1 - \exp(-E_{P_A}/P_A)]\right\}^{1/2}$

Long-term yield will more often be approximated observationally, at catchment-scale, by integration of the streamflow. It must be remembered, however, that this method assumes that all runoff, both surface and subsurface, exits the catchment as observable streamflow. This assumption fails at both extremes of wetness; in the arid case because the groundwater runoff does not enter the stream channel due to low water table, and in the very wet case because neither surface nor groundwater runoff is confined to the stream channels.

2. Soil Moisture Accounting

a. Short-term

Various empirical expressions have been proposed for evaluating the short-term yield as a function of soil moisture. Prominent among these are:

Budyko (1971)

$$\frac{Y}{P} = \frac{\theta}{n} \left\{ \mu^2 \left[1 - \left(\frac{E_p}{P}\right)^2 \right] + \left(1 - \frac{E_p}{P} \right)^2 \right\}, \quad P/E_p > 1$$

$$\frac{Y}{P} = \mu \frac{\theta}{n} \quad\quad\quad P/E_p \leq 1$$

(38)

where (Budyko, 1972), $Z_r = 1$ m, n = 0.15, and

$$\mu = \begin{cases} 0.2 & \text{latitude} > 45° \\ 0.4\text{-}0.8, & 0 \leq \text{latitude} \leq 45° \end{cases}$$

(39)

Arakawa (1972), n Z_r = 10 cm

$$\frac{Y}{P} = [1 + \left(\frac{D}{P}\right)^3]^{1/3} - \frac{D}{P}$$

(40)

where

$$D = (1 - \theta/n)\, \rho_w\, n\, Z_r,$$

(41)

Gates and Schlesinger (1977), $\theta_{fc} Z_r = 30$ cm

$$\frac{Y}{P} = \begin{cases} \frac{\theta}{2\theta f_c}(1 - E_T/P) & , \quad \theta/\theta_{fc} < 1 \\ \\ 1 - E_T/P & , \quad \theta/\theta_{fc} \geq 1 \end{cases} \qquad (42)$$

Shukla (1977), $\theta_{fc} Z_r = 10$ cm

$$\frac{Y}{P} = 0.1 + 0.9(\theta/\theta_{fc})^{1.5} \qquad (43)$$

Other moisture accounting methods, based upon physical principles have been incorporated in complete water balance models to be discussed later.

b. Long-term Average Annual

In a statistical-dynamic approach which will be described more fully later, Eagleson (1978) derived the long-term average annual yield efficiency, Y_A/P_A, as a function of the long-term average near-surface soil moisture, $\bar{\theta}$, and parameters of the local climate, soil and vegetation.

E. Surface Temperature - (short-term)

1. Rainfed Systems

Deardorff (1978) has summarized the primary methods used to calculate the landsurface temperature, T_g. The three of these most commonly used are now presented with reference to Figure 15.

    a   For an <u>insulated surface</u> (Manabe, et al., 1974), the surface heat flux $G \equiv 0$ and thus from Eq. (5)

$$R_n(T_g, t) - H(T_g, t) - \rho_w L_v e_T(T_g, t) = 0 \qquad (44)$$

which can be solved for T_g (in principle).

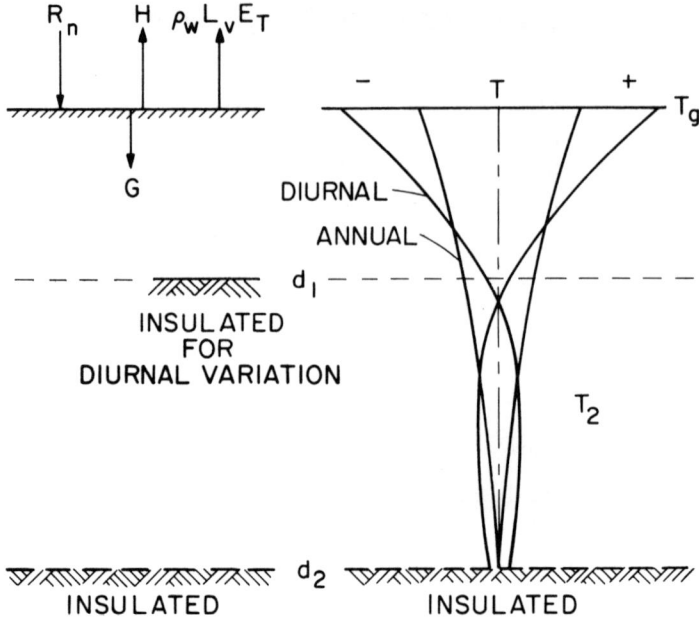

Figure 15

THE FORCE-RESTORE METHOD FOR SURFACE TEMPERATURE

b. For a <u>bottom-insulated single soil layer</u> (Arakawa, 1972)

$$\rho_s c_s d_1 \frac{\partial T_g}{\partial t} = \pi^{1/2} G(T_g, t) \tag{45}$$

where

$\rho_s = \rho_s(\theta)$ = density of soil-water system

$c_s = c_s(\theta)$ = specific heat of soil-water system

d_1 = depth of single soil layer

and the $\pi^{1/2}$ arises from imposition of a harmonic (i.e., diurnal) variation of G at the top of the bottom-insulated layer. Choosing d_1 where the amplitude of the temperature oscillation is e^{-1} times its surface value (Carslaw and Jaeger, 1959, p. 81)

$$d_1 = 0 \ (10 \text{ cm}) \quad (46)$$

$$= (\lambda_s \tau_1 / \rho_s c_s)^{1/2}$$

where

λ_s = thermal conductivity of soil-water system

τ_1 = one day

c. For a <u>bottom-insulated two-layer soil</u>, Bhumralkar (1975) and Blackadar (1979) present the "force-restore" method in which the "forced" system of Eq. (45) is restored by heat flux between the diurnal layer, d_1, and a bottom-insulated deep layer defined by the penetration, d_2, of the annual surface heat flux oscillation. This equation is

$$\frac{\partial T_g}{\partial t} = 2\pi^{1/2} G(T_g, t)/(\rho_s c_s d_1) - 2\pi(T_g - T_2)/\tau_1 \quad (47)$$

where

T_2 = temperature of second layer which is given by Eq. (47) in which $T_g = T_2$ and $d_1 = d_2$

Choosing d_2 in the same manner as d_1

$$d_2 = (365 \ \lambda_s \ \tau_1 / \rho_s \ c_s)^{1/2} = 0(10^2 \text{ cm}) \quad (48)$$

Deardorff (1978) demonstrates clearly the superiority of the force-restore method for this thermal application. In another paper (Deardorff, 1977), he applies the method to gravity-free soil moisture flux to determine the surface moisture concentration. He suggests expressing the bare soil evaporation rate in terms of this surface moisture rather than the customary bulk moisture content of a deeper soil layer.

2. Snowfed Systems

The "force" method (method b of the previous section) is also applicable to snowpack surface temperature determination assuming no heat transfer between soil and snow. This flux is normally small compared to the heat transfer at the air-snow interface. This method is not normally used, however. Instead (U.S. Army Corps of Engineers, 1956; Anderson, 1973), the snowpack is assumed to be isothermal as in method "a" of the previous section and the heat storage of the pack is followed by continuous accounting until the pack is isothermal at 0°C. Additional heat inputs cause melt and the declining liquid water storage capacity of the pack must be traced until filled. Beyond this time, additional heat inputs cause runoff.

F. Surface Water Losses - (arbitrary time scale)

The volume of moisture (rain and snow) which can be retained on natural surfaces by capillary forces (surface retention) or gravity forces (surface detention) is primarily a function of the texture of the surface.

Empirical relations are usually relied upon for estimation of these volumes. Collections of these empirical relations are given by Wigham (1973), Blake (1975) and others. A complex model of the interception process has been presented by Rutter, et al. (1972).

The actual water loss from these processes depends not only upon the supply of precipitation but also upon the interstorm opportunity for evaporation.

G. Thickness of Soil Moisture Layer - (arbitrary time scale)

In soil moisture or heat accounting, where the flux divergence is equated to storage change, it is necessary to define the size of the container in order to find the concentration of the stored substance. The majority of the heat and the dynamically-important moisture is concentrated in a layer near the surface. Using the Philip (1969) infiltration theory, Eagleson

(1978, pg. 727) has shown that the penetration depth of gravitationally-assisted capillary waves of soil moisture varies with the soil properties and with the average storm duration and spacing. Using typical values of these parameters, this capillary penetration depth is on the order of one meter. This dimension, larger in sandy soils, smaller in clays, is assumed to determine the root depth, Z_r, of the local natural vegetation.

We have already seen that, using wet soil thermal conductivities, the annual thermal penetration depth is also on the order of one meter. Because of the strong dependence of the thermal penetration on the soil moisture content, the two systems are closely coupled in this dimension.

MACROSCALE WATER BALANCE MODELS

A. Modeling Philosophy

Hydrologic models are often classified according to the degree to which process physics forms the basis for the model structure (Gupta, 1973). Three model categories are identified:

1. Empirical - where paired observations of input and output variables are related at the macroscale by some analytical means such as linear regression to obtain a prognostic relationship which is typically endowed with unwarranted generality.

2. Phenomenological - where only the primary behavioral characteristics of interest are represented (such as storage, flux, or random arrivals), by macroscopic conceptual elements. These elements focus on capturing only the appropriate storage time constant, flux path resistance or average arrival rate, rather than upon reproducing the full microscale physics.

3. Dynamic - where the one-dimensional microscale equations of fluid mechanics and thermodynamics are applied to spatial and temporal

idealizations of the real system. Such behavioral descriptions preserve generality and hence generate insight, but they are in terms of physical parameters that are observable only at the microscale. In extension to macroscale applications, the parameters and the system structure become "effective" in the sense that given the proper values they will explain the observed behavior, but they are no longer directly observable.

Most investigators will agree that scientific progress has carried us past category 1 except for expediency, and proponents of categories 2 and 3 argue their respective merits. However, the heterogeneity of most macroscale hydrologic systems rapidly draws the idealization of category 3 close to the conceptualization of category 2 and the boundary of these two categories is substantially blurred.

We will now review examples of the principal macroscale water balance models of these three categories. Only one of these models attempts to address the dynamics of vegetation water use so we will briefly treat the modeling of this process in a separate section. **A summary of many other models is presented by Fleming (1975).**

B. Empirical

    1. Thornthwaite

Thornthwaite and Mather (1955) were among the first to suggest means for computing the time variation (monthly or daily in their case) of the various components of the water balance. This they did by a continuous accounting of soil moisture in a single layer taking precipitation as the input. Outflow due to evapotranspiration was calculated according to Eq. (34), and that due to percolation was taken as a constant percentage of the soil moisture in the previous time interval. In this model, surface runoff can only occur when the entire soil layer is saturated. The primary virtue of this model is its simplicity, since as input it requires only atmospheric temperature and precipitation. The sole system parameters are the latitude, the field capacity of the soil (in depth units), the total porosity of the soil (also in depth units), and the percolation percentage.

This scheme has been prepared for machine computation by Willmott (1977), although its simplicity makes hand computation practical.

2. Lettau

Calling the process "evapo-climatonomy," Lettau (1969) combines Eqs. (2) and (5) with empirical assumptions about the dependence of runoff and evapotranspiration on soil moisture in order to arrive at monthly estimates of the latter and of the separate water balance components. Important features are his retention of system forcing by both the precipitation and the net radiation, and the use of only three system parameters. The assumptions leading to the critical evapotranspiration and runoff functions seem fairly arbitrary. However, they seem to lead to a reasonable reconciliation of both the heat and the moisture balances of large areas (Lettau, et al., 1979; Hare, 1980).

C. Phenomenological

1. Linsley

The Stanford Watershed Model of Crawford and Linsley (Linsley and Crawford, 1960; Crawford and Linsley, 1966) forms the basis for most digital simulations of streamflow in the U.S.A. and in many other nations as well. A flow chart of an early version of this model is reproduced here as Figure 16. The model uses two soil moisture zones, a shallow upper zone which reflects the rapidly changing moisture content of the surface soil that governs surface depression storage, and a thicker lower zone that provides more slowly acting regulation of infiltration, evapotranspiration and percolation to groundwater. These critical fluxes are described in terms of empirical indices that are keyed to assumed distributions of the spatial variability of soil capacity for infiltration and evaporation.

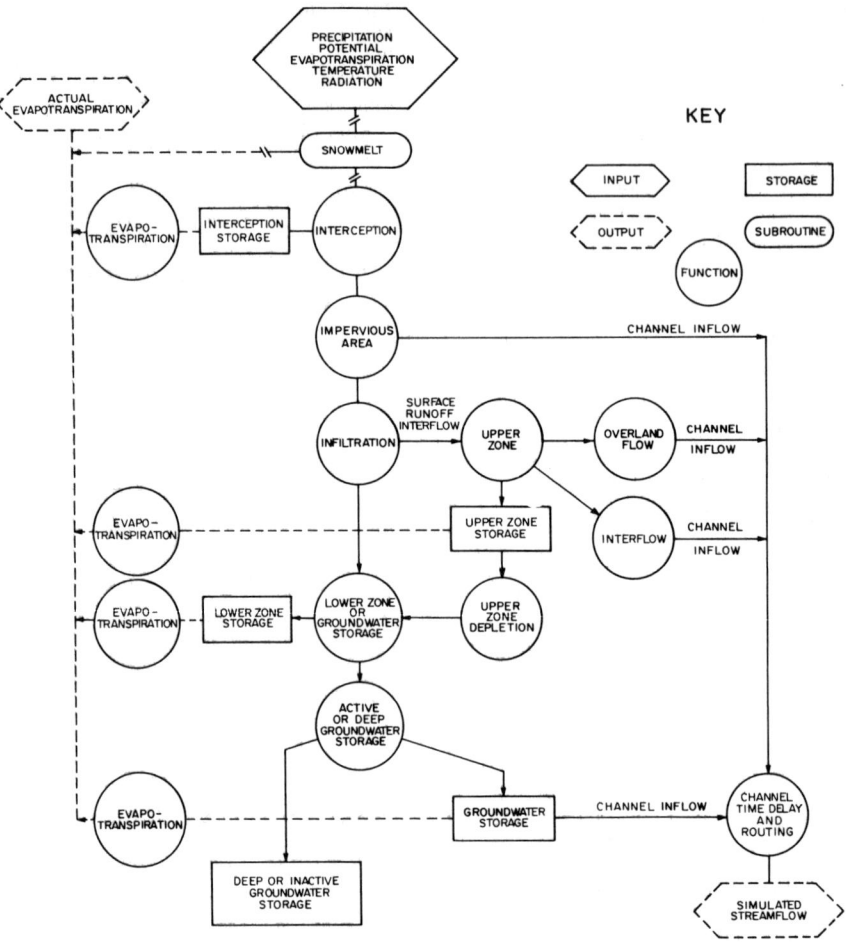

Figure 16

FLOWCHART OF THE STANFORD WATERSHED MODEL IV
(from Crawford and Linsley, 1966)

These assumed distributions, which are a clever way of accounting for spatial uncertainties in both the input variables and in the system parameters, are sketched in Figure 17. The quantities b and r are empirical indices.

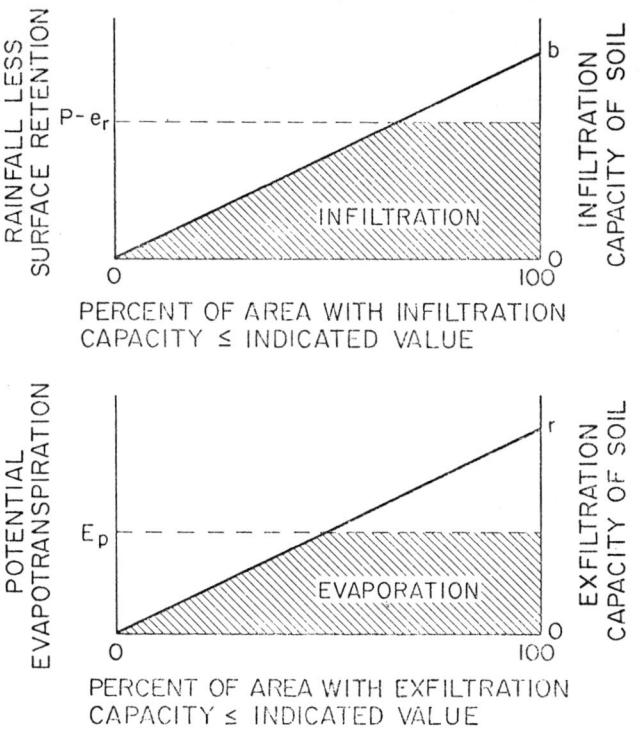

Figure 17

THE "SOURCE-AREA" METHOD OF COPING WITH SPATIAL VARIABILITY
(after Crawford and Linsley, 1966)

This model requires inputs of precipitation (15 minutes) and potential evapotranspiration (daily) and produces 15 minute values of the water balance components. Ten parameters control the water balance portion of the model. Explicit accounting for vegetation occurs only in the value of one of these parameters, the nominal upper zone storage capacity.

2. Holtan

The U.S. Department of Agriculture (Holtan, et al., 1974) developed a model which is in wide use for small agricultural watersheds. It gives explicit attention to vegetation growth through use of an empirical, time-varying crop growth index to get E_T from E_p. However, its soil moisture fluxes are estimated from a "series of empiricisms." The model has 27 parameters for the water balance portion (i.e., exclusive of the hydrograph generation).

3. Peck

For large-area river forecasting purposes, the U.S. National Weather Service has developed a model (Peck, 1976) based upon the "Sacramento Model" (Burnash, et al., 1973). It differs from the Stanford Watershed Model primarily in its separation of the soil water of both the upper and lower zones into two categories: a) tension water which is "closely bound to the soil particles" and b) free water which is "free to move." The tension moisture contributes only to evapotranspiration while the free water contributes to percolation and to runoff as well as to evapotranspiration. The moisture flux functions are empirical and there is no explicit accounting for vegetation. The many parameters of this model (there are 17 for soil moisture alone) are "lumped" in the sense of being assigned effective values for the entire catchment without explicit effort to account for areal variability.

D. Dynamic

1. Sasamori

Using a radiative-convective model of the atmosphere, Sasamori (1970) coupled the one-dimensional primitive equations for instantaneous moisture and heat content of both the atmosphere and the soil, across a bare soil interface, using the thermodynamic relation of Philip (1957)

$$S_g = \exp[g\, \psi_g(\theta)/RT_g] \tag{49}$$

where

S_g = saturation ratio of the surface atmosphere

g = gravitational constant

$\psi_g(\theta)$ = surface soil moisture potential

R = gas constant

He used this complicated numerical model to evaluate the energy balance components under contrasting conditions of soil wetness, however, it could as easily have been used to evaluate the water balance. The model contains 7 physical parameters and does not account for snow.

2. Deardorff

The heat flux and moisture flux as modulated by a vegetation layer are parameterized in a useful way by Deardorff (1978). The surface temperature and surface moisture are predicted by the "force-restore" method discussed earlier, and the transpiration is estimated by a Monteith-type (1965) relation. The critical moisture fluxes remain linear in the moisture content, however, and the effect of gravity is absent. The parameterization is arranged to handle macroscale areas with mixed fractions of vegetation and bare soil. There are six soil parameters and six vegetation parameters.

3. Eagleson

In order to incorporate the vital short-term soil moisture dynamics into the long-term average water balance, Eagleson (1978) introduced a statistical dynamic approach. He adapted the Philip (1969) infiltration equation to the case of exfiltration including a distributed sink to represent the extraction of soil moisture by plant roots. The infiltration capacity (i.e., wet surface) and exfiltration capacity (i.e., dry surface) he used were, respectively

$$f_i^*(t; \theta_o) = \frac{1}{2} S_i(\theta_o) t^{-1/2} + \frac{1}{2} [K(n) - K(\xi_o)] \quad (50)$$

and

$$f_e^*(t; \theta_o) = \frac{1}{2} S_e(\theta_o) t^{-1/2} - \frac{1}{2} K(\theta_o) - Mk_v e_p \qquad (51)$$

where

θ_o = uniform soil moisture in the root zone at the start of the flux episode

M = vegetation canopy density

S_i, S_e = infiltration and exfiltration sorptivities of soil

f_i^*, f_e^* = infiltration and exfiltration capacities

k_v = plant coefficient (see Eq. (32))

t = time

When Eqs. (50) and (51) are compared with the idealized potential infiltration rate (i.e., the rainfall rate) or the potential evapotranspiration rate as sketched in Figure 18, the infiltration or evaporation volume for a single storm or interstorm event is clearly defined. Using representative probability density functions for the independent, point (i.e., station) atmospheric variables, rainfall intensity, potential evapotranspiration rate, storm duration and interstorm duration, Eagleson (1978) derived the pdf's of the resulting point storm infiltration and interstorm evaporation given the initial moisture content, θ_o. The remaining soil moisture flux, the percolation out of the bottom of the root zone, was assumed constant at

$$p = K(\bar{\theta}) \qquad (52)$$

where $\bar{\theta}$ is the root zone space-time average moisture content,

Eagleson (1978) next introduced the average number of storms per wet season and found the average of the flux volumes just discussed in order to obtain the long-term average annual fluxes on the right-hand side of Eq.

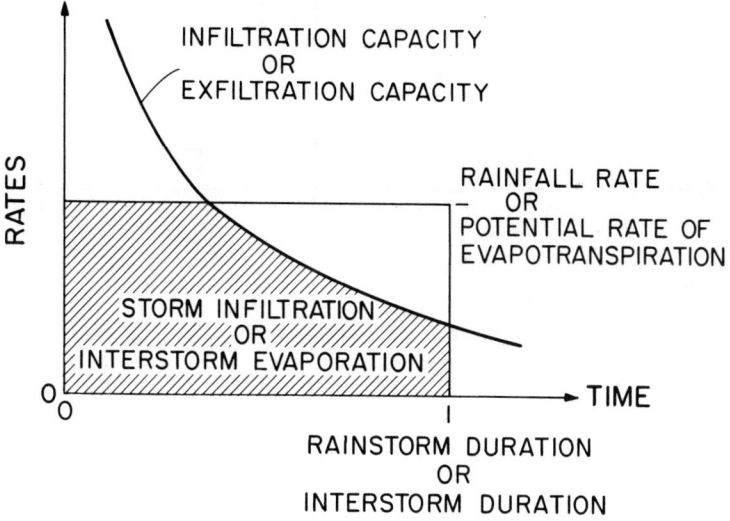

Figure 18

EVALUATION OF SINGLE-STORM SURFACE MOISTURE FLUX
(after Eagleson, 1978)

(9), the water balance equation. A major expedient assumption, that the initial soil moisture condition, θ_o, is equal to the space-time average value, $\bar{\theta}$, at the beginning of every event (see Figure 19), finally leads to the average annual local water balance

$$1 - \xi(\bar{\theta}) \exp[G(\bar{\theta})] = \bar{\beta}[E(\bar{\theta}), M, k_v] \frac{\bar{E}_{P_A}}{\bar{P}_A} + m_\tau K(\bar{\theta})/\bar{P}_A \qquad (53)$$

$$\underbrace{\hphantom{1 - \xi(\bar{\theta}) \exp[G(\bar{\theta})]}}_{\text{Infiltration}} \quad \underbrace{\hphantom{\bar{\beta}[E(\bar{\theta}), M, k_v] \frac{\bar{E}_{P_A}}{\bar{P}_A}}}_{\text{Evapotranspiration}} \quad \underbrace{\hphantom{m_\tau K(\bar{\theta})/\bar{P}_A}}_{\text{Groundwater}}$$

where

m_τ = average length of the rainy season

and where $\xi(\bar{\theta})$, $G(\bar{\theta})$ and $E(\bar{\theta})$ are dimensionless functions of climate (i.e., long-term average atmosphere) and soil characteristics. With negligible water table effects, the water balance is defined in terms of only 4 independent system parameters, 3 soil and 1 vegetation.

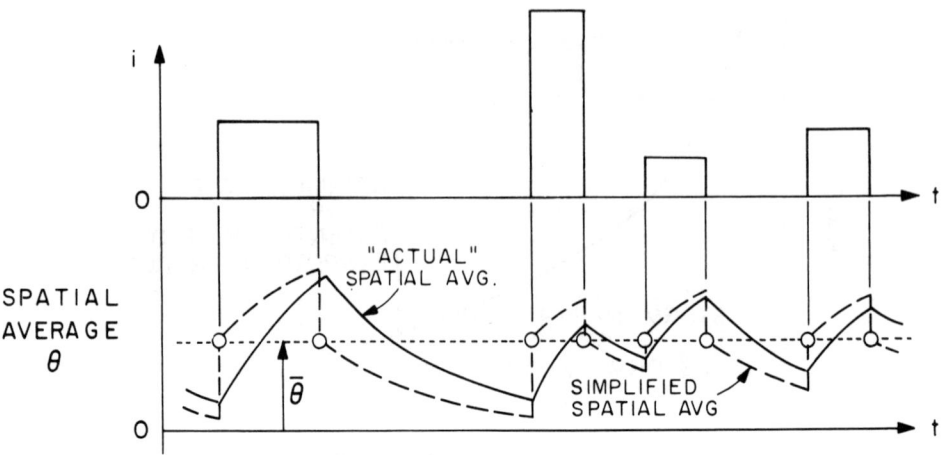

SOIL MOISTURE REPRESENTATION

Figure 19

SOIL MOISTURE REPRESENTATION IN EAGLESON WATER BALANCE MODEL
(from Eagleson, 1978)

Equation (53) has solution sets which display a maximum average soil moisture at an intermediate vegetation density. Nature selects among these sets by means of a second $\bar{\theta}$ vs. M relationship arising from the plant physiology. Eagleson (1978) bypasses the physiology by assuming, as discussed earlier, that the solution for maximum $\bar{\theta}$ is ecologically optimal because it implies minimum plant stress. With this assumption $M = M_o$ and the average evapotranspiration efficiency, $\bar{\beta}[\ ]$, is as shown in Figure 20 for various plant coefficients.

Equation (53) can also be used as a first-order approximation to the annual water balance. It can be extended to macroscale through the use of areally-averaged atmospheric inputs and of effective (i.e., lumped) physical parameters (Chan and Eagleson, 1980).

One of the beauties of the statistical-dynamic approach to the behavior of uncertain natural systems is that it expresses the statistics of

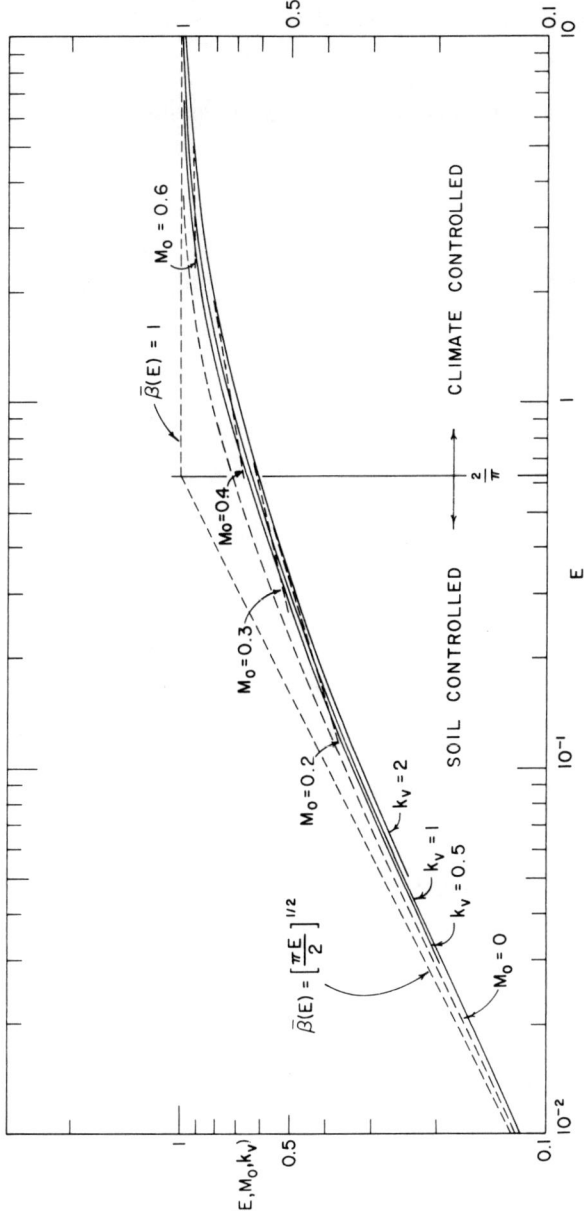

Figure 20

EVAPOTRANSPIRATION EFFICIENCY IN THE EAGLESON (1978) WATER BALANCE MODEL

the dependent variable not only in terms of the system parameters, but also in terms of the statistics of the independent variables. In dimensionless form, this can lead to significant physical insight.

MACROSCALE HEAT BALANCE MODELS

A. Vowinckel

A combined heat and water budget by Vowinckel and Orvig (1972) focusses upon solution of the former for the surface temperature, T_g. It leads to evaluation of all heat fluxes at short-time intervals but also allows the water budget to be evaluated. The model contains 5 soil layers, each of which contributes to evapotranspiration through an empirical relation. It also incorporates photosynthesis empirically. There are 20 parameters.

B. Anderson

A snow accumulation and ablation model has been prepared by Anderson (1973) for use in the U.S. National Weather Service's River Forecast System. It incorporates routines for snowpack accumulation, heat exchange at the air-snow interface, snowpack heat storage, liquid water retention and transmission and heat exchange at the soil-snow interface. It includes assumed distributions of spatial variations, but does not include sublimation or interception of snow by vegetation. These omissions may be important under certain conditions.

A preceding publication (Anderson and Crawford, 1964) does incorporate these phenomena, however, and a flow chart from that source is included here as Figure 20. The model has nine parameters.

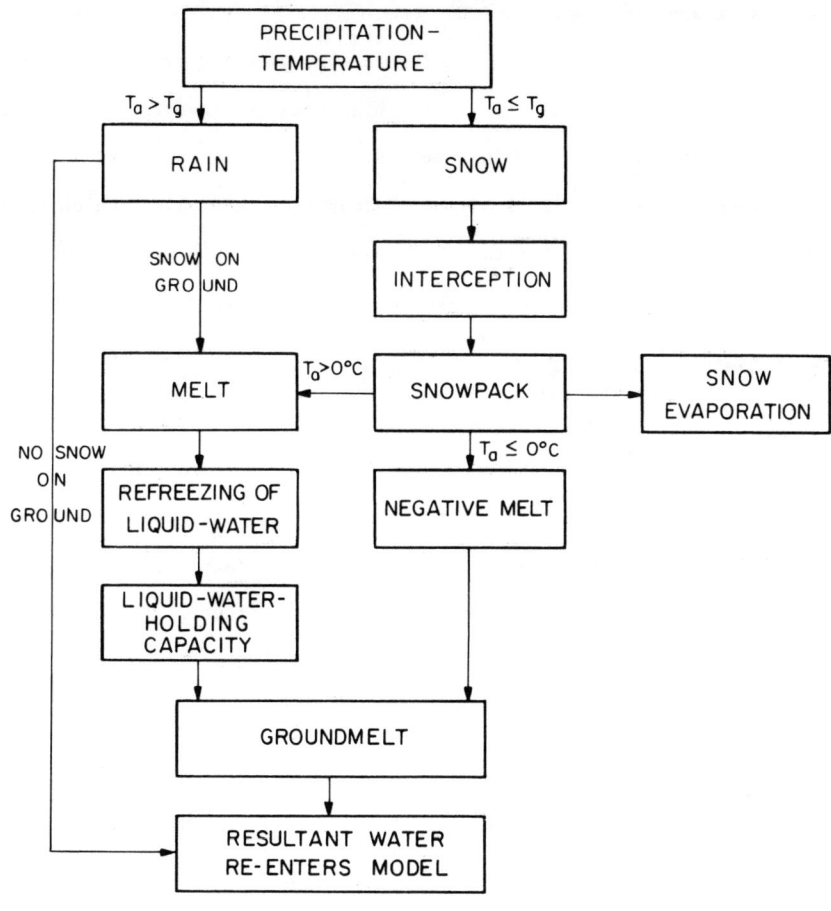

Figure 21

SNOWMELT FLOWCHART
(from Anderson and Crawford, 1964)

MACROSCALE ECOLOGICAL MODELS

Models which focus on the vegetation are more numerous and range from the sinple, largely empirical model of Saxton, et al. (1974), through the phenomenological approaches of Cowan (1965) and Federer (1979) discussed earlier, to the dynamic models of Paltridge (1970), Huff, et al. (1977), and Federer and Lash (1978). A summary of many others is given by Halldin (1979).

A. Paltridge

The Paltridge (1970) model is a complicated representation of the short-term behavior of a vegetal cover which is notable, however, for its inclusion of seasonal growth. It has 28 parameters. While it is too complex for use with GCMs, it may provide a basis of simulated "natural truth" against which to compare proposed simpler parameterizations. Many other simple growth models exist (see Baier, 1977, for a review) but most are heavily empirical.

B. Huff

The terrestrial ecology model reported by Huff, et al. (1977) is a mixture of empiricism and conceptualization. Evaporation is handled by the Monteith (1965) equation, but infiltration is estimated from an empirical cumulative infiltration-capacity curve. The vegetation system requires 19 parameters while the soil description demands 7 others. The complexity of this model probably eliminates it for use with GCM's.

C. Federer

Federer and Lash (1978) report the development of a water budget model for small, forested watersheds. It operates on a daily time interval and includes provision for snowfall and accounts for slope-aspect effects on the thermal budget. There are 18 parameters, four of which apply only to the snowmelt capability.

SUMMARY

The one-dimensional microscale dynamics of coupled heat and moisture flux at landsurfaces have been reviewed for rainfed and snowfed systems. The system has been simplified in the rainfed case by expressing the atmospheric water vapor transport capacity independently of the surface state variables, temperature and soil moisture concentration. This gives a soil moisture system which is driven by the atmospheric water vapor demand in humid climates and by the precipitation intensity in arid climates. No such decoupling of the hydro-thermal system is possible in the snowfed case. Six physical elements of the hydro-thermal balance are isolated.

Two basic parameterization problems are associated with adaptation of the one-dimensional, microscale dynamics for use at macroscale:

1. Parameterization of the processes which trade off physical fidelity in exchange for economy in data requirements (and in computation) for <u>homogeneous</u> systems of large areal extent.

2. Modification of these formulations to account for heterogeneities in both the climatic inputs and the physical parameters.

Macroscale parameterizations of the six elements of the hydro-thermal balance are reviewed as are their use in representative, lumped parameter, water balance, heat balance and ecological models. These reviews indicate that while the first of the above two problems appears essentially solved, the relative performance of the various existing macroscale models needs to be studied in terms of the current GCM needs. The second of the above problems remains largely unaddressed and unsolved, however.

In particular, the unsolved problem concerns the spatial heterogeneity of the landsurface parameters, such as vegetation type and density, and soil properties. Is it important to take explicit account of this variation, or at the extreme macroscale of the GCM grid, does the natural variability become merely a noise which is best averaged by the usual expedient lumping into "effective" parameter values? One hopes and even expects that the latter may be the case.

It also concerns parameterization, to sub-grid hydrologic scale, of atmospheric inputs that are defined only at grid points. This is a serious, unsolved problem that deserves immediate attention.

Finally, there are many unanswered questions related to the role of vegetation in the total scheme of things. Is it really important to incorporate the biophysics of plant behavior into landsurface models or is the vegetation canopy merely a buffering layer that by thermal (and hydraulic?) inertia only modulates the extremes of bare soil behavior? Here we hope for the latter but must expect the former largely because of the dependence of short-wave albedo upon vegetation state. If vegetation biodynamics <u>are</u> important, how can they best be parameterized within GCM computational restrictions?

REFERENCES

ANDERSON, E. A. (1964). The synthesis of continuous snowmelt runoff hydrographs on a digital computer. Technical Report No. 36, Dept. of Civil Engineering, Stanford University, Palo Alto, Ca., 103 pp.

ANDERSON, E. A. (1973). National Weather Service River Forecast System - snow accumulation and ablation model. NOAA Technical Memorandum NWS HYDRO-17, U.S. Dept. of Commerce, Washington.

ARAKAWA, A. (1972). Design of the UCLA General Circulation Model. Numerical simulation of weather and climate, Technical Report No. 7, Dept. of Meteorology, UCLA, 116 pp.

BAIER, W. (1973). Crop-weather models and their use in yield assessments. WMO Technical Note No. 151 (WMO-No. 458), Geneva, 48 pp.

BARTON, I. J. (1979). A parameterization of the evaporation from nonsaturated surfaces. Journal of Applied Meteorology, $\underline{18}$, pp. 43-47.

BHUMRALKAR, C. M. (1975). Numerical experiments on the computation of ground surface temperature in an atmospheric general circulation model. J. Appl. Meteorol., $\underline{14}$, pp. 1246-1258.

BLACKADAR, A. K. (1976). Modeling the nocturnal boundary layer. In Proceedings of the Third Symposium on Atmospheric Turbulence, Diffusion and Air Quality, pp. 46-49, American Meteorological Society, Boston, Ma.

BLAKE, G. J. (1975). The interception process. In Prediction in Catchment Hydrology, T.G. CHAPMAN and F.X. DUNIN (Eds.), Australian Academy of Sciences, pp. 59-81.

BOUCHET, R. J. (1963). Evapotranspiration réelle et potentielle, signification climatique. International Association for Scientific Hydrology, General Assembly of Berkeley, Pub. No. 62, Gentbrugge.

BRUTSAERT, W. and H. STRICKER (1979). An advection-aridity approach to estimate actual regional evapotranspiration. Water Resources Research, $\underline{15}$, 2, pp. 443-450.

BUDYKO, M. I. (1956). Teplovoi Balans Zemnoi Poverkhnosti, Gidrometeoizdat, Leningrad; Heat Balance of the Earth's Surface, translated by N.A. Stepanova, U.S. Weather Bureau, Washington, D. C., 1958.

BUDYKO, M. I. (1971). Climate and Life. (English translation edited by D. H. MILLER, 1974), Academic Press, N. Y., 508 pp.

BUDYKO, M. I. (1972). Hydrologic and cryospheric coupling. Chapter 9 of Parameterization of Sub-Grid Scale Processes, GARP Publications Series No. 8, WMO-ICSU, pp. 92-101.

BURNASH, R. J. C., R. L. FERRAL and R. A. McGUIRE (1973). A Generalized Streamflow Simulation System: Conceptual Modeling for Digital Computers. U.S. Dept. of Commerce, National Weather Service and State of California, Dept. of Water Resources, Sacramento.

CARSLAW, H. S. and J. C. JAEGER (1959). Conduction of heat in solids. Oxford University Press, Fairlawn, N. J. (2nd ed.).

CHAN, S.-O. and P. S. EAGLESON (1980). Water balance studies of the Bahr El Ghazal swamps. Technical Report No. 261, R. M. Parsons Laboratory, Dept. of Civil Engineering, M.I.T., 317 pp.

CORBY, G. A., A. GILCHRIST and P. R. ROWNTREE (1977). United Kingdom Meteorological Office Five-Level General Circulation Model. Methods in Computational Physics, $\underline{17}$, pp. 67-110.

COWAN, I. R. (1965). Transport of water in the soil-plant-atmosphere system. Journal of Applied Ecology, $\underline{2}$, pp. 221-239.

CRAWFORD, N. H. and R. K. LINSLEY (1966). Digital simulation in hydrology: Stanford Watershed Model IV. Tech. Rept. No. 39, Stanford University Dept. of Civil Engineering, 210 pp.

DEARDORFF, J. W. (1977). A parameterization of ground-surface moisture content for use in atmospheric prediction models. Journal of Applied Meteorology, $\underline{16}$, 11, pp. 1182-1185.

DEARDORFF, J. W. (1978). Efficient prediction of ground surface temperature and moisture with inclusion of a layer of vegetation. Journal of Geophysical Research, $\underline{83}$, C4, pp. 1889-1903.

EAGLESON, P. S. (1970). Dynamic Hydrology. McGraw-Hill Book Co., N. Y. 462 pp.

EAGLESON, P. S. (1978). Climate, soil and vegetation, Parts 1-7. Water Resources Research, $\underline{14}$, 5, pp. 705-776.

EAGLESON, P. S. and T. E. TELLERS (1981). Evidence of ecological optimality in natural soil-vegetal systems. Technical Report of R. M. Parsons Laboratory, Dept. of Civil Engineering, M.I.T. (in preparation).

FEDERER, C. A. (1977). Leaf resistance and xylem potential differ among broadleaved species. Forest Science, $\underline{23}$, pp. 411-419.

FEDERER, C. A. (1979). A soil-plant-atmosphere model for transpiration and availability of soil water. Water Resources Research, $\underline{15}$, 3, pp. 555-562.

FEDERER, C. A. and D. LASH (1978). BROOK: A hydrologic simulation model for Eastern forests. Res. Rept. No. 19, Water Resource Research Center, Univ. of New Hampshire, Durham.

FLEMING, G. (1975). Computer simulation techniques in hydrology. Elsevier, NY, 333 p.

GARP (1978), JOC Working Group on Land Surface Processes, Report of Dublin Meeting, May 8-12, 1978, Geneva, 32 pp.

GATES, W. L. and M. E. SCHLESINGER (1977). Numerical simulation of the January and July global climate with a two-level atmospheric model. Journal of the Atmospheric Sciences, 34, pp. 36-76.

GUPTA, V. K. (1973). A stochastic approach to space-time modeling of rainfall. Technical Reports on Natural Resource Systems, No. 18, The University of Arizona, Tucson, Arizona.

HALLDIN, S. (editor) (1979). Comparison of forest water and energy exchange models. Elsevier Scientific Publishing Co., NY, 255 p.

HARE, F. K. (1980). Long-term annual surface heat and water balances over North America: A reconciliation of precipitation, runoff and temperature fields. Canadian Geographer, 24, pp.

HILLEL, D. (1971). Soil and Water - Physical Properties and Processes, Academic Press, N. Y., 288 pp.

HOLTAN, H. N., G. J. STILTNER, W. H. HENSON and N. C. LOPEZ (1975). USDAHL-74 Revised Model of Watershed Hydrology. Technical Bulletin No. 1518, Agricultural Research Service, U.S. Dept. of Agriculture, Washington, D. C., 99 pp.

HUFF, D. D., R. J. LUXMORE, J. B. MANKIN and C. L. BEGOVICH (1977). TEHM: A Terrestrial Ecosystem Hydrology Model. Oak Ridge National Laboratory, Environmental Sciences Division Publication No. 1019, Oak Ridge, Tennessee, 152 pp.

KALININ, G. P. (1968). Global Hydrology (English Translation by N. KANER, 1971). Israel Program for Scientific Translations Ltd. IPST Cat. No. 5756, 311 pp. (Available from U.S. Dept. of Commerce, NTIS, Springfield, VA. 22151, Cat. No. TT 70-50054).

LANGBEIN, W. B. and others (1949). Annual runoff in the United States. Circular No. 52, U.S. Geological Survey, Washington, D. C.

LETTAU, H. (1969). Evaporation climatonomy, a new approach to numerical prediction of monthly evapotranspiration, runoff, and soil moisture storage. Monthly Weather Review, 97, 10, pp. 691-699.

LETTAU, H., K. LETTAU and L.C.B. MOLION (1979). Amazonia's hydrologic cycle and the role of atmospheric recycling in assessing deforestation effects. Monthly Weather Review, 107, 3, pp. 227-238.

LINACRE, E. T. (1968). Estimating the net-radiation flux. Agricultural Meteorology, 5, pp. 49-63.

LINSLEY, R. K. and N. H. CRAWFORD (1960). Computation of a synthetic streamflow record on a digital computer. Pub. No. 51, International Association of Scientific Hydrology, pp. 526-538.

LOWRY, W. P. (1959). The falling rate phase of evaporative soil moisture loss: a critical evaluation. Bull. Amer. Meteor. Soc., 40, p. 605.

MANABE, S. (1969). Climate and the ocean circulation 1. The atmospheric circulation and the hydrology of the earth's surface. Monthly Weather Review, 97, 11, pp. 739-774.

MANABE, S., D. G. HAHN and J. L. HOLLOWAY, JR. (1974). The seasonal variation of the tropical circulation as simulated by a global model of the atmosphere. Journal of the Atmospheric Sciences, 31, pp. 43-83.

McNAUGHTON, K. G. (1976-a). Evaporation and advection I: evaporation from extensive homogeneous surfaces. Quarterly Journal of the Royal Meteorological Society, 102, pp. 181-191.

McNAUGHTON, K. G. (1976-b). Evaporation and advection II: evaporation downwind of a boundary separating regions having different surface resistances and available energies. Quarterly Journal of the Royal Meteorological Society, 102, pp. 193-202.

MONTEITH, J. L. (1965). Evaporation and environment. Symposium of the Society for Experimental Biology, No. 19, pp. 205-234.

MONTEITH, J. L. (1973). Principles of Environmental Physics. American Elsevier Pub. Co., Inc., N. Y., 241 pp.

MORTON, F. I. (1965). Potential evaporation and river basin evaporation. Proc. ASCE, Journ. Hydraulics Division, HY6, 4534, pp. 67-97.

MORTON, F. I. (1968). Evaporation and climate - a study in cause and effect. Scientific Series No. 4, Inland Waters Branch, Canadian Department of Energy, Mines and Resources, Ottawa, Canada, 32 pp.

MORTON, F. I. (1971). Catchment evaporation and potential evaporation - further development of a climatologic relationship. Journal of Hydrology, 12, pp. 81-99.

NOVÁK, V. (1980). The estimation of the influence of the soil water content on evaporation from bare soils. Vodohospodársky Časopis, 28, 4, pp. 419-434 (in Czechoslovakian).

OL'DEKOP, E. M. (1911). Ob isparenii s poverknosti rechnykh basseinov (On evaporation from the surface of river basins). Tr. Meteorol. Observ. Iur'evskogo Univ. Tartu, 4.

PALTRIDGE, G. W. (1970). A model of a growing pasture. Agricultural Meteorology, 7, pp. 93-130.

PECK, E. L. (1976). Catchment modeling and initial parameter estimation for the National Weather Service River Forecast System. NOAA Technical Memorandum NWS HYDRO-31, Office of Hydrology, Washington, D. C., 24 pp plus appendices.

PENMAN, H. L. (1948). Natural evaporation from open water, bare soil and grass. Proc. Roy. Soc. (London), Ser. A. 193, pp. 120-145.

PHILIP, J. R. (1957). Evaporation, and moisture and heat fields in the soil. Journal of Meteorology, 14, pp. 354-366.

PHILIP, J. R. (1969). Theory of infiltration. In Advances in Hydroscience, 5, V. T. CHOW (Ed.), Academic Press, N. Y., pp. 215-296.

PIERCE, L. T. (1958). Estimating seasonal and short term fluctuations in evapotranspiration from meadow crops. Bulletin of the Amer. Meteor. Soc., 39, pp. 73-78.

PIKE, J. G. (1964). The estimation of annual run-off from meteorological data in a tropical climate. Journal of Hydrology, 2, pp. 116-123.

PRIESTLEY, C. H. B. and R. J. TAYLOR (1972). On the assessment of surface heat flux and evaporation using large-scale parameters. Monthly Weather Review, 100, 2, pp. 81-92.

RASMUSSEN, E. M. (1968). Atmospheric water vapor transport and the water balance of North America II large-scale water balance investigations. Monthly Weather Review, 96, 10, pp. 720-734.

RASMUSSEN, E. M. (1977). Hydrological application of atmospheric vapor-flux analyses. Operational Hydrology Report No. 11, WMO - No. 476, WMO, Geneva, 50 pp.

RIJTEMA, P. E. (1965). An analysis of actual evapotranspiration. Agricultural Research Report 659, Wageningen, Netherlands.

RUTTER, A. J., K. A. KERSHAW, P. C. ROBINS and A. J. MORTON (1972). A predictive model of rainfall interception in forests, 1. Derivation of the model from observations in a plantation of Corsican pine., Agric. Meteorol., 9, pp. 367-384.

SAXTON, K. E., H. P. JOHNSON and R. H. SHAW (1974). Modeling evapotranspiration and soil moisture. Transactions, American Soc. of Agricultural Engineers, pp. 673-677.

SHREIBER, P. (1904). Über die beziehungen zwischen dem niederschlag und der wasserführung der flüsse in mitteleuropa. Z. Meteorol., 21, pt. 10.

SHUKLA, J. (1977). Personal communication.

SHUTTLEWORTH, W.J. (1979). Evaporation. Report No. 56, Institute of Hydrology, Wallingford, Oxon, England, 61 pp.

SLATYER, R. O. (1967). Plant-Water Relationships. Academic Press, N.Y., 366 pp.

SLATYER, R. O. and I. C. McILROY (1961). Practical Microclimatology, UNESCO, Paris, 310 pp.

SOLOMON, S. (1967). Relationship between precipitation, evaporation, and runoff in tropical-equatorial regions, Water Resources Research, 3, 1, 163-172.

SOMERVILLE, R.C.J., et al. (1974). The GISS model of the global atmosphere. Journal of the Atmospheric Sciences, 31, pp. 84-117.

TANNER, C. B. and M. FUCHS (1968). Evaporation from unsaturated subsurfaces: a generalized combination method. Journal of Geophysical Research, 73, 4, pp. 1299-1304.

TANNER, C. B. and W. L. PELTON (1960). Potential evapotranspiration estimates by the approximate energy balance method of Penman. Journal of Geophysical Research, 65, 10, pp. 3391-3413.

TELLERS, T. E. and P. S. EAGLESON (1980). Estimation of effective properties of soils from observations of vegetation density. R. M. Parsons Lab. Report No. 254, MIT Dept. of Civil Engineering, 126 pp.

THOM, A. S. and H. R. OLIVER (1977). On Penman's equation for estimating regional evaporation. Quarterly Journal of the Royal Meteor. Soc., 103, pp. 345-357.

THORNTHWAITE, C. W. and J. R. MATHER (1955). The water balance. Publications in Climatology, 8, 1, Laboratory of Climatology, Centerton, N. J., 86 pp.

TURC, L. (1954). Le bilan d'eau des sols. Relation entre les précipitations, l'évaporation et l'écoulement. Ann. Agron. 5, pp. 491-596.

U.S. ARMY CORPS OF ENGINEERS (1956). Snow Hydrology. North Pacific Division, Portland, Oregon (available from NTIS as PB 151660).

VAN BAVEL, C.H.M. (1966). Potential evaporation: the combination concept and its experimental verification. Water Resources Research, 2, 3, pp. 455-467.

VAN DEN HONERT, T. H. (1948). Water transport in plants as a catenary process. Discuss. Faraday Soc., 3, pp. 146-153.

VOWINCKEL, E. and S. ORVIG (1972). EBBA: An energy budget programme. Department of Meteorology, McGill University, Publication in Meteorology No. 105, Montreal, 50 pp.

WIGHAM, J. M. (1973). Interception- Section IV of Principles of Hydrology. D. M. GRAY (Ed.), Water Information Center, Inc., Port Washington, N. Y.

WILLMOTT, C. J. (1977). WATBUG: A Fortran IV algorithm for calculating the climatic water budget. Water Resources Center, University of Delaware, Newark, Del., 55 pp.

NOTATION

A	net flux of advected energy
A_g	horizontal area of ground
B	factor
C	heat flux to snowpack from underlying ground
C_s	coefficient related to surface roughness
c	boundary curve
c_p	specific heat at constant pressure
c_s	specific heat of snowpack or of soil
D	factor
d	zero plane displacement of wind velocity distribution
$d_{1,2}$	depths of diurnal and annual thermal layers
E_a	rate of advective enhancement of potential evapotranspiration
E_p	potential evapotranspiration rate = atmospheric vapor transport capacity
E_{p_A}	annual potential evapotranspiration
E_{p_o}	limiting or "equilibrium" value of potential rate of evapotranspiration
E_{p_s}	potential rate of evaporation from a bare soil surface
E_{p_v}	potential rate of transpiration of a vegetation canopy
E_T	actual rate of evapotranspiration
E_{T_A}	annual evapotranspiration
e_a	vapor pressure of atmosphere at screen elevation
e_r	rate of storage of rainfall in surface retention
e_s	rate of evaporation from bare soil surfaces
e_v	rate of transpiration by vegetated surfaces
e_{v_o}	maximum rate of transpiration for a given plant species
e_a^*	saturated atmospheric vapor pressure at temperature, T_a
e_s^*	saturated vapor pressure at surface temperature

f_e^*	exfiltration capacity of soil
f_i^*	infiltration capacity of soil
G	net heat flux into the ground
g	gravitational constant
H	sensible heat flux from soil to atmosphere
I	rate of infiltration into soil matrix
i_e	rate of generation of rainfall excess
K	hydraulic conductivity of soil
K_a	atmospheric heat conductance
K_s	surface heat conductance
k	Kármán constant
k_v	plant coefficient
k_θ	factor
L	latent heat
L_f	latent heat of fusion
L_s	latent heat of sublimation
L_v	latent heat of vaporization
M	vegetated fraction of the surface
M_o	equilibrium density of natural vegetation canopy
M_s	mass of snowpack
m_s	change of storage of meltwater within snowpack
N	cloudcover
n	porosity of soil
$\vec{n}$	unit normal vector
P	rate of rainfall
P_A	annual precipitation
P_o	a particular value of precipitation
P_s	rate of snowfall
p	rate of downward percolation of soil moisture to the water table

Symbol	Description
$\vec{Q}$	total horizontal water vapor flux per unit length of atmospheric column
q_a	specific humidity of atmosphere at screen elevation
$q^*(\ )$	saturated atmospheric specific humidity at ( ) temperature
R	gas constant
R_g	rate of net groundwater runoff
R_n	net radiation flux to landsurface
R_s	surface runoff rate
R_{g_A}	annual groundwater runoff
R_{s_A}	annual surface runoff
r_a	atmospheric diffusion resistance
r_c	surface diffusion resistance
r_p	plant diffusion resistance
r_s	soil diffusion resistance
S	rate of water storage on and in ground
S_e	exfiltration sorptivity of soil
S_g	atmospheric saturation ratio at the surface
S_i	infiltration sorptivity of soil
S_s	atmospheric saturation ratio at evaporating surface
T	temperature
T_a	atmospheric temperature at screen elevation
T_g	temperature of ground
T_p	snowpack temperature
t	time
v_s	rate of storage of rainfall excess in surface depression
W	total water vapor content of atmospheric column with unit area
W_s	windspeed at screen elevation
x	downwind coordinate of distance
Y	rate of water yield

Symbol	Description
Y_A	annual water yield
Z_r	thickness of root zone
Z_w	depth to water table
z	vertical coordinate of distance
z_a	screen elevation
z_o	surface roughness dimension
α	coefficient
β	evapotranspiration efficiency $\equiv E_T/E_p$
γ	psychrometric constant
Δ	slope of saturated vapor pressure-temperature curve at $T = T_a$
θ	volumetric soil moisture content
θ_o	uniform initial soil moisture in the root zone
θ_w	soil moisture as a percent of dry weight
θ_{fc}	volumetric water content of soil at field capacity
λ_s	thermal conductivity of soil-water system
μ	factor
ρ_a	mass density of atmosphere at screen elevation
ρ_s	mass density of soil-water system
ρ_w	mass density of water
τ_1	period of one day
ψ_g	water potential at soil surface
ψ_ℓ	water potential in the leaf stomatal cavities
ψ_r	water potential at the root surface
ψ_s	water potential in the soil matrix
$E(\ )$	dimensionless evapotranspiration function
$G(\ )$	dimensionless infiltration function
$\xi(\ )$	dimensionless surface runoff function
$(\ )^o$	normalized $(\ ) = (\ )/\overline{E}_{P_A}$
$\langle(\ )\rangle$	spatial average of $(\ )$
∇	divergence operator

ACKNOWLEDGEMENTS

Thanks and credit are due to the many who have contributed constructive criticism and helpful suggestions to the preparation of this review. These include E. Bobínski, J. Deardorff, D. A. de Vries, J. C. I. Dooge, R. A. Feddes, C. A. Federer, R. A. Freeze, K. I. Itten, M. E. Jensen, K. Ya. Kondratyev, G. Kukla, H. H. Lettau, P. C. D. Milly, J. Němec, S. Orvig, E. Peck, J. Rauner, W. J. Shuttleworth, W. M. Washington and D. Z. Yeh.

The review was prepared with the support of the U.S. National Science Foundation under NSF Grant No. ATM-7812327.

SESSION IV

LAND SURFACE GLOBAL DATA SETS

USE OF REGIONAL AND GLOBAL SOILS DATA FOR CLIMATE MODELLING

M.J. Gardiner

Head, National Soil Survey of Ireland and previously Technical Officer,
World Soil Resources Office, FAO, Rome

I. INTRODUCTION

The purpose of this paper is to evaluate the adequacy of global soils data for climate modelling. Because of the horizontal variability of the land surface, such models aim at incorporating the most critical features of soil behaviour without requiring detailed information on soil characteristics. It is necessary, therefore, to decide what these critical features are, how they may be functionally related to other soil characteristics and whether adequate global data on them are available.

Soil moisture capacity, surface runoff, saturated permeability and land surface albedo seem to be the most critical features of soil behaviour in climate modelling. These have large spatial variations over the globe and, therefore, their functional dependence on more easily mappable soil characteristics such as depth, texture, structure, consistence and organic matter content must be used. It is necessary to ascertain whether these functionally related soil characteristics are included in the very large body of soils data available and whether they can be collated and interpreted for large regions and on a global basis.

It is appropriate, therefore, to examine soil classification in terms of its fundamental principles, its evolution over the years, the role of climate in soil formation and classification, the relationship between taxonomic and mapping units, the availability of soil maps of great regions and on a global basis, and their possible uses and limitations for climate modelling.

2. SOIL CLASSIFICATION

2.1 Fundamental Principles

The modern concept of soils as independent natural bodies with genetic horizons was introduced about 1870 by the Russian School led by Dokuchaiev (1). Soil was subsequently regarded as a portion of the earth's crust with properties reflecting the effects of local and zonal soil forming factors. The soil "individual" was soon defined as a soil series consisting of a collection of soil pedons essentially uniform in differentiating characteristics and in arrangement of horizons in the soil profile.

Below the series level, the soil type is usually differentiated on the basis of texture, while the soil phase takes account of any features which interfere with agriculture e.g. slope, depth, stoniness etc. At the higher abstraction levels soil series are often placed into Great Soil Groups if they show the same general profile characteristics. At the highest level, soils can be grouped into Zonal, Intrazonal and Azonal Orders; the first includes soils reflecting the influence of climate and vegetation, the second soils reflecting the dominating influence of some local factor of relief or parent material and the third, those soils without well-developed soil characteristics.

However, although it is claimed for most soil classification systems that soils are classified on their own properties, a great diversity exists in the ways that soils are seperated and named. As a result, it is very difficult to have an overall view of the worlds soil resources without comparitive studies of the different schemes in use. The degree of equivalence of soil units used in different countries, the range of variability in soil characteristics in comparable soils and the levels of generalization at which similar soils have been distinguished all need correlative investigations.

2.2 Evolution of Different Systems

Smith (2) has reviewed the evolution of soil classification schemes in the USA. He has pointed out that over the 80 years that the United States soil survey has been in operation, four systems of soil classification have evolved. The first was that of Whitney (3) which started with a narrowly defined group of soils that had similar textures. These he called soil types. A number of soil types derived from the same parent material constituted his soil series.

The second United States system was that of Marbut (4) who concluded, after studying and translating Glinkas work, that geology was less important to soil classification than climate and that Great Soil Groups could be recognised on the basis of climatic differences. There was an immediate problem because Marbuts concept left room only for zonal soils (i.e. soils mainly influenced by climate and vegetation). Therefore soils which were strongly influenced by hydromorphic conditions or by strongly calcareous parent materials had no good place in Marbuts scheme.

The third evolutionary stage of the USA classification scheme was marked by its publication in 1938 (5). This followed in part the Russian classification of Siberzev and had as its highest categories the Zonal, Intrazonal and Azonal soils. By this time, in the USA some 2000 soil series and about 6000 soil types had been recognized.

In 1951 the USDA attempted to incorporate the knowledge of as many individuals as possible throughout the world in order to develop a comprehensive system of soil classification. A series of approximations emerged over the years until the publication of the scheme itself in Soil Taxonomy in 1975 (6).

Here a new system of nomenclature was attempted using meaningful Greek and Latin roots. Names were connotative as for example Aridisols indicated the property of dryness while Histosols indicated organic soils normally saturated with water. This nomenclature proved very useful for international communication and soil correlation especially since translation of the names of the taxa of the four highest categories was not required in order to fit the main European languages.

The highest category of this system (Soil Orders) has 10 taxa, replacing the Zonal, Intrazonal and Azonal orders of the previous system. The Entisols and Inceptisols represent mostly differences in degree of horizon development, whilst Vertisols, Aridsols, Mollisols, Spodosols, Alfisols, Ultisols and Histosols represent differences in the kinds of processes that have been dominant. Oxisols represent a combination of both the kind and degree of weathering. One important deviation from the previous system was the elimination of seperate Orders for the hydromorphic and halomorphic soils.

A review of large region soil maps and their legends by Kovda et al in 1966 (7) provides useful information on the evolution of soil science in the USSR. They reported that in the early part of this century small scale soil maps had mostly been constructed on the basis of the concept of soil zonality. This zonal-geographic approach can still be traced in the existing world soil classification systems. Although it was feared at that time that horizontal zonality of soils might not be universal and that such an approach might not be adequate to explain the distribution of the soils of the world, Kovda had earlier stressed that soil regimes evolve with time and that these changes lead to alterations in the soils themselves. He called for the establishment of the principle of "differences in age (hetero-ageous) of the soils of the world". According to this principle hetero-ageous geomorphological surfaces even situated in the same climate will have different soils. However, Kovda et al later concluded that the most up to date data on global soil cover and the earths geomorphology indicated that there is a great similarity in the history of the soil forming processes of different continents.

2.3   Relationship between taxonomic and mapping units

Depending on their purpose, soil surveys range in scale from detailed to exploratory. Individual taxonomic units cannot always be shown even on detailed soil maps and usually not at all on small scale maps.

Smith (2) stated that there are always problems in developing a soil classification system which can by used for small scale soil maps in particular. Thus, for example, small scale soil maps of the U.S.A. use associations of taxa in the higher categories i.e. Great Groups, Suborders or Orders. This has serious drawbacks since, obviously, the higher the classification category the less can be said about that particular soil.

3.   CLIMATE IN SOIL FORMATION AND CLASSIFICATION

3.1 In Soil Formation

The importance of climate was recognized by early soil scientists who saw how it influenced the rate at which weathering, leaching and podzolization take place. They also saw the correlation between climate and soil distribution on a continental basis. Russian soil scientists (1) described a zone of podzols across northern Russia in a region characterized by cool moist climatic conditions and a coniferous forest vegetation. Further south, Brown Forest Soils occurred under deciduous forest where the climate was warmer. Still further south Chernozems occurred under still warmer conditions with a grassland vegetation.

About the same time it was found that in the U.S.A. analagous climatic regions seemed to have similar soils to those of the U.S.S.R. They concluded, therefore, that climate rather than parent material was the dominant soil forming factor and that given sufficient time to develop, zonal soils would eventually dominate irrespective of the other factors of soil formation.

3.2 Limitations of the Zonal Concept

As can be seen, the zonal concept of soil formation and distribution assumed major significance and was taken into account in most soil classification schemes. However, the limitations of the zonal concept must be recognized. These have been summarized by Eyre (8) who pointed out that even at its inception the zonal classification had to be qualified by various provisos. Just as all natural vegetation is not climax vegetation, many natural soils are not climatic climax soils. This is due to many factors.

First, weathered material must remain in situ for a long time. Due to relief and erosional forces this often does not happen. Soil material may either be removed or deposited as e.g. on alluvial plains. Soils subject to this persistent truncation or deposition were often referred to as permanently immature and were classed as azonal soils.

Apart from these there are other important anomalous soils within the zonal concept. For example, many soils occur in badly drained positions and as a result have either a permanent or intermittant high water table. Thus the natural soil forming process for the region e.g. leaching, cannot take place. Similarly anomalous soils develop on certain parent materials. For example in a humid region on soils formed from highly calcarreous materials leaching losses are offset. Such areas within regions of podzolised soils may carry soils which are alkaline in reaction. These are referred to as intrazonal soils.

Nevertheless the zonal concept of soils is a useful framework for a global classification provided that its limitations are understood. In the analogy with climax vegetation, climate is taken as dominant, all other factors are subsidiary. Thus for small regions where climate is relatively uniform, small scale maps based on the zonal concept mostly show homogeneous areas of certain zonal soils. However, because these areas may have a varied lithology and relief the soil distribution pattern is much more complicated than can be shown.

The zonal soil concept may have one other major limitation. It assumes that climatic conditions have remained constant throughout the whole period of soil evolution. It is well known that during the past 10,000 years, there have been profound climatic changes especially in middle and high latitudes. It is interesting to note that in the recent U.S.A. system of soil classification provision is made to distinguish soils which have acquired some of their horizons prior to the last glaciation. Some of the present properties of these soils are due to their present environment whilst others are due to a much earlier regime. The formative element pale has been used to name such soils at the Great Group level e.g. Paleudalfs.

3.3 In Soil Classification

In the 1975 (6) U.S.A. soil classification system, soil moisture and temperature regimes were used to define the taxa at the Suborder level. Furthermore the taxa names imply the soil moisture regimes. For example in the Vertisols the defined suborders are Uderts, Usterts, Xererts and Torrerts, respectively meaning moist (Ud), warm and dry (Ust), alternate moist and dry conditions (Xer), and very dry (Torr).

Smith (2) has stressed, however, that climate per se has not been used to define taxa, but rather the soil moisture and temperature regimes have been used. These are closely correlated with climate but the correlation is not perfect. For example two soils with the same rainfall may have very different moisture regimes due to the loss of water by runoff from one or its accumlation on another.

The moisture regimes were defined in terms of both the groundwater level and the presence or absence of water held at a tension of $<$ I5 bars in the moisture control section by periods of the year. The moisture control section is considered to lie between I0 and 30 cm if the texture is fine loamy, coarse silty, fine silty, or clayey and to extend from 20 to 60 cm if the texture is coarse loamy and from 30 to 90 cm if the texture is sandy. The classes of moisture regimes recognized were I) Aquic 2) Aridic 3) Torric 4) Udic 5) Ustic 6) Xeric.

Aquic implies a reducing regime that is virtually free of dissolved oxygen because the soil is saturated by ground water or by water of the capillary fringe. The aridic and torric are used for the same moisture regime but in different categories of the taxonomy. In the udic regime, the soil moisture control section is not dry in any part for as long as 90 days (cumulative). The ustic regime is intermediate between the aridic and the udic regimes whilst the xeric regime is typified in Mediteranean climates, where winters are moist and cool and summers are warm and dry.

The following classes of soil temperature regimes are used: I) Pergelic 2) Cryic 3) Frigid 4) Mesic 5) Thermic 6) Hyperthermic. Soils with a pergelic temperature regime have mean annual temperature lower than $0^{\circ}C$. The frigid regime has soils whose mean annual temperature is lower than $8^{\circ}C$, mesic is between 8° and $15^{\circ}C$, thermic is between I5 and $22^{\circ}C$ while the hyperthermic has mean annual soil temperatures over $22^{\circ}C$.

The USSR has also attempted to build relevant aspects of climate into its soil classification scheme. For the purposes of developing a world soil map, Kovda et al proposed (7) a " parallel evolutionary row" scheme of soil classification based on the soils evolutionary history in certain energy conditions. This potential energy of soil formation (E) is determined firstly by the solar radiation balance of the earths surface and secondly by the completeness of its possible utilization depending on present water resources i.e. the coefficient of humidity. It is, therefore, a function of the radiation balance (Q) and of the moisture coefficient (a) i.e. $E=f(Q,a)$.

In this proposed scheme of world soil historic genetic rows, the energetic rows included were: A - humid tropical, B - tropical arid-humid, C - tropical arid, D - subtropical arid, G - subboreal-humid H - subboreal cryogenic, I - subboreal-arid-humid, J - subboreal arid, K - boreal-humid, L - boreal cryogenic, M - polar humid cryogenic, N - polar humid-arid cryogenic.

These changes of energy potential could be expressed in a system of thermal belts, which are usually characterized by the radiation balance in calories per unit surface area or by the sum of active temperatures. Because it was for a world soil classification scheme only the main humidity facies (Table I) within thermal belts (Table 2) were proposed for consideration (7).

Table I : Humidity Facies*

I	Cryogenic	1.0 (with permafrost)
2	Humid	1.0
3	Arid-humid	0.5 - 1.0
4	Arid	0.5

Table 2 : Thermal Belts*

Nos	Belts	Radiation balance ccal/cm^2/year	Annual sum of temps. higher than + 10°C
1	Polar	0 - 25	600
2	Boreal	25 - 35	600 - 2400
3	Subboreal	35 - 50	1800 - 4000
4	Subtropical	50 - 75	3200 - 7000
5	Tropical	75 - 100	7000 - 8000

*After Kovda et al 1966 (7)

4. SOIL MAPS OF GREAT REGIONS

In reviewing global land resources for agricultural development Dudal (9) pointed out that although a number of world soil maps had been compiled at scales varying from 1:20,000,000 to 1:100,000,000, they were based on general information rather than on actual surveys. Furthermore, the distribution of soils shown differed with various schools of thought so that correlation and interpretation on different systems and source material met with great difficulties.

However, knowledge of the worlds' soils increased markedly from the early fifties onwards as soil survey activities expanded considerably. The 6th Congress of the International Society of Soil Science (ISSS) held in Paris in 1956 recommended that special attention be given to the classification and correlation of the soils of great regions. As a result, soil maps for Africa, Austrailia, Asia, Europe, South America and North America at scales ranging from 1:5,000,000 to 1:10,000,000 were presented at the 7th ISSS Congress in Madison,

U.S.A. in 1960. Although these maps made a significant contribution to our knowledge of global soils, nomenclature, survey methods, legends and systems of classification varied so widely that comparisons between major regions remained difficult (9).

5. SOIL MAP OF THE WORLD (FAO/UNESCO)

Prior to the recent publication of the FAO/Unesco Soil Map of the World the appraisal of soil resources on a global basis and at reasonable scale, had engaged the minds of scientists for many years. At the 1960 Meeting of the ISSS it was resolved to prepare through international cooperation, a soil map of the world at scale 1:5 million.

5.1 Project History

It was hoped to overcome the rather confused situation which existed prior to this because of the use of different classification systems, through which some similar soils were known by different names and conversely, the same name had been applied to different soils. This therefore required a major soil correlation effort. First drafts of maps and legends were presented at the 8th ISSS Congress in Bucharest in 1964 and international agreement on the legend was reached at the 9th ISSS Congress in Adelaide in 1968. The publication was completed in 1978 and the entire coverage of 19 maps (10) was presented at the 11th ISSS Congress at Edmonton, Canada in that year.

5.2 Cartographic Representation

The map comprises 19 sheets of 80 x 100cm, (Fig. 1). The maps covering a continent or large region are accompanied by an explanatory volume and there are 10 of these. Map units consist of associations of soils roughly equivalent to Subgroups.

Each soil association is composed of dominant and subdominant soil units, the latter estimated to cover at least 20% of the delimited area. Important soil units which cover less than 20% of the area are referred to as inclusions. Each association is represented on the map by a symbol for the dominant soil, a number for the particular association, a figure for the textural class and a letter indicating the slope class. For example Lc6-3b represents Chromic Luvisol Association number 6 and it has fine texture on rolling to hilly topography. The figure 6 refers to the description legend on the back of the map in which the full composition of the association is outlined. For example Lc6 might represent Chromic Luviosls and Calcaric Gleysols with inclusions of Eutric Regosols.

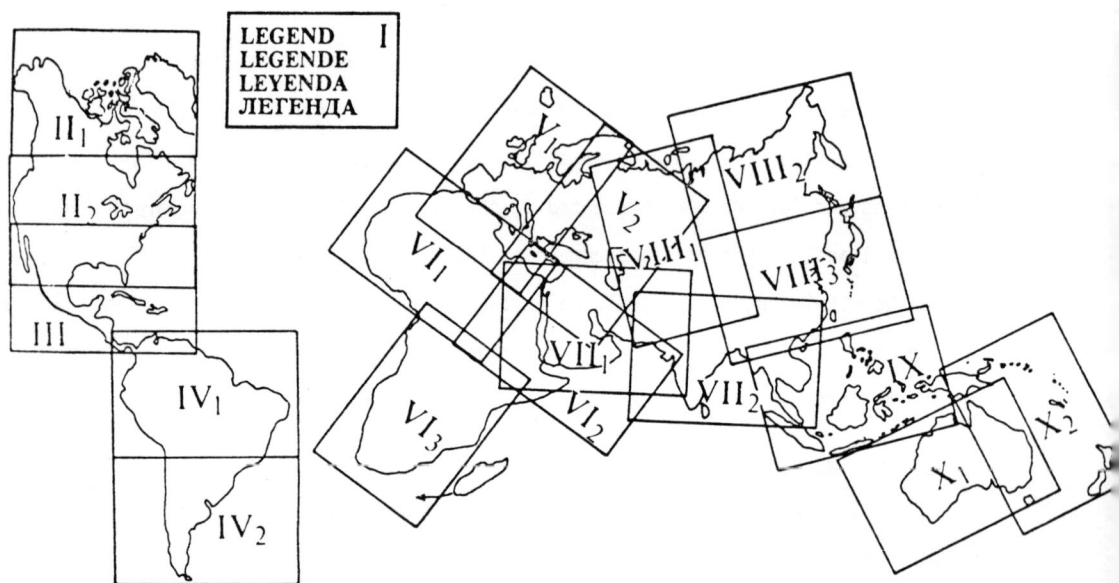

Figure I : Map sheet layout for the FAO/Unesco Soil Map of the World : After (IO), 1974.

The legend contains 106 units (Table 3) which are clustered into 26 major groups corresponding generally to either orders or suborders of previous systems. As an example, sixteen of these major groups are represented on the simplified representation of the soil map of Africa (Fig. 2).

Table 3: FAO/Unesco Soil Map of the World Units and Major Groups

J FLUVISOLS
- Je Eutric Fluvisols
- Jc Calcaric Fluvisols
- Jd Dystric Fluvisols
- Jt Thionic Fluvisols

G GLEYSOLS
- Ge Eutric Gleysols
- Gc Calcaric Gleysols
- Gd Dystric Gleysols
- Gm Mollic Gleysols
- Gh Humic Gleysols
- Gp Plinthic Gleysols
- Gx Gelic Gleysols

R REGOSOLS
- Re Eutric Regosols
- Rc Calcaric Regosols
- Rd Dystric Regosols
- Rx Gelic Regosols

I LITHOSOLS

B CAMBISOLS
- Be Eutric Cambisols
- Bd Dystric Cambisols
- Bh Humic Cambisols
- Bg Gleyic Cambisols
- Bx Gelic Cambisols
- Bk Calcic Cambisols
- Bc Chromic Cambisols
- Bv Vertic Cambisols
- Bf Ferralic Cambisols

L LUVISOLS
- Lo Orthic Luvisols
- Lc Chromic Luvisols
- Lk Calcic Luvisols
- Lv Vertic Luvisols
- Lf Ferric Luvisols
- La Albic Luvisols
- Lp Plinthic Luvisols
- Lg Gleyic Luvisols

Q ARENOSOLS
- Qc Cambic Arenosols
- Ql Luvic Arenosols
- Qf Ferralic Arenosols
- Qa Albic Arenosols

E RENDZINAS

U RANKERS

T ANDOSOLS
- To Ochric Andosols
- Tm Mollic Andosols
- Th Humic Andosols
- Tv Vitric Andosols

V VERTISOLS
- Vp Pellic Vertisols
- Vc Chromic Vertisols

D PODZOLUVISOLS
- De Eutric Podzoluvis
- Dd Dystric Podzoluvisols
- Dg Gleyic Podzoluvisols

P PODZOLS
- Po Orthic Podzols
- Pl Leptic Podzols
- Pf Ferric Podzols
- Ph Humic Podzols
- Pp Placic Podzols
- Pg Gleyic Podzols

W PLANOSOLS
- We Eutric Planosols
- Wd Dystric Planosols
- Wm Mollic Planosols
- Wh Humic Planosols
- Ws Solodic Planosols
- Wx Gelic Planosols

Z SOLONCHAKS
- Zo Orthic Solonchaks
- Zm Mollic Solonchaks
- Zt Takyric Solonchaks
- Zg Gleyic Solonchaks

S SOLONETZ
- So Orthic Solonetz
- Sm Mollic Solonetz
- Sg Gleyic Solonetz

Y YERMOSOLS
- Yh Haplic Yermosols
- Yk Calcic Yermosols
- Yy Gypsic Yermosols
- Yl Luvic Yermosols
- Yt Takyric Yermosols

X XEROSOLS
- Xh Haplic Xerosols
- Xk Calcic Xerosols
- Xy Gypsic Xerosols
- Xl Luvic Xerosols

A ACRISOLS
- Ao Orthic Acrisols
- Af Ferric Acrisols
- Ah Humic Acrisols
- Ap Plinthic Acrisols
- Ag Gleyic Acrisols

N NITOSOLS
- Ne Eutric Nitosols
- Nd Dystric Nitosols
- Nh Humic Nitosols

F FERRALSOLS
- Fo Orthic Ferralsols
- Fx Xanthic Ferralsols
- Fr Rhodic Ferralsols
- Fh Humic Ferralsols
- Fa Acric Ferralsols
- Fp Plinthic Ferralsols

K KASTANOZEMS
- Kh Haplic Kastanozems
- Kk Calcic Kastanozems
- Kl Luvic Kastanozems

C CHERNOZEMS
- Ch Haplic Chernozems
- Ck Calcic Chernozems
- Cl Luvic Chernozems
- Cg Glossic Chernozems

H PHAEOZEMS
- Hh Haplic Phaeozems
- Hc Calcaric Phaeozems
- Hl Luvic Phaeozems
- Hg Gleyic Phaeozems

M GREYZEMS
- Mo Orthic Greyzems
- Mg Gleyic Greyzems

O HISTOSOLS
- Oe Eutric Histosols
- Od Dystric Histosols
- Ox Gelic Histosols

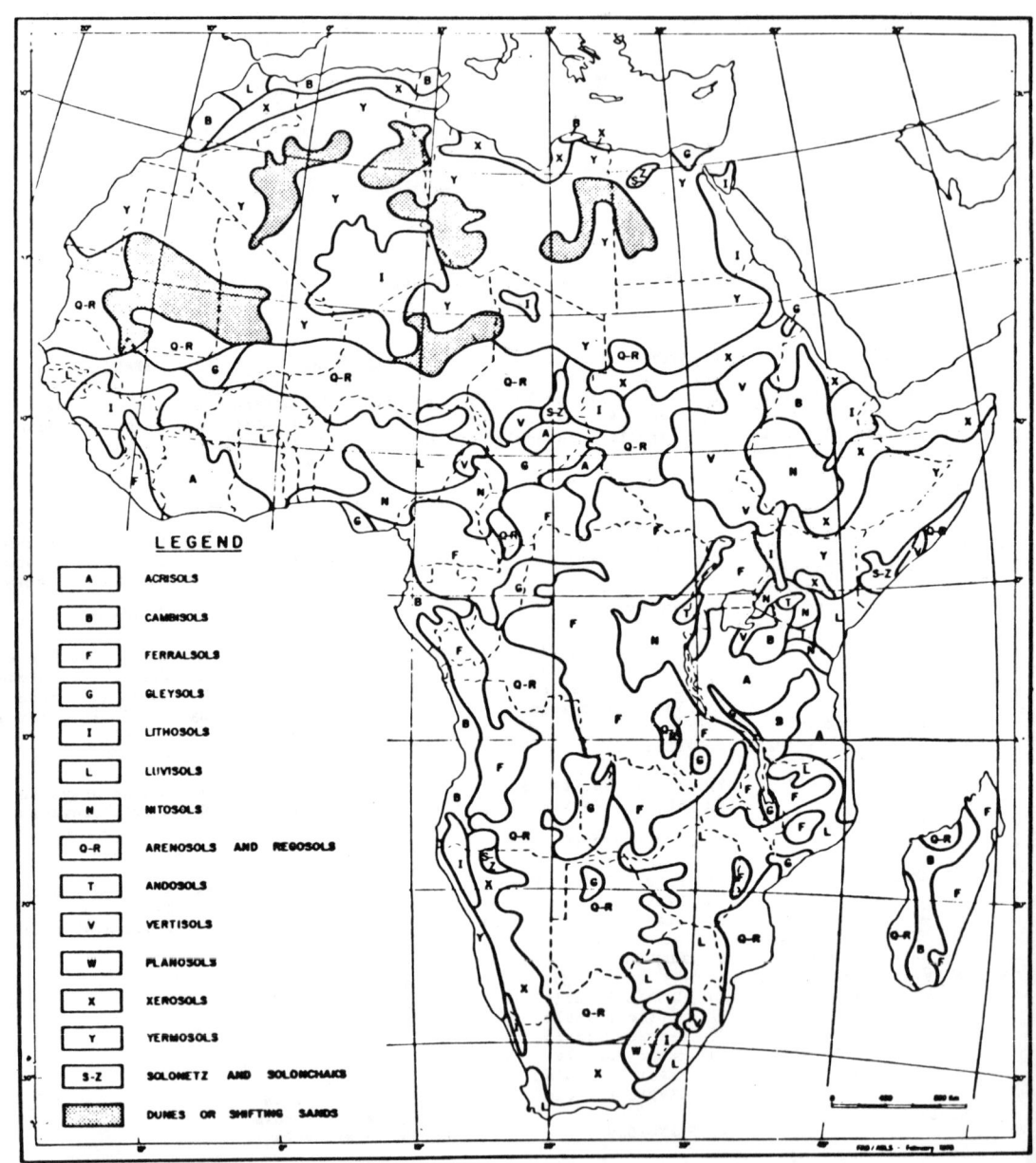

Figure 2 : Soil Associations of Africa : After Dudal, 1978 (9).

6. LIMITATIONS OF THE SOIL MAP OF THE WORLD (FAO/UNESCO) FOR CLIMATE MODELLING

6.1 Reliability of the Map

Dudal (9) has shown the sources of information as well as the level of soil survey coverage on which the map was based (Table 4).

Table 4* - Soil Survey Coverage (percent)

	Class I	Class II	Class III
Africa	7.5	38.0	54.5
Asia	19.0	49.0	32.0
Australasia	11.0	61.0	28.0
Europe	76.3	23.7	-
North & Central America	28.0	16.0	56.0
South America	14.6	45.9	39.5
World	21.0	40.0	39.0

*After Dudal (9), 1978

Class I represents those areas where systematic field surveys have been carried out whereas in Class II and III areas, soil boundaries have been derived from interpretation of general information on landforms, geology, climate and vegetation, and from scattered soil studies. It appears that only about one fifth of the worlds soils have been surveyed, the highest coverage being in Europe and the lowest in Africa. The information for China was supplied by the Soils Institute of Nanking and is therefore, realistic and reflects the knowledge acquired in the country itself (11).

6.2 Number of Mapping Units

A second problem in using the Soil Map of the World for Climate Modelling may arise because of the large number of mapping units e.g. Africa 1509, N. America 596, giving a world total of approximately 5000.

However, it is possible to group the mapping units into broad soil regions so as to reduce their number. In a recent study (12) aimed at estimating the worlds food production potential, the resulting number of broad soil regions was for Africa 33(Fig. 3), for South America 27(Fig. 4), and for the world was 222 (Table 5).

Table 5* Number of Broad Soil Regions by Continent

Continent	No. of Broad Soil Regions
South America	27
Austrailia	25
Africa	33
Asia	54
North America	34
Europe	49
World Total	222

*After Buringh et al. (1975)

It does not necessarily follow that, for climate modelling, a similar number of broad soil regions would suffice, but as an initial step the number would probably not be greatly different.

6.3 Composition of Mapping Units

A third problem in the use of the Soil Map of the World for climate modelling may arise because of the composition of the mapping units. The list of soil associations will show that widely different soil units may be present in any mapping unit. For instance, gently rolling Dystric Cambisols, flat Dystric Gleysols, and flat Dystric Histosols may be included within the one mapping unit where it is not possible to show them seperately because of small map scale.

Erikson (13) has also touched on this problem in discussing the properties of land surfaces and their influence on atmospheric processes. In particular, he dealt with 1) surface roughness 2) surface reflectivity 3) heat storage capacity 4) water storage capacity and

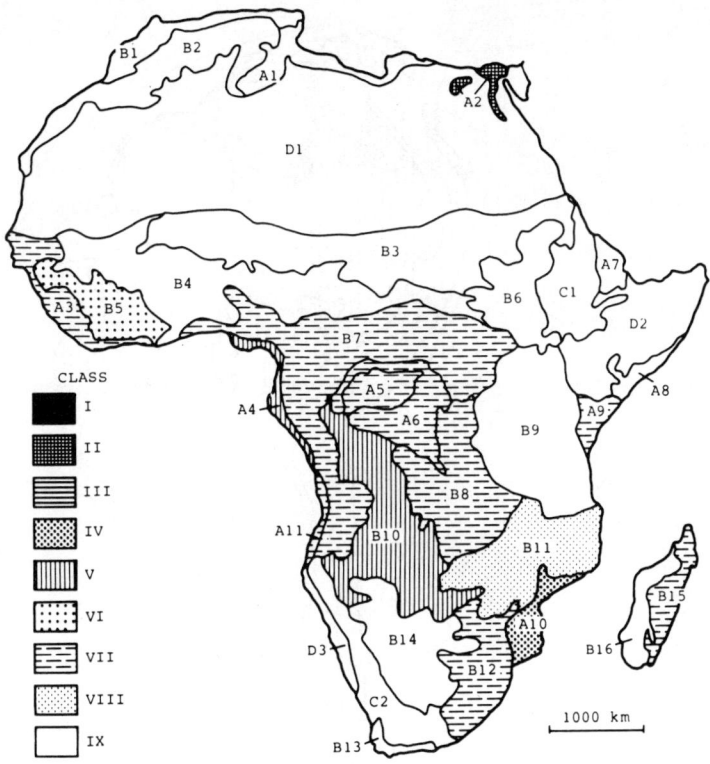

Figure 3 : Broad Soil Regions of Africa : After Buringh et al. 1975 (I2)

5) runoff. He stated that each grid point in general circulation models may represent an area of 250 by 250 km or more and that such an area is normally a composite of various surface forms such as grasslands, bogs, forests etc.

He pointed out, however, that averaging for particular land units of the various critical features of soil behaviour for climate modelling would overcome the problem and would provide sufficient information. This approach could also be applied to the map units of the Soil Map of the World and in fact has been used successfully in a recent FAO agro-ecological zones study for Africa (I4). The following was the method used.

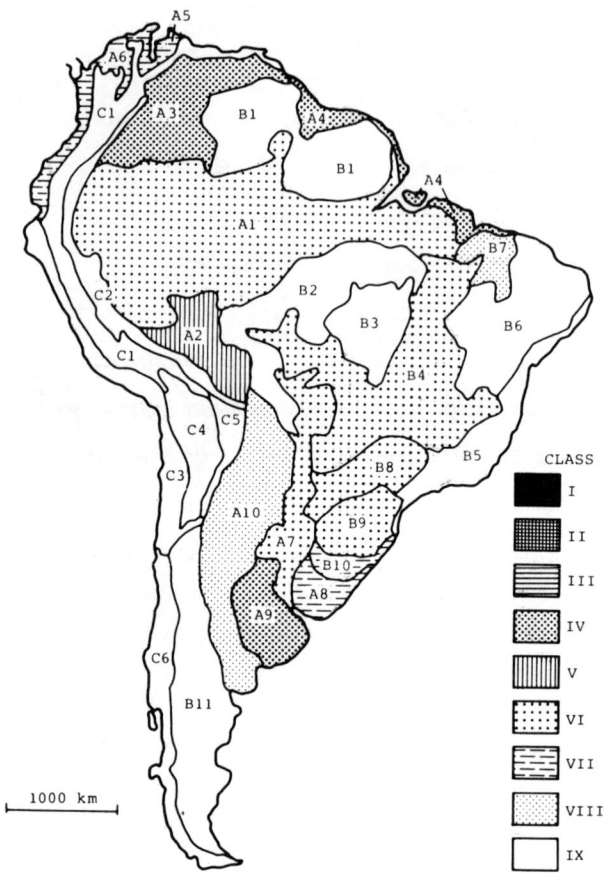

Figure 4 : Broad Soil Regions of South America : After Buringh et al. 1975 (12).

6.4 Agro-ecolological Zones Project of FAO

Although the extent of each soil association is given in the explanatory volumes, the areas covered by each soil unit within the soil associations are not available. The areas for each soil unit were calculated on the basis of the number of soil units in each association as shown in table 6.

Table 6* RELATIVE DISTRIBUTION OF DOMINANT SOIL, ASSOCIATED SOIL(S) AND INCLUSION(S) EXPRESSED IN PERCENTAGE OF THE AREA OF THE MAPPING UNITS

Dominant Soil	Associated soil (s)		Inclusion(s)	
Percentage of area	Number of soil units	Percentage of area	Number of soil units	Percentage of area
100	0	0	0	0
70	1	30	0	0
60	1	30	1	10
60	2	20 + 20	0	0
50	2	20 + 20	1	10
30	3	20 + 20 + 20	1	10
50	1	30	2	10 + 10
40	1	30	3	10 + 10 + 10
50	1	30	4	5 + 5 + 5 + 5
40	2	20 + 20	2	10 + 10
30	2	20 + 20	3	10 + 10 + 10
30	3	20 + 20 + 20	2	5 + 5
25	3	20 + 20 + 20	3	5 + 5 + 5
24	3	20 + 20 + 20	4	4 + 4 + 4 + 4

*After FAO Agro- ecological Zones Project (14)

The distribution of soil units was then computerised. The initial stage recorded the extent and composition of each mapping unit which was computer listed by countries. A climatic inventory was superimposed and by using a 2 mm grid count, the extent of each mapping unit in each major climatic division was estimated.

The second stage of the computer programme converted these basic data input into the extents of individual soil units in each major climatic division.

A complete breakdown of the extents of individual soil units by countries, by slope, texture and phases, by major climatic divisions and by lengths of growing period zones is available at FAO.

6.5 Critical Features of Soil Behaviour

Detailed and semi-detailed soil maps from which the Soil Map of the World has been compiled are usually accompanied by explanatory volumes with soil descriptions and analytical data. Features such as soil depth, texture, structure, consistence, pH, cation exchange capacity, free iron and organic matter contents on a soil profile basis are usually included. Generally, however, they do not contain measurements of moisture capacity, surface runoff, albedo or permeability.

But these features are accessory to those that are diagnostic and therefore can be either measured directly or arrived at through conversion studies based on the diagnostic characteristics of the individual soil units. For example in recent work in the Netherlands (I5), moisture retention and hydraulic conductivity curves have been derived on the basis of the soil texture triangle and supporting literature data. The recently published Atlas of Soil Reflectance Properties (I6) for the U.S.A. shows that precise measurements of the visible and near infrared reflectance, (apart from eliminating the subjectivity previously associated with the determination of soil colour) were also characteristic for each major soil. In addition, the soil spectra often revealed textural, structural, mineralogical and other significant soil differences.

6.6 Climatic Phases for the Soil Map of the World

Even though certain climatic parameters are taken into account in the evolution of soil classification systems, the maps themselves may not reflect an accurate assessment of soil climate. For this reason certain attempts to produce soil climate maps of large regions may be of interest.

Dudal (I7) pointed out that the variation in the climatic criteria used in the definitions of soil categories and their use at different levels was an additional difficulty to the correlation of soil units in the FAO/Unesco Soil Map of the World project. Although the units could show a similar morphology, they might need to be seperated because of differences in the climatic regime. He drew attention also to two other important aspects. Soils with similar morphology and chemical composition may occur in different climatic zones. This could result from weak soil development not yet reflecting the influence of climate on soil formation (e.g. certain alluvial soils) or the dominant influence of some factor of soil formation other than climate (e.g. the occurrence of podzols on quartz sands in different climatic belts).

This also applies to soils formed in successive weathering cycles thereby reflecting climatic conditions which no longer prevail - hence their occurrence in different zones.

Dudal (I7) therefore proposed the use of climatic phases which were intended to have a pedogenetic as well as an ecological sifnificance. He stressed that their introduction would be justified where soils are known to differ although this difference cannot be expressed in terms of measurable characteristics other than soil temperature and soil moisture. Furthermore the use of climatic phases would open the way to establish relationships between agricultural potential and the soils in their present environment, thus increasing the applicability of small scale soil maps.

In submissions from individual countries for the compilation of the Soil Map of the World climatic data could be expressed in terms of the climatic subdivisions currently used in the country or region or else in terms of subdivisions of climatic phases proposed within the World Soil Resources Office. These phases were I) Polar 2) Boreal 3) Temperate 4) Hot 5) Humid tropical 6) Savanna 7) Andine 8) Desertic 9) Xeric I0) Udic I2) Mediteranean.

The definitions were based largely on the work of Papadakis (I8) and involved the ratio of monthly rainfall to monthly temperatures (R/T). For example, the definitions for (a) Humid tropical and (b) Desertic are respectively:

(a) Mean annual temperature of the warmest month above 20°C., annual range of temperature below 10°C, no more than 4 months have an R/T ratio below 2; average of all monthly R/T ratios above 2. (b) All months with mean temperature above 5°C having an R/T ratio below 2; average of all these monthly ratios is below I.

Although consideration was given to these proposals to insert "climatic variants" on the Soil Map of the World, it was eventually decided that the seperation of such variants would require general agreement on a climatological classification. This appeared to be beyond the scope of the project. Consequently only the permafrost and intermittant permafrost boundaries have been shown on the map. However, it was felt (10) that the climatic maps included in the explanatory volumes supplied first approximation data for the agricultural potential of soils in terms of moisture and temperature characteristics.

7. SOIL CLIMATE MAPS

In discussions on soil climate in the U.S.A. (19) it was pointed out that historically, most classification systems for climate have emphasised the aerial biosphere. But climatic types have also been identified and their boundaries determined empirically by noting the relationships of kinds of vegetation, soils and drainage features. None of these systems, however, account for the interaction between aerial climate and soil climate.

Soil climate relates to aerial climate, but the responses are affected mainly by the water content, depth, surface cover, landscape position of the soils and human interference. The interactions are often indirect, complex and difficult to evalute. Soil climate classifications are relatively new and data have not generally been available on scales comparable with those of aerial climate data.

The attempt to make a provisional classification of soil climate for North America (19) integrated the available data with current concepts of soil temperature and moisture, the relationships of climate and vegetation to soils and recognized regional climatic seperations which have existed for some time.

The soil temperature and moisture regimes used for the U.S.A. were similar to those described earlier in this paper whilst for Canada (20) they involve definitions of a growing season (above $5^\circ C$) with mild (above $5^\circ C$) and thermal (above $15^\circ C$) periods, and a dormant season (below $5^\circ C$) with cool (above $0^\circ C$) and frozen (below $0^\circ C$) periods based on soil temperature measurements. Correlative work among soil scientists and climatologists of the U.S.A. and Canada resulted in broadly correlated criteria for categorizing temperature and moisture regimes as well as the nomenclature used.

Seven soil temperature regimes, ranging from arctic (MAST below -$7^\circ C$) to hyperthermic (MAST above $22^\circ C$) express the relationshop of soil temperatures to length, magnitude and intensity of heat conditions during various seasons of the year.

Ten moisture regimes used evaluate the duration and amount of soil moisture, ranging from peraquic (with free water surfaces) through aquic (saturated) to moist and dry regimes.

The soil climate of the dominant and sub-dominant soils in the map units of the North America sheet (Fig. 5) of the FAO/Unesco Soil Map of the World were described in terms of these regimes. The resulting soil climate map (Fig. 6) shows the distrubution of soil climates expressed in terms of dominant soil temperature and moisture regimes. It was stressed, however, that the framework of the classification used was preliminary and the nomenclature and combinations of parameters and codings used provisional.

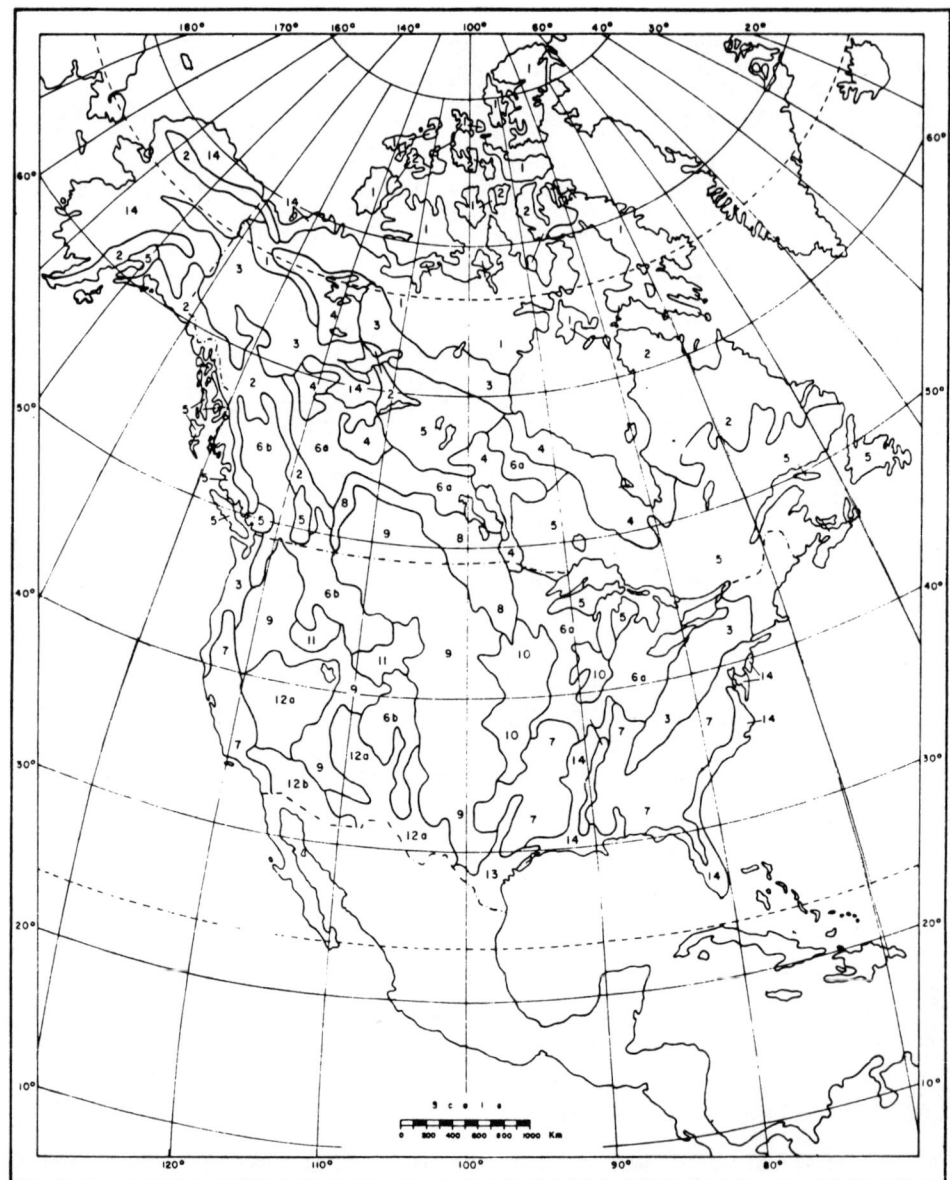

Figure 5 Major soil regions of North America
: After (17), 1975.

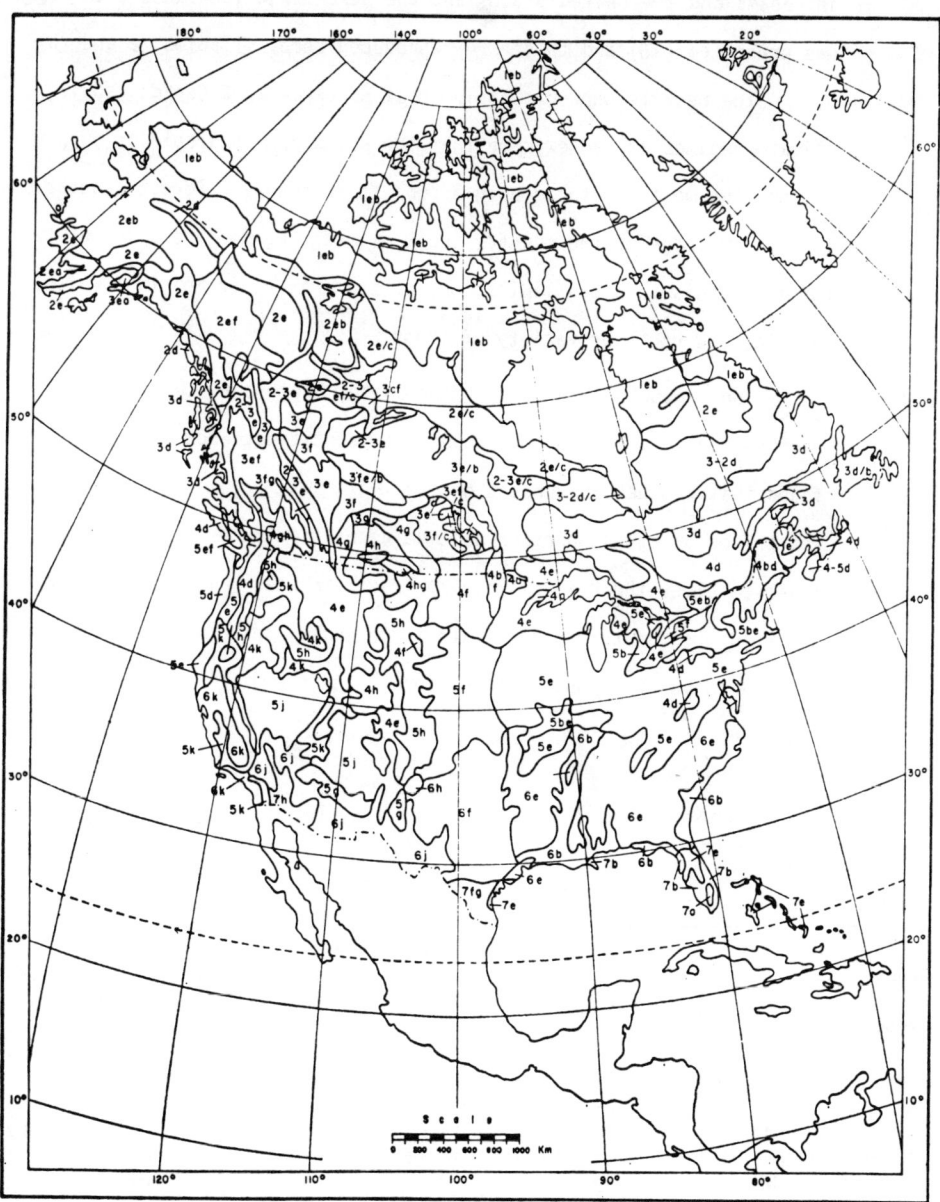

Figure 6 - Soil climate regimes of North America
: After (17), 1975.

A generalized summary of the extent of each kind of soil climate in Canada and the United States and the percentage land area each represents was presented (19) and these are reproduced here. (Tables 7a and 7b)

The results showed that in about 56 percent of Canada, soil climate regimes are so extreme that the soils are incapable of supporting any but the most limited growth of forest or tundra - forest vegetation. In the remainder of the country soil climate regimes are mild but two thirds of this area has limitations of terperature or moisture which restrict the range or variety of crops.

Only about 15 percent of the United States has soil temperature regimes too cold to support productive vegetation. However, it was found that another 30 percent has soil climatic regimes either too wet or too dry to support productive vegetation unless measures were taken to alter the soil climate for specific uses. In the remainder of the United States soil climate is generally conducive to the production of a wide range of crops, pasture, forage and forest.

TABLE 7a – EXTENT OF SOIL TEMPERATURE AND SOIL MOISTURE REGIMES IN CANADA

Soil moisture regimes \ Soil temperature regimes	Arctic	Subarctic	Cryoboreal	Boreal	Mesic	TOTAL	PERCENT
			Square kilometres				
Perhumid		77 466	1 497 169	196 452	26 363	1 797 450	19.6
Humid	2 759 836	1 823 564	786 793	55 605	49 007	5 474 805	59.6
Subhumid			244 717	43 218	11 976	299 911	3.3
Semiarid			32 267	96 924	666	129 857	1.4
Subarid				115 346		115 346	1.3
Complexes							
Perhumid and humid			120 821	231 839		352 710	3.8
Humid and subhumid		457 547	397 578	5 245	34 469	894 839	9.7
Subhumid and semiarid			75 459	31 478		106 937	1.2
Semiarid and subarid				11 541		11 541	0.1
TOTAL	2 759 836	2 358 577	3 154 804	787 698	122 481	9 183 396	100.0
PERCENT	30.0	25.7	34.4	8.6	1.3	100.0	

TABLE 7b - EXTENT OF SOIL TEMPERATURE AND SOIL MOISTURE REGIMES IN THE UNITED STATES

Soil moisture regimes \ Soil temperature regimes	Arctic	Subarctic	Cryoboreal	Boreal	Mesic	Thermic	Hyper-thermic	TOTAL	PERCENT
	..........................Square kilometres..........................								
Peraquic							7 000	7 000	0.1
Aquic				202 800	1 009 400	354 430	88 000	1 654 630	18.0
Subaquic									
Perhumid		112 300	61 100	83 200	48 900			305 500	3.3
Humid	254 300	667 300		642 900	124 800	894 500	56 200	2 640 000	28.9
Subhumid				151 600	420 400	505 800		1 077 800	11.7
Semiarid				146 700	100 200			246 900	2.7
Subarid				124 600	867 700	14 700		1 007 000	11.0
Arid					259 100	357 000	85 600	701 700	7.6
Xeric				259 000	298 100	114 900	. .	672 000	7.3
Complexes									
Aquic and humid					224 900			224 900	2.5
Aquic and subhumid				34 000				34 000	0.4
Humid and aquic		59 600	23 400					83 000	0.9
Humid and subhumid		442 500						442 500	4.8
Subhumid and semiarid						70 900		70 900	0.8
Arid and semiarid				7 000				7 000	0.1
TOTAL	254 300	1 281 700	84 500	1 651 800	3 353 500	2 312 230	236 800	9 174 830	100.0
PERCENT	2.8	14.0	0.9	18.0	36.5	25.2	2.6	100.0	

8. SUMMARY AND CONCLUSIONS

8.1 Soil Classification and the Concept of Soil Zonality

As soil science developed in the early part of this century, the correlation between climate and soil distribution was recognized. This led to the zonal concept of soil formation namely that given sufficient time, zonal soils would dominate irrespective of the other soil forming factors. Soon large - region soil maps and their legends were constructed using the concept of soil zonalilty. This zonal-geographic approach could also be traced in the evolving world soil classification systems.

But the limitations of the zonal concept were soon encountered. Soils subject to either erosion or deposition obviously failed to reach the zonal climax stage and had to be classified as azonal.

Possibly more important were the soils affected by poor drainage or dominating parent material characteristics (e.g. highly calcareous) which, therefore, had to be classified as intrazonal. A third limitation was the fact that profound climatic changes which had taken place especially in middle and high latitudes over the past 10,000 years, could not be taken into account.

8.2 Soil Maps of Great Regions

There are many problems associated with a system of soil classification suitable for small-scale soil maps. The greater the generalization of the map, the less can be said about the soil category. Despite this, a number of world and large region soil maps have been attempted since the beginning of this century. However, they all had serious drawbacks due to differences in soil nomenclature, survey methods, legends and soil classification systems.

8.3 FAO/Unesco Soil Map of the World

Intensive soil correlation under the FAO/Unesco Soil Map of the World Project (which was greatly encouraged by the International Society of Soil Science) eventually led to the production of a world soil map (1:5 million scale) in 1978 (10). Although based on the major existing soil classification systems it is not a soil classfication system in itself (17)., It is a set of soil definitions which facilitated the production of the world soil map and obtained international agreement. However, because it was based on the major world soil classification systems it includes the zonal-geographic concept together with considerations of the limitations of that concept.

The legend comprises 106 soil units which have been clustered into 26 major groupings. The map units are geographic associations usually composed of dominant and associated soils $\geq 20\%$. Provision is also made for soils of less than 20% occurrence (inclusions).

There are 19 map sheets each measuring 80 x 100 cm. An explanatory volume accompanies each group of sheets comprising a large region or continent and there are 10 volumes.

8.4 Use of the FAO/Unesco World Soil Map for Climate Modelling

Since this is the only world soil map at reasonable scale which has been developed through international cooperation and soil correlation it constitutes the best global framework for assessing the soil factor in climate modelling.

Although not based on critical features of soil behaviour for climate modelling, these features are accessory to those that are diagnostic in the system. For example, the moisture regime is used in defining classes in the very high categories of world classification systems. There is the added advantage that the moisture regime of most soils is determined by the present climate. Therefore a small-scale map can be interpreted in terms of the many accessory characteristics that are common to the soils that have a common climate. These characteristics include the amount, nature and distribution of organic matter, the base status of the soil and the presence or absence of salts.

However, because only moisture and temperature regimes were used in world soil classification systems, climate per se has not been used to define taxa. Therefore, these limitations must be realized.

It is for these reasons that attempts have been made to produce soil climate maps e.g. North America (15) or that there was provision in the FAO/Unesco Soil Map of the World scheme for the inclusion of climatic phases. These phases are broad subdivisions intended to have a pedogenetic as well as an ecological significance. Their introduction onto the soil map of the world scheme was encouraged where soils were known to differ although these differences could not be expressed in terms of measurable characteristics other than soil moisture and soil temperature.

8.5 Other limitations of the FAO/Unesco World Soil Map

Because it is based on current field knowledge of soils, the FAO/Unesco Soil Map of the World is the most reliable global soils inventory available. Nevertheless, it must be realized that only one fifth of the worlds soils have actually been surveyed and the information for certain third-world countries in particular is unsatisfactory.

The map scale and the composition of map units are other important limitations. Due to local topographic or parent material influences, intrazonal and azonal soils can form an intricate pattern with zonal soils. Because of the map scale these often cannot be shown except as associated or included soils in the mapping unit. This shortcoming has been surmounted in the FAO/Unesco Agro-Ecological Zones study for Africa (I4).

The method used consisted of an assessment of the percentage composition of the components of each mapping unit. They were computer-listed by countries and the basic data were converted into the extent of individual soil units in each major climatic division and length of growing period zone. By adopting a similar approach it should be possible to use the FAO/Unesco Soil Map of the World as a basic soils framework for climate modelling. This may require, as in the FAO Agro-ecological zones study an averaging, for the mapping units of elements which affect critical parameters such as soil moisture capacity, albedo, surface runoff and saturated permeability.

REFERENCES

1. Glinka, K.D. (1927)
   Dokuchaiev's Ideas in the Development of Pedology and Cognate Sciences Acad. of Sci., U.S.S.R. Russ., Pedol. Invest., I, 32p Leningrad.

2. Smith, G.D. (1966)
   Soil Classification in the United States, Meeting of the FAO/Unesco Map of the World Advisory Panel, Moscow. World Soil Resources Report No. 29, FAO, Rome.

3. Whitney, M.(1909)
   Soils of the United States. U.S. Dept. Agr. Bull. 55. 243p.

4. Marbut, C.F. (1935)
   Soils of the United States. In U.S. Dept. Agr., Atlas of American Agriculture, pt. 3, Advance Sheets No - 8, 98p.

5. Baldwin, M., Kellogg, C.W., and Thorp, J. (1938)
   Soil Classification. In Soils and Man, Yearbook of Agriculture, 1938, pp. 979-1001, U.S. Dept. Agr.

6. U.S.D.A. (1975)
   Soil Taxonomy. A basic system of soil classification for making and interpreting soil surveys. U.S. Government Printing Office, Washington, D.C.

7. Kovda, V.A., et al. (1966)
   An attempt at Legend Construction for the 1:5,000,000 World Soil Map Meeting of the FAO/Unesco Soil Map of the World Advisory Panel, Moscow, World Soil Resources Report No. 29, FAO, Rome.

8. Eyre, S.R.

   Vegetation and Soils - A World Picture, Arnold Ltd., London.

9. Dudal, R. (1978).

   Land Resources for Agricultural Development. IIth ISSS Congress, Edmonton, Canada.

10. FAO/Unesco (1971-1978)

    FAO/Unesco Soil Map of the World 1:5 million. North America, South America, Mexico and Central America, Europe, Africa, South Asia, North and Central Asia, Australasia, Unesco, Paris.

11. Dudal, R. (1980)

    Personal Communication.

12. Buringh, P., et al. (1975)

    Computation of the Absolute maximum Food Production of the World, Agric. University, Wageningen, Netherlands.

13. Erikson, E. (1975)

    Properties of Land Surfaces and their influence on Atmospheric Processes. GARP Publications Series, No. 16, W.M.O. Geneva.

14. FAO (1978)

    Report on the Agro-ecological Zones Project, Vol.I. Methodology and Results for Africa. World Soil Resources Report No. 48.

15. Rytema, P.E. (1969)

    Soil moisture forecasting Nota 513, Land and Water Management Research, Wejeninyer, Netherlands.

16. Stoner, E.R. et al. (1980).

    Atlas of Soil Reflectance Properties. Research Bulletin 962, Agricultural Experiment Station, Purdue University, West Lafayette, Indiana, U.S.A.

17. Dudal, R. (1966)

    Problems of International Soil Correlation, Fifth Meeting of the FAO/Unesco Advisory Panel for the Soil Map of the World, Moscow.

18. Papadakis, J. (1966)

    Climates of the world and their agricultural potentialities. Av. Cordoba, 4564, Buenos Aires.

19. FAO/Unesco (1975)

    Soil Map of the World, Vol. II, North America, Unesco, Paris.

20. - (1970)

    Proceedings of the 8th Meeting of the Canada Soil Survey Committee, Ottawa.

LAND SURFACE PROCESSES : VEGETATION

by
Alain Perrier
Station de Bioclimatologie-Télédétection (INRA)
Route de Saint-Cyr, 78000 VERSAILLES, France

I. INTRODUCTION

Land surface processes, as it is shown in several other review papers, are strongly dependent upon the nature of the surface and upon the physical and biological characteristics of the interface. Although land represents only around 29.2% of the earth's surface (149×10^6 km^2) including fifteen million square kilometers of ice-covered surface and around one million of water surface, its effects on general circulation and climate become quickly dominant when air masses move over the continents. Among the true land surfaces (around $133 \; 10^6$ km^2) it is possible to find all the transitions between complete bare soil such as sandy, gravel or rocky surfaces (respectively erg, reg or hamada desert) to well-vegetated surfaces such as the tropical rain forest which includes several layers of plants.

It is important to note that lands are very often divided into different patches introducing a great number of transitions:

- first, according to space between complete bare soil and soil fully covered by vegetation,

- second, according to time vegetative surfaces evolving with the climatic seasonal variations and biological cycles of plants.

One of the most suitable examples of these transitions according to time but also space is illustrated by intensive agricultural lands. In such a case, with a winter wheat crop, e.g., the land surface evolves and during six months may be put more or less in the same category as bare soil; then, for a few months, in a crop category partially covering the soil and finally during the next few months in the category of a well-covered surface. In fact, why, for studies concerning land surface processes, must more attention be given to a vegetated land surface than to bare soil? Just let us remember that a plant acting as an intermediate interface between soil and air is very active in capturing and dissipating energy through its spatial structure and also in pumping water through a more or less deep zone of soil defined by the depth explored by its root system (Rutter et al., 1975; Hadas et al., 1973). In an opposite way, most often, a soil surface dries up naturally, after a few hours or a few days, inducing a strong decrease in evaporation even though the dry zone is only a few millimeters or some centimeters thick (Fuchs and Tanner, 1967; Fuchs et al., 1979); this dry zone or "mulch" is directly correlated to the water and energy balance of the surface. But a vegetated surface which has its own active pumping system may maintain high evapotranspiration, reducing water loss sometimes by biological reactions when the atmospheric demand increases strongly (atmospheric demand is defined by the evaporation induced by given climatic conditions if the interface is saturated). Furthermore, this drastic modification in the exchange between soil and air by a vegetal cover is due not only to its adaptation for removing water from the soil but also to its structure (Slatyer and Gardner, 1965; Whittaker and Marks, 1975) which changes energy and momentum absorption.

Obviously, when exchanges at the lowest level of the biosphere are considered (Evans, 1963; Monteith, 1975) these considerations show that it is impossible to generalize upon the distribution of land surfaces and their modifications due to land farming. Finally, to characterize these surfaces in order to approach energy balance and water exchange some dominant parameters must be considered in more detail.

II. NATURAL AND AGRICULTURAL LAND SURFACE DISTRIBUTION

In all global atmospheric circulation models which use relatively large scale frames (about 100×100 km^2 to 500×500 km^2 even up to 1000×1000 km^2) land surfaces must be characterized at same scale; this is particularly difficult for a country or zones with high human density. Human activities modify strongly what we consider as natural land by introducing agriculture which cuts up the landscape into small pieces of land very often widely contrasting.

Nevertheless, according to ecological studies, land surfaces where no annual or biannual human intervention occurs are classified into uniform areas which can be considered as more or less fundamental units with common characteristics; each unit being called an ecosystem including biotic (living organisms) and abiotic (environmental air and soil conditions) factors (Duvigneaud, 1974). For ecological purposes and for our aims ecosystems are depicted by means of:

- their own specific structure with listed species and index of frequency, sociability; from a botanical viewpoint this analysis leads to the notion of smaller units called associations (phytosociology) generally associated in higher amalgamation orders and classes (Braun-Blanquet, 1964; Ozenda, 1963-1966);

- their spatial structure with horizontal distribution and vertical stratification including root zone; this description, more important than the previous one for land surface processes (heat and mass transfer), is categorized by the notion of vegetal formation (forests, savannas, maquis or scrubs, steppes) generally in good correlation with abiotic factors (Walter, 1979);

- their temporal variations which are cyclic when they are in relation to climate (abiotic factors) and which evolve slowly and regularly when they are affected by a transient process, the system evolving towards an equilibrium generally called a climax.

2.1 NATURAL LAND SURFACE DISTRIBUTION

Among the classifications of the geo-biosphere, climatic zones appear as a strong abiotic factor which defines zones or biomes called zonobiomes. In fact, they are more or less modified by other abiotic factors such as soil type (Gardiner, 1981) and orography. These modifications lead to some specific zonotions called pedobiomes and orobiomes but on an international scale quite a good agreement exists between zonal vegetation and zonal soil type due to the threefold interaction between climate, soil and vegetation. Many zonobiomes are transient zones and we can consider them as zonoecotomes.

Two dominant classes may be considered among zonobiomes if we must predominantly take account of the structure which acts upon momentum, energy, and, in particular, mass transfer:

- zonobiomes which include formations with a complete closed canopy (cf. paragraph 2.1.1) or with well-covered soil surfaces (in such formations radiative energy is absorbed mostly by leaves and different parts of plants and not by the soil these plant surfaces represent 80 to 100% of the evaporating surface, strongly modifying the microclimate inside the canopy, so that the soil acts only as a nutrient support and water reservoir;

- zonobiomes which include all formations covering more or less the soil surface (cf. paragraph 2.1.2) including as one extreme, desert (complete bare soil); in such a surface, in accordance with the percentage of bare soils, energy balance and evaporation evolve quickly after rainfall because the soil is a drying surface which also acts by microadvection on the evaporating surfaces of plants.

Even with this dichotomy which describes essentially spatial structure, we must superimpose cyclic variations (temporal variation) because if some zonobiomes maintain closed vegetation throughout the year or a constant percentage of active vegetation (cf. paragraph 2.1.1.1 evergreen canopies) others (cf. paragraph 2.1.1.2, 2.1.1.3 deciduous canopies and woodland or grassland) like, in particular, agricultural lands have a great fluctuation in vegetation cover.

2.1.1 Zonobiomes with closed vegetation

2.1.1.1 Evergreen canopies

(a) Sempervirent forest (equatorial zonobiome) or evergreen forest

This kind of forest located in countries with a strong rainy climate (from 2000 to 7000 mm per year) and with small variations in air temperature (around $27^\circ C \pm$ a few degrees), more or less extends in surface area around the equator in relation to oceanic and orographic effects (Figure 1) (Golley and Medina, 1975; Allen and Lemon, 1976; Bernhard-Reversat et al., 1978). It is characterized by a very dense canopy with many leaves: the uppermost one with the highest trees (30 to 60m) is always discontinuous; the intermediate zone (10 to 30m) is practically continuous with a great number of species, and the bottommost one is an herb and shrub layer (Longman and Jenik, 1974). Among these layers lianas, epiphytes and hemi-epiphytes increase the leaf density of the canopy; the total leaf area is often between 7 to 10 m^2 of leaves per m^2 of soil (Kira, 1978; Odum et al., 1963; Ogawa et al., 1965). The particularity is the continuous growing period with some specific periodicity which never affects simultaneously all the canopy, always maintaining a high leaf area index. Most of the root system (75%) is generally found in the first twenty to fifty centimeters and approximately more than 95% in a layer 1.3m thick (Huttel, 1975).

For these evergreen forest surfaces water balance is positive on a year-long basis (drainage is an important term) and most often it is positive throughout the year, if the climatic zone is not a transient one, and when no strong dry season appears. For example, in a zone such as South Ivory Coast (Bernhard-Reversat, 1972; Lemée et al., 1975) even with a dry season, water available in the soil never reaches values below 50% of soil water capacity. But, when a sufficiently strong dry season appears, the actual evapotranspiration estimated in the evergreen forest is affected and its values decrease from 1300 to 1200 mm and even 1100 mm according to the water capacity of the soil (respectively 500, 300 and 200 mm); this evapotranspiration reduction underlines a water stress (for these data the mean potential evaporation calculated from Turc's formula is around 1200 mm and the mean maximal evapotranspiration measured on well-watered grass around 1300 mm; the corresponding mean annual rainfall is about 1800 mm).

(b) Dense ombrophile forest

It is a zonobiome that covers small surfaces with warm temperate humid climate in the southern hemisphere (Figure 1). This evergreen forest or sempervirent forest is determined by the strong influence of the ocean and grows in a climate defined by a mean temperature between 10 to $18^\circ C$ but with no temperature below $0^\circ C$ and with heavy rains (starting from more than 1500 or 2000 mm). It is composed of few species (principally Eucalyptus, Nothofagus) but with several strata; the uppermost one may be very dense and consists of very high trees (40 to 100m in Tasmania which is the highest forest). This kind of forest, with a positive water balance, is very specific but similar to the evergreen rain forest in its behaviour towards land surface processes. In fact, this very local forest disappears quickly when rainfall decreases to be replaced by sclerophyllous sempervirent forest.

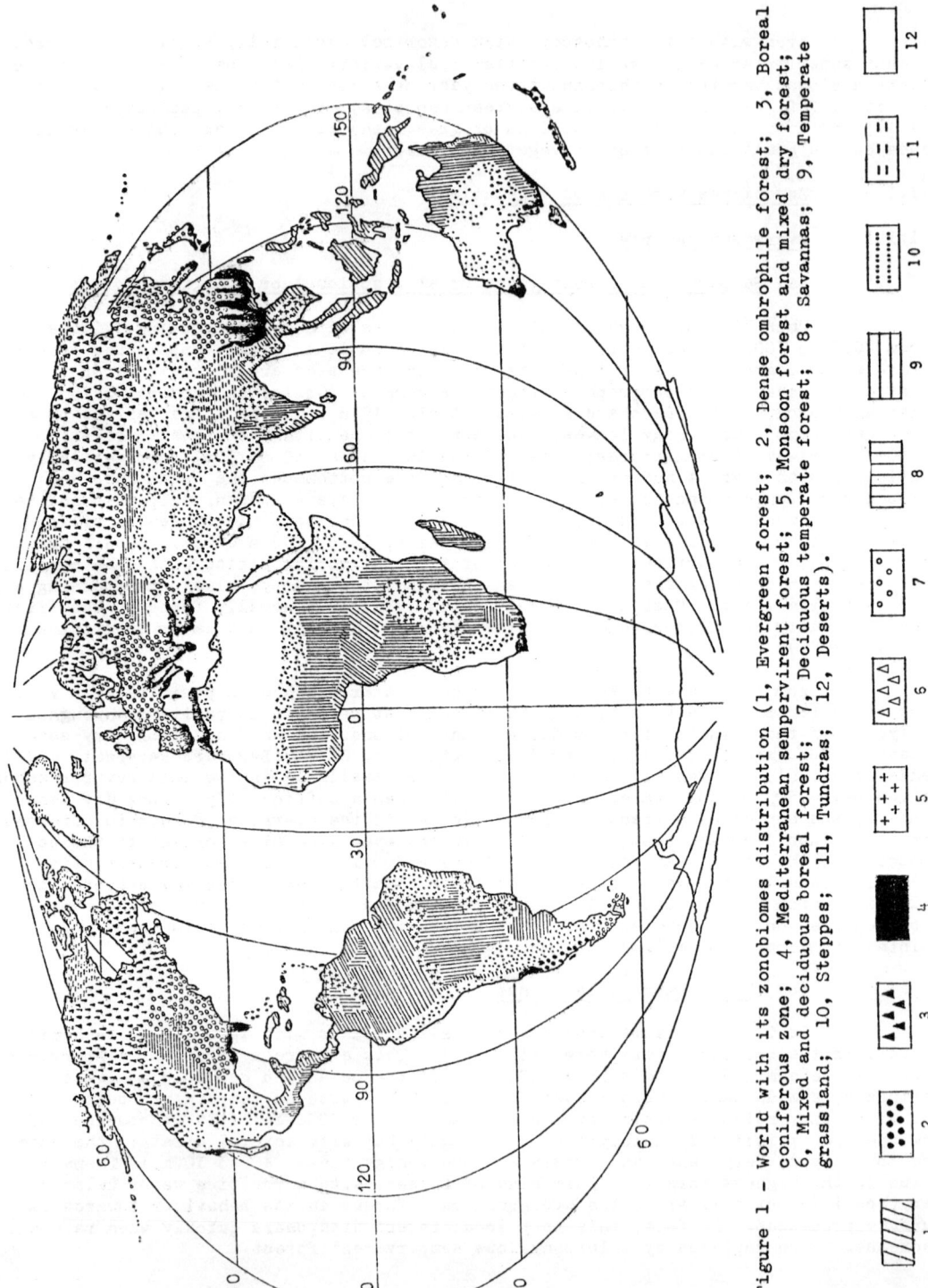

Figure 1 - World with its zonobiomes distribution (1, Evergreen forest; 2, Dense ombrophile forest; 3, Boreal coniferous zone; 4, Mediterranean sempervirent forest; 5, Monsoon forest and mixed dry forest; 6, Mixed and deciduous boreal forest; 7, Deciduous temperate forest; 8, Savannas; 9, Temperate grassland; 10, Steppes; 11, Tundras; 12, Deserts).

(c) <u>Sempervirent boreal forest or boreal coniferous zone</u>

This evergreen forest covers a large part of the land surface of the northern hemisphere (Figure 1). This forest with a reduced number of species is characterized by a short growing period, a function principally due to root zone temperature and air temperature (Albrektsson, 1976; Axelsson, 1976; Jarvis et al., 1976). The climate, by its low temperature, modulates the density of the forest, for below a mean temperature of $-10°C$ needles enter into dormancy. If during the year only a few days with a mean temperature above $10°C$ occur, the tundra (discontinuous canopy) is progressively colonized by coniferous trees and if this number of days above $10°C$ reaches one, two, three, or four months, the forest becomes more and more dense with a leaf area index evolving from 2 to 10 (Spruce forest of the Taïga). In such a zone the ground frost and very often the ground water table is a limiting factor to the root growth; therefore root zone stays in a very shallow zone (20 to 40 cm) and for this reason temporary dry conditions may occur associated with reduced evapotranspiration. Water interception which reaches values of 50% (Sweden coniferous forest) (Halldin et al., 1979; Leyton et al., 1967) compared to 30% for the deciduous forest and 10% for the evergreen forest, attempts to produce frequently high evapotranspiration (because such saturation after rainfall corresponds to the physical definition of potential evaporation in a canopy). Thus, the canopy acts as a regulator of evapotranspiration when it is dry; regulation becomes strongly effective only when the part of the remaining 50% of the rainfall, stored in the layer of soil explored by the root system, has evaporated. As it was shown for the previous biomes analysed, water balance of a sempervirent boreal forest is also positive over the year (drainage). Of course, some short dry seasons during summer appear according to climate (essentially latitude) and water losses evolve from 250 mm in the northern part to 450 mm in the southern part of the coniferous ecosystems.

(d) <u>Mediterranean sempervirent forest</u>

This sclerophyllous woodland (Figure 1) belongs to a zonobiome characterized by rainy winters but dry summers with an annual rainfall of around 800 to 2000 mm (Walter, 1979). If the strata of trees are less and less continuous, according to the length of the dry season and human forest destruction in general, the shrub layer and more often a mixed layer of scrub and herb completes the soil cover (Lossaint and Rapp, 1978; Di Castri and Mooney, 1973; Mooney and Parsons, 1973). The leaf area index of the forest is about 4 or 5 sometimes 6 with Eucalyptus forests and that of the scrub and herb part generally much less, around 2 and 3 producing a mean value of only 2 or 3 for tree leaves according to the density of trees.

Interception (Lossaint and Rapp, 1978) is also non-negligible with 30% of the local rainfall (example of quercus ilex forest) in a climate zone with total precipitation of 770 mm; this interception maintains for some time evaporation to a potential value. But sclerophyllous vegetation induces a high resistance to water loss, chiefly during the dry season, reducing quite strongly actual evapotranspiration. From Turc's method the calculated value of actual to potential evapotranspiration gives a ratio of 0.56. Therefore, for at least several years, water balance is generally positive even in a drier zone (around 800 mm of annual rainfall), but the water balance may be negative in such a country during a particular year.

2.1.1.2 <u>Deciduous canopies</u>

A partial or complete loss of leaves is a response of plants to climatic stresses such as cold or drought. Thus, with a clear seasonal rhythm, forests during a period are reduced to a tangle of trunks and branches which intercept radiation and modify turbulence, exchanging only sensible heat whereas at the bare soil surface beneath this forest skeleton, latent heat exchange takes place.

Several zonobiomes can be considered among these vegetation types which do not present a closed canopy throughout the year (Figure 1).

(a) Monsoon forest and mixed dry forest

If with evergreen forests the shedding of leaves appears most often regularly throughout the year, sometimes a periodicity affects the forest, particularly when the dry season becomes stronger, so that the leafless state of trees related to water conditions spreads throughout the forest during a noticeable time period. In some transient zones between evergreen forest and a total leafless state, intermediate leafless state occurs progressively in the mixed dry forest according to the duration of the dry season; thus, these forests may be considered as a zonoecotome between the evergreen forest and the savanna.

In fact, this leafless period during a dry season, even with high total precipitation of 1000 or 2000 mm in a monsoon region, modifies drastically for a period the structure of the forest, even though the soil surface layer with herbs and scrubs maintains a complete soil coverage. Effectively, during the dry season, a strong reduction of evapotranspiration affects all the zone because of scleromorphic leaves, xerophyl scrubs and dry herbs.

For this forest, according to water balance and during the wet period, the complete leaf area index of all layers may approach around 6 to 10 and perhaps sometimes 15 (Ogawa et al., 1961).

(b) Mixed and deciduous boreal forest

When the climate of the northern part becomes strongly continental the boreal forest evolves into a deciduous boreal forest (composed of few species such as Larix) (Overbeck, 1975; Elhai, 1968). When the climate becomes more and more temperate, the mixed forest takes the place of the boreal forest and evolves progressively towards the temperate deciduous forest. This mixed forest is typically a zonoecotome. Some of these mixed forests particularly in the mountains (orobiomes) contain a lot of different species. Alike are forests along the Pacific coast which are characterized by a great number of species, a very luxuriant canopy and most of the time, look like a tropical forest.

(c) Deciduous temperate forest

Even though most of these forests are at present monospecific the number of species in this zone is important. This type of forest presents a tree stratum with a shrub stratum, a bush layer and/or a herb layer. This more or less high forest, between 15 to 30m, is quite homogeneous and covers completely the soil for over six months (Lemée, 1978; Rauner, 1976). During this period the leaf area index of the trees is about 4 to 6 and that of the total vegetation reaches 6 to 8 (Kestemon 1973; Heath et al., 1966).

This forest intercepts generally only 10 to 15 per cent of rainfall because of the leafless period and does not affect much the value of the forest evapotranspiration.

Rainfall is quite well distributed throughout the year and we can observe positive water balance with drainage during the winter and spring periods and a reduction of evapotranspiration according to soil depth and water availability during the summer. The contribution of the herb and bush layer to evapotranspiration is always small during the leafy period, but dominant in the early spring as this layer covers partially the soil for more than a month before tree leaves appear.

2.1.1.3   Woodland and grassland

Opposite to forests as high and dense canopies covering more or less completely the soil, we can find in many places grasslands, particularly in temperate zones with a continental climate. But there are also many mixed zones with a strong interaction at all levels between trees and dense herb layers.

(a) Savannas

This kind of formation (Figure 1) covers great surfaces around evergreen rain forests, monsoon forests, and mixed dry forests. When the duration of the dry period increases and when the precipitation does not exceed 1000 to 1500 mm, the tree density decreases, crowns become discontinuous and grassland develops its dense canopy (Bourlière, 1978; Lamotte, 1978; Beard, 1953; Hopkins, 1974). Trees are evergreen or deciduous and when they are scarce, the tree biomass may evolve from a dense to a clear savanna. In canopies such as these, the woody savanna represents 80 to 100% of the total biomass and the leaf area index is only around 1 to 1.5 and 0.4 to 0.2 (Lamotte, 1975).

In fact, because of the dry season, the biomass of roots is important. For herbs (roots localized around the first fifty centimeters) and for trees (between 50 cm and 2m or more) the biomass reaches about 50% of the aerial biomass and can represent more than 100% for herbs (Bille and Poupon, 1972a, 1972b, 1974). This developed root system is a good adaptation to the dry season. But even with this adapted system during the dry season most of the herbs dry up; most of them dry up totally but some still remain alive if the dry season is not too long.

So, in function of the rainy season and the growing period of the herb layer (leaf area index varying between 2 and 4), evapotranspiration must be high, but it decreases quickly with water deficit because of stomatal regulation. Trees play a role as radiation screen and windbreaker, maintaining longer, better conditions for herbs.

(b) Temperate grassland

This formation (Figure 1) correlated with continental temperate climate associated with cold winters and hot and very often dry summers is described generally by two periods of dormancy: first, during the long cold period but most of the aerial part remains dead and dry covering the soil and secondly during the dry period, some of the aerial parts remaining alive (Ripley and Redmann, 1976; Ricou, 1978; Lauenroth, 1979; Delting, 1979).

This formation may be tall, more than one or even two meters, or low (around 50 cm) but always very dense (leaf area surface of 3 to 6), the root biomass, which exceeds the leaf biomass, explores intensively the first twenty centimeters and often reaches more than one meter.

With a total rainfall of around 300 to 500 mm and even up to 1000 mm the water balance is positive but negative for more or less long periods during the summer, particularly if the root zone is shallow.

Water interception may be high at around 20 to 90% according to intensity and duration but does not play a role as important as that in oceanic or monsoon regions.

During most of the time of the snowy period and particularly when water deficit occurs during the dry season, actual evapotranspiration is under stomatal regulation which reduces quite strongly the potential evaporation.

2.1.2 Zonobiomes with open vegetation

These zonobiomes with more and more discontinuous vegetation are due to climatic stress combined with human and fire agressions (San Jose and Medina, 1975). We find once more the two dominant stresses which are:

- drought inducing steppes to evolve towards hot deserts;

- cold inducing boreal forests to evolve towards cold deserts (ice surface); these zonobiomes represent a high percentage of the earth's land surface (Figure 1).

(a) Steppes

Steppes are defined by the percentage of soil covered by vegetation including plants and trees with xerophyllum or succulent characteristics.

All xerophytes are adapted to limit water loss and develop a strong root system.

Thus, for steppes developing with rainfall of around 300 mm and less, the runoff is generally important, up to 40%, and after rain, if the soil surface dries out quickly, grass and vegetation may spread out maintaining some evapotranspiration.

The heterogeneity due to trees, bushes, scrubs, and grasses increases turbulence but evapotranspiration stays at quite low values. Most arid zones are halobiomes with special species.

(b) Tundras

With a very cold climate (few months with a mean temperature above $0^{o}C$), little precipitation (around 200 mm) and strong wind during the winter having a strong mechanical action, plant growth depends upon the thaw of the upper layer of soil (Bliss and Wielgolaski, 1973; Lewis and Callaghan, 1975).

Dispersed plants, small trees, bushes, grasses, and mosses are found and the vegetation develops during the summer time (Linacre, 1975). Because the thawed soil does not reach more than 0.5m, drainage does not exist and excess water runs off and creates swamps.

For many months (autumn, winter, spring) only sublimation may take place and during the summer most of the surfaces remain wet and evapotranspiration is probably not far from potential evaporation.

(c) Deserts

Without vegetation or only a few species from place to place, hot deserts are localized in climatic zones with rainfall totals of less than 50 mm to sometimes near 0 mm. When it rains, the runoff is important and the surface dries out immediately, so that evaporation becomes quickly negligible (Evenari et al., 1971; Logan, 1960).

With cold deserts characterized by a long period of permanent snow or ice with a process of low sublimation and generally only little precipitation, evaporation even near potential for long periods remains low (Kuhn, 1981).

All natural zonobiomes are strongly related to climate and particularly air temperature and precipitation (Figure 2) which allows classification by climatic diagrams (Gaussen, 1954).

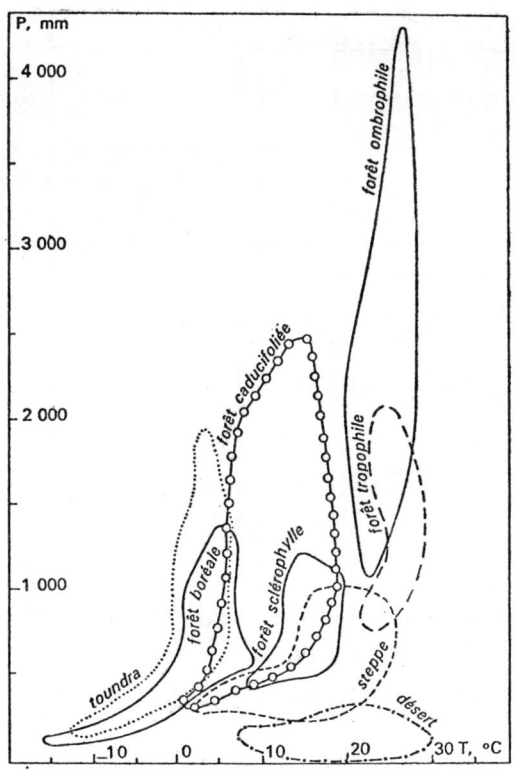

Figure 2 - Principal biome diagrams related to precipitation and temperature (after Lieth, in Rey, 1960).

2.2   AGRICULTURAL LAND SURFACE

Agriculture changes continuously and repetitively the previous description. Some human interventions are small or scarce enough that equilibrium has time to re-establish and delete human alteration. But since the last few millenia the impact has been so high that irreversible degradation can be seen everywhere.

Thus, most of these natural land surface distributions, especially in temperate zones, are completely or partially modified by intensive agriculture. At the present time more than 10% of land surfaces may be cultivated permanently with annual and perennial crops with a more or less intensive exploitation. About 25% of the land surface is grassland and grazing surfaces. Probably about 50% of these surfaces is not natural grassland and consequently results from human modification of some natural land (Table 1).

Table 1. Land surface distribution in 10^6 km^2 (from FAO data)

	Africa	North America	South America	Asia	Europe	USSR	Oceania	Total
Tillable land and permanent agricultural land	2.05	2.55	0.84	4.61	1.47	0.47	2.33	14.32
Grassland and pasture land	8.45	3.72	4.13	4.99	0.93	4.64	3.74	30.60
Forest land	7.10	8.15	9.08	6.24	1.40	0.81	9.10	41.88
Non-agricultural land	12.40	7.90	3.80	11.30	1.10	2.60	7.10	46.20
Total	30.00	22.32	17.85	27.14	4.90	8.52	27.27	133.00

It is possible to assess that about 20% of natural land has been actually modified by agriculture (without counting forestry).

We must consider several characteristics of agricultural land which affect land surface processes (Monteith, 1976):

- those regarding the structure of the crop (height, density, water adaptation),

- the annual cycle of plant growth with periods of partial soil coverage, complete soil cover and a period of maturation (drying phase),

- farming intervention such as irrigation which modifies intensively water balance and consequently, latent heat exchange (Yaron et al., 1973).

Therefore, for example, most cereals which represent (Table 2) around half of a per cent of arable land (7.2 10^6 km^2) are not permanent crops. They cover the soil for a few months (3 or 5) acting during this time as a real vegetative surface with nevertheless perhaps one month covered by more or less dried canopy, and beyond this short period soil surface is predominant.

In order to properly use natural and agricultural land surface description for land surface processes, a description of vegetal cover properties and of some energy balance equations which use these properties is useful. A certain understanding of main exchange modifications due to vegetal cover may then explain what kind of relations and data sets give the best adequacy for simulation problems and the best argument to verify such programming models.

Table 2. Approximative land surface areas devoted to crops (10^6 km^2)

Cereals	7.2
Proteaginous plants	1.0
Starches (potatoes-manioc)	0.4
Sugar plants (sugarbeet, sugarcane)	0.3
Tea - Coffee	0.3
Citrus trees - Vineyards	0.1
Others (vegetables-fruit trees)	1.5
Potential and non-permanent agricultural land	2.5

III. PROPERTIES OF VEGETAL COVERS

With regard to heat and mass exchanges a canopy must be considered as a porous medium which includes several kind of elements (leaves, stems, branches, trunks). This medium is closely related to the soil by its root system. Of course leaves and active roots represent the most important living part of the plant and act dominantly upon exchanges. Consequently a canopy will be described as a porous medium taking into consideration all the spatial distribution of each class of its elements, but of course the complexity of such a description very often leads scientists to select a much simpler model which takes into account, throughout the volume occupied by the vegetation, only the distribution of leaf surfaces.

3.1 STATIC PLANT CANOPY DESCRIPTION

3.1.1 Geometrical properties

With quite homogeneous canopies such as most agricultural crops it may be possible to define a vegetal cover (plant cover) by a mean crop height z_h and at each level ($0 < z < z_h$) by the percentage of exchange surfaces per unit of volume for each kind of element (Figure 3a, b).

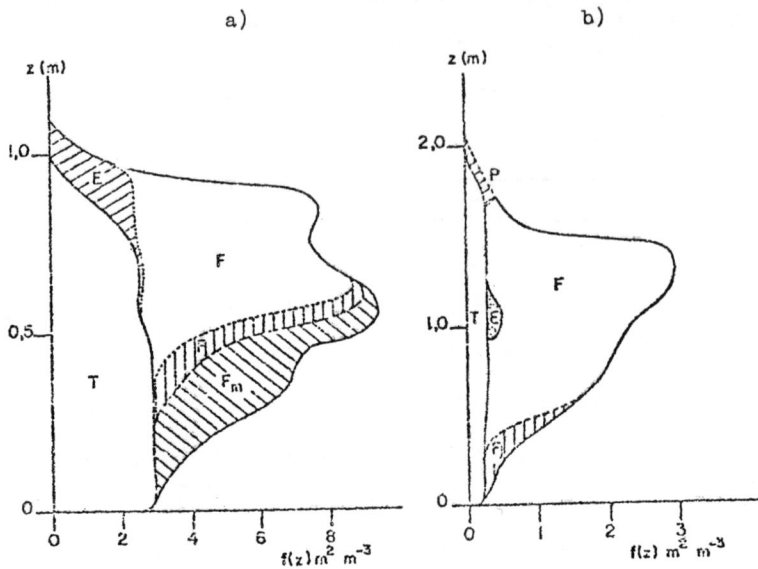

Figure 3a, b - Distribution of surface density of vegetation elements (f)z versus height z for a wheat and maize crop (F, green leaves; Fj, yellow leaves; F_m, dead leaves; T, stems; E, ears; P, tassels).

As mentioned above, this leads to a leaf density profile f(z) and for most practical purposes to a cumulative leaf area profile F(z) per unit of soil surface calculated from the top of the canopy (z_h) to the soil surface (the total leaf area per unit of soil surface is F(z) = 0 and is called leaf area index) (Figure 4). This description is sometimes completed by that of the frequency of the distribution of the azimuthal and inclination angle of the normal direction of each small element acting as an exchanger (Table 3) (Verhagen et al., 1963). Such investigations were done by many authors with different techniques (Warren Wilson, 1960) but essentially on agricultural crops and natural grassland and very scarce are experimentations on natural stand covers, particularly on forests where such measurements are very difficult to perform (Figure 4).

The difficulty is much greater when we want to obtain such results with roots but partial descriptions are now available in a number of papers describing such crops or biomes (Figure 5).

3.1.2   Biological properties

When we go further in detail through a canopy in order to analyse the three dominant transfers: radiative transfer, momentum transfer and turbulent heat and mass transfer, the previous description does not suffice to permit analytical results.

Table 3. Distribution of leaf density (m²/m³) for each layer (0.3 m) and class of leaf inclination (10°) in the case of a maïze (hybrid INRA F7EP1 at "La Minière")

Layers \ Classes	0-10	10-20	20-30	30-40	40-50	50-60	60-70	70-80	80-90	Total
0 - 30	0.013	0.016	0.037	0.011	0.003	0.006	0.005	0.008	0.010	0.109
30 - 60	0.035	0.040	0.010	0.037	0.055	0.070	0.047	0.010	0.028	0.332
60 - 90	0.042	0.016	0.021	0.037	0.103	0.136	0.082	0.068	0.042	0.547
90 - 120	0.062	0.040	0.026	0.101	0.041	0.159	0.103	0.022	0.024	0.578
120 - 150	0.107	0.041	0.106	0.083	0.036	0.118	0.094	0.051	0.008	0.644
150 - 180	0.067	0.057	0.113	0.046	0.127	0.135	0.074	0.056	0.021	0.696
180 - 200	0.003	0.002	0.003	0.004	0.006	0.008	0.004	0.002	0.001	0.033
Cumulative value	0.329	0.212	0.316	0.319	0.371	0.632	0.409	0.217	0.134	2.939

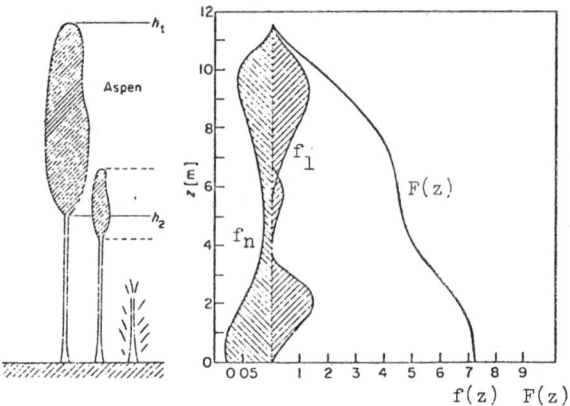

Figure 4 - Complex deciduous forest with main layer aspen (Populus tremula), then regrowth and bush hazel layers and others; non-leaf area density $f_n(z)$, leaf area density $f_i(z)$ and cumulative total area $F(z)$ from the top to the bottom versus height (after Rauner, 1976).

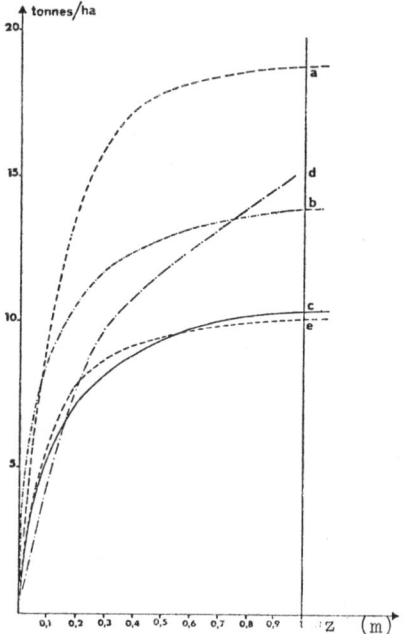

Figure 5 - Cumulative root biomass versus depth for grassland ((a) grassland savanna "Loudetia simplex"; (b) grassland savanna "Andropogonées"; (c) bush savanna "Andropogonées; (d) woody savanna "Andropogonées"; (e) never burnt bush savanna) (after Cesar in Lamotte, 1978).

3.1.2.1 Radiative properties

For radiative purposes, properties of reflection and diffusion as well as transmission of leaves with their angular variations (Anderson, 1966; Chartier, 1966) are very useful as is emissivity. Such details can be often avoided if some general mean coefficients are established for a given crop (Udagawa et al., 1969).

The use of an all-inclusive extinction coefficient (k) is often precise enough for many purposes (see paragraph 4.1) (Figure 6) and expressions such as the following are currently used (Impens et al., 1969; Impens et al., 1970):

$$Rg = Rg_o \, e^{-k \, F(z)} \tag{1}$$

$$Rn = Rn_o \, e^{-k_0 \, F(z) + k_1 \, F(z)^2} \tag{2}$$

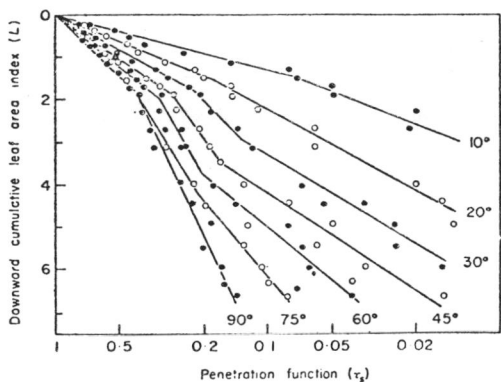

Figure 6 - Penetration function of direct solar radiation in a Sorghum crop whose depth is indicated by the cumulative leaf area index. (After Laisk, 1965.)

3.1.2.2 Momentum properties

For momentum transfer $K(z)$, an analysis based on momentum absorption and momentum balance (Thom, 1969, 1971) shows the importance of the drag coefficient $C_D(z)$ and wind speed $U(z)^2$ (Inoue, 1963; Uchijima, 1966; Cowan, 1968)

$$\frac{d\tau(z)}{dz} = \frac{1}{2} \rho \, C_D(z) . f(z) . U(z)^2 \tag{3}$$

and the variation of this coefficient with the angular position through the flow. Such analyses lead to the complete equation (Perrier, 1975)

$$K(z) = k^2 L(z)^2 \frac{d U(z)}{dz} \qquad (4)$$

$$U''(z).U'(z) + \left[L'(z)/L(z)\right]\left(U'(z)\right)^2 - \left[C_D(z).f(z)/4k^2.L(z)\right]U(z)^2 = 0 \quad (4a)$$

which describes wind speed $U(z)$ through the canopy and turbulent diffusivity $K(z)$ (Cionco, 1965; Takeda, 1965) which appears as mainly influenced by geometrical properties acting by means of parameters such as leaf density $f(z)$, drag coefficient $C_D(z)$ and mixing length $L(z)$, properties which also depend on the stability or instability of air.

3.1.2.3 Heat and mass exchange properties

For heat and mass transfers which are strongly interdependent, relationships giving the energy balance of each layer of canopy, describing transfer equations through the porous media, and formulating exchanges chiefly at leaf surfaces, may also be reduced to two general equations:

- enthalpy flux equation $H(z)$ (with $H(z) = \left(L\ M/R\ T\right).P\left(T_D(z)\right) + \rho\ c_p\ T(z))$:

$$Rn(z) + S + K(z) \frac{dH(z)}{dz} = 0 \qquad (5)$$

- air temperature deficit equation (with $Y(z) = T(z) - T_D(z)$) :

$$K(z).Y''(z) + K'(z).Y'(z) - h(z).f(z).\beta(z).Y(z) + \left(1-\beta(z)\right)\frac{R(z)}{\rho.c_p} = 0 \qquad (6)$$

These equations show the classical role played by geometrical properties (leaf density $f(z)$) and previous properties relative to radiative $(R(z))$ and momentum $(K(z))$ transfers but also underline two other parameters:

- the mean exchange coefficient h of one element due to its drag coefficient and the boundary layer developing around it (Raschke, 1956; Bouchet, 1963),

- the mean stomatal resistance r_s which defines the resistance to vapour flow from the inner site of the leaf where there is saturation to the leaf surface. This coefficient is always combined with the exchange coefficient h and appears in a classical form already used by Bouchet (1963), Monteith (1965), Perrier (1975), Thom (1975):

$$\beta = 1/\left[1 + \left(\gamma/(P'+\gamma)\right) h\ r_s\right] \qquad (7)$$

This parameter also defines potential evaporation ($r_s = 0$) as compared to actual evapotranspiration:

$$ET = \beta\ EP \qquad (7a)$$

It is often called an aridity coefficient or index which varies from 1 to 0.

But if the exchange coefficient $h(z)$ depends on leaf structure only the stomatal resistance $r_s(z)$, also a function of stomatal structure, is strongly related to biological mechanisms which can make this term vary greatly.

3.2 TIME MODIFICATION OF THESE PROPERTIES

This evolution is time dependent through the impact of the plant growth but with a strong feedback due to plant reaction and adaptation to climatic and soil factors which evolve with time.

Thus, plant growth modifies structure (height, leaf density profile, leaf area index, root system) as well as leaf properties because new organs are never similar to the previous ones. The most sensible parameter in its range of variations and in its time dependency is stomatal resistance which varies according to radiation, water deficit in the plant, humidity in the air, and temperature.

Two periods must be regarded with special attention:

- the first stage of the plant during which the soil is being covered by the vegetation,

- the last stage, or maturation period, when the crop begins to dry out.

As we have seen earlier, this time evolution affects all agricultural land, but to a lesser extent pasture land which covers the soil throughout the year, and some of the natural biomes defined (Figure 7). We can consider only the evergreen forest as an apparently more or less stable biome approaching a certain equilibrium, throughout the year with more or less fixed structure and constant leaf area.

3.3 SPATIAL DISTRIBUTION OF LANDS

If many zonobiomes which cover surfaces of more than 1000 km^2 may be considered as quite homogeneous, in many places human activities and agricultural land have divided the zone. A second factor of heterogeneity appears clearly in all zones which are transient relatively to climatic gradients; the consecutive low climatic variations result from latitude and modification by oceanic zones, orographic zones, and all local effects.

Heterogeneity appears more clearly in agricultural land which very often superimposes for diversification reasons a strong contrast even in a large climatic and edaphic homogeneous zone.

Thus, for many places, consideration of how means must be assessed is essential. If integration to attain a greater zone is feasible (map of small region), the possibility of estimating directly some mean value for the previous properties with a significant value for bigger surfaces would be worthwhile. Some examples of this problem will be developed further on.

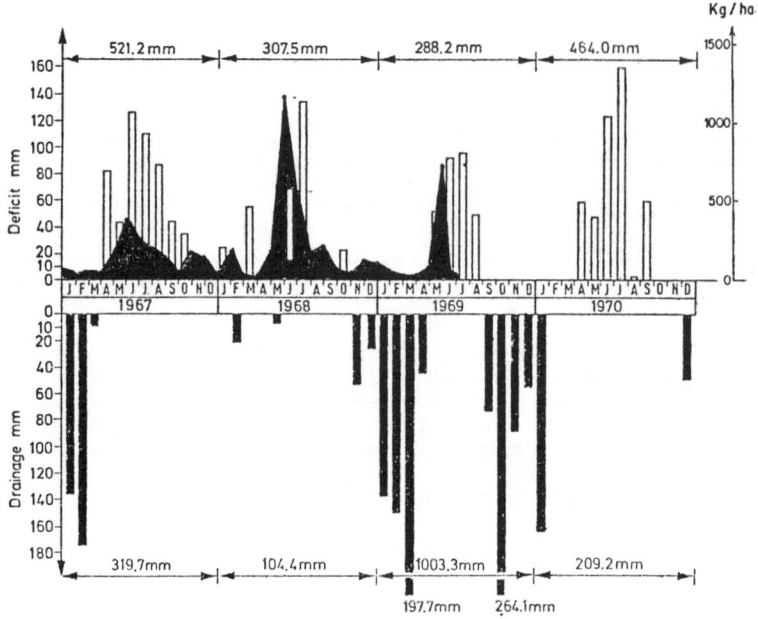

Figure 7 - Soil water deficit and drainage monthly values in a Mediterranean sempervirent forest and corresponding leaf fall (Kg/ha - black area -). This latter gives an indication of the annual fluctuation of the total leaf area index (1.5 T ha^{-1} ~ 20% LAI). (After Lossaint, 1973.)

IV. PARAMETERIZATION OF ENERGY BALANCE TERMS

Land surface processes must be regarded from a definite point of view, and data sets will be adequate for use in land surface processes only if there are choices strongly correlated to this problem. Thus, let us approach these most useful data through an analysis of the parameterization of energy balance terms without entering into the details of processes as do Brutsaert's and Fritschen's papers (1981)

4.1 STATIC CONCEPTION OF POINT ENERGY BALANCE

For a given canopy, with sufficient area to avoid too much perturbation due to the fetch effect, but not too big a surface to accept climatic data such as air temperature and dewpoint temperature of the air as independent of the field concerned, the classical energy balance equation may be written as follows:

$$Rn + S = H + LE \qquad (8)$$

where heat capacity of the vegetation and energy used for biological activities have not been taken into account.

Each term is written as a function of climatic data and canopy parameters:

Soil heat flux: $\quad S$ = heat flux at the soil surface $\hfill (9)$

Net radiation: $\quad Rn = (1 - a)Rg + \epsilon Ra - \epsilon \sigma Ta^4 + 4 \epsilon \sigma T^3(Ta-Ts)\hfill (10)$

(Rg, global radiation and Ra, atmospheric radiation, are the two terms of incident radiation; a and ϵ define total radiative properties of the canopy, albedo and emissivity; Ta and Ts* the air temperature and radiative surface temperature of the canopy (°K).

Net radiation Rn may be written as climatic net radiation Rn* with a correction term:

$$Rn = Rn^* + f(Ta - Ts^*) \qquad (10a)$$

Sensible heat flux:

$$H = \rho\, c_p \left[h\right]_{z_h}^{z_r} \left[T(z_h) - T(z_R)\right] \qquad (11)$$

(ρ volumetric mass of air; c_p, heat capacity of air; $\left[h\right]_{z_h}^{z_r}$ exchange coefficient between surface canopy, z_h, and the reference level, z_R, including forced and natural convection; $T(z)$ air temperature at level z; $T(z_h)$ may be defined as canopy surface temperature, °C).

Latent heat flux:

$$LE = (L\,M\,/\,R\,T) \left[h\right]_{z_h}^{z_R} \left[P\left(T_D(z_h)\right) - P\left(T_D(z_R)\right)\right] \qquad (12)$$

(L, latent heat of water; M, molar weight of water; R, constant of perfect gas; $P[T_D(z)]$, saturation vapour pressure for temperature $T_D(z)$; $T_D(z)$, dewpoint temperature at level z).

Latent heat flux may also be regarded as a reduced flux compared to a saturated surface due to mean structure and stomatal resistance of the canopy (r). This reduced flux corresponds to a non-saturated surface with a dewpoint temperature at the top of the canopy $T_D(z_h)$ more or less removed from the temperature at this level $\left[T(z_h)\right]$. Latent heat flux is then written as follows:

$$LE = (L\,M\,/\,R\,T)\, \frac{P[T(z_h)] - P[T_D(z_h)]}{r} \qquad (12a)$$

with r the mean crop resistance to water vapour.

Following Penman's classical calculation (1948), the previous relationships (equation 8 to 12a) allow one to calculate evapotranspiration but only if two hypotheses specify that:

- climatic data are independent of the studied surface,
- fluxes are conservative between the two levels chosen (z_h and z_R).

Actual evapotranspiration is given by:

$$LE = \left(P'/(P'+\gamma)\right)(Rn + S) \frac{1 + \left(\gamma/(P'+\gamma)\right) h\, r_c}{1 + \left(\gamma/(P'+\gamma)\right) h\, r} \qquad (13)$$

(Rn may be replaced by climatic net radiation Rn* if γ the psychrometric constant is transformed by addition of the term $4\,\epsilon\,\sigma\,T^3$ and if we suppose the approximate relationship Ts* $\sim$ T(z_h)).

In this relationship r_c is a climatic resistance defined from climatic variables

$$r_c = (L\,M/R\,T)(P'+\gamma)\left[T(z_R) - T_D(z_R)\right]/(Rn + S) \qquad (13a)$$

Thus, if we write the evapotranspiration term as Priestley and Taylor (1972):

$$ET = \alpha\left(P'/(P'+\gamma)\right)(Rn + S) \qquad (14)$$

this α term may be defined according to equations (7) and (13):

$$\alpha = \frac{\beta(crop)}{\beta(climate)} \qquad (15)$$

with $\alpha = 1$ if there is an equilibrium between air potential at reference level and that at surface level (Bouchet and Perrier, 1973; Thom, 1975) (Figure 8),

$\alpha < 1$ which is an "Island effect", air potential at reference level being less than that at surface level (air at reference level is wetter than at the surface),

$\alpha > 1$ which is an "Oasis effect", air potential at reference level being greater than that at surface level (air at reference level is drier than at the surface).

4.2 DYNAMIC CONCEPTION AND EQUILIBRIUM

If we consider air and surface exchanges in a dynamic way, it is necessary to integrate, as was mentioned but not elaborated in the previous paragraph, the action of climatic data on surface exchanges, and the reverse action of exchanges on climatic data. Obviously, it is impossible to consider a given region with an air mass passing over without taking into account the modifications in air characteristics which are well described by the air temperature rising during the morning.

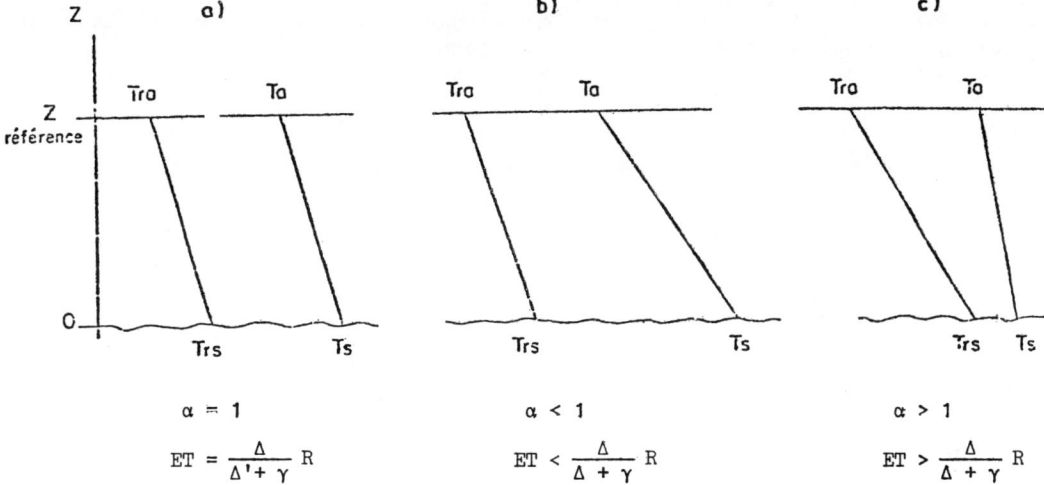

Figure 8 - Schematic profile of air and air dewpoint temperature in three cases:
(a) Equilibrium between air potential at reference level and surface level.
(b) "Island effect".
(c) "Oasis effect".

For this purpose a very simple analysis may give us a good idea of what happens on a regional scale. This simple example is based on:

- constant incoming radiative fluxes (Rn),

- constant surface characteristics defined as previously by their exchange coefficient h and their surface resistance r,

- a constant conductive heat flux at the soil surface which may be associated to the incoming radiative fluxes in order to give a constant available net energy $R=(Rn + S)$,

- a defined convection boundary layer of height z_M which may be considered as quite homogeneous particularly as regards the thin surface sub-boundary layer (height r_R). The interface at height z_M may be as a first hypothesis assimilated to a zero surface flux.

Under these conditions two dynamic equations must be added to the previous equations 8, 11 12 and 13:

$$d(T(t) - T_i)/dt = H / \rho\, c_p\, z_M \qquad (16)$$

$$d(T_D(t) - T_{Di})/dt = (\gamma/P') LE / \rho\, c_p\, z_M \qquad (17)$$

which define for the convective layer the modification of air mass characteristics as a function of surface exchanges.

These equations lead to a specific relationship for air temperature deficit $[T(t) - T_D(t) = Y(t)]$ which modifies the climatic resistance r_c (equation 15) and the surface exchanges themselves (Perrier, 1980):

$$Y'(t) - \beta_o Y(t) + \beta_o Y_f = 0 \qquad (18)$$

$$\beta_o = \beta \, h/z_M \qquad (19)$$

(β, surface aridity index)

This interaction of surface and air characteristics, by means of this result (equation 18 and 19), allows for three conclusions:

- $Y(t)$ varies more or less quickly according to β_o to a finite value Y_f well defined by the net available energy and the surface resistance (exchange coefficient h acting only on time lag of the equilibrium $1/\beta_o$):

$$Y_f = \left(\gamma/(P' + \gamma)\right) R \, r/\rho \, c_p \qquad (20)$$

- this finite value corresponds to a finite value for r_c which is obviously r, and consequently, to a defined value for evapotranspiration corresponding to $\alpha = 1$ (equation 14). Thus, this particular value of evapotranspiration which may be called climatic or limit evapotranspiration is also an equilibrium or a critical value called ET_o to which the regional evapotranspiration must naturally tend,

- if according to Bouchet (1963) we try to examine through these dynamic relationships what he defines as some logical correspondance:

$$ET + EP = 2EP_o \qquad (21)$$

we can assume that according to relation (7a, $ET = \beta EP$), the surface aridity index being β:

$$ET + EP = EP(1 + \beta) \qquad (22)$$

and if the finite value is always reached after a given time "t_o" (equation 18) then ET is always equivalent to ET_o with EP given by ET_o/β, and we can write (Figure 9a):

$$ET + EP = ET_o \left[\frac{1 + \beta}{\beta}\right] \qquad (23)$$

But as is shown by equation (19) when β decreases from one ($r = 0$ to zero ($r \to \infty$), the time "t_o" necessary to accede to the limit value or equilibrium tends to become also infinite. Consequently, for a reasonable time over a regional scale (a few hours for instance to $\sim z_M/h$), the previous equation (23) may be rewritten more precisely as a function of this time "t_o"; air temperature deficit $Y(t_o)$ is given (eq. 18) by:

$$Y(t_o) - Y_i = (Y_f - Y_i)e^{-\beta_o t_o} \qquad (24)$$

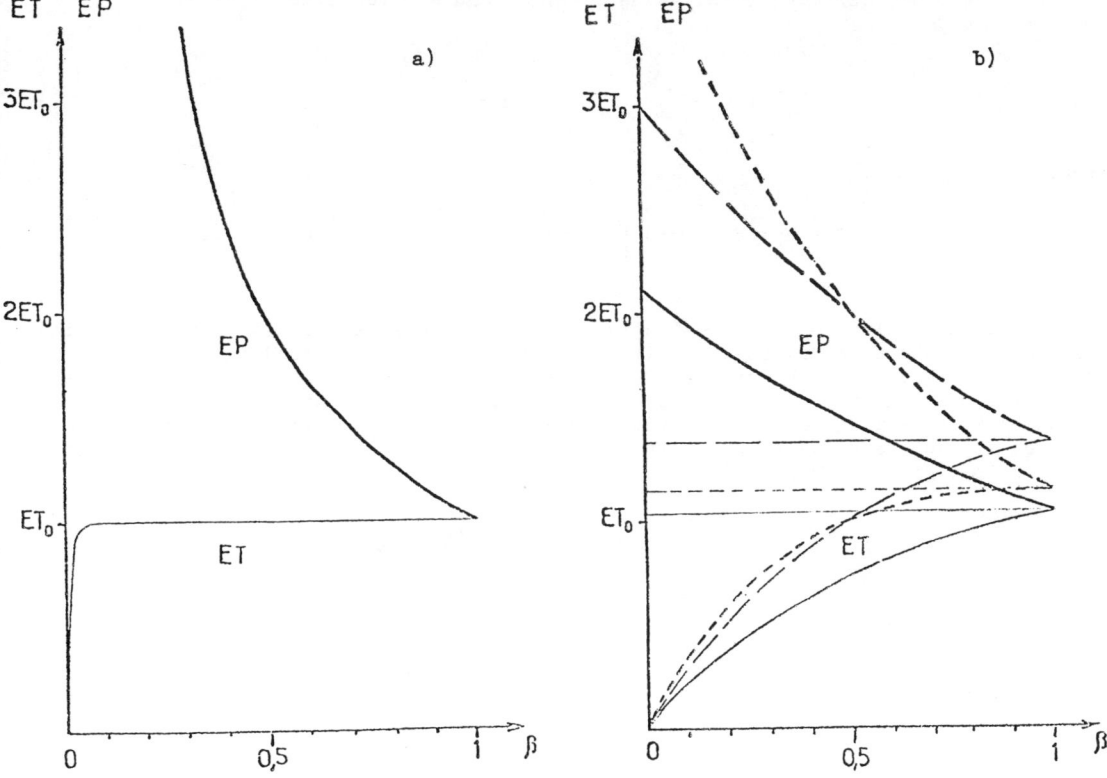

Figure 9 - Correlative variation at regional scale between actual evapotranspiration ET and potential evaporation EP versus the surface aridity index β (limit or equilibrium evaporation ET_o).
(a) at a perfect equilibrium ($t \to \infty$ equation 24b)
(b) under different initial conditions of the climatic data defined by β_i (climate):
- $\beta_i = 0.9$ with $t_o = 1$ (———)
- $\beta_i = 0.5$ with $t_o = 1$ (— —)
           with $t_o = 2$ (-----)

and the climatic resistance r_c which acts in ET and EP expressions varies also as a function of time according to the following relationship:

$$r_c(t) = r_{ci} + (r - r_{ci})e^{-\beta_o t_o} \qquad (24a)$$

(r_{ci} is the value of r_c for the initial conditions: $t = 0$, with $Y = Y_i = T_i - T_{Di}$, which also permits one to define the value of a corresponding β_i(climate) according to equation (7)).

Thus, for an initial value Y_i or β_i and a given time "t_o":

$$EP = ET_o \left[\frac{1}{\beta} + \frac{1-\beta_i}{\beta_i} - \frac{1-\beta}{\beta} \right] e^{-\beta h t_o/z_M} \qquad (24b)$$

and:

$$ET + EP = ET_o (1+\beta) \left[\frac{1}{\beta} + \frac{1-\beta_i}{\beta_i} - \frac{1-\beta}{\beta} \right] e^{-\beta h t_o/z_M} \qquad (25)$$

a relationship which is in perfect agreement with Bouchet's formula (1963) (equation 21) as it is shown in Figure 9b, only if the time chosen is around z_M/h. In fact, the greater this time, the greater the divergence (Figure 9a) particularly for the low value of aridity β of the surface. If Bouchet assumes that EP_o is well approximated by $0.5(1 - a)Rg$ (Rg global radiation and a albedo), the previous calculation defines EP_o as the value of ET_o (limit evapotranspiration) which is very often quite near the one given by Bouchet:

$$\left[P'/(P' + \gamma) \right] R \sim 0.5(1-a)Rg$$

4.3 CONCLUSIONS

Evapotranspiration can always be expressed as a function of the limit evapotranspiration (equation 14):

$$ET = \alpha\, ET_o$$

with a coefficient α which represents the ratio between the crop surface aridity β and the climatic aridity β_c (eq. 15). Most often if on a local scale the divergence between these two values theoretically allows for a great variation of α ($0 < \alpha < 1/\beta_c$), in fact, on a regional scale, after a few hours, the evolution of air characteristics ($\beta_c \rightarrow \beta$) towards an equilibrium with the surface tends to maintain α near 1.

As Priestley and Taylor's analysis (1972) has shown the maximum value of α is around 1.3 but only with water surfaces or plant covers on a well-watered soil. With a positive water balance, α may vary between 1 and 1.3 according to the values of the real exchanges with the high atmosphere which take place at the level of the boundary chosen ($z > z_M$).

But with a bare soil and a crop with a negative water balance the surface resistance r increases quickly and the corresponding aridity β decreases.

Three kinds of evapotranspiration calculation can be performed in order to introduce soil and plant surface characteristics:

(i) A more or less complete calculation of evapotranspiration (equation 13), with a determination of Priestley and Taylor's coefficient from a specific calculation of the aridity coefficient β for a crop as a function of the crop resistance r, the exchange coefficient h and from the mean climatic aridity coefficient β_c.

(ii) A calculation based on some empirical expressions of Priestley and Taylor's coefficient α starting from its theoretical definition ($\alpha = \beta(\text{crop})/\beta(\text{climate})$) On a regional scale, as has been demonstrated, the value of this coefficient cannot be far from unity (equilibrium). If some divergence around the value appears, it may only be related to soil water stress and consequently to soil water availability for the root zone. If the mean soil water capacity W_c (kg m^{-3}) is defined as the maximum water storage per unit of soil volume, the relative water content for a useful depth of z (mm) will be:

$$RW = \left[W_c z - (\Sigma ET - \Sigma P) \right] / W_c z \leq 1$$

(ΣET is the cumulative evapotranspiration and ΣP the cumulative precipitation in kg m^{-2} or mm of water and if $W_c < \Sigma P - \Sigma ET$, drainage occurs).

- For a bare soil with a useful depth z_s and with an always drying top layer at the surface, the coefficient α may be approached by:

$$\alpha = 1 - (\Sigma ET - \Sigma P)/W_c z_s \quad (\text{with } \alpha \leq 1) \qquad (26)$$

- For a crop with a root zone z_r and active water absorption by the root system:

$$\alpha = 1.3 - 0.6\left[1 - (\Sigma ET - \Sigma P)/W_c z_s\right] \qquad (27)$$

(for $\alpha > 1$ or $RW \geq 0.5$)

and:

$$\alpha = 1 - (\Sigma ET - \Sigma P)/W_c z_s \qquad (27a)$$

(for $\alpha < 1$ or $RW \leq 0.5$)

(iii) An estimation of ET could be directly performed from the general mean equations (21 or 25 and Figure 9a, b) by means of only a calculation of potential evaporation (EP = ET_0/β_i(climatic)) (Dooge, 1981):

$$ET = 2 ET_0 - EP = ET_0\left(2 - \frac{1}{\beta_i}\right) \qquad (28)$$

V. MAIN EXCHANGE MODIFICATIONS BY A VEGETAL COVER

This living cover induces modifications in space and time which interact at all levels of the energy balance terms. Thus, the main terms which appear in the description of the energy balance given in the previous paragraph are also more or less canopy dependent. Let us consider how strongly the vegetation cover acts by means of these main data or parameters.

5.1 NET RADIATION

This term depends on the canopy because of its radiative properties as well as its leaf density.

- Albedo for a given crop is more or less constant, radiative properties being determined (Figure 10b), but in fact it varies strongly (between 10 to 20%) as a function of soil cover with plant spacing and leaf density as well as a function of the annual variation in plant growth (Figure 10a) which introduces quite high temporal variations among crops, herbaceous covers, and deciduous forests (Seo and Yamaguchi, 1972). With vegetation, the contrast to soil increases as the soil surface dries.

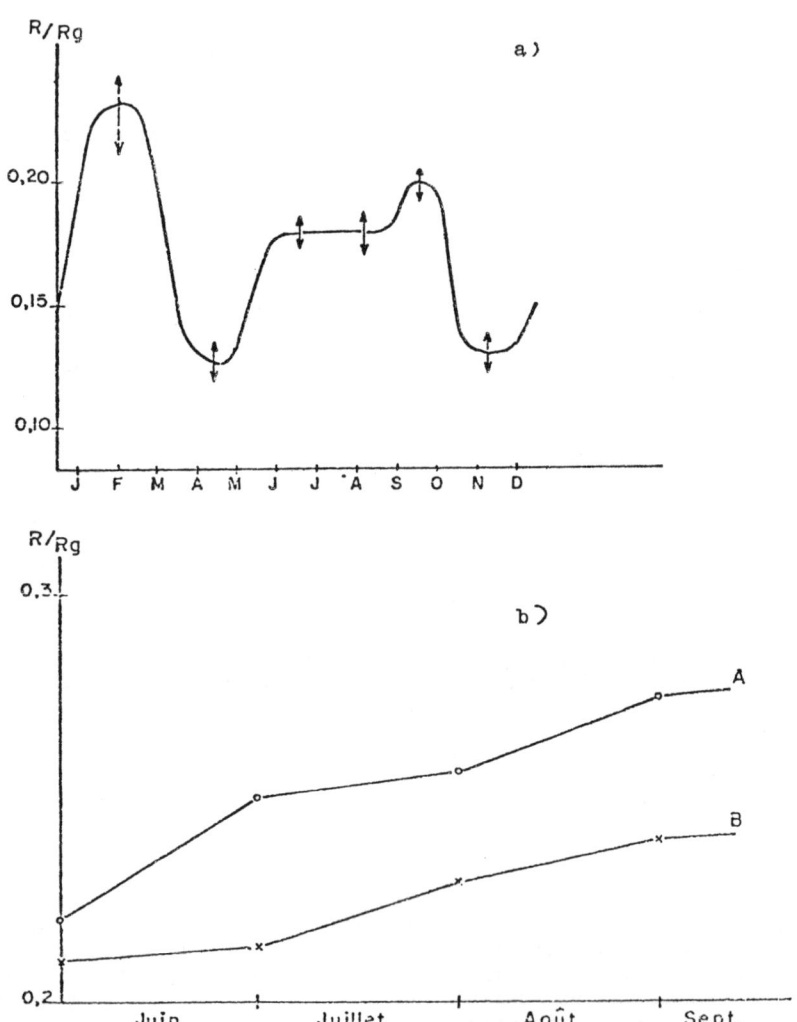

Figure 10 - Variation of forest and crop albedo during the growing period.
(a) Albedo of a beech forest with a snow-covered soil during winter and leaves during spring and summer (after Federer).
(b) Albedo of a maize crop versus time dependent LAI increase for two different species (after Davies).

- Emissivity and temperature also modify the net radiation but with a weak effect; only very dry surfaces introduce a significant divergence (6 W m^{-2} °C^{-1}).

- The last but not least important effect is due to the leaf net radiation absorption through the canopy (equation 2). The energy absorbed at the top of the canopy or at the bottom does not have the same distribution of sensible and latent heat and the energy collected in a more or less confined environment (bottom) yields more sensible heat.

Thus, with the same total leaf area but with a distribution of leaves presenting a maximum at the top or a minimum at the bottom, we get 20% differences in latent heat according to the crop resistance r (Table 4).

Table 4. Evapotranspiration versus crop resistance for top and bottom leaf distributions (Rn = 500 W m^{-2}, $Y(z_R)$ = 8°C, U = 3 m s^{-1})

r (s m^{-1})		0	100	200	400	1000
LE (W m^{-2})	Maximum leaves at the top	569	354	276	208	149
	Maximum leaves at the bottom	481	311	252	198	150

But the net radiation at the soil surface, which is generally the dryer part of the canopy, may also change significantly the fluxes. With low leaf density, the net radiation at the soil surface increases and the part of the sensible heat flux for the whole canopy increases. More commonly, conductive heat flux in soils increases and the available energy in the canopy R (R = Rn + S) is reduced by the same quantity.

In fact with a low plant cover and small values of leaf area index the soil surface characteristics play a more important role.

5.2   MOMENTUM BALANCE

The drag coefficient of all the elements in a stand of vegetation is one of the basic parameters which influence momentum absorption (eq. 3) but do not change too much the heat and mass exchanges. Greater effects are induced by the canopy structure with its leaf density profile $f(z)$ and some functions as the mixing length $L(z)$ in a porous medium (eq. 4a). They act strongly on sensible and latent heat exchange h through parameters such as height displacement D, roughness element z_o and friction velocity U_*.

The exchange coefficient is expressed as a function of two heights (reference level z_R and top of canopy z_h) with the common aerodynamic expression (Webb, 1965; Saito, 1964) (Figure 11a):

$$\left[h \right]_{z_h}^{z_R} = \left\{ \text{Log} \frac{z_R - D}{z_h - D} \right\} / k\, U_* \tag{29}$$

with the friction velocity defined by the wind profile

$$U_* = k\, U(z_R)/\text{Log}\left[(z_R - D)/z_o\right] \qquad (30)$$

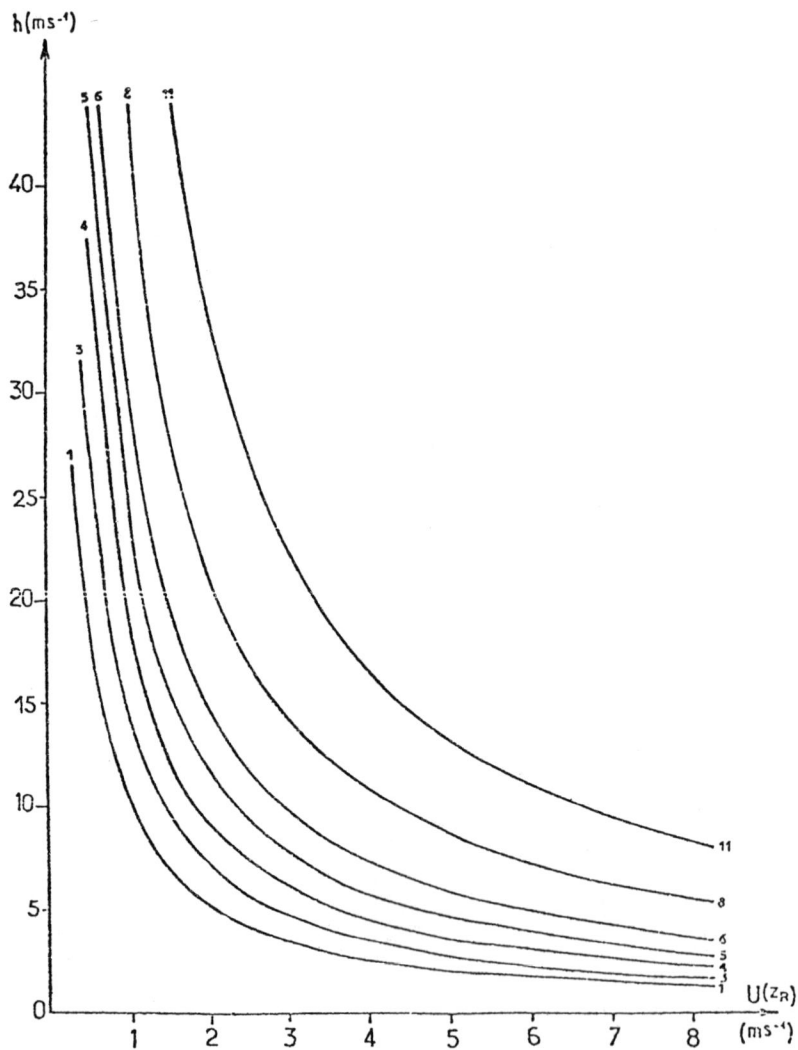

Figure 11a - Typical variation of the exchange coefficient (h) between the crop surface (height z_h) and reference level ($z_R = 2z_h$) versus windspeed at reference level for different LAI values (1 to 11).

In fact many authors have defined D and z_o as a function only of the height of the canopy z_h (Maki, 1975a, b).

$$D = 0.75\ z_h \quad \text{or} \quad D = 1.04\ z_h^{0.88}$$

$$z_o = 0.10\ z_h \quad \text{or} \quad z_o = 0.06\ z_h^{1.08}$$

which is quite good for many agricultural crops with a dense canopy. For natural stand vegetation with sometimes low, sometimes high density, and with a great variation in height, this relationship may be completely erroneous.

For a well-defined canopy, equations (3, 4, 4a) give in relation to leaf density profile $f(z)$, the variation of height displacement D and roughness parameter z_o (Figure 11b). With some simple hypotheses on homogeneous leaf distribution, an analytical solution may be proposed (Perrier, 1976) with LAI = z_h f:

$$D/z_h = 1 - \left[2/(z_h\ f)\right]\left[1 - e^{-(z_h\ f)/2}\right] \tag{31}$$

$$z_o/z_h = \left[1 - D/z_h\right]\left[(z_h\ f)/2\right] e^{-(z_h\ f)/2} \tag{32}$$

$$z_o/z_h = \left[1 - e^{-(z_h\ f)/2}\right] e^{-(z_h\ f)/2} \tag{32a}$$

These mean expressions can be extended to a profile of leaf density (f) if the product (z_h f) is corrected by a coefficient (a) which describes how many leaves are above, ΔLAI (sup), or below, ΔLAI (inf), the middle of the canopy, hence (z_h f), will be replaced by a(z_h f)

with a = 2 ΔLAI (sup)/LAI if (a > 1), or LAI/2 ΔLAI (inf) if (a < 1) (33)

These relationships similar to the proposed general relationships (Tani, 1963; Takeda, 1966; Lettau, 1969; Kondo, 1971; Seginer, 1974) allow for different kinds of vegetation cover and in particular for each kind of defined biomes, the calculation of the surface exchange coefficient.

On a regional scale if surfaces are heterogeneous with dispersed rows of trees or windbreaks, the expression of the roughness parameter becomes preponderant but the values given by equations (31) (32) remain in good accordance with analyses of Seguin (1973) and Iqbal et al. (1977).

5.3 MEAN CROP RESISTANCE

It is the most commonly used term to analyse evapotranspiration from surfaces (Rel. 13) but it is possible to recombine equations (8 to 12) in order to describe latent heat flux starting from an equivalent surface humidity parameter which is the value of air temperature deficit at the top of the canopy $[Y(z_h) = T(z_h) - T_D(z_h)]$ and also express the strength of water retention at the surface (soil or crop) as well as that of the mean crop involved. Evapotranspiration is thus rewritten as follows:

$$LE = \left[P'/(P' + \gamma)\right] R \left[1 + \rho\ c_p\ h\left(\frac{Y(z_R) - Y(z_h)}{R}\right)\right]$$

and Priestley and Taylor's coefficient appears as equal to one modified by a positive or negative ratio which translates, as was previously mentioned (4.1), the divergence of air potential between the reference level at the top of the crop as compared to the total available energy. In fact this term is useful only for direct measurement or for the use of a model which can directly predict this parameter.

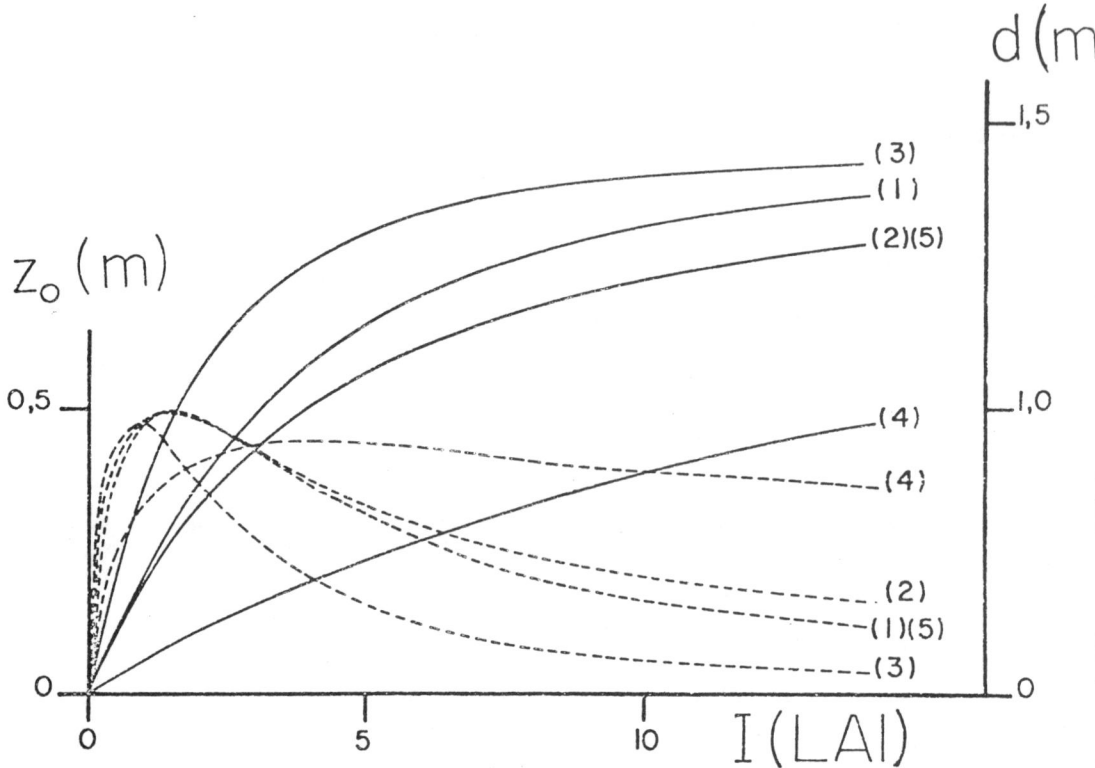

Figure 11b - Theoretical variation of roughness parameter (z_o) and crop height displacement (d) as function of crop LAI for several leaf density profiles. (1) constant leaf density profile; (2) parabolic leaf density profile (maximum for $z_h/2$); (3) maximum leaf density at the top (3 $z_h/4$); (4) maximum leaf density at the bottom ($z_h/4$); (5) measured maize leaf density.

The concept of mean crop resistance is easier to handle and is directly related to the physiological response of the plant to climate and to water storage in soil or water potential in the plant.

Most often the variations of canopy resistance with climatic factors are proportional (Katerji, 1977):

- to the reverse value of net radiation (or total radiation 1/Rg; Figure 12a),

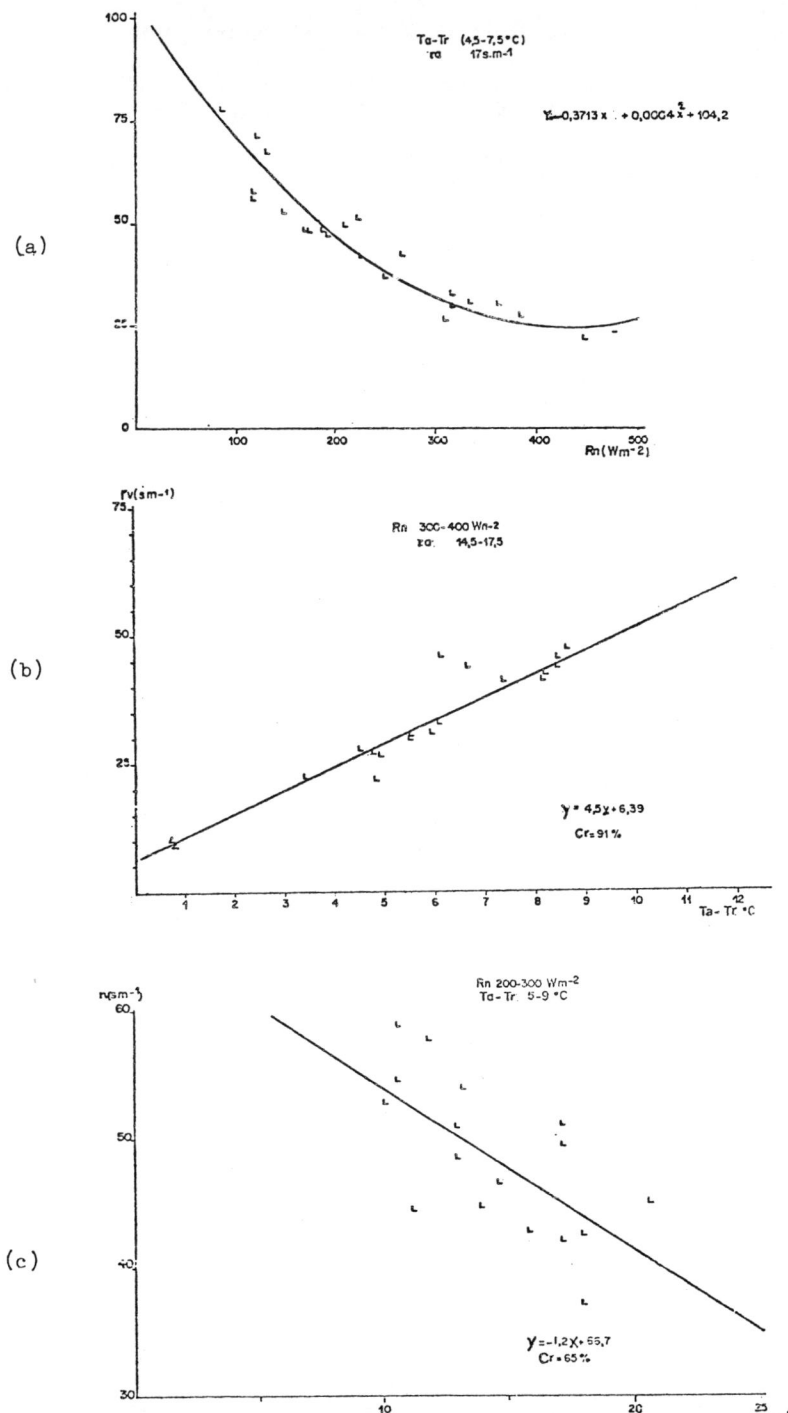

Figure 12 - Wheat crop resistance versus net radiation (a), air deficit (b) and wind speed plotted as a function of the crop aerodynamic resistance (c) (after Katerji 1977).

- to air deficit temperature ($Y = T - T_D$; Figure 12b)

- to wind speed vaguely according to equations (29) and (30) U proportional to $h = 1/r_a$ with r_a the aerodynamic resistance of the canopy (Figure 12c),

- and to the water potential in the plant which is strongly dependent on soil water potential (Figure 13) and climate.

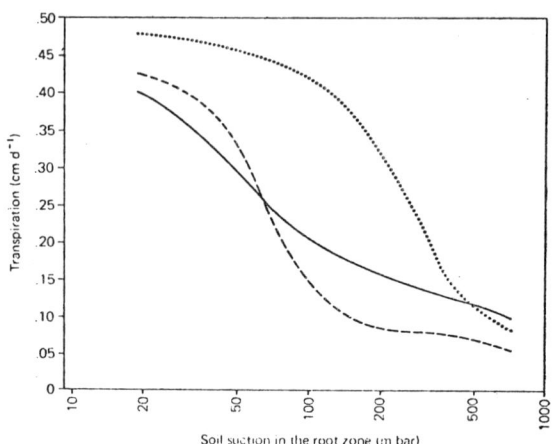

Figure 13 - Transpiration as a function of soil water suction in the root zone for spruce forest (———) mixed beech forest (....) and meadow (- - -). (After Item, 1979.)

These classical variations are the result through some complex integration of all the variations in stomatal resistances r_s at the level of each element (West and Gaff, 1976). For most species stomatal resistances also vary as a function of these terms and in the same way. As an example, for a pine forest (Halldin et al., 1979) the formula for stomatal resistance is given by the following relationship:

$$r_s = (a + b\ Rg)/Rg + (Y + c)/d$$

(Rg, global radiation; Y, air temperature deficit; a, b, c, d, constants).

One simple way to approach crop resistance as a whole is to link mean leaf resistance $\overline{r_s}$ for the canopy to the total leaf area index (Monteith, 1965) and write the expression for crop resistance as follows:

$$r = \overline{r_s}/LAI \qquad (34)$$

or in a more precise way:

$$\frac{1}{r} = \int_0^z \frac{f(z)}{r_s(z)}\,dz \qquad (34a)$$

This is the approximation which gives quite good results for a well-covered soil, soil surface having little action as compared to vegetation. But if the soil surface is characterized by a dry surface and consequently by an equivalent resistance to water diffusion equal to r_{ss}, and above all if the soil surface is not completely covered with vegetation, the previous relationship may be transformed as follows:

$$\frac{1}{r} = c \frac{LAI}{r_s} + (1 - c) \frac{1}{r_{ss}} \qquad (35)$$

or

$$\frac{1}{r} = c \int_0^z \frac{f(z)dz}{r_s(z)} + (1 - c) \frac{1}{r_{ss}} \qquad (35a)$$

where c is the percentage of soil covered with vertical projection on leaves on the surface.

We must note that the soil surface resistance (Gardner and Fireman, 1958; Nerpin et al., 1968; Feodoroff and Rafi, 1962) increases rapidly as a function of the received available energy. This kind of evolution is given in Table 5 which underlines the evolution of the thickness of the surface dry layer (Vernet, 1963) according to cumulative evaporation and the relationship between dry layer thickness and resistance (Perrier, 1973). In fact according to the water diffusivity of the soil and the value of evaporation (Gardner and Hillel, 1962) this dry layer evolution may thicken quicker (with low diffusivity and high evaporation) or slower with the reversed conditions (Murakami, 1969).

Table 5. Increase of the dry layer thickness at soil surface versus cumulative evaporation (after Vernet 1964) and corresponding soil surface resistance (after Perrier 1973).

$\sum$ ET (mm)	0	5	7	10	15	20	25	30	35	40	50	60
x (mm)	0	0	1	3	11	19	26	33	40	46	58	70
r (s m^{-1})	0	0	30	100	400	670	850	1150	1300	1400	1700	-

5.4 MEAN CROP VARIATION

One of the main variations is due to species; an illustration (Table 6a, b) gives some idea of the range of variations and some precise values of leaf resistance (see also Körner, 1979). According to the previous calculation of mean crop resistance, all modifications in stomatal structure as a function of species, modifying leaf resistance (Penman et al., 1951; Bange, 1953), change the mean crop resistance values. But even with similar values of leaf resistance it is clear that the leaf area index (equation 34, 34a) superimposes its modification at the scale of the whole canopy, reducing the mean crop resistance value and inducing different evapotranspiration (Figure 14).

Table 6a (after Rutter, 1975). Some examples of leaf stomatal resistances for several plant species (Herbaceous or Trees).

	$s\,cm^{-1}$		
	r_e	r_l(min)	
HERBACEOUS PLANTS			
Pelargonium zonale	75	0·75[b]	Williams and Amer (1957)
Wheat	17	0·25[b]	Milthorpe (1961)
Trifolium repens	16	0·33[b]	Shepherd (1964)
Cotton	32	0·9–1·3	Slatyer and Bierhuizen (1964)
Turnip, sugar beet	35–40	1·5–1·7	Gaastra (1959, 1963)
8 Crop species		0·7–2·3	El-Sharkawy and Hesketh (1965)
Phaseolus vulgaris		0·5–1·5	Kanemasu and Tanner (1969)
Vicia faba		0·5–2·0	Kassam (1973)
Circaea lutetiana	41–140	9·2–16·1	Holmgren et al. (1965)
Lamium galeobdolon	17–40	7·6–11·3	Holmgren et al. (1965)
TREES			
Betula verrucosa	56–83	0·9–1·4	Holmgren et al. (1965)
Populus tremula		2·3	Holmgren et al. (1965)
Acer platanoides	85–140	4·7–13·9	Holmgren et al. (1965)
Quercus robur	150–460	6·7–14·6	Holmgren et al. (1965)
Pinus sylvestris		ca. 2·0	Rutter (1967)
Citrus sinensis		1·0–3·0	Kriedemann (1971)
Populus sargentii		ca. 4·0	Parkhurst and Gates (1966)
Liquidambar styraciflua Liriodendron tulipifera Magnolia grandiflora Quercus velutina		4·0–6·0[c]	Knoerr (1967)
Picea sitchensis		3·0–8·0	Ludlow and Jarvis (1971)
Pinus resinosa, young leaves		2·7–4·7	Waggoner and Turner (1971)
Pinus resinosa, 1-year-old leaves		7·0–9·0	Waggoner and Turner (1971)

[a] All values based on projected leaf area.
[b] These values were determined from stomatal geometry, and are included for comparison with r_e.
[c] These values include some external resistance.

Table 6b (after Jarvis et al., 1976). Some examples of leaf stomatal resistances for several plant species (Herbaceous or Trees).

Species	Needle age	Minimal r_s (s m^{-1})	Method	Reference
Pinus halepensis	—	420	1	Whiteman and Koller (1967)
	—	1 000	2	Kaufmann (1973)
Pinus strobus	—	1 200	3	Gates (1966)
Pinus resinosa	—	780	3	Gates (1966)
Pinus contorta	Current year	290	1	Dykstra (1974)
Pinus resinosa	Current year	310	2	Waggoner and Turner (1971)
	1 year	430	2	Waggoner and Turner (1971)
	2 year	860	2	Waggoner and Turner (1971)
	3 year	890	2	Waggoner and Turner (1971)
	4 year	1 100	2	Waggoner and Turner (1971)
Pinus ponderosa	Current year	950	2	Running (1974)
Pinus sylvestris	—	ca. 200	4	Rutter (1967)
	—	400	2	Robins (In Stewart and Thom, 1973)
Tsuga heterophylla	Current year	450	2	Running (1973)
Sequoia sempervirens	--	1 500	2	Kaufmann (1973)
Thuja occidentalis	—	2 700	3	Gates (1966)
Abies grandis	Current year	240	2	Running (1974)
Picea mariana	—	1 700	3	Gates (1966)
Picea Breweriana	Current year	140	2	Running (1974)
Picea engelmannii	—	2 400	2	Kaufmann (1973)
	—	400	2	Kaufmann (unpublished)
Picea sitchensis	Current year	120	1	Ludlow and Jarvis (1971), Neilson et al. (1972)
	1 year	480	1	Ludlow and Jarvis (1971), Neilson et al. (1972)
	2 year	650	1	Ludlow and Jarvis (1971), Neilson et al. (1972)
	3 year	1 400	1	Ludlow and Jarvis (1971), Neilson et al. (1972)
	Current year	140	2	W. R. Watts (unpublished)
	1 year	220	2	W. R. Watts (unpublished)
	2 year	330	2	W. R. Watts (unpublished)
	Current year	290	2	Running (1974)
Pseudotsuga menziesii	Current year	120	2	Running (1974)
	Current year	300	1	Leverenz (1974)

Method: 1 assimilation chamber; 2 diffusion porometer; 3 leaf energy balance; 4 loss in weight.

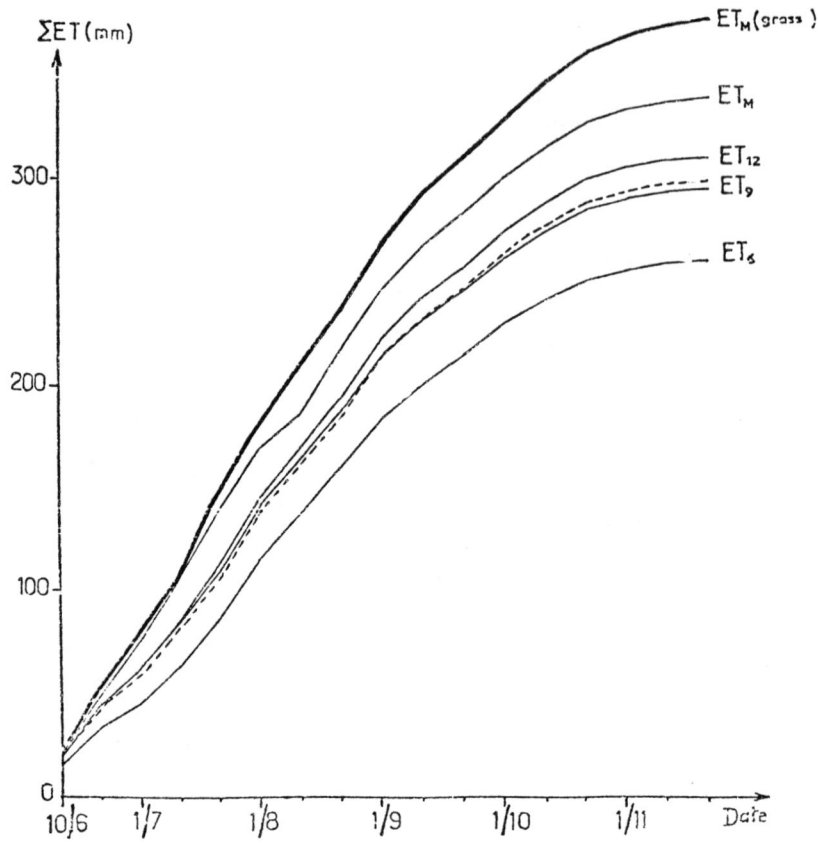

Figure 14 - Cumulative evapotranspiration during the growing period underlining:

- effect of crop resistance between grass ET_M(grass) and a well-watered maize crop ET_M(maize),
- effect of water supply between an irrigated maize crop ET_M(maize) and a non-irrigated one ET_9 (90,000 feet ha^{-1} crop density),
- effect of leaf area index on actual evapotranspiration (120,000 feet ha^{-1}, ET_{12}; 90,000 feet ha^{-1}, ET_9 with replication and 60,000 feet ha^{-1}, ET_6).

From the ratio between ET and EP on a monthly scale and from the exchange coefficient of the stand some mean values of crop resistances are proposed (Table 7).

If diurnal crop resistance variation (Figure 15) is always observed (Waggoner and Turner, 1971; Berger, 1973), the effect of high values during the beginning and the end of the day affects but little the cumulative value of evapotranspiration for the whole day because the available energy remains low during these two periods and often dew during the morning cuts down the value of stomatal resistan

Table 7. Some stand or crop resistances ($r\ s\ m^{-1}$) calculated from the ratio ET/EP and the exchange coefficient estimate.

Alfalfa	40	Citrus	250
Barley	70	Coniferous forest	200-300
Cotton	130	Deciduous forest	100-150
Maïze	80	Tropical forest	100-300
Potato	70	Grassland (temperate)	100
Rice	80	Grassland (subtropical)	200
Sugarbeet	50	Tundra	400
Sunflower	40		
Wheat	60		

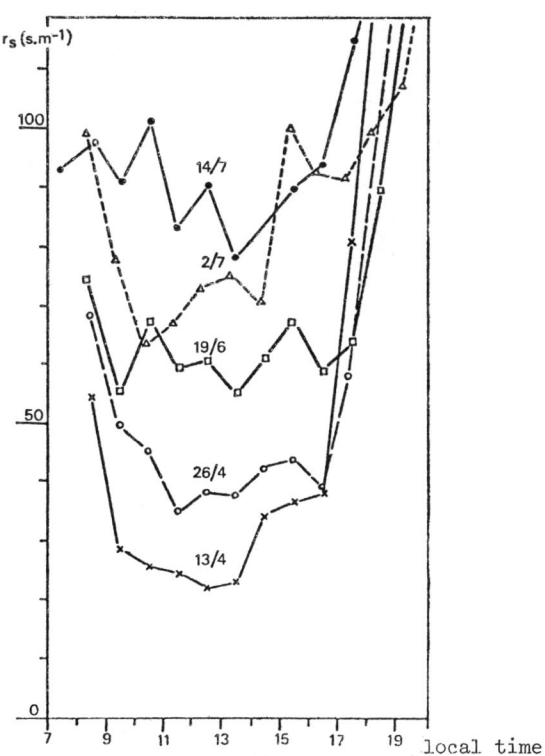

Figure 15 - Diurnal variation of a wheat crop resistance on several dates during the growing period.

The mean diurnal crop or stand resistance value, as was mentioned, varies with daily parameters of climate but remains around some mean value which follows plant growth and plant stage. For most of crop land and natural grassland the maturation period is accompanied by a great increase in the mean crop resistance (Figure 16).

Figure 16 - Increase of the daily mean resistance value of a wheat crop over the growing period.

If direct crop resistance measurements are always quite scarce nowadays, there are many investigations regarding direct leaf resistance measurements by diffuse porometers, and a great number of values for all kinds of species is available allowing the calculation of crop or stand resistance (Turner and Begg, 1973; Pereira and Kozlowski, 1976; Turner and Heichel, 1977; Pereira and Kozlowski, 1977).

VI. EVALUATION OF DATA SETS

6.1 DATA SETS CONCERNING VEGETATION PARAMETERS

Data sets are useful, firstly to estimate parameters needed in all models to assess land surface fluxes by means of some of the previous relationships described above, and secondly to directly verify from measured variables some data outputs from general atmospheric circulation models (Anderson, 1979).

6.1.1   Repartition of vegetation

Vegetation distribution according to climate with a description of stand type, form and species, with specific emphasis on interaction with climatic area, with relief, with soil type, with human modification and use, is the work of botanists, ecologists, geographers, agronomists, etc.

But if one tries to analyse potential natural vegetation (Figure 1), and in this respect the analysis given by Walter (1979), Kuchler (1964), Schnell (1970, 1977) is more or less relevant to this kind of description, vegetation describing the real and actual occurrence is of course the most useful thing for land surface processes. In such a description we not only find some wide natural zones correlated to potential natural vegetation described but also semi-natural zones which are much more commonly encountered and of course cultural zones representing nowadays a great percentage (Table 1). Thus, if biogeography (Elhai, 1968; Lacoste, 1978) is always more potential than actual, new investigations issued from IBP tend to describe more precisely natural terrestrial ecosystems in their present forms (Lamotte and Bourlière, 1978; Monteith, 1976).

One of the most accessible data sets are maps of every scale and dealing with most countries. A first attempt to collect information on it has been made by Kuchler (1967) in a three-volume book: International Bibliography of Vegetation Maps. Unesco (1973) has tried to develop some standardized vegetation classification and a map of world vegetation (scale around 1/1,000,000) is in preparation likewise some maps of the continents (Africa, South America, scale 1/5,000,000). For many countries (North America, Europe, North Africa, Japan, Madagascar, etc.) maps already exist, developed by special state agencies or bureaux, with scales between 1/50,000 to 1/1,000,000. For example, France is covered by a map of vegetal cover (1/1,000,000) and most of France by vegetation maps (1/200,000 - Service de la Carte de Végétation - P. Ray - Toulouse); more than that, phytosociology and ecology maps (1/20,000, Centre d'Etudes phytosociologiques et écologiques - Montpellier) are planned and some are already available; some particular parts of France are detailed (scale 1/50,000 to 1/400,000) such as the Alps (P. Ozenda, 1963) and some experiments works have been extended to foreign countries (Tunisia, India, Madagascar). Developments have improved rapidly particularly with new techniques of remote sensing (National Academy of Sciences and National Research Council, 1976) but at present no data collection in digital form is available.

6.1.2   Land use

Land use for agriculture modifies surfaces and if an enormous quantity of data is collected at the level of small regions through several kinds of organizations, only a little is used to establish and publish statistical data or mean land use on a bigger scale.

But in the U.S.A., e.g., maps are in preparation by the U.S. Geological Survey (scale 1/250,000) (Anderson et al., 1976). Some data are available through the FAO production yearbook (FAO statistics series No. 22, 1979) or in the World Atlas of Agriculture (scale 1/5,000,000 to 1/2,500,000) (1969). New presentations are actually planned which attempt to intergrate pedology, ecology, climate and human impact to define a map of the potentiality of zones (Economic Commission for Europe).

6.1.3 Vegetation parameters

No special data set exists for these parameters such as height, density, biomass distribution in air (Halle et al., 1978) and soil (root system) (Huttel, 1975), biological behaviour, and particularly, annual cycle, but only some descriptions; in addition we have just tried to demonstrate the usefulness of data sets concerning stomatal resistances at leaf level for every kind of species and also in fact the necessity of data sets for stand or crop resistances of a whole canopy.

The second type of parameters that can improve models of land surface processes are exchange coefficients for crop or stand vegetation and measurements or determinations of all the aerodynamic parameters as described in paragraph 4 and 5.2 (Takeda, 1965; Kondo, 1971). But all the scientific investigations made on this matter have been developed for more than 20 years and have been particularly promoted by the International Biological Programme (IBP); this latter has given a lot of data which are at present available in articles in international or national reviews. Some data have been collected and pointed out (Series of Ecological Studies, 1973-1980; Mihara, 1974; Monteith, 1975, 1976; Halldin, 1979; Lamotte and Bourlière, 1978).

6.2 DATA SETS CONCERNING USEFUL VARIABLES

Among useful variables which can help the validation of atmospheric models, some are commonly available but often less precise because they are not so dominant in the model, whereas some resulting from many interactions are much more significant as test output data such as latent heat flux or actual surface temperature.

6.2.1 Climatic data sets

Each country has its own meteorological service and for most of them, data such as classical air temperature, or humidity and precipitation are available in digital form. For our purpose and as far as vegetation is concerned, the most important climatic term is the energy input or the net radiation term (equation 13, 14) and it is one of the principal terms to test for, before latent or sensible fluxes or even surface temperature (Thornthwaite, 1948).

Few net radiation data are available but some specialized institutes have recorded this variable for many years; of course global radiation is often much more available but the transfer of this variable into net radiation remains uncertain (Gates, 1965; Durand, 1974) partly due to the incoming atmospheric radiation. The direct measurement of net radiation is undoubtedly the most suitable variable to have in an analysis of surface fluxes, latent and sensible heat and any error in its estimation is directly transferred into latent and sensible heat flux calculation. This measurement is one of the most important recommendations made by the report of the Round Table on Evapotranspiration (Budapest, 1977) organized by WMO and the International Commission for Irrigation and Drainage (ICID, 1977).

6.2.2 Evapotranspiration data sets

An abundant literature dealing with evapotranspiration calculated from formulas exists throughout the world (Thornthwaite, 1942 and 1948; Blaney-Criddle, 1962; Penman, 1956; Turc, 1964). Many countries have developed such calculations, e.g., Spain (Castillo and Beltran, 1977) for many years and even some month by month frequency analyses have been performed giving a water deficit calculation illustrated by monthly maps (Lecarpentier and Scherer, 1979). In fact, for many years such simple calculations and sometimes only the climatic index have been mapped for countries, continents and even the world (1/25,000,000) (Meigs, 1952), and Atlases have been constructed (Walter and Leith, 1974). A new map has been established by Unesco (1979) on the world repartition of arid zones. Many other examples are of course available.

More sophisticated models of energy balance and fluxes have been built (Jensen, 1976; Choisnel, 1977) and some are continuously running for technical information and are regularly checked with measurements. They also give good data sets for long periods to test general atmospheric circulation models.

More precise are data sets on plant water consumption or evapotranspiration (Naito, 1969). They are very useful for irrigation and drainage purposes. Direct or indirect measurements (Slatyer and McIlroy, 1961) are used everywhere to fit calculation or measured pan evaporation data with the actual evapotranspiration of crops and to define the water requirement. A great number of data (Doovenhos et al., 1975) is collected and presented in "Crop Water Requirement" (publ. FAO) accompanied by many details concerning all varieties of crops according to their growing period. The International Commission for Irrigation and Drainage manages a collection of these data in order to eventually prepare a world map of actual evapotranspiration.

In fact, if many data are available in most agronomic and agricultural institutes, they are only listed. However some digital data sets are already available, e.g. in France a complete data set of measured evapotranspiration going back 10 to 25 years on five locations and for more than ten crops has been built on a daily basis with the corresponding complete set of meteorological data (Grebet, 1978).

6.2.3 Energy balance data sets

Much more interesting should be complete data sets of all the terms of the energy balance of natural surface (Budyko, 1958). The energy balance method has been used to determine, generally on an hourly basis over more or less long periods, evapotranspiration and all the other terms of energy balance (Sestak et al., 1971). This method allows to gather, as well as other methods (aerodynamic method and combined or mixed method), a great number of flux data useful for daily, monthly or yearly data sets. Now some are also available in a digital form; as an example the SWECON project owns a few years of measurements on pine forest (Lindroth and Noren, 1979) and we have also established (Katerji and Grebet, 1979) data sets of energy balance terms on an hourly basis in series of three or four months (vegetative period) on wheat, maize, grass, and alfalfa crop. Such data exist in some laboratories and correspond to a well-fitted data set for testing models.

6.2.4 Water balance data sets

This aspect is only mentioned because the water balance data sets are useful in calculating stand characteristics such as the mean ratio between actual evaporation and potential evapotranspiration (Table 7). Data collections of measured fluxes over natural stands such as forests or savannas are generally very difficult to perform and almost impossible for some heterogeneous vegetative covers; in these cases, water balance data (Baumgartner, 1981; Dooge, 1981) are the only way to obtain an estimation of the evapotranspiration of such surfaces.

6.2.5 Surface temperature data sets

Surface temperature is the complex result of all climatic driving variables and parameters which fix flux values in terms of the energy balance of the surfaces. Any fluctuation in these driving variables or parameters induces a modification of the surface temperature with of course a given time lag. Several models are proposed (Feddes et al., 1975; Rosema and Bijleveld, 1977; Perrier et al., 1980) in order to use the surface temperature as an indicator of fluxes. Anyway, this surface temperature is, for those who develop models, a very accurate test variable particularly if the main problem is to determine and verify the energy balance of a surface essentially governed by water balance, because the aridity parameter (defined by eq. 7 and 15) is the most active term in the modification of the surface temperature.

These data sets become very common through satellite information (HCMM, NOAA, METEOSAT, etc.) (Itten, 1981) and are available in a digital form.

VII. GAPS AND DEFICIENCIES

The first dominant gap and deficiency is probably the problem of interactions between time and space scales.

Most of the vegetation stand parameters are only known for a local area and must be adapted to a regional scale which is always a difficult task for non-homogeneous surfaces.

As an example. if we analyse the exchange coefficient by means of the wind profile on grassland (Seguin and Gignoux, 1974) but in an open area and in a similar wide area with many windbreaks (Figure 17) results show:

- a much higher value for the roughness parameter (z_{oR} = 0.8m) in the case of a wide region divided into small fields by windbreaks than in the case of open grassland (z_o = 0.015),

- a much higher friction velocity (slope proportional to $1/U_*$) in the area with windbreaks than in the open area.

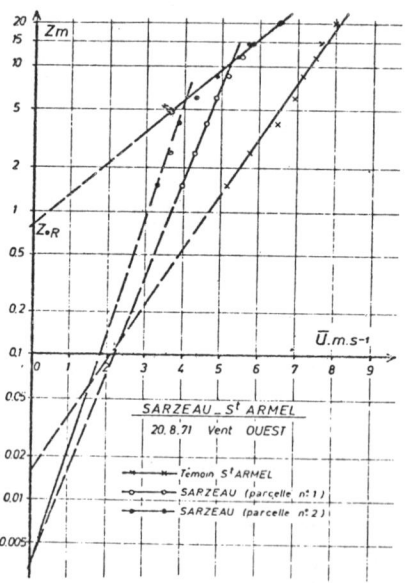

Figure 17 - Wind profile plotted according to logarithmic scale for grassland in open area (x) and in sheltered area (o, •) (grass was not mown on the open area and induces a higher roughness parameter) (after Guyot and Seguin, 1978).

In fact near the grassland surface, the roughness parameter is similar everywhere (z_o = 0.01m) with small differences due to crop height, but friction velocity is much higher in open area. In conclusion, the aerodynamic characteristics are very different on a regional scale between the two areas (upper profiles 5 or 10 to 20m) giving a lower exchange coefficient for the open area; but, in the lower layer, the observed reverse effect at the surface on the friction velocity parameter imposes a reduction in the exchange coefficient in the area with windbreaks. Finally, if we compare the total exchange coefficient h (between 0 to 20m) in the windbreak area (h_{wb} = h(0 to 8m) + H(8 to 20m)) and in the open area (h_o = h(0 to 20m) the total exchange coefficient for heat and mass transfer becomes in the sheltered area half of the exchange coefficient in open area ($h_{wb} = h_o/2$). Such a combination between local and regional scales is very hard to perform without specific data.

The second point concerned, as shown in paragraph 4.2, is this interaction between local fluxes and air mass characteristics at the scale of 100 to 400 km; this interaction imposes that the time step considered must be more than six hours, and surface fluxes must be taken, more or less, only on a daily mean value. We know that daily mean fluxes such as evapotranspiration are known by data sets on a local scale but spatial integration is always complex. Even over a small region it is very often difficult to perform and in some conditions impossible. Of course, over a great zone, this integration becomes even more impossible for homogeneous surfaces such as natural biomes with low human density. Obviously the lack of regional scale estimation values or direct measurements is the main problem of good comparisons with models. One possible way of obtaining a mean value on a regional scale should be some rough mean value in time and space, for example all the data of local evapotranspiration of grass at five points throughout France are represented (Figure 18) for many years according to dates and indicate perhaps some idea of how a mean regional value looks and evolves throughout the year.

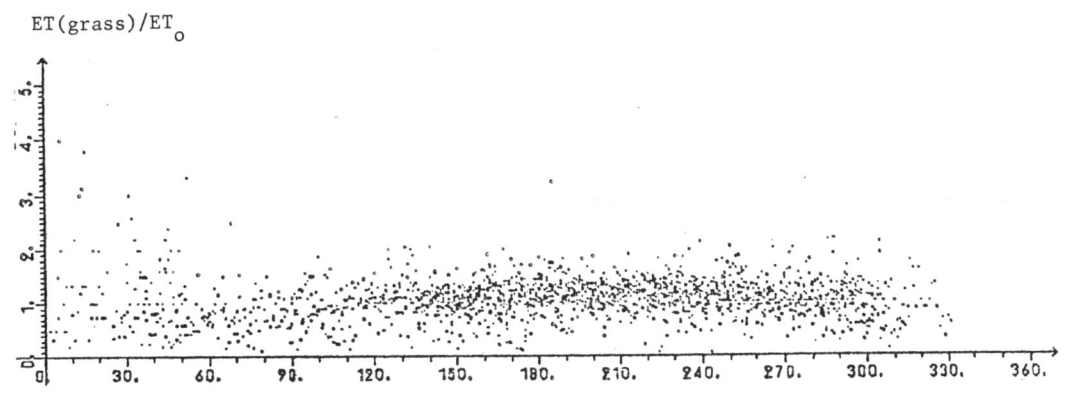

(ordinal numbers)

Figure 18 - Ratio of measured evapotranspiration of grass to limit or equilibrium evapotranspiration $\left[ET_o = \{\Delta/(\Delta + \gamma)\} R \right]$ for several dispersed locations in France (during winter evapotranspiration low values lack accuracy and induce great fluctuations in the ratio value - ratio mean value being about 1.1).

For mean daily regional values of fluxes, surface temperature can be very appropriate, particularly for space scale integration, but unfortunately probably less so for time scale which may be partly solved by using the maximum and minimum daily temperature. These two temperatures are more representative with respect to time evolution because they are quite stable for a few hours and well-correlated to daily evolution.

VIII. CONCLUSIONS

Many values describing vegetative surfaces, and consequently the resulting fluxes, are obtained most often for local points only sometimes extended to natural biomes and sometimes generalized to a country; of course very few descriptions are extended to the world scale. Most of these data are scattered throughout a great amount of scientific and technical publications and only very partial syntheses exist. But also the first difficulty is that these data are numerical data most often listed and only very rarely put in digital form for easy usage in computer comparison with models.

REFERENCES

Albrektsson, A., 1976: The amount and the distribution of tree biomass in some pine stands (Pinus silvestris) in northern Gästrikland, 26 p.

Allen, L. H. Jr. and E. R. Lemon, 1976: Carbon dioxide exchange and turbulence in a Costa Rican tropical rain forest. In "Vegetation and the atmosphere", Vol. 2, Case Studies, J. L. Monteith ed. 265-307.

Anderson, J. R. et al., 1976: A land use and land cover classification system for use with remote sensor data: U.S. Geological Survey Professional paper 964, 28 p.

Anderson, J. R., 1979: Geographical land use and assessment data. Working Group 6, Harpers Ferry, 8-11 May 1979, 127-147.

Anderson, M. C., 1966: Stand structure and light penetration. J. Appl. Ecology, 3 (1), 41.

Axelsson, B. and G. Agren, 1976: Tree growth model (PT 1) - a development paper, 79 p. (E) ISBN 041-5.

Bange G. G. J., 1953: On the quantitative explanation of stomatal transpiration. Acta Botanica Neerlandica, Vol. 2 (3), 255-297.

Baumgartner, A., 1981: Water Balance. Comm. Cgr. "Atmospheric General Circulation Models and Climate Simulation", Greenbelt, MD, U.S.A., 5 January 1981. This volume.

Beard, J. S., 1953: The savanna vegetation of northern tropical America. Ecol. Monogr. 23, 149-215.

Berger, A., 1973: Le potentiel hydrique et la résistance à la diffusion dans les stomates indicateurs de l'état hydrique de la plante. In "Réponse des plantes aux facteurs climatiques", Coll. Unesco, Uppsala 15-20 September 1970, série Ecologie et Conservation No. 5, 201-211.

Bernhardt-Reversat, F. et al., 1978: La forêt sempervirente de Basse Côte d'Ivoire. In "Structure et fonctionnement des écosystèmes terrestres" M. Lamotte and F. Bourlière, ed., 313-345.

Bernhardt-Reversat, F., 1972: Décomposition de la litière de feuilles en forêt ombrophile de Basse Côte d'Ivoire. Oecol. Plant. 7, 279-300.

Bille, J. C. and H. Poupon, 1972a: Recherches écologiques sur une savane sahélienne du Ferlo septentrional, Sénégal: Description de la végétation. Terre et vie, 26, 351-365.

_____ and _____, 1972b: Recherches écologiques sur une savane sahélienne du Ferlo septentrional, Sénégal: Biomasse végétale et production primaire nette. Terre et vie, 26, 366-382.

_____ and _____, 1974: Recherches écologiques sur une savane sahélienne du Ferlo septentrional, Sénégal: La régénération de la strate herbacée. Terre et vie, 28, 21-48.

Blaney, H. F. and W. D. Criddle, 1962: Determining consumptive use and irrigation water requirements. USDA (ARS) Techn. Bull., 1275, 59 p.

Bliss, L. C. and F. E. Wielgolaski, 1973: Primary production and production processes, Tundra Biome. Proc. Conf. Dublin 1973, Swedish IBP Comm., Stockholm, 250 p.

Bouchet, R. J., 1963: Evapotranspiration réelle, évapotranspiration potentielle et production agricole. In "L'eau et la production végétale", INRA Paris, 151-232.

_____ and A. Perrier, 1973: Bilan d'énergie et évapotranspiration à différentes échelles. Cgr. "Le soleil au service de l'homme", Unesco Paris, 29 p.

Bourlière, F., 1978: La savanne sahélienne de Fété-Olé, Sénégal. In "Ecosystèmes terrestres", Lamotte, M, Bourlière, F., 187-230.

Braun-Blanquet, J., 1964: Pflanzensoziologie. Springer (Vienne).

Brutsaert, W. H., 1981: Vertical flux of heat and moisture at a bare soil surface. Comm. Cgr. "Atmospheric General Circulation Models and Climate Simulation", Greenbelt, MD, U.S.A., 5 January 1981. This volume.

Budyko, M. I., 1958: The heat balance of the earth surface. Engl. Transl. Washington D.C., U.S.Depart. Commerce, Leningrad, Gidrometeorologicheskoeizdatel' stvo. 259 p.

Castillo, F. E. and L. R. Beltran, 1977: Agroclimatologia de Espana. Instituto Nacional de Investigaciones agrarias, Madrid, cuaderno No. 7.

Chartier, P., 1966: Etude du microclimat lumineaux dans la végétation. Ann. agron., 17, (5), 571-602.

Choisnel, E., 1976: Etude des échanges thermiques de l'homme en plein air au cas d'un environment froid. La Météorologie, 6, (5), 85-106.

_____, 1977: Le bilan d'énergie et le bilan hydrique du sol. La Météorologie, 11, (6), 103-159.

Cionco, R. M., 1965: A mathematical model for air flow in a vegetative canopy. J. Appl. Meteorol., 4, (4), 517-522.

Cowan, I. R., 1968: Mass, heat and momentum exchange between stands of plants and their atmospheric environment. Quart. J.R. Meteorol. Soc., 94 (402), 523-544.

Delting, J. K., 1979: Processes controlling Blue Grama Production on the Shortgrass Prairie. In "Perspectives in Grassland Ecology", N. French, ed., Springer-Verlag, 25-40.

Di Castri F. and H. A. Mooney, 1973: Mediterranean type ecosystems. Origin and structure. Ecol. Stud. No. 7, Springer-Verlag, 405 p.

Dooge, J. C. I., 1981: Parameterization of hydrologic processes. Comm. Cgr. "Atmospheric General Circulation Models and Climate Simulation", Greenbelt, MD, U.S.A., 5 January 1981. This volume.

Doorenbos J. and W. O. Pruitt, 1975: Les besoins en eau des cultures. Bull. Irrig. Drainage, 24, 198 p. FAO, Rome.

Durand, R., 1974: Estimation du rayonnement global à partir de la durée d'insolation. Ann. agron., 25, (6), 779-795.

Duvigneaud, P., 1974: La synthèse écologique. Doin, Paris, 296 p.

Elhai, H., 1968: Biogéographie. Colin, Paris, 407 p.

Evans, L. T., 1963: Environmental control of plant growth. Proc. Symp., Canberra, August 1962, Academic Press, 449 p.

Evenari, M. et al., 1971: The Negev: the challenge of a desert. Harvard, Cambridge, Mass., 345 p.

FAO, 1979: FAO Production Yearbook. Vol. 32, FAO Statistics Series No. 22, 45-56.

Feddes, R. A. et al., 1975: Finite element analysis of two dimensional flow in soils considering water uptake by roots. II. Field applications. Soil Sci. Soc. Amer. Proc., 39, 231-237.

Feodoroff, A. and A. M. Rafi, 1962: Evapotranspiration de l'eau à partir d'un sol nu. Les trois stades de l'évaporation. C. R. Acad. Sci. (Paris), 225, (23), 3220-3222.

Fritschen, L., 1981: The vertical fluxes of heat and moisture at a vegetated land surface. Comm. Cgr. "Atmospheric general circulation models and climate simulation", Greenbelt, MD, U.S.A., 5 January 1981. This volume.

Fuchs, M. and C. B. Tanner, 1967: Evaporation from a drying soil. J. Appl. Meteorol., 6, (5), 852-857.

Fuchs, M. et al., 1969: Evaporation from drying surfaces by the combination method. Agron. J., 61, (1), 22-26.

Gardiner, M. J., 1981: Use of regional and global soils data for climate modelling. Comm. Cgr. "Atmospheric general circulation models and climate simulation", Greenbelt, MD, USA., 5 January 1981. This volume.

Gardner, W. R. and M. Fireman, 1958: Laboratory studies of evaporation from soil columns in the presence of a water table. Soil Sci., 85 (5), 244-249.

_____ and D. I. Hillel, 1962: The relation of external evaporative conditions to the drying of soils. J. Geophys. Res., 67 (11), 4319-4325.

Gates, D. M., 1965: Radiant energy, its receipt and disposal. Meteorol. Monographs, 6, (28), 1-26.

Gaussen, H., 1954: Théorie et classification des climats et microclimats. Congr. Internat. Bot. Paris, Sect. 7 & 3, 125-130.

Golley, F. B. and E. Medina, 1975: Tropical ecology systems. Ecol. Stud. No. 11, Springer-Verlag, 398 p.

Grebet, Ph., 1978: Note interne (unpublished).

Guyot, G. and B. Seguin, 1978: Influence du bocage sur le climat d'une petite région: résultats des mesures effectuées en Bretagne. Agric. Meteorol., 19 (5), 411-430.

Hadas, A. et al., 1973: Physical aspects of soil water and salts in ecosystems (Ecological stud. No. 4) Springer-Verlag, 460 p.

Halldin, S. et al., 1970: Model for energy exchange of a pine forest canopy. In "Comparison of forest water and energy exchange models", Halldin S. ed. Elsevier (Amsterdam), 59-75.

Halldin, S., 1980: Comparison of forest water and energy exchange models. Proc. IUFRO Conf., Uppsala, 1978, Elsevier ed., 250 p.

Halle, F. et al., 1978: Tropical trees and forests. An architectural analysis. Springer-Verlag, 441 p.

Heath, G. W. et al., 1966: Studies on leaf litter breakdown. I - Breakdown rates of leaves of different species. Pedobiologia, 6, 1-12.

Hopkins, B., 1974: Forest and savanna (West Africa) Ibadan-London, 154 p.

Huttel, Cl., 1975: Root distribution and biomass in three Ivory Coast rain forests plots. Ecol. Stud. 11, 123-130.

I.C.I.D., 1977: Round Table on Evapotranspiration and Water Balance. C.R. Conf. Budapest, May 1977, La Météorologie, 6, (11).

Impens, I. et al., 1969: Extinction of net radiation in different crop canopies. Arch. Meteorol. Geophys. Bioklimatol. Ser. B, 17 (4), 403-412.

_____, 1970: Spatial and temporal variation of net radiation in crop canopies. Agric. Meteorol. 7 (4), 335-337.

Inoue, E., 1963: On the turbulent structure of airflow within crop canopies. J. Meteor. Soc. Japan, Series II, Vol. 41 (6), 317-326.

Iqbal, M. et al., 1977: A study of the roughness effects of multiple windbreaks. Bound. Layer Meteorol. 11, 187-203.

Item, 1979: Model for the water regime of forest and meadow. In "Comparison of forest water and energy exchange models", S. Halldin ed. 133-143.

Itten, K. J., 1981: Possibilities for remote sensing of surface characteristics. Comm. Cgr. "Atmospheric general circulation models and climate simulation", Greenbelt, MD, U.S.A., 5 January 1981. This volume.

Jarvis, P. G. et al., 1976: Coniferous forest. In "Vegetation and the atmosphere" Vol. II, case studies, J. L. Monteith ed., 171-236.

Jensen, M. E., 1976: On farm water management: irrigation scheduling for optimal water use. In "Conf. on salt and salinity management", 23-24 September 1976, Santa Barbara (U.S.A.), rep. 38, 54-65.

Katerji, N., 1977: Contribution à l'étude de l'évapotranspiration réelle du blé tendre d'hiver, application à la résistance du couvert en relation avec certains facteurs du milieu. Thèse Dr. Ing., Univ. Paris VII, 120 p.

_____ and Ph. Grebet, 1979: Note interne (unpublished).

Kestemont, P., 1973: Production primaire de la strate aborée d'une hêtraie à fétuques. Bull. Soc. Roy. Bot. Belgique, 106, 305-316.

Kira, T., 1978: Community architecture and organic matter dynamics in tropical lowland rain forests of Southern Asia with special reference to Posoh forest, West Malaysia. In "Tropical trees as living systems", Tomlinson and Zimmermann ed., Cambridge Univ. Press, 561-590.

Kira, T. et al., 1978: Biological production in a warm temperate evergreen oak forest of Japan. JIBP Synthesis (Tokyo) Vol. 18.

Kondo, J., 1971: Relationship between the roughness coefficient and other aerodynamic parameters. J. Meteorol. Soc. Japan, 49 (2), 121-124.

Korner, Ch., 1979: Maximum leaf diffusive conductance in vascular plants. Photosynthetica 13 (1), 45-82.

Kuchler, A. W., 1964: Potential natural vegetation of the conterminous United States (map and manual). Amer. Geograph. Soc. sp. Pub. 36, p. 116.

_____, 1967: Vegetation maps as climatic records. Biometeorology, Vol. 2, Part 2, "Ecological climatography", 953-964.

Kuhn, M., 1981: Vertical flux of heat and moisture in snow and ice. In "Atmospheric general circulation models and climate simulations", Greenbelt, MD, U.S.A., 5 January 1981. This volume.

Lacoste, A. and R. Salanon, 1969: Eléments de biogéographie et d'écologie. Nathan, (Paris), 189 p.

Laisk, A. H., 1965: Izvestiya akademii nauk Estonskoi SSR, ser. fiz., mat. i techn. nauk, 14, 107-119.

Lamotte, M., 1975: The structure and function of a tropical savanna ecosystem. Ecol. stud. 11, 179-222.

Lamotte M. and l'équipe PBI/Lamto, 1978: La savane préforestière de Lamto (Cote d'Ivoire). In "Problèmes d'écologie: Structure et fonctionnement des écosystèmes terrestres", M. Lamotte, F. Bourlière ed. 231-311.

Lamotte, M. and F. Bourlière, 1978: Problèmes d'écologie: Structure et fonctionnement des écosystèmes terrestres. Masson (Paris) 345 p.

Lauenroth, W. K., 1979: Grassland primary production: North American grasslands in perspective. In "Perspectives in grassland ecology", N.R. French ed. 3-21.

Lecarpentier C. and J. C. Scherer, 1979: Etude fréquentielle des besoins en eau d'irrigation 1946-1976. Serv. Hydraul., Minist. Agric. Paris, Darves-Bornoz éd.

Lemee, G., 1978: La hêtraie naturelle de Fontainebleau. In "Problèmes d'écologie: structure et fonctionnement des écosystèmes", M. Lamotte, F. Bourlière ed. 75-128.

Lemee, G. et al., 1975: Recherches sur l'écosystème de la forêt subéquatoriale de Basse Côte d'Ivoire. La Terre et la vie, 29, 169-264.

Lettau, H. A., 1969: Note on aerodynamic roughness parameter estimation on the basis of roughness element description. J. Appl. Meteorol. 8 (5), 828-832.

Lewis, M. C. and T. V. Callaghan, 1975: Tundra. In "Vegetation and the atmosphere" Vol. 2, case studies, Monteith J.L. ed. 399-435.

Leyton, L. et al., 1967: Rainfall interception in forest and moorland. In "Forest hydrology", W.E. Sopper et al., ed. 163-178.

Linacre, E., 1975: Swamps. In "Vegetation and the atmosphere" Vol. 2, case studies, Monteith J.L. ed. 329-444.

Lindroth A. and B. Noren, 1980: Evapotranspiration measurements at Jädraas. Instrumentation, data gathering and processing. In "Comparison of forest water and energy exchange models", Halldin S. ed. 15-26.

Logan, R. F., 1960: The central namib desert, South West Africa. Publ. 758, 162 p.

Longman, K. A. and J. Jenik, 1974: Tropical forest and its environment (Ghana) 196 p. Thetford, Norfolk.

Lossaint, P., 1973: Soil vegetation relationships in mediterranean ecosystems of Southern France. In "Mediterranean type ecosystems - origin and structure" F. di Castri, H.A. Mooney ed. 199-212.

_____ and M. Rapp, 1978: La forêt méditerranéenne de chênes verts. In "Problèmes d'écologie: structure et fonctionnement des écosystèmes terrestres", M. Lamotte, F. Bourlière ed. 129-152.

Maki, T., 1975a: Interrelationships between zero-plane displacement, aerodynamic roughness length and plant canopy height. J. Agric. Meteorol. 31 (1), 7-15.

_____, 1975b: Wind profile parameters of various canopies as influenced by wind velocity and stability. J. Agric. Meteorol. 31 (2), 61-70.

Meigs, P., 1952: La répartition mondiale des zones climatiques arides et semi-arides. C. R. rech. relatives à l'hydrologie de la zone aride, 208-215, Paris, Unesco (Recherches sur la zone aride, 1).

Mihara, Y., 1974: Agricultural meteorology of Japan. Soc. Agric. Meteorol. Japan, Univ. Tokyo Press, 215 p.

Monteith, J. L., 1965: Evaporation and environment. In "The state and movement of water in living organisms". G. E. Fogg ed. Symp. Soc. Exper. Biol. Swansea (G.B.) 8-12 September 1964, 205-236.

_____, 1975-1976: Vegetation and the atmosphere. Vol. I - Principles, 276 p. Vol. II - Case studies, 439 p., London Acad. Press.

Mooney, H. A. and D. J. Parsons, 1973: Structure and function of the Californian chaparral - an example from San Dimas. Ecol. Studies 11, 83-112.

Murakami, R., 1969: Effect of straw mulch on heat balance. J. Agric. Meteorol, 25 (2), 93-99.

Naito, Y., 1969: Studies on the transpiration and evapotranspiration amount of various crops. Bull. Tokai-Kinki Natl. Agr. Exptl. Station, 19, 49-151.

Nation. Acad. Sc./Nat. Res. Council, 1976: Research and environmental surveys from space with the thematic mapper in the 1980's, land use management. Nat. Acad. Sc., 43-51.

Nerpin, S. et al., 1968: Evaporation from bare soil and the ways of reducing it. In "Eau dans la zone non saturée" Wageningen 1966, Intern. Ass. Sci. Hydrol, t. 2, 595-602.

Odum, H. T. et al., 1963: Direct and optical assay of leaf mass of the lower mountain rain forest of Puerto Rico. Proc. Nat. Acad. Sci. Wash. 49, 429-434.

Ogawa, H. et al., 1961: A preliminary survey of the vegetation of Thailand. Nature life SE Asia 1, 21-157.

_____, 1965: Comparative ecological studies of three main types of forest vegetation in Thailand, II: plant biomass, Nature and life in SE Asia 4, 34-49.

Overbeck, F., 1975: Botanisch geologische moorkinde. 719 S Neumünster.

Ozenda, P. et al., 1963 ... 1966: Documents pour la carte de la végétation des Alpes. Lab. Biol. vég. Grenoble Lautaret, 5 tomes, Grenoble (France).

Penman, H. L., 1948: Natural evaporation from open water, bare soil and grass. Proc. R. Soc. A., 193, 120-145.

Penman, H. L. et al., 1951: Some physical aspects of assimilation and transpiration. Symp. Soc. for Exper. Biol. 5, 115-129.

Penman, H. L., 1956: Evaporation. An introductory survey. Neth. J. Agric. Sc., 4, 9-29.

Pereira, J. S. and T. T. Kozlowski, 1976: Leaf anatomy and water relations of Eucalyptus camaldulensis and E. globulus seedlings. Can. J. Bot. 54, 2868-2880.

_____ and _____, 1977: Influence of light intensity, temperature and leaf area on stomatal aperture and water potential of woody plants. Can. J. Forest Res. 7, 145-153.

Perrier, A., 1973: Bilan hydrique de l'assolement blé-jachère et évaporation d'un sol nu, en région semi-aride. In "Réponse des plantes aux facteurs climatiques" Coll. Unesco, Uppsala, 15-20 September 1970, sér. ecologie et conservation, 477-487.

_____, 1975: Methods of observation of heat and mass transfer in the lower atmosphere and in plant canopies. In "Heat and mass transfer in the biosphere. Part I: transfer processes in the plant environment" Coll. Intern. Heat and mass transfer, Dubrovnik, August 1974, De Vries D.A. and Afgan N.H. ed. 229-249.

_____, 1976: Etude et assai de modélisation des échanges de masse et d'énergie au niveau des couverts végétaux. Thèse. Doct. Sc., Univ. Paris VI, 236 p.

_____, 1980: Etude micro-climatique des relations entre les propriétés de surface et les caractéristiques de l'air: applications aux échanges régionaux. In "Météorologie et Environnement", Evry (France), 6-8 October 1980, 13 p.

Perrier, A. et al., 1980: Signification du concept d'inertie thermique pour diverses surfaces naturelles, sols ou couvertures végétales. Comm. 14ème cong. Soc. Intern. Photogrammétrie, Hambourg, juillet 1980, commission VII, groupe No. 2, 9 p.

Priestley, C. H. B. and R. J. Taylor, 1972: On the assessment of surface heat flux and evaporation using large scale parameters. Month. Weath. Rev. 100 (2), 81-92.

Raschke, K., 1956: Physikalische beziehungen zwischen wâr meibergangszahl usweines blattes. Planta 48, 200-238.

Rauner, J. U. L., 1976: Deciduous forest. In "Vegetation and the atmosphere" Vol. 2, case studies, Monteith J.L. ed. 241-263.

Ricou, G., 1978: La prairie permanente du nord ouest français. In "Problèmes d'écologie: structure et fonctionnement des écosystèmes terrestres" Lamotte M. and Bourlière, F. ed. 17-74.

Rey, P., 1960: Essai de phytocinétique biogéographique. Ed. CNRS Paris.

Ripley, E. A. and R. E. Redmann, 1976: Grassland. In "Vegetation and the atmosphere" Vol. 2 case studies, Monteith J.L. ed. 351-396.

Rosema, A. and J. H. Bijleveld, 1977: Tellus test of an algorithm for the determination of soil moisture and evaporation from remotely sensed surface temperatures. E.A.R.S. b.v., Delft, December 1977.

Rutter, A.J. et al., 1975: A predictive model of rainfall interception in forests. II. Generalization of the model and comparison with observations in some coniferous and hardwood stands. J. Appl. Ecol., 12, 364-380.

Saito, T., 1964: On the wind profile within plant communities. Bull. Nat. Inst. Agr. Sci. ser. A.11 - 67-74.

San Jose, J. J. and E. Medina, 1975: Effect of fire, organic matter production and water balance in a tropical savanna. Ecological Studies 11, 251-264.

Seo, T. and N. Yamaguchi, 1972: Albedo of several field crops. Berichte des Ohara Inst. fur Landwirtschatliche Biologie, 14 (3), 133-143.

Schnell, R., 1970-1977: Introduction à la phytogeographie des pays tropicaux. Vol. 1: Les flores, les structures. Vol. 2: Les milieux, les groupements. Vol. 3 et 4: La flore et la végétation de l'Afrique tropicale. Gauthier-Villars, Paris.

Schnell, R. and S. T. Schnell, 1977: Les applications de la météorologie et l'environnement. Bull. OMM, 26 (2), 139-141.

Seginer, I., 1974: Aerodynamic roughness of vegetated surfaces. Bound. Layer Meteorol. 5, 383-393.

Seguin, B., 1973: Rugosité du paysage et évapotranspiration potentielle à l'échelle régionale. Agric. Meteorol. 11, 79-98.

―――― and N. Gignoux, 1974: Etude experimentale de l'influence d'un réseau de brise vent sur le profil vertical de vitesse du vent. Agric. Meteorol. 13, 15-23.

Sestak, Z. et al., 1971: Criteria for the selection of suitable methods. In "Plant Photosynthetic Production, Manual of Methods", Sestak Z. et al., ed., D.W. Junk (The Hague), 1-38.

Slatyer, R. O. and I. C. McIlroy, 1961: Practical microclimatology (with special reference to the water factor in soil-plant-atmosphere relationships). Unesco Paris.

_____ and W. R. Gardner, 1965: Overall aspects of water movement in plants and soils. In "The state and movement of water in living organisms", Fogg G. E. ed. Cambridge Univ. Press, 113-129.

Takeda, K., 1965: Turbulence in plant canopies (2). J. Agric. Meteorol. 21 (1), 11-14.

_____, 1966: On roughness length and the zero plane displacement in the wind profile of the lowest air layer. J. Meteorol. Soc. Japan, ser. II, 44 (2), 101-108.

Tani, N., 1963: The wind over the cultivated field. Bull. Natl. Inst. Agr. Sci., Ser. A-10, 1-100.

Thom, A. S., 1969: The exchange of momentum, mass and heat between an artificial leaf and the air flow in a wind tunnel. Quart. J. Roy. Meteorol. Soc. 95 (405), 656-657.

_____, 1971: Momentum absorption by vegetation. Quart. J. Roy. Meteorol. Soc., 97 (414), 414-428.

_____, 1975: Momentum, mass and heat exchange of plant communities. In "Vegetation and the atmosphere", Vol. 1, principles, J.L. Monteith ed. 57-109.

Thornthwaite, C. W. and B. Holzman, 1942: Measurement of evaporation from land and water surfaces. USDA Techn. Bull. 817, 76 p.

Thornthwaite, C. W., 1948: An approach toward a rational classification of climate.

_____, 1948-1949: Micrometeorology of the surface layer of the atmosphere. Interim report No. 4, 5, 6, 7, 8.

Turc, L., 1964: Evapotranspiration potentielle mensuelle ou décadaire. Coll. Franco-Polonais de l'aménagement et de l'économie de l'eau, 251-268.

Turner, N. C. and J. E. Begg, 1973: Stomatal behaviour and water status of maize sorghum and tobacco under field conditions. I - At high soil water potential. Plant Physiol. 51, 31-36.

_____ and H. H. Heichel, 1977: Stomatal development and seasonal changes in diffusive resistance of primary and regrowth foliage of red oak (Quercus rubra L.) and red maple (Acer rubrum L). New Phytol, 78, 71-81.

Uchijima, Z., 1966: Micrometeorological evaluation of integral exchange coefficient at foliage surfaces and source strengths within a corn canopy. Bull. Nat. Inst. Agric. Sci. ser. A 13, 81-94.

Udagawa, T. et al., 1969: The penetration of direct solar radiation into a corn canopy and the intensity of direct radiation on the foliage surface. Photosynthesis and utilization of solar energy JIBP/PP - Photosynthesis level III group, 80-84.

Unesco, 1973: Classification internationale et cartographie de la végétation. 93 p. Coll. Ecologie.

_____, 1979: Map of the world distribution of arid regions. Techn. Note MAB No. 7, 55 p.

Verhagen, A.M. et al., 1963: Plant production in relation to foliage illumination. Ann. Bot. 27, 627-640.

Vernet, A., 1963: Evaporation de l'eau du sol. In "L'eau et la production végétale" INRA Paris ed. 415-428.

Waggoner, P. E. and N. C. Turner, 1971: Transpiration and its control by stomata in a pine forest. Bull. Connecticut agr. exp. sta. New Haven, 726, 1-87.

Walter, H. and H. Lieth, 1974: Klimadiagramm Weltatlas. Fischer Verlag, Iena.

Walter, H., 1979: Vegetation of the earth and ecological systems of the geobiosphere. Springer-Verlag New York, 274 p.

Warren Wilson, J., 1960: Inclined point quadrats. New Phytol., 59, 1-8.

Watts, W. R. et al., 1976: Photosynthesis in Sitka spruce (Picea sitchensis (Bong) Carr.). VII - Measurements of stomatal conductance and 14 CO_2 uptake in a forest canopy. J. Appl. Ecol. 13, 623-638.

Webb, E. K., 1965: Aerial microclimate. In "Agricultural meteorology", Meteorological monographs Vol. 6 (28) 27-44.

West, D. W. and D. F. Gaff, 1976: The effect of leaf water potential, leaf temperature and light intensity on leaf diffusion resistance and the transpiration of leaves of Malus sylvestris. Physiol. Plant. 38, 98-104.

Whittaker, R. H. and P. L. Marks, 1975: Methods of assessing terrestrial productivity. In "Primary productivity of the biosphere", Leith H., Whittaker, R. H. ed. Ecol. Stud. 14, 55-118.

Wielgolaski, F. E., 1973: Influence of climate on primary production in a Northern Mountain Area. Int. J. Biometeorol., 17, (4), 355-357.

World Atlas of Agriculture (Committee for), 1969: World Atlas of Agriculture. Novaro, Instituto Geografico de Agostine, 4 Vol., 527 p.

Yaron, B. et al., 1973: Arid zone irrigation. Ecol. Stud. Vol. 5, Springer, 434 p.

DATA ON SNOW COVER AND GLACIERS FOR THE GLOBAL CLIMATIC MODELS

V. M. Kotliakov, A. N. Krenke

1. The spreading of snow cover and glaciers.

2. The impact of snow cover and glaciers upon climate. Parameters necessary for climatic models.

3. Primary data - their types, quality and recurrency.

4. The primary available data. Their assemblage and exchange.

5. Gaps in information and ways of filling them.

1. THE SPREADING OF SNOW COVER AND GLACIERS

About 25,000 km^3 of water precipitate annually on the Earth's surface in the form of snow. About one third of it melts simultaneously in the sea, while the remainder forms snow cover on land, glaciers and sea ice. This mantle remains stable throughout a year only in the accumulation areas of glaciers (about $15 \cdot 10^6$ km^2). According to pre-satellite era estimation in the years of average snowfall still further $100 \cdot 10^6$ km^2 of our planet including sea ice are temporarily covered with snow, while in the years of excessive snowfall this value may reach to $110 \cdot 10^6$ km^2, in other words snow spreads over nearly a quarter of our planet (Shumsky, Krenke, Zotikov, 1964). On the inland territories snow covers temporarily more than $75 \cdot 10^6$ km^2 in normal years, i.e. nearly one half of its area (Table 1).

The data of Table 1 incorporate the areas of oceans with discontinuous ice cover (the computations were undertaken up to the mean perennial "ice fringe"), highlands with spots of bare rocks and territories with unstable (for several days) snow cover, such as Western Europe or Arizona.

Since 1966 the area of snow cover has been computed from satellite data (Matson, 1977; Wiesnet, Matson, 1979; Kukla, Kukla, 1974; Kukla, Gavin, 1980). In transitional zones only the areas directly covered by ice or snow are taken into account; for mountainous areas the snow coverage is assumed to be only 0.33, which is evidently underestimated for winter season. Only the areas where during the month snow prevailed over the snowless state are taken into consideration. As a result, e.g. Western Europe is considered snowless in February in the mean approximation for many years (Wiesnet, Matson, 1980). That is why the summarizing figures in the above-mentioned calculations appeared to be much lower than our previous estimations, shown in Table 1, especially for the temporary snow cover of the Northern Hemisphere, assumed to be 42-46 million km^2. But even under the assumptions made by American authors, about $100 \cdot 10^6$ km^2 remain under snow annually, among them $64 \cdot 10^6$ km^2 (25% of the area) in the Northern and $36 \cdot 10^6$ km^2 (14% of the area) in the Southern Hemisphere (Kukla, Gavin, 1980).

Taking into account that in February in the Northern Hemisphere, and in August in the Southern Hemisphere, i.e. at the moments of snow cover peaks on the land, sea ice does not reach yet the maximum area for approximately $3 \cdot 10^6$ km^2; the area directly impacted by snow and ice varies from $44 \cdot 10^6$ km^2 in August up to $96 \cdot 10^6$ km^2 in February. For continuous and stable snow and ice covers this area varies from $42 \cdot 10^6$ km^2 in August up to $83 \cdot 10^6$ km^2 in February, according to satellite information.

Table 1: The area and mass of snow cover formed annually (according to Shumsky, Krenke, Zotikov, 1964 and Kotliakov, 1968).

Snow cover	Area $10^6 km^2$	Accumulation kg/m^2	Mass of seasonal snow, 10^9 tons/km^3 of water
Northern Hemisphere			
Permanent snow on glaciers	2	250*	500
Seasonal snow on land	59	140	8,300
On pack ice	9	100	900
On seasonal sea ice	9	120	1,100
Southern Hemisphere			
Permanent snow on glaciers	14	160	2,200
Seasonal snow on land	2	150	300
On pack ice	5	180	900
On seasonal sea ice	15	200	3,000
Total	115	150	17,200

* New estimation.

According to satellite measurements the mean annual area of snow cover in the Northern Hemisphere varies around $35 \cdot 10^6$ km^2 (Kukla, Gavin, 1980), and consequently its mean duration is about six months. In the Southern Hemisphere, where one third of the snow cover is stable throughout a year, its average surface is $25 \cdot 10^6$ km^2 and mean duration exceeds months.

Lastly, it is noteworthy that the quaternary snow cover occupied 35% of the Northern Hemisphere and 24% of the Southern Hemisphere (Lamb, 1964).

Ice flow from the accumulation areas of glaciers to ablation areas exposing the old monolith ice in summer, and calves icebergs into the oceans. The distribution of glaciers in terms of their area and volume is presented in Table 2.

About 3,000 km^3 of icebergs are calved annually into the sea (about 2,500 km^3 from the Antarctic - Kotliakov, Losev, Loseva, 1977, and about 500 km^3 from the Arctic - Kotliakov, Krenke, 1980). About 8,000 km^3 of them are afloat (Nazarov, 1963), i.e. their ice survives in average for about three years. Occupying thousands of km^2 they are scattered over the water area of $63 \cdot 10^6$ km^2, i.e. nearly over a fifth of the ocean. The melting of icebergs consumes a great amount of heat (about $1 \cdot 10^{18}$ kJ/year), comparable to the heat discharge of all rivers and affects considerably the worldwide system of ocean currents.

Table 2: Dimensions of glacierization (Shumsky, Krenke, Zotikov, 1964; ICEX, 1979).

Glacierization	Area, 10^6 km^2			Volume, 10^6 km^3
	Accumulation area*	Ablation area*	Total	
Northern Hemisphere				
Continental ice sheets	1.1	0.6	1.7	2.7
Ice sheets on islands	0.2	0.15	0.35	0.2
Mountain glaciers	0.1	0.1	0.2	0.03
Southern Hemisphere				
Continental ice sheets	13.8	0.1	13.9	28.0
Mountain glaciers	0.02	0.01	0.03	0.01

* Accumulation area means the area where accumulation exceeds ablation; an ablation area - vice versa.

2. THE IMPACT OF SNOW COVER AND GLACIERS UPON CLIMATE. PARAMETERS NECESSARY FOR CLIMATIC MODELS

The impact of snow cover on climate is mainly caused by its high albedo, small heat conductivity, heat expenditures on snow melting and the relatively small roughness at its surface. Together with the high radiating emissivity of snow this induces low-surface temperatures and high temperature inversions above it. Among the first studies of the effects of snow cover upon climate were those of a well-known Russian climatologist, A. I. Voieykov (1889). Later, G. D. Rikhter (1948) analysed the impacts of snow cover on the geographical environment and showed its role in the evolution of landscapes. One of the authors of this paper (Kotliakov, 1968) has introduced the notion of "snowness" as the complex of natural phenomena connected to the presence of snow cover on the Earth's surface.

Special attention to the role of snow and ice covers in formation of the Earth's climate through the albedo is paid in the papers of M. I. Budyko (1971), Kukla (Kukla, Kukla, 1974) and others. Snow cover with regard to its albedo peculiarities and heat losses on melting has been included into the global climatic models (e.g. Manabe, Holloway, 1975; Chil, Battacharya, 1978). Experiments performed with the global model have suggested considerable impacts of the fluctuations of sea ice boundaries on the general circulation of the atmosphere (Herman, 1978).

Convective heating of the troposphere is absent, and absorbed short-wave radiation is reduced approximately three times due to the high albedo. If the area of ice and snow cover in both the hemispheres, averaged for the year, is assumed $62 \cdot 10^6$ km^2 (Kukla, Gavin, 1980), it would appear, under constant cloudiness and other assumptions, that the radiation income is lowered by $13 \cdot 10^{19}$ kJ/year due to the snow cover. This makes up more than 4% of the radiation absorbed by the Earth, and contributes to latitudinal differentiation of climate. Detailed computations of the albedo changes due to snow and ice covers were undertaken quite recently (Kukla, Robinson, 1980). Their account will help to specify the above-mentioned figure. Taking into account the typical values of cloudiness, its reflectivity, absorbtive capacity, and common values of albedo on glacier types, we concluded that continental ice sheets reflect into space an additional 54% of solar radiation coming to the upper boundary of the atmosphere; the accumulation areas of mountain glaciers reflect 33%, ice caps on relatively small islands only 16% and the tongues of mountain glaciers almost 9% of the solar radiation. The effect of mountain glaciers' tongues on climate is small because of the low albedo, while the effect of the ice caps on small islands is insignificant because of the almost permanent clouds, whose albedo is close to that of glaciers.

If the volumes of snow and ice melting are assumed to be the same as the volume of snow accumulation (17,200 km^3 in the water equivalent according to Table 1), these expenditures will make up nearly $6 \cdot 10^{18}$ kJ per year annually or about 0.2% of the solar radiation, absorbed by the Earth.

Investigations conducted in the U.S.S.R. in the 1950s showed the impact of snow cover in Eurasia on the occurrence of the Siberian anticyclone. The role of snow cover in North America is similar. The growth of thickness and duration of winter-spring snow cover in Tibet causes the lowering of summer temperatures in the troposphere, lag and attenuation of summer monsoons (Tu Cheng Yen, Xong-Shan-Chen, Cong-Bin-Fu, 1980). The annihilation of forests brings about the spreading of bare snow surfaces, the growth of albedo and climatic changes (Burroughs, 1978).

We have carried out quantitative estimations of glacier effects upon climate (Krenke, 1974). Proceeding from the value of radiation cooling and heat expenditures on melting, compared with the heat capacity, density and exchange rate of air masses, we can assume that the atmosphere above Greenland may cool by $5°C$ for June, July and August on the average due to the ice cover at the 300-500m layer. Consequently, the layer at 1,500m can cool on the average for $1°$, a figure which is in good agreement with the heights of the 500mb and 700mb surfaces. Heat transfer to the Greenland Ice Sheet in a year averages about $2 \cdot 10^{15}$ kJ/day, which is commensurate with the energy of cyclones crossing the ice sheet in 12 or 24 hours. Therefore the influence of the Greenland Ice Sheet on the Arctic baroclinic field is explained from the energetic point of view. The impact of the Antarctic Ice Sheet is still greater. According to similar estimations, the effects of ice domes much less than Greenland's on the global climate are negligibly small. The local air temperature jump between rock and ice surfaces is about $1.0-2.5°$ (Krenke, 1975).

The form of ice sheets creates additional impacts on climate. The steep vertical profile on their margins intensifies the vertical component of the wind velocity and precipitation, causing the mass-exchange to take place mainly in the marginal areas of the Antarctic (Kotliakov, 1961). The further spreading of ice sheets replaces the positive feedback between their nourishment and growth (increase in snowfall due to ice sheet orography) by the negative one (decrease in snowfall due to deviation of cyclone tracks).

The obstacles created by the ice sheets deform waves. Calculations based on diffraction formulae, show that this deformation is significant for obstacles whose dimensions equal those of the Greenland Ice Sheet, but is practically negligible for obstacles having the dimensions of Novaya Zemlya. This suggests the necessity to consider in the global climatic models those ice sheets and snow fields whose diameter exceeds 500km, and the possibility of neglecting smaller patches of snow and ice in global scale models.

The basic snowness characteristics of the territory, which are to be incorporated in the climatic models are:

(a) The surface of snow covered areas;

(b) Albedo of snow mantle spreading continuously and discontinuously over rocks and vegetation;

(c) Roughness of the snow surface, in unforested and forested areas;

(d) Timing of snow cover and the dates of its establishment and disappearance;

(e) The thickness (height) and mass of snow cover;

(f) Duration of snow-cover melting and the dates of its beginning and end.

The norms as well as the anomalies and temporary trends of the above-mentioned properties are very important for climatic models. One of these temporary trends was the increase of snowness in the Northern Hemisphere in the 1970s compared to the 1960s ($37 \cdot 10^6$ km^2 of annual average snow cover instead of $33 \cdot 10^6$ km^2) as discussed by G. Kukla (Kukla, Kukla, 1974; Wiesnet, Matson, 1980).

The area and form of glaciers may be assumed constant when solving the problems of short-term climatic fluctuations (decades and even centuries), but their variations should be taken into account for evaluations of longer fluctuations of climate. It becomes necessary to calculate or reconstruct the changes in glacier form accompanying climatic changes. This question comprises the problems of mechanics and thermodynamics of glaciers and so far has been solved by physical methods only in very rough approximations. Geological methods, using the evolution of ice sheets' form in the past to forecast the future, may serve as the alternative. The present theoretical knowledge obviously requires the following information:

(a) Topography of the glacier surface and bed;

(b) Fields of the annual net accumulation and ablation on their surfaces;

(c) Fields of the ice temperature at the attenuation level of its seasonal variations;

(d) Values of the geothermal heat flux;

(e) Fields of the ice velocity on the surface.

Some of the enumerated parameters are necessary not only for the predictions and reconstructions of glacier variations but also for the computations of their impact upon climate. Besides the above, we should know:

(a) Albedo of the glacier surface;

(b) Surface roughness of glaciers of different scales;

(c) Temperatures of the glacier surface;

(d) The dates of the beginning and end of melting on glacier surfaces;

(e) Values of the total ablation on the surface of glaciers and on their bed, if ablation exists there.

The glacier wind should be taken into account in the models, for it complicates the computations of air transformation over glaciers and changes the problem from pure thermodynamic into thermohydrodynamic. Under these conditions applications of the methods of comparative geographical analyses and statistical connexions between the indices of glacier morphology and the extent of their impacts upon climate are necessary.

3. PRIMARY DATA - THEIR TYPES, QUALITY AND RECURRENCY

The primary data on snow cover are obtained by weather stations, ground-based snow surveys, remote observations of snow (aircraft and spaceborne) and measurements of radiation of different wavelengths.

The network of weather stations measures the duration, thickness and water equivalent of snow cover over small sites. The density of observation is one point for several thousands of km^2. The advantage of these measurements is their high frequency - up to two times per day (the density of snow is measured more rarely). Snow albedo is measured at a very limited number of stations. The surface temperature is as a rule measured incorrectly because of radiation heating of instruments, and it is very difficult to establish the duration of snow melting.

Snow surveys are conducted along the routes from several to several dozens of kilometres. The routes coincide with representative but accessible landscapes (in the mountains - the valley bottoms). The thickness and water equivalent of snow cover are measured evenly along the route or in representative sites snow-survey points. The routes are distributed unevenly over large territories and do not provide continuous observations. The frequency of snow-surveys is approximately once a month or once in ten days.

The combination of routine and route observations does not provide precise data on the area of snow cover, nor on the position of the front and the area of snow-melting. These data appear to be extremely approximate.

Remote snow surveys may comprise remote aircraft and surface-based readings of stakes, established in hazardous localities in winter, automatic transmission of the data from such stakes or similar poles, airborne determination of water equivalent in snow covers, as determined by attenuation of the natural radiation of gamma rays. The latter method is successfully applied to the plains (Dmitriev et al., 1970; Carrol, 1980), but meets with navigation difficulties in the mountains.

Satellite (and sometimes aircraft) images allow the determination of the boundaries of snow cover establishment and melting, the extent of snow cover areas of synchronous snow melting (surveys in the near infrared band). Methods of automatic mapping of snow cover boundaries have been developed (Kravtsova, 1977). The resolution of the Landsat images is 80m, and the frequency of images over the same area is once per 18 days (Meier, 1975), or nine days with two Landsat in the space now. Because of cloudiness this interval may be still longer.

Meteorological satellites of the "NOAA" and "METEOR" systems (Deleur, 1980) provide small-scale observations of large territories of our planet. However, they possess low resolution (respectively 1.1 and 1.5 km) and low accuracy of the locality identification in case of operational use (Kurilova, 1975).

The data are on the area of snow cover and are used for the estimation of global climate (Kukla, Kukla, 1974; Kukla, Robinson, 1980) as well as for the computations of river discharge (Kalinin, 1974). In the U.S.A. the satellite information is efficiently used in hydrological computations for 30 river basins with their areas from 3,000 up to 100,000 km^2 (Schneider, 1980). Statistical (Rango, Solomonson, Foster, 1977) and deterministic models (Rango, Martinec, 1979) are applied.

Attempts to determine the thickness and water equivalents of snow cover directly from space images were successful for only thin sheets (Thomas, Lewis, Ching, 1978). The methods of determining snow storages and the thickness of snow cover from the passive or active (reflected radar) micro-wave radiation are now being developed. Correlation between the dielectric properties of snow sequence, determining its absorbtive and reflective capacities and its thickness and water equivalent is available. However, this correlation is disturbed, up to the change of its sign, by additional agents: the presence of liquid water in snow, its structure, the type of underlying surface and vegetation (Rango, Chang, Foster, 1979; Microwave Remote Sensing, 1980). With a certain delay, however, the water equivalent of snow can be calculated by methods of "thermal development" worked out in the U.S.S.R. (in application to spaceborne surveys - Garelik, Greenberg, Krenke, 1975 and Kotliakov, Khodakov, Greenberg, 1981). The method is based upon the equality of snow accumulation and ablation at the snow line. Ablation is determined from the totals of positive air temperatures (in grade-days) or by other heat-balance methods or is measured at specific points. Daily air temperature is interpolated from the stations, taking in mind vertical lapse rate. Moreover, the most important parameters of melting - the albedo - can also be measured on the image. Accumulation (i.e. water storage in snow) can be reconstructed from ablation values along successively depicted snow lines. In different modifications this method has been used already for certain territories and dates of surveys. This same method was earlier applied to the interpretation of terrestrial (Glacierization of the Urals, 1966) and aerial photography (Denisov, 1963).

The areas of glaciers are reliably determined from topographic maps and airborne photography. Because of the limited use of monotone-white stereo-pairs, the surface topography has so far been determined very inadequately on the basis of aircraft and terrestrial barometric levelling, combined with radio echo sounding. The errors were as large as 100m. Precise methods of geodetic surveys from satellites providing the accuracy of 1m have now been developed. A topographic map of the southern area of the Greenland Ice Cap has been already compiled (ICEX, 1979).

Topography of the bed with an accuracy of 1-2% of the glacier thickness is determined by radio echo soundings: both surface-based and with aircraft. The accuracy of these measurements is greater by an order of magnitude than the accuracy of the previous geophysical methods. Although continuous thickness profiles are available, however, the main difficulty is the identification of the position of aircraft profiles. Nevertheless navigation systems can be readily improved.

Ablation on the glacier surfaces can be estimated from the mean summer temperature (Krenke, Khodakov, 1966). On the equilibrium line height, where it is equal to accumulation it can be recalculated in accumulation (Ahlmann, 1924; Krenke, 1975). The errors of such calculation for glacier groups are reduced to a reasonable level. Interpolation between such groups helps in precipitation and run-off estimation.

Accumulation and ablation distribution through the glacier surfaces still require ground studies. Snow line is absent in the vast accumulation areas of ice sheets and the method of thermal development can not be applied there. Ground studies provide only scarce and uneven network of points. The data on a sequence of years are obtained only by digging pits and securing ice core, although new methods of satellite geodesy could provide the total values of accumulation and ablation for 5-8 years by repeated surveys of ice sheets; however such surveys have not yet been completed.

The same geodetic methods can be applied to the plotting of velocity fields on ice sheets, also unknown so far, and calculated only as a residual. The basic parameter - temperature at the glacier bed - remains yet unavailable. Computation methods provide but discrepant results.

4. THE PRIMARY AVAILABLE DATA. THEIR ASSEMBLAGE AND EXCHANGE

Snow surveys have already been performed for a century. In Russia they were pioneered by A. I. Voieykov in the 1880s. Since the beginning of the twentieth century these observations have been made in England, Sweden and Alpine countries. In the U.S.A. the first surveys were conducted in 1910. At present snow surveys are performed in nearly all the countries with snow cover. In the U.S.S.R. and the U.S.A. snow-survey routes make up many hundreds of kilometres. During the last 10-15 years remote snow surveys of different types, including gamma-surveys, have been broadly initiated in the U.S.S.R., the U.S.A., Norway and Canada.

Results of snow surveys are published in meteorological annals, and recently they are tape-recorded and stored in national meteorological centres (e.g. Asheville in the U.S.A., Obninsk in the U.S.S.R.). In the U.S.S.R. the data are kept on punch-cards of the "Minsk"-computer format and include the data on precipitation of different type, the data on the thickness, water equivalent and duration of snow cover, its area and the height of the snow line in the mountains.

The first regular spaceborne data were obtained in 1965. They include measurements of different wavelengths, radiation and albedo measurements from space. Since 1966 the weekly maps of snow and ice in the 1:50,000,000 scale (Glaciological data, 1979) have been compiled in the U.S.A. from the imagery obtained from polar-orbital and geostationary satellites. The data obtained from these maps are kept on magnetic tapes in the World Data Center A for Glaciology in Boulder, Colorado (U.S.A.). For Canada the similar maps at a scale of 1:9,000,000 are compiled weekly from NOAA and Landsat satellites. Resolution of these maps is about 10 km^2.

The World Atlas of Snow and Ice Resources (Kotliakov, 1977) is being compiled in the U.S.S.R. on the basis of all types of snow-cover observations. The Atlas is intended to show the norms and anomalies. Maps of the number of days with snow cover are available for the majority of countries. At present direct information on the maximum thickness of snow cover is available in Finland, France, German Democratic Republic, Poland, Czechoslovakia, Bulgaria, Switzerland, U.S.S.R., U.S.A., Canada and Japan. However, this information is not always presented as the mean long-term data and different averaging periods are used. Systematic information on the thickness of snow cover, the territory of Asia (except for the U.S.S.R.) and South America is unavailable.

The data on glaciers are not systematically assembled, but are gathered in course of conducting special national and international programmes. The areas of glaciers are specified through preparing the World Glacier Inventory, undertaken by the Snow and Ice Commission. The data recorded on tapes are kept in the Secretariat of the Programme at the Geographical Department of the "Eidgenossische Technische Hochschule" (ETH) in Zurich. The Glacier Inventory of the U.S.A. (except for Alaska) and glacier inventories of Scandinavia, the Alps, the Soviet Union and Africa have now been accomplished. In Central Asia (except U.S.S.R.), Canada and South America glacier inventories have been compiled only for separate areas. Inventorization of ice sheets of Greenland is now initiated (Scherler, 1980). The inventorization of Antarctica glaciation has not yet been organized.

The present knowledge of surface topography and glacier beds has been mentioned above. Maps of the glacier bed for half of the Antarctic Ice Sheet, separate ice sheets on islands and mountain glaciers have been compiled. Accumulation fields are known only for particular glaciers. Maps of such fields have also been compiled for continental ice sheets, but with greater approximation because of the scarce network observation points and routes (Kotliakov, Losev, Loseva, 1977; Mock, 1967). Fields of ice temperature are only approximate (Atlas of Antarctica, 1966; Mock, Weeks, 1966). Velocity fields are known only for separate small glaciers.

Albedo has been evaluated only at separate rare points on the glacier surface. Available information on glacier fluctuations is kept and published by the "Eidgenossische Technische Hochschule" (Fluctuation of Glaciers, 1967, 1972, 1978). Three centres assembling glaciological information have existed since 1957: Center A in the U.S.A. (Boulder, Colorado), B in Moscow, and C in Great Britain (Cambridge). Information is stored there by routine methods. Banks of data storage and processing on magnetic tapes have not been organized as yet.

5. GAPS IN INFORMATION AND WAYS OF FILLING THEM

Important data gaps can be filled in the years to come using satellite information, in particular polar-orbiting satellites.

These data gaps are: first, data on the position of the snow line and the snow-melting front on the basis of visible and near-infrared images, which will permit the use of the method of thermal development on the global scale; second, the operational use of active and passive microwave methods for estimation of water equivalent to snow; third, this is the accomplishment of satellite surveys of the ice topography of sheets; and, lastly, the plotting of the field of surface velocities of ice sheets with the help of satellite geodesy.

It is also necessary to organize the network of ground surveys of accumulation, ablation, cloudiness and air temperature over the ice covers, specially devised on a global scale. Empirical generalizations of the relations of these indices to the morphology of ice sheets are also desirable. It is expedient to establish the global network of boreholes in ice with selection of core for the reconstruction of the past glacial climates.

It is necessary to develop measurements of radiation in different wavelength bands for determining brightness temperature, albedo and roughness of glacier surfaces and snow cover.

Organization of such a programme could be conducted by the Snow and Ice Commission as the basic part of the Glaciers-Ocean-Atmosphere programme devoted to the special problem of determining the interaction between glaciers, ocean and the atmosphere.

REFERENCES

Ahlmann, H. (1924): Le niveau de glaciation comme fonction de l'accumulation d'humidité sous forme solide. Geogr. Annaler.

Atlas Antarktiki (1966) (Atlas of Antarctica), Moscow, GUGK Publishing House.

Budyko, M. I.: Klimat i Zhyisn (Climate and Life), Gidrometeoizdat, Leningrad.

Burroughs, W. J. (1978): Snow cover and climate change - seeing the wood for the snow. Weather, V.33, No. 7.

Carrol, T. R. (1980): Operational airborne measurement of snow water equivalent using terrestrial Gamma Radiation. Proceedings of the 48th annual Western Snow Conference, Laramie, Wyoming, April.

Chil, M. and K. Battaeharya (1978): An energy balance model of glaciation cycles. A review of climate models, GARP Publ. Ser. WMO, Geneva.

Deleur, M. S. (1980): Kosmicheskiye metody izucheniya snezhnogo pokrova Zemli (Spaceborne methods of studying snow cover of the Earth). Leningrad, Hydrometeoizdat.

Denisov, Yu. M. (1963): Metod rascheta raspredeleniya snezhnogo pokrova v gorakh po dannym aerofotos'yemki i temperature vozdukha (Methods, computating the distribution of snow cover in the mountains from the data of aerial photography and air temperatures). Izvestiya Academii Nauk Uzbekskoi SSR. Seriya Tekhnicheskikh Nauk, No. 6.

Dmitriev, A. V., R. N. Kogan, M. V. Nikiforov and S. D. Fridman (1970): Samoljetnaja gamma-sjemka snezhnogo pokrova (Airborne gamma-survey of snow cover). Meteorologi y Hydrologia, No. 3.

Fluctuations of Glaciers, publ. of Unesco, ed. by P. Kasser, Vol. 1, Louvain, 1967; Vol. 2, Louvain, 1972; Vol. 3, Louvain, 1978.

Garelik, I. S., A. M. Greenberg and A. N. Krenke (1975): Ispolzovaniye materialov s'emok so sputnikov dlya glyatsiologiches-kikh issledovanii (The use of satellite observations for glaciological studies). Izvestiya Academii Nauk SSSR. Seriya Geographiche-skaya, No. 1.

Glaciological Data (1979), No. 7.

Herman, G. (1978): The effect of extreme sea ice variations on the climatology of the Goddard GCM, Sea Ice Processes and Models. R. Pritchard, ed., University of Washington Press.

ICEX (1979). Ice and Climate Experiment. Greenbelt, Maryland, 1979.

Kalinin, G. P. (1974): Ot aerokosmicheskikh snimkov k prognosam y raschjetam stoka. Leningrad. Hydrometeoizdat.

Kotliakov, V. M. (1961): Snezhnyi pokrov Antarktidy i ego rol v sovremennom oledenen materika (Antarctic snow cover and its role in the present-day glaciation of the continent). Glatsiologiya, No. 7, Moscow, Academiya Nauk Publishing House.

Kotliakov, V. M. (1968): Snezhnyi pokrov Zemli i ledniki (Snow cover of the Earth and glaciers). Leningrad, Hydrometeoizdat.

_____ (Editor) (1977): Programma i metodicheskiye ukazaniya po sostavleniyu Atlasa snezhno-ledovykh resursov mira (Programme and instructions on the compilation of the World Atlas of Snow and Ice Resources). In: The Data of Glaciological Studies. Chronicle. Discussion, No. 29, Moscow.

_____, V. G. Khodakov and A. M. Greenberg (1981): Teplovoye proyavleniye snezhno-ledovykh ob'yektov kak metod kolichestvennoi interpretatsii aerokosmicheskoi informatsii. ("Thermal development" of snow and ice objects as a method of quantitative interpretation of airborne and satellite information). Izvestiya Akademii Nauk SSSR, Seriya Geograficheskaya, No. 3

_____ and A. N. Krenke (1980): Rol nazemnogo oledeneniya v vodno-ledovom balanse Arktiki (The role of glacierization in the water-ice balance of the Arctic). Izvestiya Akademii Nauk SSSR. Seriya Geograficheskaya, No. 4.

_____, K. S. Losev and I. A. Loseva (1977): Ledovyi balans Antarktidy (Ice-balance of Antarctica). Izvestiya Akademii Nauk SSSR. Seriya Geograficheskaya, No. 1.

Kravtsova, V. I. (1977): Sovremennye vozmozhnosti ispolzovaniya kosmicheskoi informatsii v glyatsiologicheskikh tselyakh (Present-day opportunities to use spaceborne information with glaciological objectives). In: The Data of Glaciological Studies. Chronicle. Discussion, No. 31, Moscow.

Krenke, A. N. (1975): Climatic conditions of present-day glaciation in Soviet Central Asia. Proceedings of the Moscow Snow and Ice Symposium, August 1971. IAHS-AISH Publ., No. 104.

_____ (1974): Climatic existence conditions for glaciers and the shaping of glacial climates. Meteorology and Climatology, Vol. 1, Geophysics series. G. K. Hall & Co., Boston, Massachusetts.

_____ and V. G. Khodakov (1966): Zavisimost' poverkhnostnogo tajania lednikov ot temperatury vosdukha (Relation of the surface melting of glaciers to air temperature). Materialy glyatsiol. issled. Khronika, Obsuzhdeniya, No. 12, Moscow.

Kukla, G. J. and J. Gavin (1980): Recent secular variations of snow and sea ice cover. Proceedings of the Riederalp Workshop. September 1978, IAHS-AISH Publ., No. 126.

_____ and H. J. Kukla (1974): Increased Surface Albedo in the Northern Hemisphere. Science, V.183, p.709-714.

_____ and P. Robinson (1980): Annual Cycle of Surface Albedo - Monthly Weather Review, Vol. 108, No. 1.

Lamb, H. H. (1964): The role of atmosphere and oceans in relation to climatic changes and the growth of ice sheets on land. In: Problems of paleoclimatology. London-New York-Sydney.

Manabe, S. and T. L. Holloway (1975): The seasonal variation of the Hydrologic Cycle as simulated by a global model of the atmosphere. Journal of Geophysical Research, Vol. 80, No. 12.

Matson, M. (1977): Winter snow-cover maps of North America and Eurasia from satellite records, 1966-1976 - NOAA Technical Memorandum NESS 84, Washington.

Meier, M. F. (1975): Application of remote-sensing techniques to the study of seasonal snow cover. Journ. of Glaciology, V.15, No. 73.

Microwave Remote Sensing of Snowpack Properties (1980): Proceedings of a NASA Workshop at Fort Collins, Colorado, May 20-22, 1980. NASA Conference Publication 2153, Washington.

Mock, S. T. (1967): Accumulation patterns on the Greenland Ice Sheet. CRREL rep. No. 233.

_____ and W. F. Weeks (1966): The distribution of 10m snow temperature on the Greenland Ice Sheet. Journal of Glaciology, V. 6, No. 43.

Nazarov, V. E. (1963): Sravnitel'naya kharakteristika ledovitosti Severnogo i Yuzhnogo polusharii (Comparative characteristics of the ice cover extent over the Northern and Southern Hemispheres). Vsesouznaya Konferentsiya po rezultatam MGG, Moscow, 1963.

Oledeneniye Ulara (Glacierization of the Urals). Akademiya Nauk Publishing House, Moscow, 1966.

Rango, A., A. T. C. Chang and J. L. Foster (1979): The utilization of spaceborne microwave radiometers for monitoring snowpack properties. Nordic Hydrology, 10.

_____ and J. Martinec (1979): Application of a snow-melt runoff model using Landsat data. Nordic Hydrology, 10, p.225-238.

_____, V. V. Salomonson and J. L. Foster (1977): Seasonal streamflow estimation in the Himalayan region employing meteorological satellite snow cover observations. Water Resources Res., V.13, No. 1.

Rikhter, G. D. (1948): Rol snezhnogo pokrova v fiziko-geograficheskom protsesse (The role of snow cover in physico-geographical process). Trudy Instituta Geografii Akademii Nauk SSSR, No. 40.

Scherler, K. E. (1980): Report on World Glacier Inventory. Status December 1980, Zurich.

Schneider, S. R. (1980): The NOAA/NESS program for operational snowcover mapping: preparing for the 1980's. Proceedings of a final workshop "Operational Applications of Satellite Snowcover Observations". Sparks, Nevada, 1979, NASA Conference Publication 2116, Washington.

Shumsky, P. A., A. N. Krenke, and I. A. Zotikov (1964): Ice and its changes. In: Geophysik of Earth, V.II, Washington.

Thomas, I. L., A. I. Lewis and H. P. Chins (1978): Snowfield assessment from Landsat-photogramm. England Remote Sensing, V.44, No. 4.

Tu Cheng Yeh, Xiong-Shan-Chen and Cong-Bin-Fu (1980): The Feedback Process of Long-scale Precipitation on the Variation of Atmospheric Circulation and Climate - The air-land interaction, Beijing.

Voieykov, A. I. (1889): Snezhnyi pokrov, ego vliyaniya na pochvu, klimat i pogodu i sposoby issledovaniya (Snow cover and its impacts upon soil, climate, weather and methods of its investigations). In: <u>Zapiski Russkogo Geograficheskogo Obschestva</u>, Vol. 18, No. 2.

Wiesnet, D. R. and M. Matson (1979): The satellite-derived Northern Hemisphere snow cover record for the winter of 1977-78. <u>Mon. Wea. Rev.</u>, 107.

_____ and _____ (1971): NOAA Satellite-derived continental snow cover data base. EDIS, 1980, Vol. 11, No. 1.

THE SHORTWAVE ALBEDO AND THE SURFACE EMISSIVITY

by
K.Ya.Kondratyev, V.I.Korzov, V.V.Mukhenberg,
L.N. Dyachenko
Main Geophysical Observatory, Leningrad

INTRODUCTION

Development of the atmospheric general circulation numerical models and climate theory makes it necessary to parameterize various physical processes taking place on the Earth's surface. The choice of the input parameters for the parameterization scheme affects significantly the results of numerical modelling of the atmospheric general circulation and climate formation. Two of such parameters are: the shortwave albedo (A) and surface emissivity (ε).

The given review paper discusses the available information on these parameters as well as the problems of parameterizing the dependence of albedo and emissivity on different factors, to use them in numerical climate models. It should be noted, however, that to parameterize reliably A and ε, much more observational data are needed in many cases.

1. The Shortwave Surface Albedo

This parameter is strongly variable and depends on wavelength (in the case of spectral albedo), Sun elevation, direct solar to scattered radiation ratio, type of the surface, moisture content, as well as microstructure of soil, and roughness (in the case of water basins), etc.

The surface albedo depends significantly on a height at which the instrument measuring the reflected radiation is located. Ground-based observations (usually made at the 2-m level) enable one to evaluate the albedo of only small areas, and not always characterize adequately the reflectivity of the extended areas. Observations of A made from aircraft, balloons or satellites are most valuable for determination of the albedo of large territories. However, when the albedo is measured at high altitudes (above 200 m), the effect of an intermediate atmospheric layer is substantial, which makes it necessary to consider the transfer function of the atmosphere.

Note, that the aircraft measurements are made along the restricted routes covering only some regions. Therefore, the existing data of aircraft measurements of the surface reflectivity are still inadequate to characterize, for instance, its seasonal variations. However, the albedo seasonal variations for large territories can be calculated. In this case the aircraft and other measurements are necessary to verify the calculation techniques.

In calculating the albedo for the summer period one should bear in mind that the albedo for large territories is determined by its values for various parts of such territories, which may differ substantially. In winter the albedo values in moderate and high latitudes are determined first of all by the snow cover contribution.

The actinometric observations at the meteorological network are mainly carried out with thermoelectric pyranometers sensitive in the 0.4-3.0 μm spectral range and installed at 1-2 m above the surface level. The albedo is calculated from the measured reflected and global radiation for three day times 9:30, 12:30, and 15:30 (mean local time).

An average day-time albedo is calculated in this case as an arithmetical mean determined from the routine observational data. But being calculated in such a way, it is not quite accurate .

TABLE 1

The snow-cover albedo from the data of the "Severny Polus-4,6,7" drifting stations [3]

No.	Snow structure	Moisture content and colour	Albedo, %		
			mean	max	min
1	Fresh snow	Dry, bright-white, pure	88	98	72
2	Fresh snow	Wet, bright-white	80	85	80
3	Drifted snow	Dry, pure, loose	85	96	70
4	Drifted snow	Wet, grey-white	77	81	59
5	Fresh or drifted from 2 to 5 days ago	Dry, pure	80	86	75
6	Fresh or drifted from 2 to 5 days ago	Wet, grey-white	75	80	56
7	Solid snow	Dry, pure	77	80	66
8	Solid snow	Wet, grey-white	70	75	61
9	Overcrystallized snow	Wet	63	75	52
10	Water snow	Light-green	35	-	28

With long days, the errors can be quite substantial in the case of the surfaces characterized by highly variable day-time albedos.

Mean-monthly, -seasonal, and -annual albedo values are determined as an arithmetical mean from the albedo values for month, season or year, respectively.

When using the albedo values, one should bear in mind that the reflectivity of natural formations exhibits in many cases a substantial angular dependence described in some publications [25, 55, 83]. According to their spectral and angular dependences of reflectivity and to typical albedo values, the basic natural formations are divided schematically into several classes [55]: snow, ice, soil, stony surface, vegetated surface, desert, and water basins.

Let us now consider the observational data characterizing the albedos of various surfaces.

1.1 The snow cover albedo.

The albedo of the snow cover varies widely (Table 1) depending on its structure and state (density, thickness, moisture content and contamination). Due to multi-year systematic observations in the Arctic and Antarctic, considerable information is available on the snow cover albedo for the polar regions. The mean-annual albedo for the Antarctic constitutes 86% [51] according to the data from all the stations located on glaciers, the seasonal variations in albedo being insignificant (within 84-89%). From the data of the american stations [64], the annual variations of the albedo in the Antarctic constitute 75-93%, in 53% of all the cases the albedo being 87-90%.

The snow cover albedos in overcast weather in the polar regions are, on the average, 3-6% higher than those in clear weather [44].

There is no common opinion in literature on the dependence of the snow cover albedo on Sun elevation. Several authors [40,51,64] believe that snow and ice albedos decrease with increasing Sun elevation. Others [29,44] presume that the albedo of a homogeneous snow-ice surface is independent of Sun elevation within the uncertainties of the measurements. This problem has to be further explored in more detail.

1.2 The ice albedo.

Investigations of the ice cover in the Arctic, Antarctic, and on glaciers, point to a great variability of the ice albedo due to melting, snow cover, and contamination [19,22,29,43, 44,50]. The ice albedo may vary from 2-20% (nilas, dark) to 90-95% (shelf ice) (Table 2).

TABLE 2

The Antarctic ice albedo [43]

No.	Type of ice	Concentration (tenths)	Snow coverage (tenths)	Albedo (%)
1	Nilas, dark	9-10	-	2-20
2	Nilas, grey, pancake	8-9	-	20-50
3	Grey ice	9	-	50-70
4	Grey-white, white ice	7-8	1-2	60-70
5	Ice containing the first-year ice	9	2-3	80-90
6	First-year and multi-year ice	9-10	2	80
7	Coastal ice	-	2-3	80
8	Shelf ice	-	-	90-95
9	Ice cap	-	-	90

The ice-covered surfaces of the Elbrus and Alps regions are characterized by average albedos listed in Table 3.

TABLE 3

The ice albedos for the Elbrus and Alps regions [6, 61]

No.	Surface	Albedo, %
1	Pure moist snow-ice	50 - 54
2	Contaminated snow-ice	46
3	Pure ice, and ice covered with tiny pellets of snow	39 - 41
4	Contaminated melting ice	26 - 33

From the data of aircraft observations /¯44_7, an ice-cap albedo for Franz Josef Land constituted 71-78%. The ice albedo depends on such interrelated factors as: surface state, ice age, contamination, snow coverage, ice concentration, and the extent to which it is destructed. Figure 1a shows the albedo vs. the ice cover destruction. As the destruction extent grows from 0 to 5 tenths, its albedo decreases from 70-80% to 10%. The albedos of the ice of different age differ only till the 2-3 tenths destruction, i.e. till the ice starts drying. The dry ice of different age destructed to more than 3 tenths, has the albedo which varies depending on the extent of the ice surface destruction (Fig.1). Variations of albedo depending on ice contamination and the extent of destruction are shown in Fig. 1b. As the ice contamination grows, its albedo decreases.

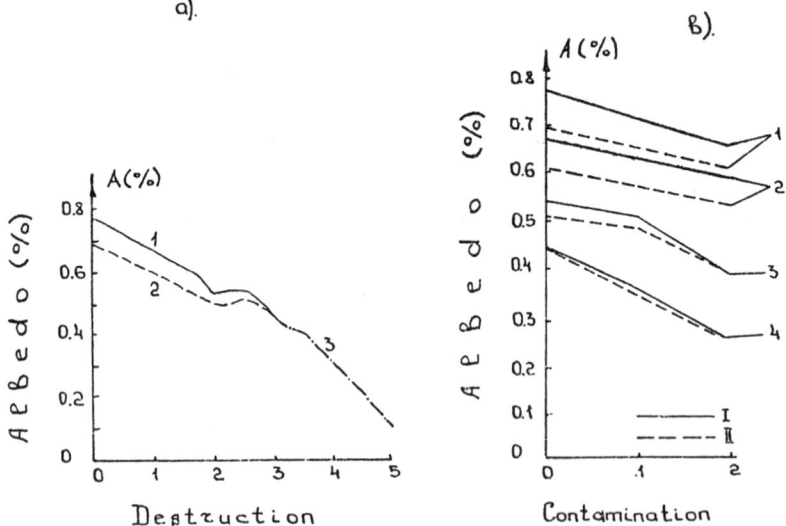

Figure 1 - The albedo vs. destruction (a) and contamination (b) of ice.

(a) 1 - the fall and two-year ice
2 - the winter and fall ice
3 - from the data of /¯29_7

(b) 1 - the fall and two-year ice
2 - the winter and fall ice

Destruction: 1 - 0, 2 - 1 tenths,
3 - 2 tenths, 4 - 3 tenths.

TABLE 4

The soil albedo.

No.	Soil characteristics	Albedo (%)	Reference
1	2	3	4
	Dry surface		
1	Chernozem (black soil, ploughed surface)	8	39
2	Alluvial soil on sea sediments	12	78
3	Loam, dark grey	12	78
4	Chernozem (smooth surface)	13	39
5	Chestnut soil (ploughed surface) light-grey	14	39
6	Iron-enriched soil on acid rocks	15	78
7	Semiarid, brown and reddish soils	17	78
8	Red ferrous soil on sand deposits	17	78
9	Chestnut soil, light-grey	18	39
10	Chestnut soil, grey-red	20	39
11	Grey sand	18-23	20
12	Weak-podzolic soil, sandy loam	18-24	46
13	Blue clay	23	39
14	Grey soil (Crimea)	25	39
15	Strong-podzolic soil (Moscow region)	26	39
16	Yellow sand	35	20
17	White sand	34-40	20
18	White sand	40	39

(to be continued)

TABLE 4 (continued)

1	2	3	4
	Moist surface		
1	Chernozem (ploughed surface)	4	39
2	Chestnut soil, light-grey (smooth surface)	6	39
3	Chernozem (smooth surface)	8	39
4	Black alluvial and hydromorphous soil	9	78
5	Chestnut soil, light-grey (smooth surface)	10	39
6	Chestnut soil, light-red (ploughed soil)	11	39
7	Blue clay	16	39
8	Weak-podzolic soil, sandy loam	16-18	46
9	Chestnut soil, grey-red (smooth surface)	18	39
10	White sand	20	39

1.3 The soil albedo.

The values of soil albedo systematized in Table 4 vary widely (4-40%) and depend not only on the type of soil but also on its colour, structure, and moisture content. The albedo of dry soil varies within 8-40%, and that of moist soil from 4 to 20% (Table 4). The soil albedo depends, in particular, on soil cultivation, i.e. its roughness. The albedo of the smooth soil is roughly 1.5 times higher than that of the ploughed soil.

The albedo of dry soils is about 1.8 times higher than that of moist soils. The albedo decreases most drastically as the soil moisture content increases from 1 to 15-20%, but further increase in the moisture content does not affect significantly the albedo value. So, for instance, the albedo of sand with moisture contents 5, 15, and 25% is 26, 16, and 14%, respectively.

The dependence of the albedo on surface coloration is clearly seen in the case of sand (Table 4). The albedo values for dry soils in low latitudes (Nigeria /¯78_7) are within 12-17%. Measurements were made at high Sun elevations (60-80°).

The diurnal change of the soil albedo can be determined by both insolation conditions (Sun elevation) and a change in the surface state (first of all, a change in its moisture content). As the Sun height increases, the soil albedo lowers /¯1_7, the greatest changes in the albedo values being observed in the morning hours at $h_\odot < 30°$.

TABLE 5

The dependence of the soil albedo (%) on Sun elevation /¯1_7

Sun elevation / Soil type	10	20	30	40	50	60	65
Rocky soil, dry	22	16	14	13	12	12	11
Loam, dry	34	29	21	20	19	18	17
Grey-green, dry	-	30	27	26	25	24	23

1.4 The albedo of vegetation.

The albedos of vegetation are different for various species of plants. For the same plant species, the albedo depends on

TABLE 6

The vegetation albedo [12,41,42,46,47,63]

Vegetation type	Albedo %	Month or h_o	Place of measurements
1	2	3	4
Meadow, green	17-18	52	Leningrad region
Meadow (fading stage)	13-14	35	Leningrad region
Grass (at a meteorological station)	21-22	45-54	Leningrad region
Grass (at a meteorological station)	19-21	50	Kiev
Thick grass in a desert	22	IV-VI	Turkmen SSR
Thin grass in a desert (light soil)	28	III-IV	Turkmen SSR
Thin grass in a desert (sand)	25	IV,57	Turkmen SSR
Lichen, dry, light	22	51-52	Canada
Lichen, dry, greyish-brown	17	51-52	Canada
Lichen, pale-green	21	51-52	Canada
Sedge in a marsh, green	18	51-52	Canada
Sedge in a marsh (with water)	12	51-52	Canada
Sedge and moss	11	51-52	Canada
Wet moss	18-19	44-53	Leningrad region
Potatoes at different vegetation stages	13-18	63-71	The Trans-Caucasus
Maize at different vegetation stages	16-23	69-70	The Trans-Caucasus
Maize at different vegetation stages	20-22	40	Estonian SSR

(to be continued)

TABLE 6 (continued)

1	2	3	4
Cotton at different vegetation stages	17-22	IV-XII	Turkmen SSR
Winter rye at different vegetation stages	21-22	VI-VII	Leningrad region
Oats at different vegetation stages	18-21	VII-VIII	Leningrad region
Winter wheat at different vegetation stages	13-21	41.7	Leningrad region
Sunflower (the stage of floscule maturing)	20	40-50	Estonian SSR

insolation conditions, the surface state, and the vegetation stage (in the course of which the height of plants changes, the crop-grass compactness grows, the colour changes, the plants drain, etc.).

Table 6 lists the data on the vegetation albedo obtained at high (40-70°) Sun elevations, $h_\odot$, when the vegetation albedo depends weakly on $h_\odot$. It follows from the analysis of the data in Table 6 that the green vegetation albedo varies mainly within 13-28%, about a mean value of 19%.

The vegetation albedo varies during a day. The diurnal change of the albedo is affected by surface roughness and insolation conditions: Sun elevation, the ratio between the scattered and global radiation, as well as spectral changes in the incident radiation. As V.G. Kastrov has shown, the latter affects less the diurnal change in albedos of soil and vegetation as compared to changes in roughness. As a rule, in a cloudless atmosphere conditions, the vegetation albedo is minimum at noon and increases as Sun lowers. Figure 2 exemplifies the dependences of the vegetation albedo on Sun elevation during a day.

Similar dependences were obtained for crops (wheat, maize, etc.). The wheat albedo increases by 2% when the Sun lowers by 10°. Such results were obtained from measurements made in different regions /⁻46,15,80_7 for plants at various vegetation stages. Note, that in the diurnal change of the vegetation albedo, an asymmetry is observed about noon, the reason of which is still unknown. Tooming /⁻46_7 believes that one of the reasons of such an asymmetry is variations of plants' physiological properties.

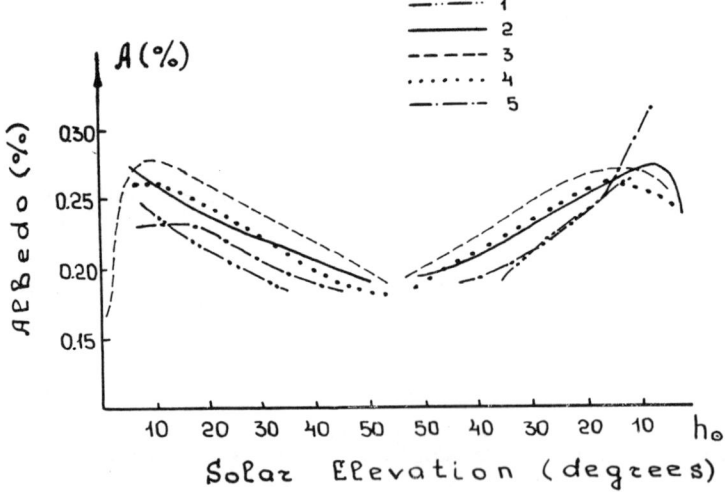

Figure 2 - The grass cover albedo vs. Sun elevation /⁻46_7:
    1 - dry grass of the year before, March;
    2 - new grass, May;
    3 - grass cover, June;
    4 - grass cover, July;
    5 - grass cover, August.

The dependence of the grass albedo on Sun elevation becomes more significant with growing compactness of the grass crop (grass in May-June). In the case of "pole" surfaces (sunflower, maize), this dependence is also distinctly observed. The albedo of bushy, thinly planted vegetation (for instance, potatoes), does not practically change within 10-50° Sun elevations.

Clouds reduce the dependence of the vegetation albedo on Sun elevation, since the absorption of the scattered radiation, the part of which in the global radiation flux increases with increasing cloudiness, is practically independent of Sun elevation. In cloudy weather, the diurnal course of the vegetation albedo is negligible.

A nomogram [12] was drawn from the data of numerous measurements over the grass cover to determine the albedo with different ratios between scattered (D) and global (Q) radiation (Fig.3). The upper curve shows the vegetation albedo vs. Sun elevation for direct solar radiation, the lower curve is for the case of overcast cloudiness.

To determine the grass albedo at different zenith solar distances and different values of D/Q , the following formula was proposed [31]:

$$A = A_o \left[1 + 2.5 \left(1.25 - \frac{D}{Q} \right)(1 - A_o) \sin \frac{3Z_o}{2} \right],$$

where A_o is an albedo value at a zenith angle of $Z_o = 0$.

Comparison of calculations made with this formula with the experimental data points to the fact that the formula approximates well the real albedo course for the grass cover with varying D/Q and Z_o.

Measurements [84] of the albedo of a pine forest (at a height of 20.3 m) have shown that the value A depends weakly on Sun elevation and humidity of the forest. So, for instance, the [A] value for the pine forest increased from 8.7% at

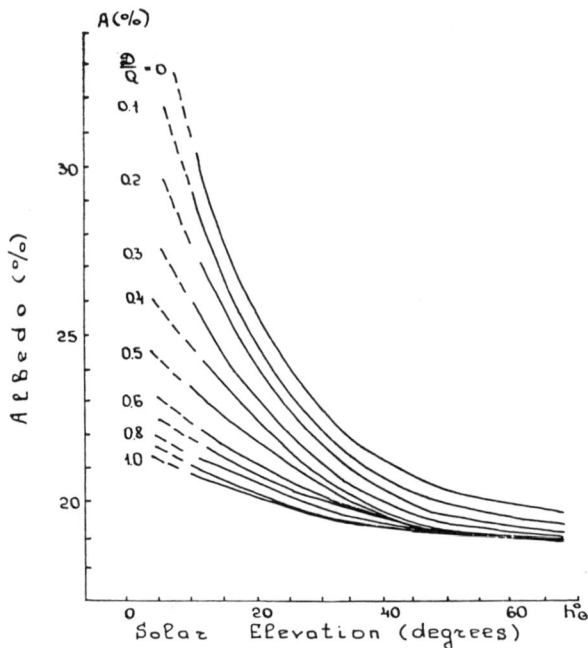

Figure 3 - The grass cover albedo vs. Sun elevation with various scattered-to global radiation ratios.

$h_\odot = 60°$ to 9.5% at $h_\odot = 10°$. With an appropriate Sun elevation and cloud amount, the albedo of a dry pine forest is 0.5% higher than that of a moist one.

Apart from the factors mentioned above, the vegetation albedo is affected by some structural parameters of the vegetated surface. Investigations made in /⁻54_7 show that the crop albedos are affected by such parameters as the leaf area index, the height of crops, the soil moisture content, the temperature and humidity of leaves. Figure 4 exemplifies the dependence of the spring wheat albedo on the enumerated parameters. The same figure shows the regression equations for the above-mentioned dependences.

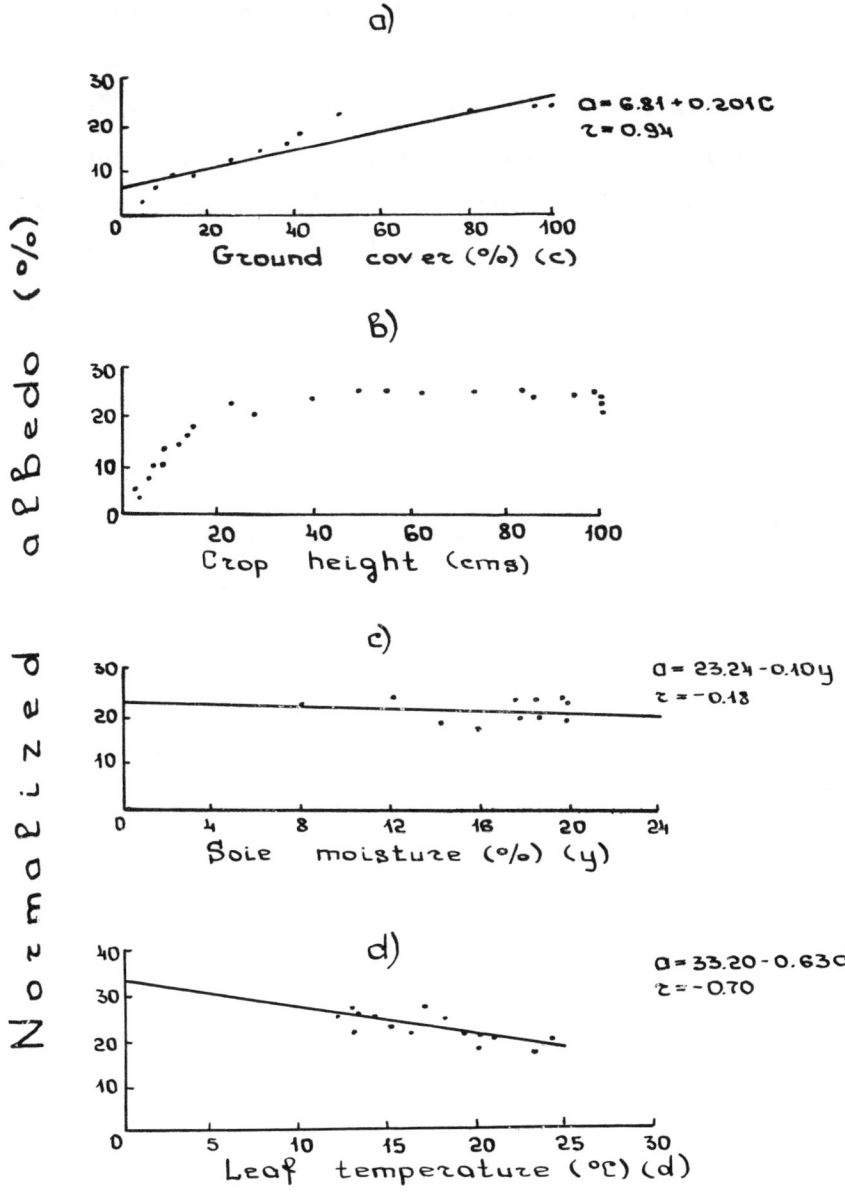

Figure 4 - Normalized albedo over agricultural crops (spring wheat) as a function of ground cover (a), crop height (b), soil moisture (c), and leaf temperature (d) [48].

As the leaf area index grows, the wheat albedo increases linearly (Fig.4a) reaching a maximum of 24-26%. When the height of plants reaches 20 cm, the albedo increases rapidly (Fig.4b). With further growth of wheat, the albedo varies negligibly.

The soil moisture content affects weakly an albedo value (Fig.4c), but the effect of the leaf temperature is greater. When the leaf temperature increases, the wheat albedo decreases linearly.

The vegetation albedo depends on the extent of leaves' drying, increasing as the moisture content in leaves lowers. This is clearly seen from the dependences obtained in [73] and shown in Fig.5. The reflectivity of a dying maize leaf increases at all wavelengths in the 0.5-2.5 um spectral region.

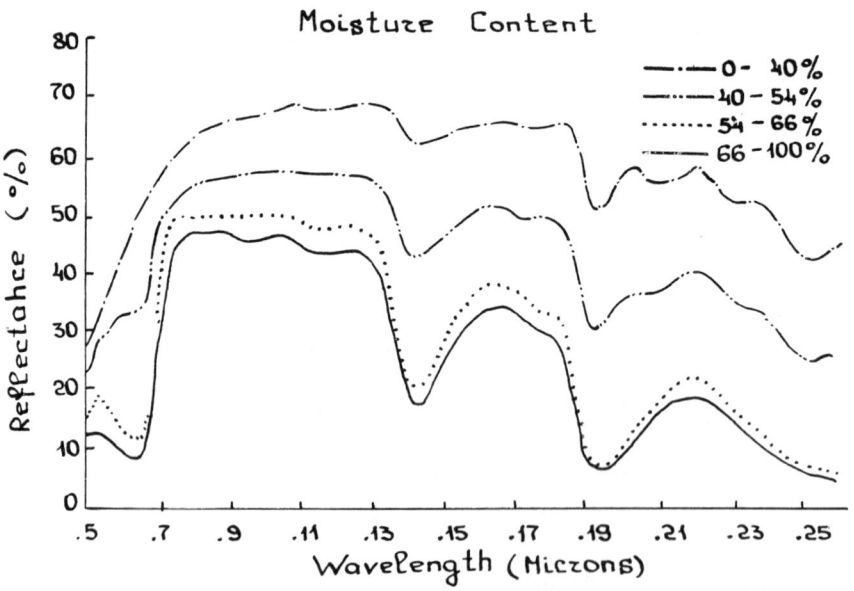

Figure 5 - Reflectance curves for corn leaves with different moisture content [73].

The plants' albedo varies significantly at different stages of vegetation. Most plants such as grass, potatoes, tomatoes, etc., change their colour from green to greyish-brown-green as they mature. When dying, they become dark, and their albedos decrease. Some crops (maize, sunflower and cereals) change their colour from green to yellow when maturing, and the albedo can increase in this case.

It has been observed /⁻5_7 that the albedo of maturing cereals varies differently in arid and humid regions. In the latter case a maximum value falls on the stage of earing - blossoming (Table 7). At the stage of ripeness, the albedo of cereals decreases, since the leaf area index diminishes /⁻15_7. In arid regions the albedo grows due to yellowing and clearing at the tubing-to-ripeness transition stage.

TABLE 7

The albedo (%) of spring and winter cereals in different vegetation phases /⁻15_7.

Regions	Vegetation phases		
	Tubing	Earing-blossoming	Ripeness
Humid	18	19 - 22	18 - 21
Arid	16	17 - 20	20 - 24

Figure 6 /⁻54_7 shows seasonal variations in the albedo of the grass cover and cereals, from which it is seen that in May the A values for the grass cover reach a maximum (25%).

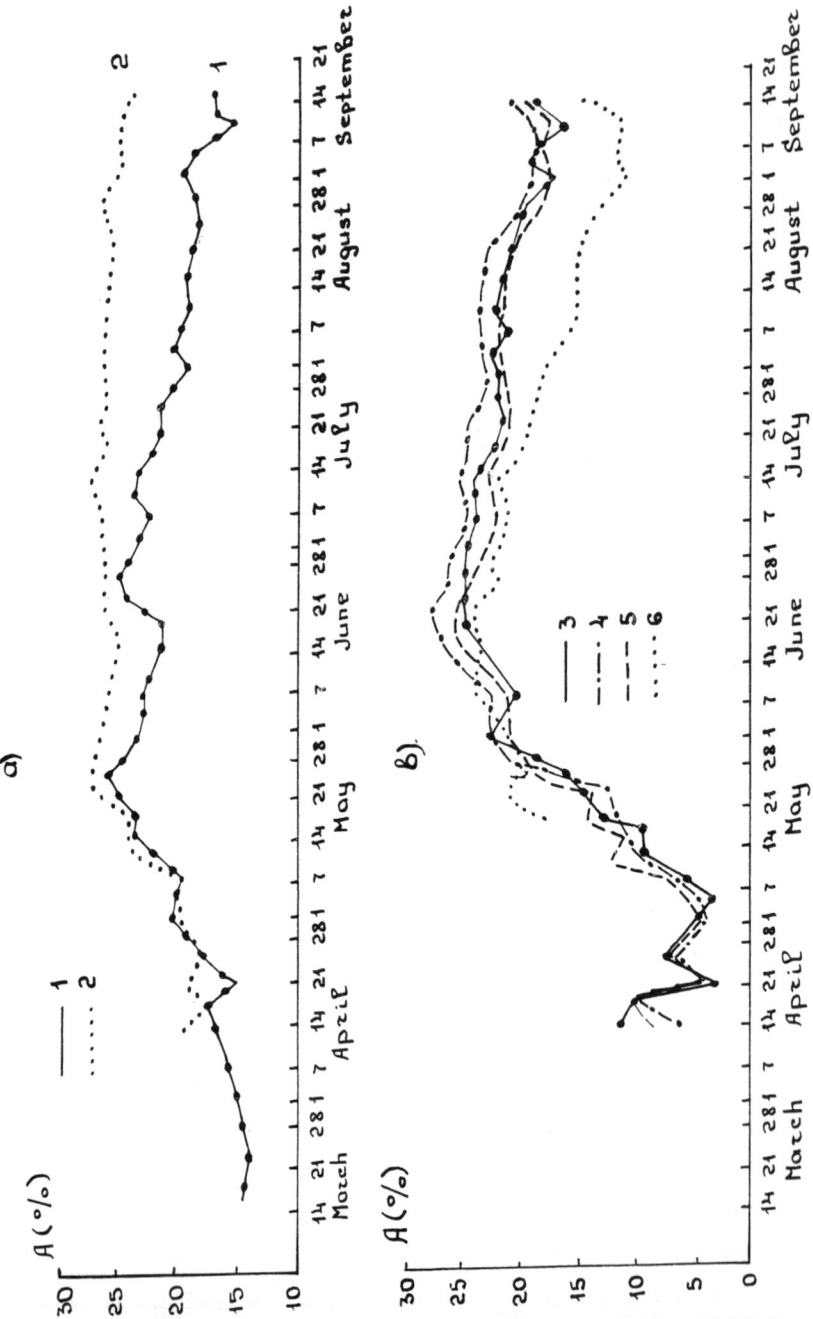

Figure 6 - Seasonal variations of the albedos of grass (a) and crops (b) [54].

1 - field grass; 2 - golf-playing field; 3 - spring wheat;
4 - barley I; 5 - barley II; 6 - winter wheat.

Then the field grass albedo (grazing grass) decreases slowly during the entire summer (curve 1). The albedo of the grass cover on the golf-playing field (curve 2) remains practically constant throughout the summer period since the field was maintained in a good state.

The albedos of the wheat and barley in May (Fig.2b) increase drastically from a value typical of a bare ground to a maximum in June, and then it decreases slowly till harvesting, after which the albedo somewhat increases.

1.5 The sea surface albedo.

The sea surface albedo depends on several factors: Sun elevation, cloud amount, roughness, surface contamination, and characteristics of water basins (depth, water transparency, etc.). The dependence of the albedo on Sun elevation and cloud amount can be considered similar for different types of water surfaces - oceans, lakes, reservoirs, but the albedo variability due to water transparency and roughness is rather specific.

1.5.1. The albedo of oceans and seas.

Generalized 10-year ship observations for various oceans have made it possible to obtain the averaged data on the albedo for the global radiation depending on Sun elevation and cloud amount in conditions of the 3-4-point roughness. Table 8 /⁻9_7 gives the calculated mean-daily albedos vs. noon Sun elevation for the ocean surfaces in the Northern Hemisphere. Analyzing these data, one can trace the effect of clouds on the ocean albedo, which depends non-linearly on Sun elevation. At 90°-40°

Sun elevations the albedo in clear-sky conditions is lower than with clouds, and at 30°-40° these values coincide. As Sun lowers still further, the reverse dependence is observed.

TABLE 8

The ocean albedo (%) for the global radiation depending on Sun elevation and cloud amount.

Cloud amount (tenths)	$h_\odot$								
	10	20	30	40	50	60	70	80	90
0 – 2	21	16	12	10	8	7	6	6	6
3 – 7	20	15	12	10	8	8	7	7	7
8 –10	18	14	12	10	9	8	8	7	7

The albedo of deep-sea areas almost coincides with the ocean albedo within 60°-20° Sun elevations. The difference is observed only at the 10° Sun elevation. For a shallow sea site, the albedo is 2-3% higher than that for a deep one.

Mean-monthly values of the ocean albedo for both hemispheres are shown in Table 9. These values were calculated from noon Sun elevations using the Table 8 data.

Sea state affects substantially an ocean albedo. The dependence of an ocean albedo on sea state (in points) is given in Table 10 [14]. The albedo was calculated from systematic ship observations made in conditions of the 8-point cloud amount and turbid atmosphere (the turbidity factor T = 2.5-3.2). It follows from these data that in conditions of the 9-point roughness (when the entire ocean surface is covered with foam), an albedo equal to 7% is independent of Sun elevation. At $h_\odot > 34°$

the ocean albedo increases with increasing roughness, and at $h_\odot < 34°$ the pattern reverses, i.e. the ocean albedo decreases with its increasing roughness.

TABLE 9

Mean-monthly (%) ocean albedos /⁻9_7

Month	I	II	III	IV	V	VI	VII	VIII	IX	X	XI	XII
70°N	-	21	16	12	10	9	10	11	14	19	24	-
60	20	16	13	10	8	8	8	9	11	15	19	21
50	16	13	10	8	8	8	8	8	9	12	15	16
40	13	11	9	8	7	7	7	7	8	10	12	13
30	10	9	8	7	7	7	7	7	7	8	10	11
20	8	8	7	7	7	7	7	7	7	8	8	9
10	8	7	7	7	7	7	7	7	7	7	8	8
0	7	7	7	7	7	7	7	7	7	7	7	7
10	7	7	7	7	8	8	9	8	8	7	7	7
20	7	7	7	8	8	9	8	8	7	7	7	7
30	7	7	8	8	10	11	10	9	8	7	7	7
40	7	8	8	10	12	13	13	11	9	8	7	7
50	8	8	10	12	15	17	16	14	11	9	8	7
60	8	9	12	15	19	21	20	17	13	10	9	8

1.5.2 The albedo of reservoirs, lakes and ponds.

Systematic observations made on various lakes and reservoirs reveal the fact that single measurements give quite different albedo values at the same Sun elevations. So, for instance, for Sevan lake at $h_\odot > 30°$ an albedo value may vary from 4 to 14%. At $h_\odot < 30°$ these variations become even greater (5-25%).

TABLE 10

The ocean albedos (%) at different Sun elevations ($h_\odot$) and sea state (V) [14]

$h_\odot$	V (points)									
	0	1	2	3	4	5	6	7	8	9
3	69.8	54.3	42.3	32.6	25.2	19.4	15.0	11.5	8.9	7.0
10	34.0	28.7	24.0	20.0	16.9	14.1	11.7	9.9	8.3	7.0
20	14.1	12.9	12.0	11.1	10.2	9.4	8.8	8.1	7.5	7.0
30	8.2	8.0	7.9	7.7	7.6	7.4	7.4	7.3	7.2	7.0
34	7.0	7.0	7.0	7.0	7.0	7.0	7.0	7.0	7.0	7.0
40	5.5	5.6	5.8	5.9	6.1	6.2	6.4	6.6	6.8	7.0
50	4.5	4.7	5.0	5.2	5.4	5.7	6.0	6.3	6.6	7.0
60	4.0	4.3	4.5	4.8	5.1	5.4	5.8	6.2	6.6	7.0
70	3.6	3.9	4.2	4.5	4.8	5.2	5.6	6.0	6.5	7.0
80	3.3	3.6	3.9	4.2	4.6	5.0	5.4	5.9	6.4	7.0
90	3.0	3.3	3.6	4.0	4.4	4.8	5.2	5.8	6.4	7.0

The analysis of mean-monthly albedos for interior reservoirs shows that in summer (June, July, August) an albedo of reservoirs and large lakes does not exceed 6-10%. From measurements on four reservoirs and five lakes, an average albedo in summer is 8%. In September and October the lake albedos exceed those in summer by 2-5%. The albedos of small lakes and ponds are somewhat higher than those of large lakes and vary within 11-16%.

1.6 The spectral albedo of natural formations.

Natural formations differ considerably as to their spectral reflectivity. The wavelength dependences of the spectral albedo for the main types of natural formations are shown in Fig.7.

The snow cover has a maximum, independent of the wavelength, reflectivity in the 0.4-0.8 μm spectral region (Fig.7, curves 1,2). At wavelengths exceeding 0.8 μm, the snow spectral albedo (SA) decreases sharply, being selective in the water absorption bands. Maximum spectral albedos correspond to dry fresh snow. As snow ages and becomes contaminated, its albedo decreases throughout the spectrum.

Soil surfaces are characterized by a monotomous increase in SA from the UV to the IR, i.e. from 0.4 to 1.0 μm. With further increasing of the wavelength the SA varies weakly, decreasing slightly in water-absorption bands (Fig.7, curves 3,4,5). Maximum SA are observed through the entire spectrum for sand, loam, and minimum ones for chernozem.

The green vegetation has a rather specific spectral albedo. The SA values in the 0.40-0.55 μm spectral region increase monotonically with wavelength reaching a small maximum (5-15%) at λ = 0.55 μm. In the 0.55-0.69 μm region, the SA values decrease, which is caused by the presence of the fundamental chlorophyll

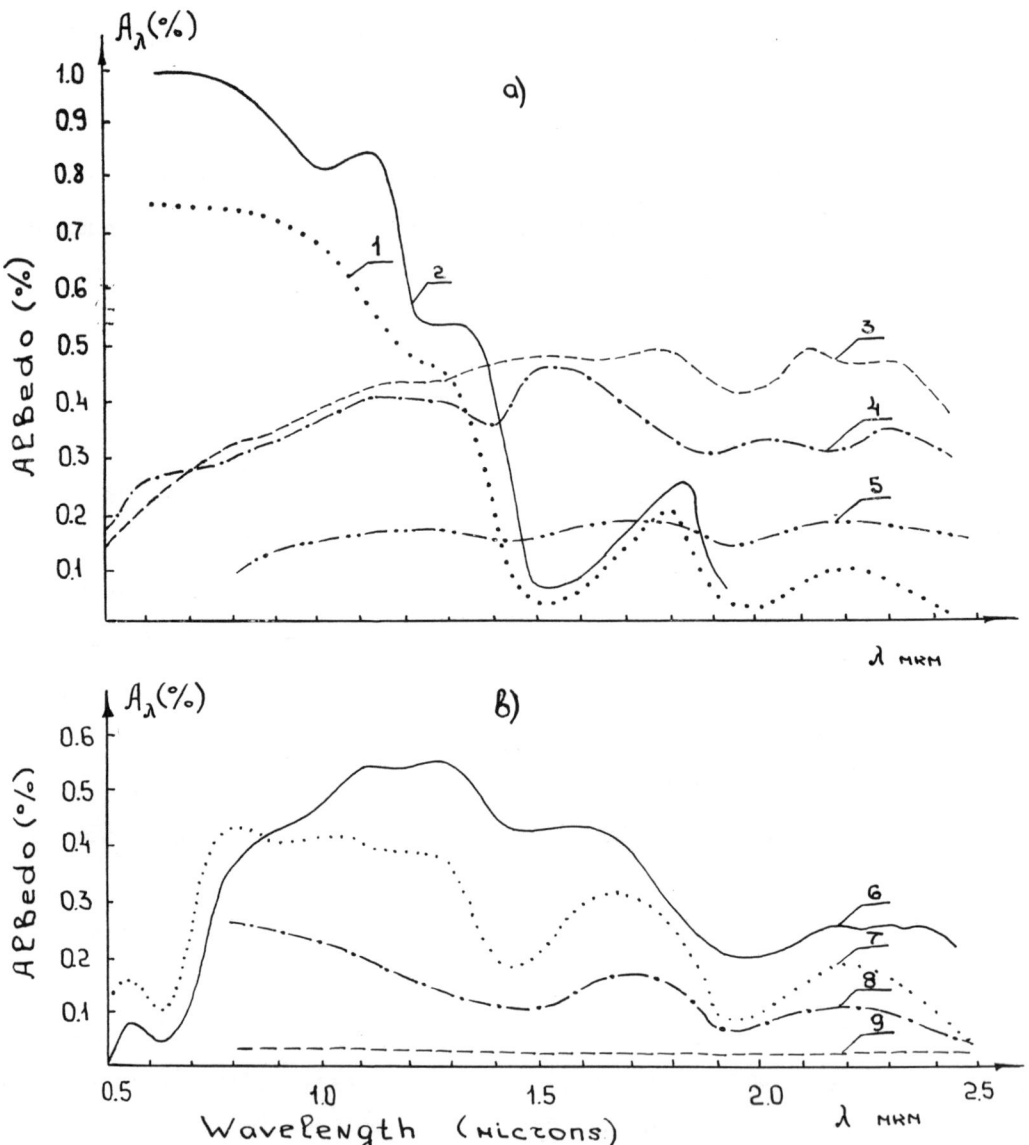

Figure 7 - The wavelength dependence of the spectral albedo for some natural formations:

1 - new snow, $h_0 = 20-25°$ [45];
2 - snow cover [58];
3 - sandy loam, $h_0 = 25-30°$ [47];
4 - dry sand, $h_0 = 40-45°$ [47];
5 - chernozem, sandy loam [16];
6 - blossoming meadow, $h = 40-45°$ [47];
7 - green leaf [54];
8 - green coniferous forest [16];
9 - water surface [16].

absorption band. Beginning from 0.69 μm, SA increases sharply with wavelength, reaching a maximum in the 0.8-1.2 μm spectral region (Fig.7, curves 7,8). In the IR (after 1.2 μm), characterized by water absorption bands, SA decreases slowly.

The spectral characteristics of the forest albedos (Fig.7, curve 8) differ from the respective dependences of the grass cover by smaller values of SA.

Water surface has a neutral course of the spectral albedo with increasing wavelength in the 0.5-2.5 μm region (Fig.7, curve 9) The water rich in phytoplankton has a small maximum of SA in the 0.55-0.57 μm region. The presence of particles suspended in water affects a spectral change of albedo and its value.

1.7 A technique for determination of the albedo for large territories.

For climatic description, an indirect (calculational) technique is used to determine albedos for large territories, which enables one to better estimate the seasonal variations of albedo in different geographical regions. The authors of /⁻34-36_7 calculated the albedo as a weighted average for a territory using a technique according to which the land area under consideration was divided into coordinate squares with a side of 2-4°. For each square, forest percentage was determined using forest maps /⁻17_7 as related to the area of the square. To calculate a mean-monthly weighted albedo of a square, the mean multi-year dates of appearing and melting of the snow cover were taken from climatological reference books as well as the dates of establishment and destruction of the stable snow cover, the dates of 10° air temperature transitions (corresponding to the beginning and the end of vegetation season), and then an albedo was assumed characteristic for each period with a certain state of the underlying surface. More laborious was determination of albedo in transient seasons when the state of the underlying surface changed several times during one month.

Mean albedo values for the basic types of land (Table 11) used in calculating a mean local albedo, were determined by averaging the published data and the observational data (both ground-based and aircraft) as well as by generalizing the systematical ground-based actinometric observations of albedo. Such a calculation technique is somewhat schematic and therefore the results considered below do not take into account several microfeatures of individual natural surfaces (the effect of soil texture and humidity, etc.).

Comparison of calculated albedos with those measured at nine actinometric stations (Riga, Khabarovsk, Omsk, Toronto, Yakutsk, Odessa, Aral sea, Resolute, Gus) has shown that a minimum difference between them (1-2%) is observed in summer. For winter, with the stable snow cover, these differences amount to 7-9%, and for transient seasons with the unstable snow cover to 10%. All this points to the fact that the calculated data are sufficiently reliable, and therefore the technique described above can be used to determine the global land albedo.

For the territory of the Northern Hemisphere, the albedos were calculated by squares. The territory of tropical forests, with

TABLE 11

Mean albedos for the main types of natural land surfaces

No.	Type of the surface	Albedo
1	The stable snow cover in high (more than 60°) latitudes	80
2	The stable snow cover in mid-latitudes (less than 60°)	70
3	The forest with the stable snow cover	45
4	The unstable spring snow cover	38
5	The forest in spring with the unstable snow cover	25
6	The unstable snow cover in autumn	50
7	The forest in autumn with the unstable snow cover	30
8	The steppe and forest in the period between snow cover melting off and a 10°-transition of the mean-diurnal air temperature	13
9	The tundra in the period between melting off and appearance of the snow cover	18
10	The steppe, decidious forest in a period between a 10°-transition of spring air temperature and the appearance of the snow cover	18
11	The coniferous forest in the period from a 10°-transition of spring air temperature and the appearance of the snow cover	14
12	The forests shedding their leaves in a dry season; savannas, semi-deserts in dry seasons	24
13	The same in humid seasons	18
14	Desert	28

a small annual change of albedo was not divided into squares. From a vegetation map in the geographical atlas [7], a zone was determined typical of the region under consideration, and using a mean-monthly value of precipitations [33], the boundaries of dry and humid periods were established, with their respective albedo values. The boundary of the dry period was assumed to be an amount of precipitation less than 20-40 mm depending on air temperature [13].

Based on the data obtained, the monthly maps were drawn of a land albedo spatial distribution in the Northern Hemisphere [34] and that of the continents [35]. The shaded mountain regions failed to be estimated because of the lack of the reliable initial meteorological data. The albedos for ocean surfaces shown in maps were taken from [4]. Investigations carried out by several authors [18,32] have shown that mean-monthly albedos assumed in calculations have been underestimated by about 1% as compared to the data in Table 9. Mean-monthly albedos of limited reservoirs are, on the average, 1% higher than for the oceans at respective latitudes.

1.8 Characteristic features of the geographical distribution of the surface albedo.

Figures 8-11 show the maps of the global surface albedos for January, April, July, and October. The isolines were drawn by 10%-steps, and for the forest regions and desert regions additional isolines were drawn, corresponding to albedos of 15 and 25%.

In January (Fig.8) a maximum albedo (80%) is observed in high latitudes of the Northern Hemisphere. South of 65°N with increasing forest-covered areas, the albedo decreases down to 45% in the regions of unbroken forests. Approaching steppe regions, the albedo again increases and reaches 70%. In marine regions, with a stable snow cover, the albedo varies within 44-25% for the regions with and without forests, respectively. In low latitudes of the Northern Hemisphere and on the continents of the Southern Hemisphere, where there is summer and, hence, no snow (apart from high-mountain regions and the Antarctic), the albedo varies from 18 to 24%, which corresponds to A observed in humid and dry tropical and sub-tropical regions. In desert regions of both hemispheres the albedo is 28%. In the Antarctic the surface albedo constitutes 80-82%.

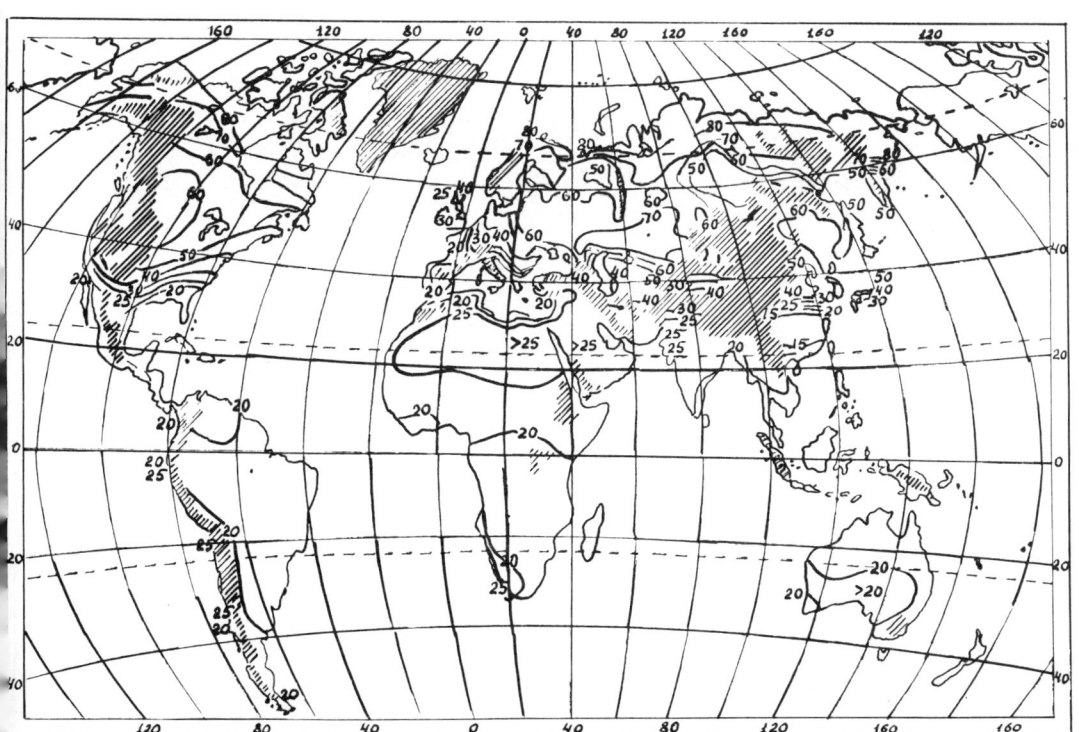

Figure 8 - Surface albedo (%). January.

In April (Fig.9), in high latitudes of the Northern Hemisphere (to 70°N) a region of maximum albedos restricted by the 80% isoline, remains almost the same, since here the stable snow cover is preserved. Down to low latitudes the albedo decreases. In the West Europe (except its northern part) and in the southern regions of the European territory of the USSR, the snow cover is absent and the albedo constitutes about 15%. Further to east, the albedo somewhat increases. In the Northern America, in April, the albedo changes drastically with latitude (from 80 to 20%), which is associated with different distribution of the snow cover over the territory. South of 50°N, the albedo is 15-20% due to lacking snow cover and the beginning of plants' vegetation. In deserts the albedo constitutes 28%.

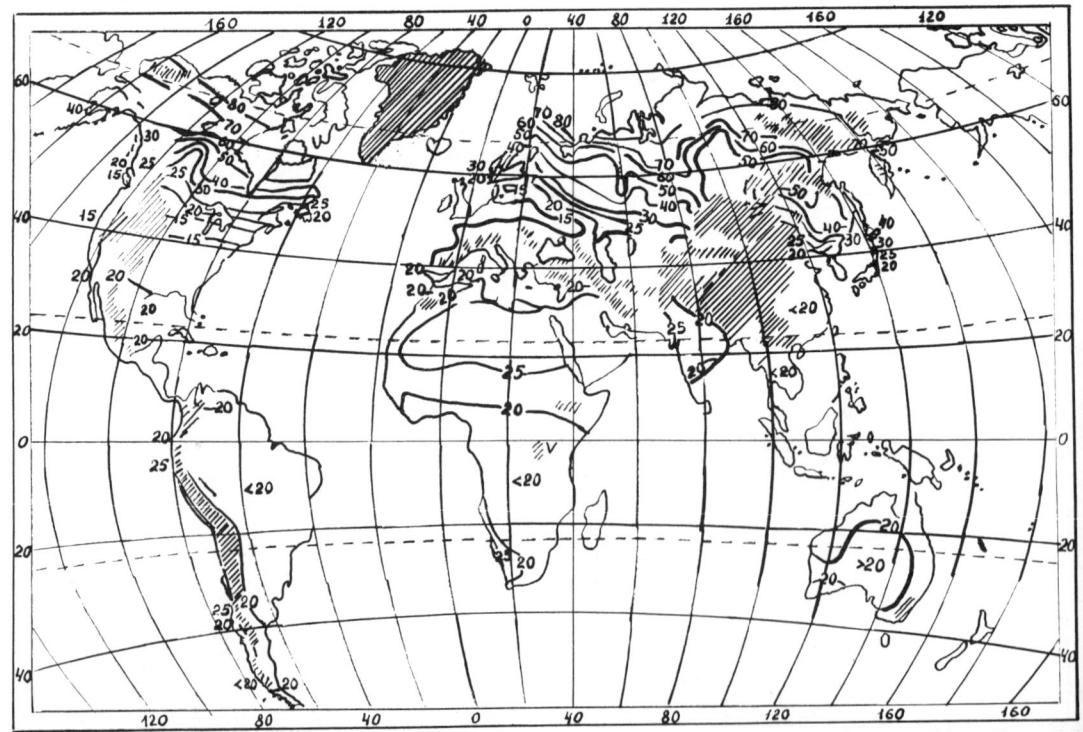

Figure 9 - Surface albedo (%). April.

In the Southern Hemisphere, in April the albedo of arid tropical regions is 24% and of humid ones 18%. A maximum (28%) is observed in deserts. In the Antarctic the albedo amounts to 90%. In July (Fig.10), albedo variations are negligible since there is no snow over the entire territory in question. In coniferous areas the albedo is 14-15%, in steppe and decidious areas the albedo varies from 18 to 20%, in dry steppe and semi-steppe regions from 22 to 24%, in deserts it is 28%.

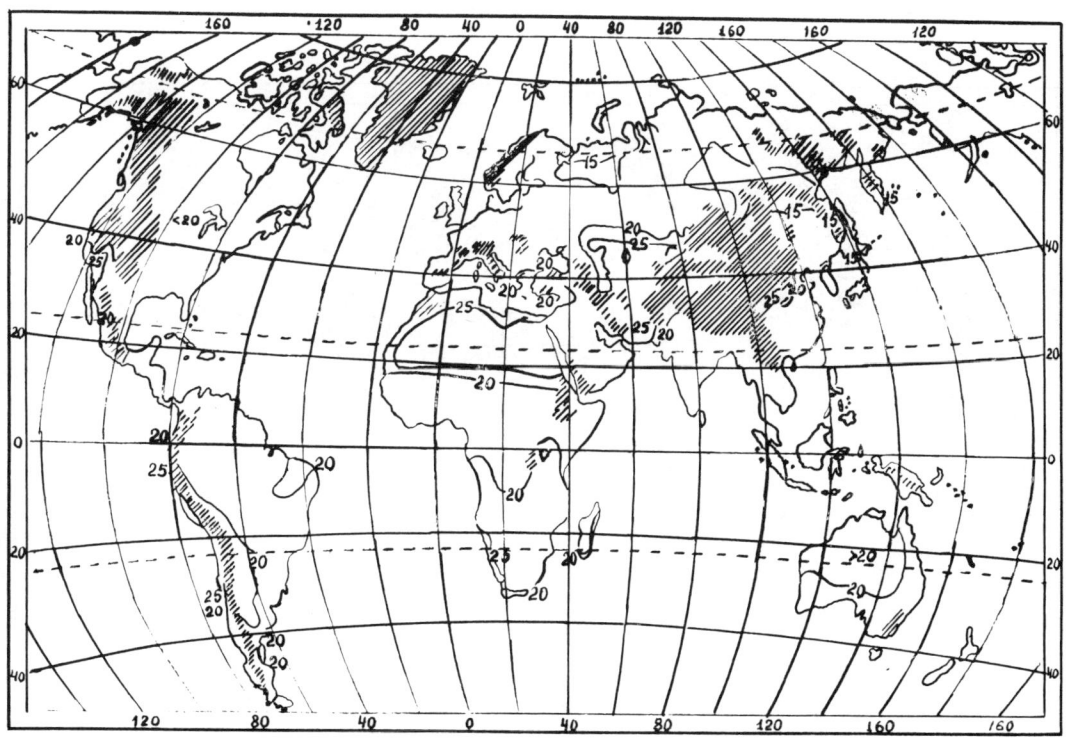

Figure 10 - Surface albedo (%). July.

In October (Fig.11) in high northern latitudes, the stable snow cover is established and up to 70° the albedo is maximum (80%). Over the main part of Europe, the snow cover is absent, and the albedo varies from 15 to 20%. The snow cover is observed only north and north-east of Europe (higher than 60°N), and for this reason, the albedo increases from 20 to 50% north of 60°, the albedo isolines being drawn from north-west to south-east, which is associated with climatic conditions in Europe. In Asia, the isolines stretch almost along latitudes, and the albedo varies from 80% in the north to 20% in mid-latitudes.

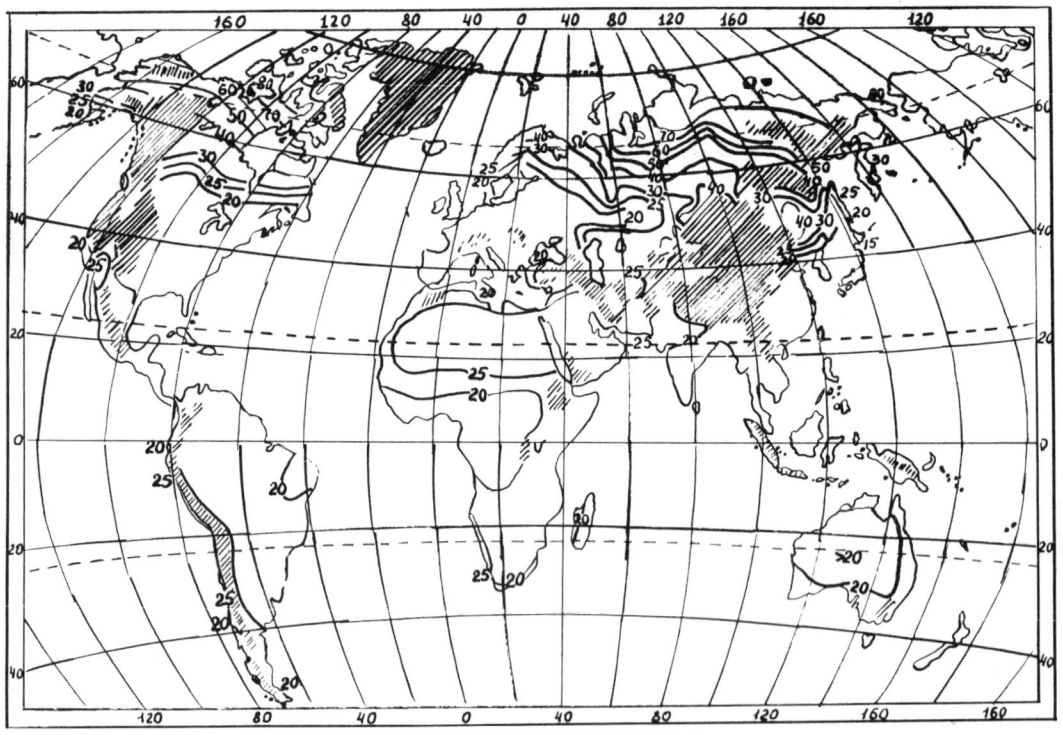

Figure 11 - Surface albedo (%). October.

In the North America, in October, the Arctic zone covered with stable snow has a maximum albedo (80%). Down to south the albedo increases rapidly due to a change in the snow cover state. South of 50°N in the North America the albedo does not exceed 20%, except arid and desert regions, where it reaches 24-28%, respectively. In the South America, Africa, and Australia variations of albedo depending on surface are negligible (from 18 to 28%). In the Antarctic the albedo constitutes about 85%.

1.9 Comparison of mean-monthly albedos.

Let us compare the maps of the mean-monthly albedo distributions with the observational results obtained by several authors. Comparison with the albedo maps drawn by Berlyand /⁻2_7 for the European territory of the USSR has shown a satisfactory agreement. Higher albedos obtained by Berlyand for the summer tundra and winter forests are caused by the fact that the surfaces taken in calculations have been assumed to have albedos of 25 and 50%, respectively. Recent investigations /⁻37,60_7 have shown that an albedo value assumed by Berlyand for the summer tundra are overestimated by 7%, and for the winter forest by 5% (Table 11).

The albedo maps drawn by different authors for individual regions of the Northern Hemisphere (Ukraine /⁻11_7, Azerbaijan /⁻52_7, Georgia /⁻49_7, Middle Asia /⁻30_7, Poland /⁻71_7, Hungary /⁻57_7, China /⁻53_7) give a more detailed picture of albedo variations over the territories under consideration. On the whole, one can speak about their satisfactory agreement with the maps shown here. However, for some regions with snow-covered surfaces the albedos were underestimated by 8%. This could be associated with the fact that in drawing the global maps, less detailed information had been used about the snow cover for some comparatively small territories.

Using the aircraft data on albedos of different surfaces, Kung, Bryson and Linshow [74] drew the seasonal maps of albedo for the North America. The albedo distribution is shown numerically.

Comparison of these maps with the global ones has shown that in the tundra, despite a variety of figures, an agreement with global maps can be considered satisfactory, and in densely populated regions where forests and uncultivated areas are broken ones, the global data not taking account of these moments turned out to be overestimated by 5-7%.

Possey and Clapp [81] have drawn the global albedo maps for January, April, July and October. These calculations were based on albedo values for different types of surfaces given in a monograph by Budyko [4], and in the Smithsonian meteorological tables [82]. It should be noted that albedo values for individual surface types assumed by american authors, differ considerably from those assumed by the authors of this review. So, for instance, an albedo for a snowless deciduous forest was assumed to be 18%, and for a coniferous forest 14%. For a period with snow cover, 12 and 6%, respectively, were added (without sufficient reasons) to the above-mentioned values. Measurements made by some authors [60,74] show that the albedo values given for a winter forest (30 and 20%) are characteristic of territories with the unstable snow cover, and for regions with the stable snow cover they are underestimated. The 7-10% albedos used in [81] for the tropical and equatorial forests are underestimated since they have been obtained with the help of the photometric technique. This has been confirmed by the observational results obtained on the Barbados $/^-75\_7$ and in the South China $/^-53\_7$ where an albedo of coniferous forests was 16-18%, and an albedo of humid equatorial forests in Nigeria $/^-78\_7$ was 12-13%. As a result, in the mid-latitude forest zone on Possey and Clapp's maps the winter albedos are underestimated by 15-20 %, and in the equatorial and tropical zones they are underestimated by 8-10%. However, in the latter case the difference will be less if an albedo value of 18% used by the authors for equatorial coniferous forests is considered to be overestimated. According to observations discussed in $/^-78,79\_7$, it is expedient to take 12-13% for the albedo.

Baumgartner et al. $/^-57\_7$ based their calculations of the mean-annual, January and July albedos' global distribution on the values taken from Bartman $/^-55\_7$, Budyko $/^-4\_7$, Davies $/^-60\_7$, Jackson $/^-67\_7$, Kung $/^-74\_7$, Lockwood $/^-76\_7$ and other authors. Comparison with our maps gives a satisfactory agreement, though in some regions a difference is observed.

Hummel and Reck $/^-66\_7$ calculated a model of the albedo global distribution considering 49 surface types. An albedo for each surface type is determined from noon measurements in clear sky conditions $/^-39,74,\text{etc.}\_7$. Comparison of these data with ours reveals higher values on our maps. This is apparently explained by the fact that the authors of $/^-66\_7$ used the albedo values observed at noon, when $[A]$ is minimum, and they did not take into account the cloud effect, while our data were obtained in moderate cloud conditions. There is no doubt that leaving clouds out of account leads to underestimation of the observed albedo values.

Figure 12 compares the albedo values on our maps (A_1) and those of other authors (A_2). It is seen from this figure that the greatest difference is observed in albedos typical of the unstable snow cover. With the stable snow cover, the difference does not exceed 7-8%, and in a snowless season 1-2%. The comparison results enable one to draw a conclusion that the above-mentioned global maps represent the basic features of global distribution of surface albedo adequately enough.

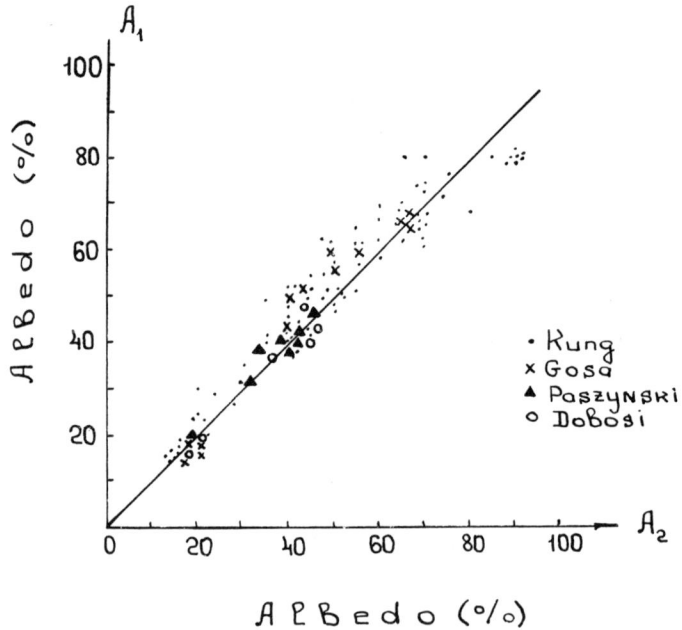

Figure 12 - Comparison of albedo values for large territories obtained by different authors.

1.10 The mean-zonal albedo distribution.

In climate modelling, of great interest is the information on the mean-zonal distribution of albedo. Several authors /‾36,56,66,72_7 have calculated the mean-zonal albedos of the

Earth's surface for every month and a year, on the average. In /¯36_7 the land albedos were obtained by averaging the values taken from the 10°-step maps. For water surfaces the albedos were taken according to the Table 9 data. In /¯2_7 a 2°-step by latitude was taken for calculations of the mean-zonal albedos. Figure 13 compares the latitudinal changes of mean-annual albedos obtained by different authors. The curves of the A latitudinal change are mainly similar, though there is some difference in calculation techniques used by different authors. However, there are some quantitative differences in high latitudes where the albedos are high. Between 30°N and 55°S, a zonal albedo varies weakly, and between 60-80°N and 58-70°S albedos increase sharply. The data of Baumgartner /¯53_7 turned out to be close to our mean-zonal albedo values.

1.11 The annual change of albedo.

Due to annual variations of the underlying surface, there are several types of the annual change of albedo over the global land surface. Table 12 illustrates an annual course of albedo in different landscape-climatic conditions. The table shows mean-multiyear albedo values which may exhibit considerable year-to-year variations depending on climate conditions and surface state.

In the tundra of the Northern Hemisphere an annual course of albedo is characterized by a sharp decrease in June-September and remains almost constantly high in October-May. An annual amplitude of albedo reaches 60%. Similar annual course is observed in mid-latitudinal forest and steppe zones. However, the annual change is less (30% for forests, and 50% for steppe).

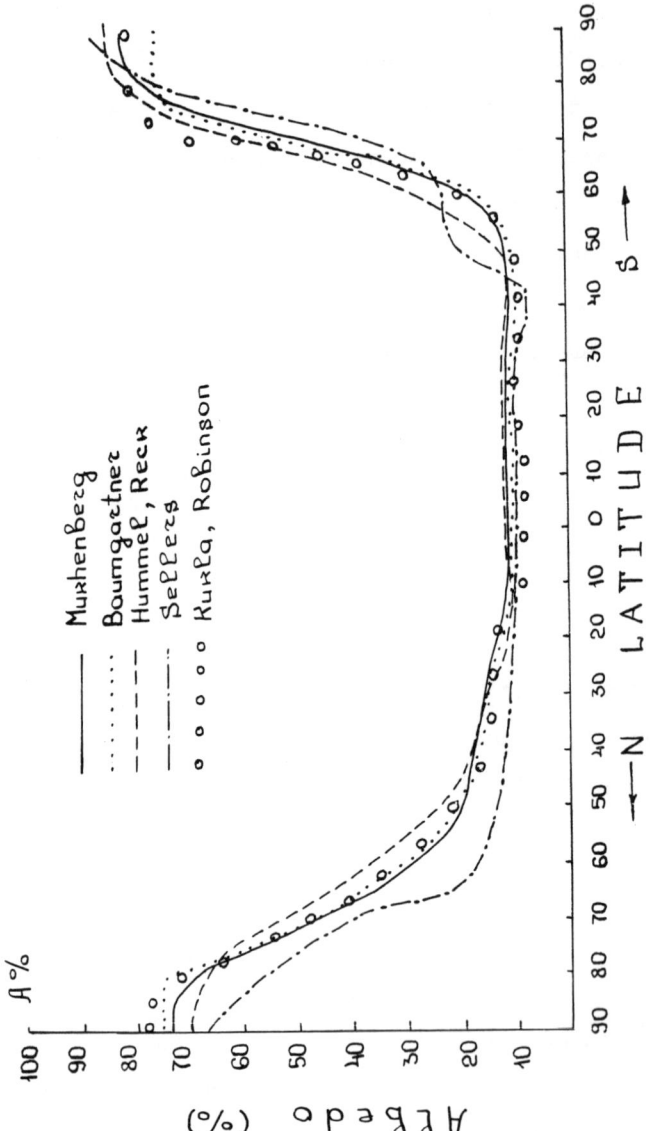

Figure 13 - Latitudinal change of the surface albedo from the data of different authors.

TABLE 12

The annual change of albedo.

Landscape-climatic zones	I	II	III	IV	V	VI	VII	VIII	IX	X	XI	XII
The tundra	80	80	80	80	80	32	18	18	45	79	80	80
The mid-latitude forests	46	46	46	45	14	14	14	14	14	37	46	46
The winter steppe with snow cover	70	70	70	31	18	18	18	18	18	20	51	70
The winter desert with snow cover	44	44	24	27	28	28	28	28	28	28	28	34
The western coast of Europe	21	22	20	18	18	18	18	18	18	18	18	20
The savannas	18	18	20	24	24	24	24	24	24	20	18	18
The tropical deserts	28	28	28	28	28	28	28	28	28	28	28	28
The equatorial coniferous forests	18	18	18	18	18	18	18	18	18	18	18	18

In marine-climate conditions (e,g. the western coast of Europe), where the unstable snow cover with several days' life time is observed, the summer-winter difference of albedo is small (2-5%), without a noticeable annual change.

In the equatorial zone there is practically no annual change of albedo. It should be noted that during some years the spatial-temporal distribution of albedo may differ considerably from that observed above, since year-to-year variations of albedo for one month may substantially differ. So, for instance, for Minsk the 10-year variations of albedo in March constituted from 74% to 16%, in Kiev from 62 to 16%.

The available observational data on albedo make it possible to draw a conclusion that in extra-tropical latitudes in transient seasons as well as in winter with unstable snow cover, the 10-year mean albedo values can be estimated to an accuracy of 30%. In summer, the 10-year mean albedo values may be considered as mean-climatological ones to an accuracy of 10%.

2. Surface Emissivity.

All natural surfaces emit thermal radiation in the IR spectral region 4-100 um. The amount of this emission is governed by surface temperature and emissivity (ε). Measurements of ε for different natural formations were taken both in laboratories [62] and natural conditions [24,68]. Unfortunately, such measurements are substantially restricted as compared to measurements of the shortwave albedo, which makes it impossible to obtain an ample information on emissivity of natural formations. In addition, the data on ε for the same surface type obtained by different authors do not always agree [73]. This is probably explained by the technique applied, and, mainly, by differences in spectral characteristics of radiation sensors used in these measurements. Radiation sensors do not always cover the entire 4-100 μm spectral region.

In natural conditions, ε is usually determined through pyrheliometric observations. In laboratories, an indirect technique is used based on measurements of reflectivity of opaque samples. In this case it is possible to study spectral emissivities of different objects.

Table 13 lists the data on integral emissivity of some natural formations obtained by different authors both in laboratory and in natural conditions. It follows from the analysis of these data that emissivities of natural formations vary within relatively narrow limits (0.85-0.99).

TABLE 13

Emissivities of various natural formations.

Type of natural formations	ε	Reference
Chernozem (black soil)	0.87	24
Sand	0.89	62
Sand	0.95	68
Loam	0.97	68
Yellow loam	0.85	24
Humus loam	0.99	68
Clod	0.98	62
Limestone	0.91	62
Gravel	0.91	62
Rye field	0.93	24
Lilac bushes	0.91	24
Bean leaf	0.96	73
Tobacco leaf	0.97	73
Snow	0.995	62
Ice	0.97	27
Water	0.96	27
Vegetation	0.97	85

Data on emissivity of various soils measured in the spectral regions 5-15 μm and 10.4-12.6 μm in the South States of America (Alabama, Georgia, and Florida) are shown in Table 14 [86].

TABLE 14

Emissivities (ε) of various soils for three American States [86]

Soil type	Spectral region	
	5-15 μm	10.4-12.6 μm
Brown loam, brown chert	0.97	0.98
Shale and clay	0.97	0.98
Brown loam, grey shale, and loam	0.86	0.98
Brown loam, brown gravel, loam	0.94	0.98
Grey and brown loam	0.90	0.96
Brown loam, brown sand, and gravel	0.94	0.97
Grey and brown loam	0.93	0.96
Brown sand, and loam	-	0.96
Grey sand	0.91	0.97
Brown and grey loam, swamps, acid sands	0.90	0.96
Acid sands, and muck, poorly drained soil	0.96	0.97
Phosphate and acid sands	0.96	0.97
Dark sands	0.97	0.98
Loam, peat, and swamp land	0.97	0.98

The data in Table 14 show that soil emissivities vary slightly from mean values 0.94 and 0.97 for broad and narrow spectral regions, respectively. Reliable data on emissivities of various natural formations in the 9-12 μm spectral region were obtained by Gaevsky [7] (Table 15).

TABLE 15

Emissivities of soils /⁻7_7

Surface type	ε
Fine-grained sand, dry	0.949
Saturated fine-grained sand	0.962
Dry loamy sand	0.954
Saturated sandy loam	0.968
Dry peat	0.970
Saturated peat	0.983
Thick green grass	0.986
Thin green grass on humid sandy loams	0.975
Coniferous needles	0.971
Fresh snow	0.986
Contaminated snow	0.969

Emissivity of a humid surface is usually 1.5% higher than that of a dry one.

Detailed studies and analysis of emissivities of vegetation and soil have been carried out in /⁻73_7. Based on the data obtained, several authors /⁻24,69,73_7 have come to the conclusion that $\varepsilon = 0.95$ can be assumed as a mean relative emissivity of the Earth's surface. Apparently, such a conclusion can be accepted with a comment that on the average, relative emissivities of underlying surfaces vary within 0.90-0.99. Note, however, that the problem of emissivity measurements for various natural surfaces is far from an exact solution at the present time. Further efforts are needed to improve the measurement techniques and to make these measurements on large scale.

CONCLUSION

The data on surface albedo and emissivity compiled in this review paper illustrate inadequateness of the information available. The following problems are most urgent:

1. Broadening of the network for regular (all-the-year-round) ground-based observations of albedo in such locations which are representative from the point of view of characterizing albedo variability for typical underlying surfaces.

2. Wider application of aircraft for observations of albedos and emissivities over large territories.

3. Development and application of techniques for retrievals of albedo for large territories from satellite data [26,65].

4. Improving the measurement techniques to obtain ample information about emissivities of typical surfaces, and to reveal its variations depending on different factors.

5. Development of parameterization techniques for albedos and emissivities of typical surfaces depending on basic governing factors which can be used in numerical modelling of atmospheric general circulation and climate.

References

1. Barashkova E.P., Gaevsky V.L., Dyachenko L.N., Luchina K.M., Pivovarova Z.I. Radiative regime of the USSR territory.- Leningrad, Gidrometeoizdat, 1961, 528 pp. (in Russian).

2. Berlyand T.G. Climatic studies of the thermal balance.- Trudy GGO, 1967, issue 218, p.89-100 (in Russian).

3. Bryazgin N.N. On the surface albedo of drifting ice.- Problemy Arktiki i Antarktiki, 1959, No.1, p. 33-39.(in Russian).

4. Budyko M.I. Thermal balance of the Earth's surface. Leningrad, Gidrometeoizdat, 1956, 255 pp. (in Russian).

5. Budyko M.I. Climate and Life.- Gidrometeoizdat, 1971, 470 pp. (in Russian).

6. Voloshina A.P. The summer thermal balance of high-mountain glaciers.- "Nauka" Publ House, Moscow, 1966 (in Russian).

7. Gaevsky V.L. Surface temperature over large territories.- Trudy GGO, 1951, issue 26(88). (in Russian).

8. Geographical Atlas for Teachers.- GUGK SSSR, Moscow, 1955 (in Russian).

9. Girdyuk G.V., Ivanova I.I., Kirillova T.V. Cloud effect on the ocean albedo.- Trudy GGO, issue 326, 1975, p.136-139 (in Russian).

10. Goisa N.I. Airborne measurements of large territories' albedo.- Ukrainsky Fizichesky Zhurnal, 1959, vol.4, No.6, p.793-802 (in Russian).

11. Goisa N.I. A brief characteristic of the Ukrainian territory's albedo.- Trudy Ukr. NIGMI, issue 31, 1962 (in Russian).

12. Goisa N.I. Some regularities of the diurnal and annual courses of the radiative balance of an underlying surface and its constituents.- Trudy Ukr.NIGMI, issue 31, 1962 p.60-81 (in Russian).

13. Goltsberg I.A. Estimation of the global water provision of crops.- Trudy GGO, issue 192, 1966 (in Russian).

14. Gushchin G.K. Ocean surface albedo. Oceanography and marine meteorology.- Trudy Dalnevostochnogo NIGMI, issue 30, 1970, p.244-251 (in Russian).

15. Efimova N.A. The albedo of the agricultural areas.- Trudy GGO, issue 307, 1974, p.43-56 (in Russian).

16. Ivanyan G.A. Spectral brightness coefficients for natural formations in the 0.8-2.5 μm region.- Problemy Fiziki Atmosfery, LGU, 1975, p.17-21 (in Russian).

17. Kirillova T.V. Radiation balance for water basins of different depths and size.- Trudy GGO, 1965, issue 267 (in Russian).

18. Kirillova T.V. Ocean albedo.- Trudy GGO, 1972, issue 282, (in Russian).

19. Climate of polar regions (translation from English, edited by E.P. Borisenkov), Leningrad, Gidrometeoizdat, 1973, 96 pp. (in Russian).

20. Kondratyev K.Ya. Radiative solar energy.- Leningrad, Gidrometeoizdat, 1954, 600 pp. (in Russian).

21. Kondratyev K.Ya. The spectral albedo of natural underlying surfaces.- Meteorologia i Gidrologia, 1960, No.5 (in Russian).

22. Kondratyev K.Ya. Actinometry.- Leningrad, Gidrometeoizdat, 1965, 692 pp. (in Russian).

23. Kondratyev K.Ya. Radiative Factors of the Present-Day Global Climate Change.- Leningrad, Gidrometeoizdat, 1980, 279 pp. (in Russian).

24. Kondratyev K.Ya. Radiative heat exchange in the atmosphere.- Leningrad, Gidrometeoizdat, 1956, 419 pp. (in Russian).

25. Kondratyev K.Ya., Korzov V.I. Studies of the angular anizotropy of the shortwave radiation reflection from different types of the underlying surface and continuous cloudiness.- In book: "Radiative Processes in the Atmosphere and on the Earth's Surface". Leningrad, Gidrometeoizdat, 1979, p.146-151 (in Russian).

26. Kondratyev K.Ya. et al. Studies of natural formations in combined observations from "Soyuz-7" spacecraft and flying laboratories.- Problemy Fiziki Atmosfery, Leningrad State University, 1972, No.10, p.3-19 (in Russian).

27. Kozelkin V.V., Usoltsev I.F. Fundamentals of the Infrared Technique. "Mashinostroenie" Publ. House, 1967, 307 pp. (in Russian).

28. Kuzmin P.P. Radiation reflected from sea surface and absorbed by water layers of different depths.- Meteorologia i Gidrologia, Nos. 7-8, 1939 (in Russian).

29. Kuznetsov I.M., Timerev A.A. Ice albedo changes depending on its surface state as measured from aircraft.- Problemy Arktiki i Antarktiki, 1972, issue 40, p.71-77 (in Russian).

30. Lopukhin E.A. Macroalbedo of some landscapes in the Middle Asia.- Trudy Sredneaziatskogo NIGMI, issue 11(26), 1963 (in Russian).

31. Makarova N.M., Mikirov A.E., Smerkalov V.A. Generalized dependence of land and sea surfaces' albedos on Sun elevation.- Trudy IPG, issue 17, 1973, p.203-210 (in Russian).

32. Meteorological conditions over the Pacific ocean.- "Nauka" Publ. House, "Pacific Ocean", vol.1, 1966 (in Russian).

33. World Agroclimatological Reference Book.- Leningrad-Moscow, Gidrometeoizdat, 1937 (in Russian).

34. Mukhenberg V.V., Land surface albedo in the Northern Hemisphere.- Scientific Reports of the Institute for Geology and Geography of the Lit.SSR, 1962, vol.13, p.175-179 (in Russian).

35. Mukhenberg V.V. The surface albedo over the USSR territory.- Trudy GGO, 1963, issue 139, p.43-60 (in Russian).

36. Mukhenberg V.V. Global land surface albedo.- Trudy GGO, 1967, issue 193, p.37-43 (in Russian).

37. Nekrasov I.A. Annual course of the albedo in the interior areas of Chukotka.- Priroda, No.12, 1957 (in Russian).

38. Pleshkova T.T., Vegetation cover albedo.- Trudy GGO, 1955, issue 46 (108), p.120-124 (in Russian).

39. Radiative characteristics of the atmosphere and Earth's surface.- Ed. by Kondratyev K.Ya., Leningrad, Gidrometeoizdat, 1969, 563 pp. (in Russian).

40. Rusin N.P. Meteorological and radiative regimes of the Antarctic.- Leningrad, Gidrometeoizdat, 1961, 447 pp. (in Russian).

41. Sitnikova M.V. The results of measurements of the albedo of different underlying surfaces.- Trudy Sredneaziatskogo NIGMI, 1964, issue 18, p.56-58 (in Russian).

42. Climatological Reference Book for the USSR. Part I, issue 1-34, Gidrometeoizdat, 1966-1968 (in Russian).

43. Climatological Reference Book for the Antarctic (Solar radiation, radiative balance, sunshine). Ed. by Dolgin I.M., Mirshunova M.S., Petrov L.S., Gidrometeoizdat, 1976, vol.1, 131 pp. (in Russian).

44. Timerev A.A. Reflecting properties of the polar regions' surfaces.- Trudy AANII, 1976, vol.328, p.106-115 (in Russian).

45. Tolchelnikov Yu.S. Optical properties of the landscape as applied to surveying.- "Nauka" Publ. House, 1974, 252 pp. (in Russian).

46. Tooming Kh. Diurnal and seasonal variations of the albedo of some natural surfaces of the Estonian SSR.- Issledovania po Fizike Atmosfery, 1960, No.2, p.115-161 (in Russian).

47. Chapursky L.I. et al. Spectral brightness of clouds and landscape objects in the visible and near IR spectral regions.- Trudy GGO, 1968, issue 221 (in Russian).

48. Tsutskiridze Ya.A. The albedo of some cultural plants and other natural surfaces.- Trudy Tbilisskogo NIGMI, issue 8, 1963 (in Russian).

49. Tsutskiridze Ya.A. Radiative and thermal regimes of the Georgia territory.- Gidrometeoizdat, 1967, 163 pp.(in Russian)

50. Chernigovsky N.T., Marshunova M.S. Climate of the Soviet Arctic.- Leningrad, Gidrometeoizdat, 1965, 198 pp.(in Russian).

51. Chernigovsky N.T. The albedo of the Antarctic surface.- Information Bulletin of the Soviet Antarctic Expedition, 77, 1970, p.68-72 (in Russian).

52. Shikhlinsky E.M. Radiative balance of Azerbaijan.- Trudy Azerb. Geograph.Soc., Baku, 1960 (in Russian).

53. Chen-Tsyan-Sui. Measurement of the surface albedo in China and its distribution.- Acta Geographica Sinica, 1964, No.2, vol.30, p.85-93.

54. Ahmad S.B., Lockwood J.G. Albedo.- Progr. Phys. Geogr., 1979, 3, No.4, p.510-543.

55. Bartman F.L. The reflectance and scattering of solar radiation by the Earth.- Techn. Report, NASA Contract No.NASr 54(03), Washington, D.C., 1967.

56. Baumgartner A., Mayer H., Metz W. Globale Verteilung der Oberflächenalbedo.- Meteorol. Rundsch., 1978, Bd.29, H.2, s.38-43.

57. Borhidi Attila, Dobozi Z. A felszini albedo területi eloszlasa Magyarorszägon. - Idöjárás 71(3), 1967.

58. Choudhury B.J., Chang A.T.C. Two-stream theory of reflectance of snow.- IEEE Trans. on Geoscience Electronics, 1979, vol. GE-17, No.3, p.63-68.

59. Climatological data (1950-1966) National Summary. Asheville (US Dept. of Commerce, Weather Bureau).

60. Davies I.A. Albedo investigation in Labrador-Ungava.- Arch. Meteorol. Geophys. und Bioklim., 1963, Bd.13, Nr.1, s.376-384.

61. Dirmhirn J., Trojer E. Albedountersuchungen auf dem Hintereisferner.- Arch. Meteorol., Geophys. und Bioklim., 1955, Ser.B, Bd.6, H.4.

62. Falckenberg G. Die Absorptionskonstanten Einiger Meteorologisch Wichtiger Korper Für Infrarote Wellen.- Met. Z., 1928, s. 334-337.

63. Fritz S. The albedo of the ground and atmosphere.- Bull. Amer. Meteorl. Soc., 1948, vol.29, No.6.

64. Hoinkes H.C. Studies of solar radiation and albedo in the Antarctic.- Arch. Meteorol., Geophys. und Bioklim., 1960, Bd.10, H.2, s.175-181.

65. Henderson-Sellers A. Albedo changes - surface surveillance from satellites.- Climatic Change, 1980, vol.2, No. p. 275-282.

66. Hummel J.R. and Reck R.A. A global surface albedo model.- J. Appl. Meteorol., 1979, vol.36, No.3, p.238-253.

67. Jackson C.I. Estimates of total radiation and albedo in sub-arctic Canada.- Arch. Meteorol., Geophys. und Bioklim., 1960, Ser. B, Bd.10, H.2, s. 193-199.

68. Kawabata Y., Oomori F., and Hattori T. Radiation properties of some soils.- J. Agric. Meteorol., Tokyo, 1951, vol.6, Nos.3-4.

69. Kondratyev K.Ya. Radiation in the Atmosphere.- N.Y. Academic Press, 1969, 912 pp.

70. Kondratyev K.Ya. Radiation Processes in the Atmosphere. WMO Monograph, No.309, Geneva, 1972, 214 pp.

71. Kozlowska-Szczesna T., Passyuski Ja. Wstepne opracowanie mapy albedo dla Polski.- Przegl. Geogr., 37, No.2, 1965.

72. Kukla G., Robinson G.D. Annual Cycle of Surface Albedo.- Mon. Weath. Rev.,, Jan. 1980, 3371, p.135-150.

73. Kumar R. Radiation from plants - reflection and emission: a review.- 1972, Febr., Purdue University, Research Project No.5543, 87 pp.

74. Kung E.C., Bryson R.A., Linschow D.H. Study of continental surface albedo on the basis of flight measurements and structure of the Earth's surface over North America.- Mon. Weath. Rev., 1964, vol.92, No.12, p.543-564.

75. Lin Sien Chis. Albedos of natural surfaces in Barbados.- Quart. J. Roy. Met. Soc., 1967, vol.93, No.395, p.116-120.

76. Lockwood J.S. World climatology: an environmental approach. - London, Edward Arnold Ltd., 1974, 330 pp.

77. Monthly climatic data for the world, 1957-1967. vol.10-20, Asheville (US Dept. of Commerce, Weather Bureau).

78. Oguntoyinbo J.S. Reflection coefficient of natural vegetation, crops and urban surface in Nigeria.- Quart. J. Roy. Met. Soc., 1970, vol.96, No.409, p.430-441.

79. Oguntoyinbo J.S. Land-use and reflection coefficient (albedo) map for southern parts of Nigeria.- Agr. Meteorol., 1974, vol.13, No.2, p.227-237.

80. Piggin J., Schwerdtfeder P. Variations in the albedo of wheat and barley crops.- Arch Meteorol., Geophys. und Bioklim., 1973, Bd.21, No.4, p.365-391.

81. Posey J.W., Clapp P.F. Global distribution of normal surface albedo.- Geophisica Internacional, 1964, vol.4, No.1, p.33-48.

82. Smithsonian Meteorological Tables.- Smithsonian Institution, Washington, 1951, 527 pp.

83. Salomonson V.V. Anisotropy in reflected solar radiation.- Atmos. Sci., 1968, paper 128, 143 pp.

84. Stewart J.B. The albedo of pine forest.- Quart. J. Roy. Met. Soc., 1971, vol.97, No.414, p.561-564.

85. Fichs M. and Tanner C.B. Infrared thermometry of vegetation.- Agron. J., 1966, No.58, p.597-601.

86. Taylor S.E. Measured Emissivity of Soils in the Southeast United States.- Remote Sens. Environm., 1979, vol.8, No.4, p.359-364.

WATER BALANCE

by

A. Baumgartner
Professor for Bioclimatology and Applied Meteorology
University of Munich, Federal Republic of Germany

1. INTRODUCTION

From two standpoints water is an important element in atmospheric circulation and climate models: water exchanges have to be predicted and the terms of the water budget are needed as parameters in prediction models. Of specific interest are water balances; they are most valuable, not only for the quantitative control of water-, vapour- or energy-exchange processes at the earth-atmosphere interfaces, but also for the validation of model runs.

Water balance is an expression for mass conservation, applied to the water cycle on earth, in the atmosphere or to the earth/atmosphere system. It includes water in solid, liquid or vapour form. In the water balance of continents Precipitation P can be stated as the driving force. Runoff or Discharge D as well as Evaporation E and Storage S are dependent variables. The water balance of the oceans may be related primarily to evaporation and that of the atmosphere or of the soil to the water mass of the reservoirs.

For surfaces, as the earth atmosphere interface, the balance equation has the form

$$P + D + E = 0.$$

A flux to the surface is defined as positive, therefore

$$P = D + E.$$

For reservoirs of water in the atmosphere, hydrosphere or lithosphere the changes dS of the water storage S have also to be included and

$$P = D + E + dS,$$

but for multiannual time-intervals dS = 0 and the storage term S can be neglected or initially applied as a constant rate.

2. SCALES OF WATER BALANCES

Water balances of surfaces or volumes can be evaluated for the globe, for atmosphere, sea and land, for oceans and continents, regionally and locally. The time scale of the assessment increases generally with the space scale. Sufficient precision results for global water balance in multiannual averages. Annual, seasonal, monthly or daily evaluations are more difficult due to the increasing weight and influence of storage. For a soil column daily or hourly estimations are routines.

The quantities of the balance terms may be given as water-heights with the units mm, cm or m, as water-volumes (cm^3, m^3, km^3) as exchange-rates (mm/d, cm/a or m^3/s, km^3/a) or as exchange densities, fluxes or source power ($m^3/km^2 d$, $km^3/10^6 \cdot km^2 a$).

3. MAGNITUDE OF WATER RESOURCES

The volume of water that is available to the planet earth amounts to $1,384 \times 10^9 \text{km}^3$. The partition to reservoirs is given in Table 1.

Table 1: Water volumes of the globe in solid, liquid and gaseous form

	Volume 10^3km^3	%
Oceans	1,348,000	97.39
Polar- and sea-ice, glacier	27,820	2.01
Ground water, soil moisture	8,062	0.58
Lakes and rivers	225	0.02
Atmosphere	13	0.001
Total	1,384,120	100.00

The total water mass is $1,384 \times 10^{18}$ tons. The ratio to the volume of the globe is 1 : 777 or 0.00129. In a homogeneous distribution the earth could be covered by a water layer of 2,715m; however, more than 97% of total water volume is in the oceans, covering 2/3 of earth surface. The average water depth of all oceans is 3,730m. With an exchange rate of less than 1 m/a, a molecule of sea water has a residence time of about 4,000 years.

Fresh water volume is $36,020 \times 10^6 \text{km}^3$ or 2.6% of the total resource, comprising $27.82 \times 10^6 \text{km}^3$ ice in polar ice caps, icebergs and glaciers, $8.06 \times 10^6 \text{km}^3$ groundwater and soil moisture and $0.23 \times 10^6 \text{km}^3$ surface water on land in lakes and rivers. The soil moisture ($60,000 \text{ km}^3$) is a quick-exchange-reservoir for evaporation on land. In a homogeneous distribution it would cover the land of the globe with 40cm or 400mm corresponding to a half annual precipitation.

The atmospheric water reservoir amounts to $13 \times 10^3 \text{km}^3$; the mass is 13×10^{12} tons. Total depletion by precipitation would result in a 2.5cm layer of water on earth. From mean annual precipitation on earth, that is 97cm, it follows that the average residence time of a water molecule in the atmosphere is about 10 days. The precipitable water of atmosphere W has a strong meridional gradient and seasonal variations. Table 2, according to Kessler (1965) provides information on magnitudes.

Table 2: Precipitable Water (W) of Atmosphere (mm) and vapour pressure (e) at earth surface (mb)

	North						South					
φ	90	70	50	30	10	0	10	30	50	70	90°	
e	3.5	5.7	9.3	18.1	16.0	26.7	24.2	15.2	7.6	3.5	3.4	mb
W	5	7	14	24	40	43	41	24	12	5	4	mm

1 mb of e $\simeq$ 1.5 mm of W

Average	Hemispheres		Globe
	M	S	
January	19	25	22 mm
July	34	20	27 mm

It is postulated that the water resources on earth remain constant. Climate fluctuations may influence water exchange rates, the intensity of the exchange processes, the spatial distribution and the ratio of the reservoirs.

4. BALANCE REQUIREMENTS

The following requirements are necessary criteria for representing water balance as well as for the evaluation of specific balance terms and for the closure of the equation systems:

Trivial conditions are generally

$P = E + D$, $E = P - D$ and $D = P - E$.

Similar assumptions exist for storage changements

$dS = P - (E + D)$.

Another relationship is given by the connexion of evaporation-ratio E/P and runoff-ratio D/P

$E/P + D/P = 1$

and analogues in the case of storage changes

$E/P + D/P + dS/P = 1$,

where dS/P is storage change ratio. S/P may be called storage ratio.

4.1 Earth, globally

For the globe, with the statement of conservation of mass and long-term periods the equality of global Precipitation P_g and global Evaporation E_g must be valid

$P_g - E_g = 0$.

It follows that no advective term exists and the water vapour flux within the atmosphere over the sea D_S corresponds to the water flow in streams and groundwater from land to sea D_L

$D_S - D_L = 0$.

Independent integration of precipitation- and evaporation-distribution on earth and thorough estimates of water vapour advection in the air by radiosonde data and of water discharge on earth are the basis of global water balance. Maps of the resultant flux of water vapour in air are presented in the UdSSR World Water Balance (1974).

For shorter periods the time lags must be accounted for in the recycling of precipitated water in the form of water vapour or runoff. The deposited water on earth returns partly after the integration period. The decrease of the storage follows exponentially with time as indicated by Eagleson (1978) and is regionally differentiated.

4.2 Land

Here land means the integration of all continents. The main balance requirement is the equality of precipitation over Land P_L with evaporation E_L and water runoff D_L, thus

$$P_L = E_L + D_L.$$

The water discharge D_L originates from surface water runoff $D_{L,o}$ and from groundwater runoff $D_{L,i}$ to the sea

$$D_L = D_{L,o} + D_{L,i}.$$

4.3 Sea

Here the integral of all oceans is defined as Sea. First of all the trivial equation of continuity

$$E_S = P_S + D_S$$

is to be observed. Evaporation from sea surface is the source for precipitation P_S at the sea surface and for the water vapour-advection from sea to the atmosphere over the land of the globe. The closure of the system is given by the equality of D_S and D_L due to conservation requirements

$$D_S = D_L.$$

It can be seen easily that the water balance of the sea is the crucial point of all investigations in the past, present and future. It is also clear that D_S can be supplemented by D_L, therefore

$$E_S = P_S + D_L \quad \text{or}$$

$$P_S = E_S - D_L \quad \text{and}$$

$$P_S = E_S - P_L + E_L.$$

However, $D_L = P_L - E_L$ is a relatively small magnitude against P_S or E_S, as can be seen in Chapter 8 or in Figure 2.

4.4 Individual oceans

The water balances of individual oceans, as North Atlantic or Pacific, are defined by precipitation on ocean surfaces P_o, evaporation from ocean surfaces E_o and inflow of water D_c from the surrounding continents. The ocean currents OC between the water bodies should not be considered here.

Water cycles on continents

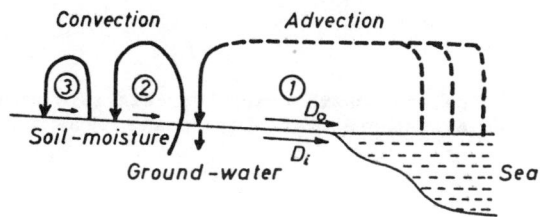

	Water - source
① Global cycle:	Atmosphere
② Regional cycle:	Soil-water and ①
③ Local cycle:	Surface-water and ①+②

Figure 1 - Global, regional and local water cycle on continents. After Lettau (1979).

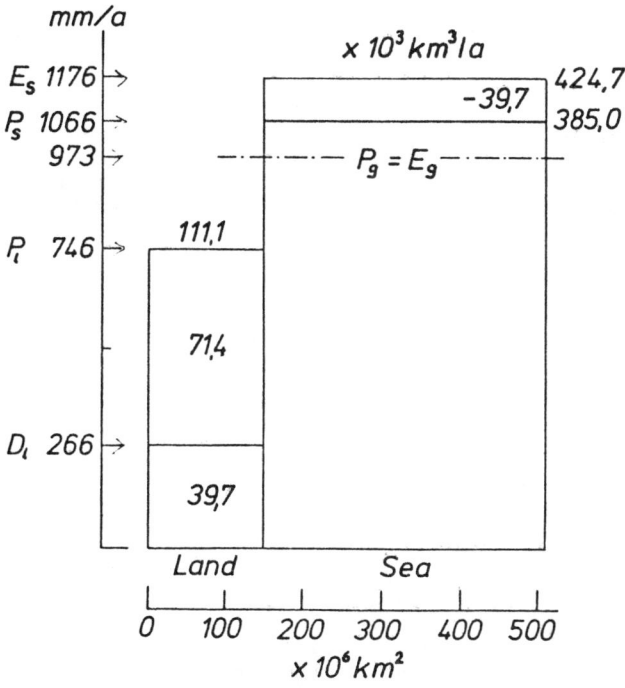

Figure 2 - Water depth and volumes in the global water balance. After Baumgartner and Reichel (1975).

Of importance for the water budget is the relation of P_o to E_o, which is different from the situation for the sea of the globe, where $P_S^o < E_S^o$. It is necessary to distinguish between

$$P_o \lessgtr E_o.$$

In the case of $P_o > E_o$ the atmosphere over the ocean is a sink of water vapour and in the case of $P_o < E_o$ the atmosphere over that ocean acts as a water vapour source.

The net gain or loss of water by an individual ocean dS_o is quantified by the difference

$$D_o = E_o - P_o$$

and by the water inflow D_c. Two different cases occur

$$dS_o = D_c - D_o > 0 \quad \text{Surplus of water}$$
$$dS_o = D_c - D_o < 0 \quad \text{Deficit of water.}$$

The balance requirement for individual ocean therefore is

$$dS_o = D_c - E_o + P_o.$$

The water excesses or deficits S_o by the hydrological cycle explain (by external causes) ocean currents which balance out inequalities of the individual water balances of the oceans. They may be called "equilibrium currents". Certainly they are superposed on and relatively small compared to the well-known ocean water drifts with internal or other causes. The equalization currents among the oceans obtained from water budget considerations include only parts of the total hydrospherical current system. The links, however, between the hydrological cycle and the ocean water cycle must still be explored. The water masses of both drifts have to be compared and the influence of D_c and $P_o - D_o$ on the salt content of surface-ocean-water investigated. Estimations are given in Figure 6.

4.5 Continents and basins

For the continents, naturally in the long-term average, participation P_c, evaporation E_c and runoff or discharge D_c are balanced:

$$P_c = E_o + D_c.$$

There are fluctuations of water storage in surface waters as streams, lakes, ponds, swamps, soil moisture and groundwater through climatic trends. The magnitudes of the water retention play an increasing role in integration periods shorter than one year.

Within the continents precipitation P may exceed E, be equal to or smaller than E. If P > E then the discharge D alternately may be directed peripherally (externally) to the coasts or centrally (internally) to the interior of lakes, swamps, etc. In humid areas P > E. In arid areas, when P = E or E > P, the suction of the air may be just satisfied or not and no runoff may exist. In the case of E > P the discharge D will be consumed in the central area.

The recycling of precipitated or discharged water in the continents initiated by Lettau (1973) to identify three coupled processes, see Figure 1,

(a) Global cycle; driving force is water vapour advection from sea to land;

(b) Intercontinental or regional cycle; water source is precipitated water in soil and groundwater;

(c) Local cycle; source is quickly evaporated surface water or soil moisture in the upper soil horizon.

Lettau et al. (1969) offered a rational way to separate the cycles with the following concept of evapoclimatonomy: in the climatological transform of the balance equation for cycle 1, where vapour advection A is the driving force, the water exchanges w are defined by

$$\partial w/\partial t = E - P - A,$$

where w is the precipitable water content of atmosphere. A similar equation stands for the recycling processes 2 and 3, where exchangeable soil moisture m is the driving force, and D is runoff,

$$\partial m/\partial t = P - D - E.$$

The addition of the two equations results in

$$\partial w/\partial t + \partial m/\partial t = - (D + A).$$

The ratio of the reservoirs w/m is about 1 : 4. The moisture content at and in the ground depends on the flushing rates of runoff and evaporation as well as on the intensity of vapour advection in air. Lettau separated out the parts of D and E which depend on stored moisture m by introducing five new parameters:

(a) Flushing rates r_D and r_E for runoff and evaporation;

(b) Dimensionless characteristics d_p and e_p, defining the input-dependent parts of runoff and evaporation, and

(c) Threshold value P_D'.

The parameters are determined by

$$D = D' + r_D \cdot m$$

$$D' = d_p (P - P_D'), \text{ with } D' = 0 \text{ if } P \leq P_D'$$

$$E = E' + r_E \cdot m$$

$$E' = e_p (P - D')$$

With the aid of the new parameters the reduced input f and the dimensionless time scale τ are defined

$$f = (P - D' - E')/(r_D + r_E)$$

$$d\tau = (r_D + r_E) \cdot dt$$

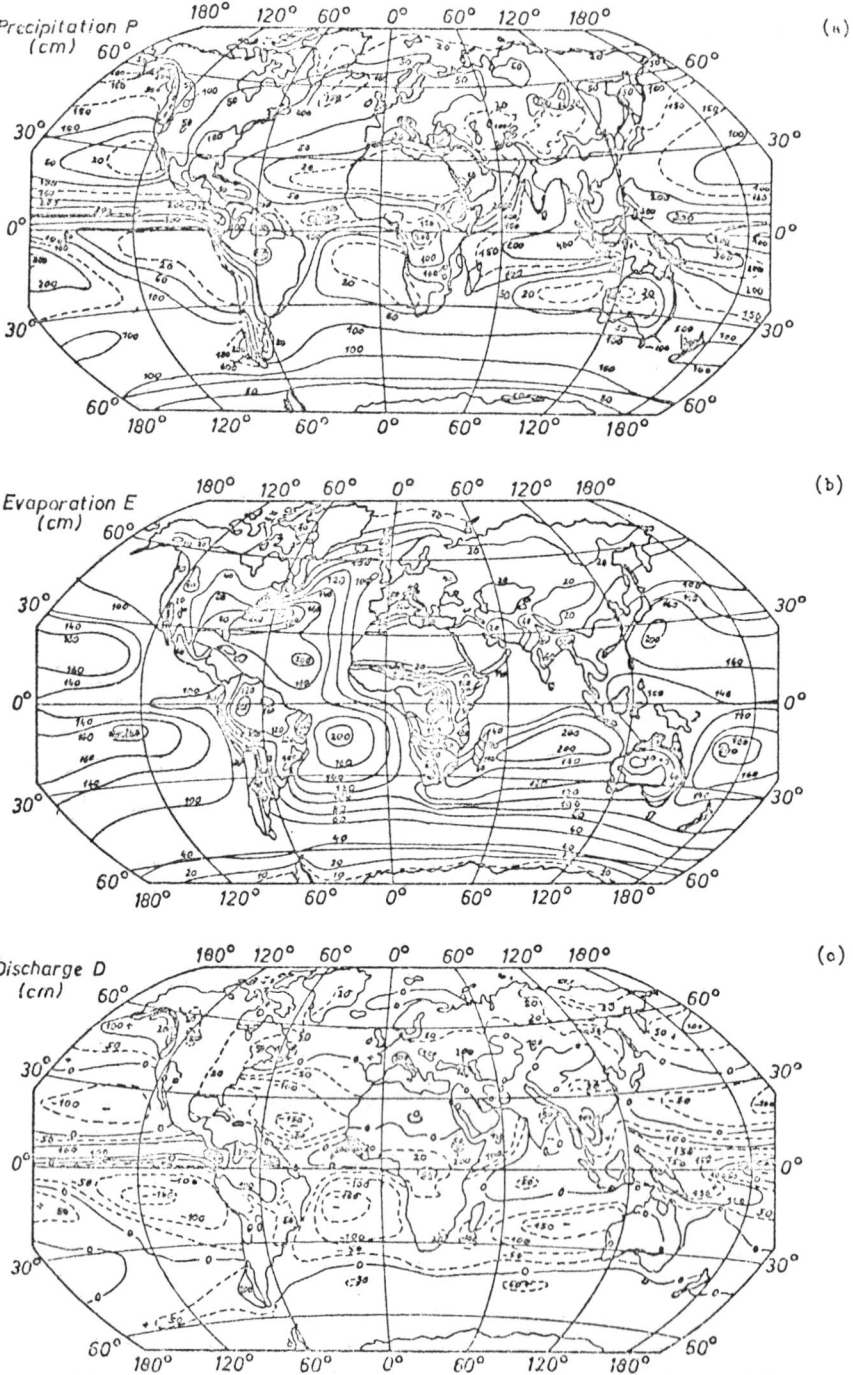

Figure 3 - Global water balance: (a) distribution of precipitation (P); (b) distribution of evaporation (E); (c) distribution of discharge (D). By Baumgartner and Reichel (1975).

The new form of balance equation

$$f = m + \partial m/\tau$$

has been solved, assuming nonlinear flushing rates for the depletion of m. The starting m is arbitrary, "it becomes determined either by prescribing or by a requirement of climate stability demanding that the multi-annual mean of $\partial m/\partial t$ approaches zero". In applying the above method Lettau and Molion (1979) estimated, that at the western border of the Amazonas basin 88% of the precipitation are recycled evaporation water. Due to the long travel distance of air masses the advected water vapour will be precipitated on the continent at least a second time from the air.

In the UdSSR World Water Balance, (Korzum 1974), moisture cycling coefficients and moisture-use-coefficients have been evaluated for the continents.

The establishment of regional water balances for catchments, basins, river or stream systems is based on areal means $\bar{P}$ evaluated from point measurements P_i and from runoff gauging D. D has the character of an areal mean $\bar{D}_b = D/F$, related to the whole gauging area F; however, the averaging area of $\bar{P}$ and $\bar{D}$ is different. Basin-evaporation is evaluated from $\bar{E}_b = \bar{P}_b - (\bar{D}_b + \bar{S})$. In regional systems, as for countries, recycling of used water, e.g. groundwater for irrigation, industry-wastewater, etc. has to be taken into account.

4.6 Soil column

The water balances have to be adapted to the space scales. For a soil column precipitation has to be separated into the P offered from atmosphere to earth-atmosphere interface and the effective precipitation P_e at the soil surface which infiltrates into the soil. The difference $P - P_e = I$ is a part of the evaporation water loss and could be substantial, e.g. for forests: 20-40% of P. It is the quickest flow in the recycling process. Runoff D is split up into surface runoff D_o and internal runoff D_i as percolation or transmission water to the groundwater floor. In the case of a vegetative soil cover the available water store is most important down to the root-zone for evapotranspiration. The root-zone can be used as a control layer for the water balance of a soil column in the form

$$\partial m/\partial t = P_e - (E + D_i).$$

5. ENERGY RELATIONSHIPS

Water budgets and exchanges must be considered in relation to energy budgets. The energy potential of water balance terms in aggregate changes are

$$P' = r \cdot P, \quad E' = rE, \quad D' = rD \quad \text{and} \quad S' = r \cdot S,$$

where r stands for the specific transformation heat of about 2,540 J/g water. From the energetic evaluation of the water balance

$$P' = D' + E' + S'.$$

In connexion with the energy balance

$$Q = H + E' + B$$

it therefore follows

$$E' = Q - (H + B) = P' - (D' + S').$$

There are relationships between net radiation Q and precipitation heat flux P', of sensible heat flow H and discharge heat flow D', and of soil heat flux B and storage heat flow S'. In analogy to the ratio E/P, D/P and S/P some dimensionless parameters are obtained by normalizing on net radiation Q. These are P'/Q, E'/Q, D'/Q and S'/Q. The evaporation ratio can be expressed in the energetic form

$$E/P = k \, (Q/P'), \text{ with } k = E'/Q.$$

In a similar way the runoff ratio can be evaluated from

$$D/P = 1 - E/P = 1 - E'/P'.$$

It can be concluded, e.g., that runoff withdraws water from the evaporation process and is therefore also an important part for heat balance of land. Runoff regulated by land use, etc. are anthropogenic influences on climate.

The comparison of hydrological and energetic balance results in the climatonomy equation, introduced by Lettau (1969)

$$(1 - D/P) \cdot (1 + H/E') = Q/P'.$$

It combines directly water balance terms as runoff ratio D/P with heat balance terms as Bowen-Ratio H/E' and Budyko-Ratio Q/P'. The climatonomy equation is a very sensitive tool for critical control of water and energy balance evaluations. The so-called radiation-dryness-index Q/P' is also an aid for classification of hydroclimate: $(Q/P') < 1$ wet, humid climate, $(Q/P') > 1$ to 3 insufficient humid, $(Q/P') > 3$ dry, arid climate. The quantities are strongly correlated to vegetation types: 0.3 tundra, 0.3 - 1.1 forests, 1.2 - 2.3 savanna, steppe, prairie, 2.3 - 3.4 semi desert, > 3.4 desert.

In the energy cycle of the atmosphere, latent heat E' for the precipitated and again evaporated water volume of $496 \cdot 10^3 km^3/a$ (corresponding to $496 \cdot 10^{18}$ g/a $\approx 5 \cdot 10^{20}$ g/a) is of the order $E' = 1.3 \cdot 10^{21}$ kJ/a. Related to the earth surface of $A = 510.0 \times 10^6 km^2$ the annual average energy flow from the atmosphere to earth surface and back is

$$E'_g = 81 \text{ W/m}^2$$

The energy potential of the precipitable water of the atmosphere is $3.2 \cdot 10^{19}$ kJ. Information for the meridional distribution and transports can be extracted from Chapter 7 or in Sellers (1965).

The evaporated water gets potential energy by the vertical flow within the atmosphere, which will be released with falling precipitation or with runoff from elevated earth surfaces to sea level. The theoretical water power of the globe is after Siebinger (1950) of the order 50 GW and the energy is about $44 \cdot 10^3$ TWh.

6. PARAMETERIZATION OF BALANCE TERMS

Some methods for parameterization are already given in the foregoing section using balance and searching for terms as a residual or by the aid of dimensionless numbers.

Well-known are the formulae for evaporation. They can be classified after the methods, but will not be repeated here. Only some empirical experience shall be described. Very useful for parameterization is the Dalton-type, especially for water surfaces the evaporation of which is called potential evaporation

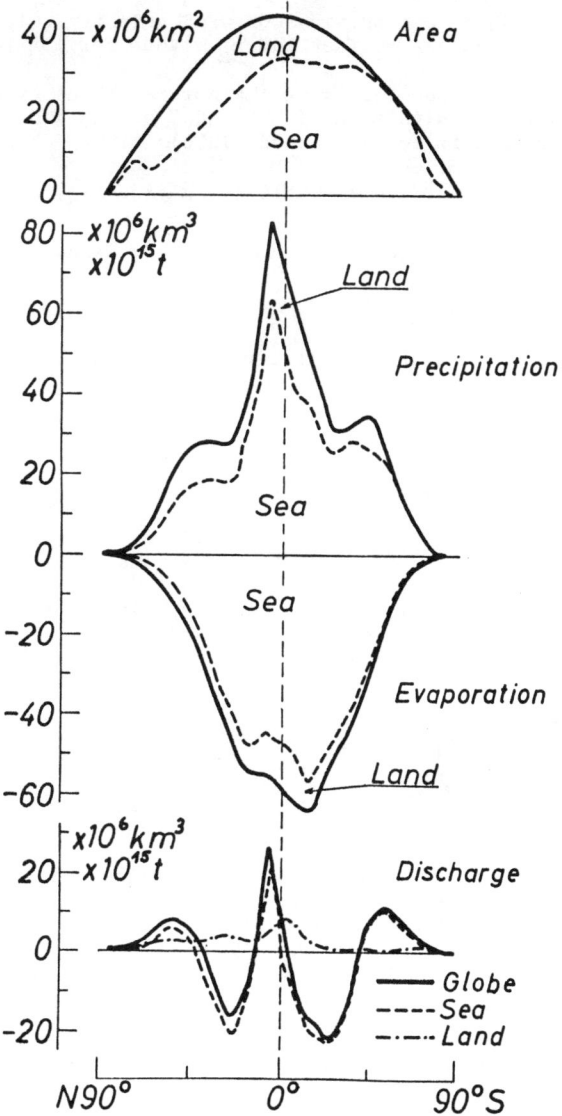

Figure 4 - Meridional distribution of 10°-latitudinal means of area (A), precipitation (P), evaporation (E) and discharge (D) for land, sea and globe. After data from Baumgartner and Reichel (1975).

$$E_p = c(E_o - e_a),$$

where c is a function of exchange conditions or wind velocity (u), E_o is saturation vapour pressure of the evaporating surface and e_a stands for the vapour pressure of air

The function f (u) has the following character: $f(u) = \sqrt{u}$, $f(u) = u^p$, with $p \approx 0.5$, $f(u) = a + b \cdot u$, with an a of 0.2 - 0.7 and b of 0.1 - 0.4. To the first group the formulae of Trabert is to be counted (1896)

$$E_p = c \cdot (1 + \alpha T) \cdot \sqrt{u} (E_o - e_a) \approx 0.41 \sqrt{u} (E_o - e_a).$$

To the second group belongs the equation of Brockamp and Wenner (1963)

$$E_p = 0.543 \cdot u^{0.456} \cdot (E_o - e_a).$$

The disadvantage of E→0 with u→0 is not involved in the third group of equations, e.g.

Richter (1973) $\quad E_p = (a + b \cdot u_{200}) \cdot (E_o - e_a)$

Kohler (1955) $\quad E_p = (0.419 + 0.214 \cdot u)(E_o - e_a)$

WMO (1966) $\quad E_p = (0.0173 + 0.1245 u)(E_o - e_a)$

For dimensions see original papers. Other equations relate E to the energy transfer to the earth surface, like those of Penman (1956) and Budyko (1963) or use temperature-relationships like Thornthwaite (1948). Wundt (1939) related evaporation to precipitation and temperature. For areal average $\overline{E}$ is given by

$$\overline{E} = \left[\overline{P}/(0.95 + P/T')^{-3}\right] \cdot 0.65 \cdot \overline{P}.$$

T' is a parameter for Temperature T. The formula is illustrated in Figure 5. Evaporation cannot increase proportional to P, as radiation energy limits vaporization.

A specific problem is the evaporation from unsaturated land surfaces. In circulation models it seems to be successful to first evaluate potential evaporation from radiation, temperature, saturation deficit and mass exchange. The qualitative distribution is corrected in a second trial on the basis of balance requirements and reduced to real evaporation values. The problem has been investigated extensively by Albrecht (1950, '62, '65).

Runoff can be related to the balance terms E, P, S or to precipitation and air temperature. The discharge from regions follows after Kalweit (1953) in Europe

$$\overline{D} = -95 + 0.8 \overline{P} - 23 \overline{T}.$$

A refinement was introduced by Liebscher (1974) for annual values by separating annual precipitation into winter-p_w and summer-p_s

$$\overline{D} = -454 + 0.93 \overline{P} - 24 \overline{T} + 151 (P_s/P_w).$$

In the runoff prediction models, e.g. the SSARR model (1976), a soil moisture index is evaluated, that is the effective part of precipitation P, which is available for runoff. The fact that long-term averages of the hydrologic ratio or dimensionless parameters are characteristic for specific basins, catchments or regions, can be used for estimations in short-term periods. An example is the application of $(\overline{D}/\overline{P}) \cdot P$ or $(\overline{E}/\overline{P}) \cdot E$ to get D or E for short periods.

7. EVALUATION OF DATA SETS, ACCURACY, COMPLETENESS

This question must be answered differently according to the scale of balances in time and space and with the specific needs for circulation models.

Starting with space, it is well known that meteorological networks produce directly precipitation data, at about 120,000 stations in the world, which are available in daily, monthly and annual sums in Weather Reports and Climatological Yearbooks. Network density is inhomogeneous on land and very scarce on sea, it is not arranged to grid net values of the models. With a density of 1 precipitation-station for 100 km^2, for example in Europe, about 100 data are available for areal average in $1°$ fields. With the method of network-optimizing, characteristic stations can be selected to make averaging rational.

A grid net of $1°$ distance is sufficient for accuracy, however, the measured values themselves are incorrect due to wind and other influences on the measuring instruments. Experiences and balance-discrepancies have forced UdSSR researchers for their World Water Balance, Korzum (1974), to adjust the precipitation data. The adjustments ΔP in % are for latitudes (φ)

```
φ  90 - 80 - 70 - 60 - 50 - 40 - 30 - 20 - 10 - 0
N  25   20   31   24   35   10    5    6    6 %
S   -    -    -   14    7    7    6    7    4 %
```

Baumgartner and Reichel (1975) also have corrected the precipitation values for northern latitudes but not in that rigorous magnitude as for the middle latitudes of the Northern Hemisphere.

The main problem are precipitation data for the sea. Available are weather ship data at ocean stations, drift stations, measurements on islands and the observations from commercial ships. The precipitation amount is gained from frequency distributions of precipitation events and correlations between precipitation at fixed stations and the frequencies. The accuracy is inconsistent and discussed recently by Dorman (1980). Average frequencies are mapped in the US-Navy-Marine Climatic Atlas of the World (1955-69). On the ocean grid, net width could be from the point of hydrological balances in the order of $4°$ or more.

Hydrology networks produce runoff data for about 20,000 selected gauging stations in the world. The transfer of runoff data for grid net is not well developed, it cannot be done linearly in the gauging area. Altitude-dependent transformations have to be developed. The hydrological network is very inhomogeneous throughout the world. The density is in Europe 1 station per 2,000 km^2. The runoff estimations for the great streams must be revised. Averages are difficult to obtain due to great variances and due to the hidden influences of incorrect measurement or of anthropogenic influences on the catchments. Data are available from hydrological yearbooks but data exchange is not well developed. Subsurface water data are generally lacking or very crude. This is true also for data of soil and groundwater content.

There is no network for the uniform evaluation of actual evaporation. A dense network exists for pan-evaporation. However, these data are no real aid to get grid net values for the circulation models.

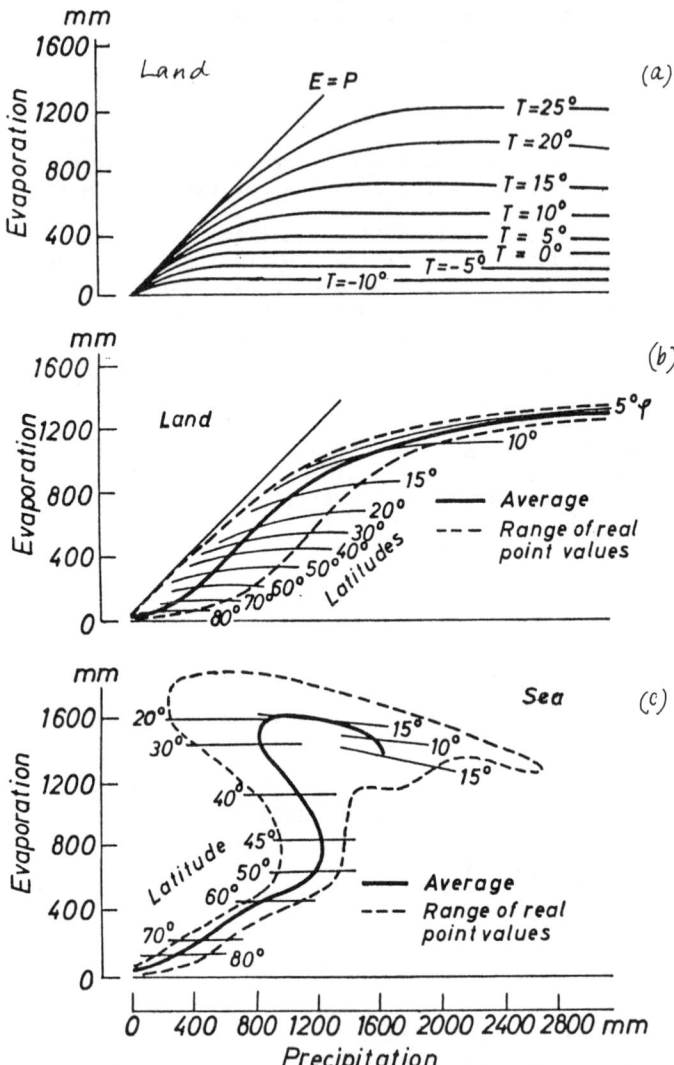

Figure 5 - Correlation of evaporation E on precipitation, temperature and latitude. (a) After Wundt (1939), (b) and (c) after Baumgartner and Reichel (1975).

$\bar{E}$-values, more dependent on large-scale atmospherical processes are quasi-invariant compared with P, D and S, especially at the ocean surfaces.

Concerning the time scale, series of precipitation data are available for a one hundred year period, longer series are collected in World Weather Records, Clayton (1927). Runoff data series are rare for periods of one hundred years. Acceptable accuracy is increasing; the local data have to be controlled in the hydrological balance by integration over the whole basin area.

It has to be observed that areal averages of precipitation are related to the projection area of a terrain, while runoff is collected at the gauging station from real area. If altitude variations of the balance terms have to be taken into account then vertical gradients have to be developed separately for each region. Examples for altitude variations for the Alps are given below:

$dP/dz = + 70$ mm/100 m·a

$dD/dz = + 100$ mm/100 m·a

$dE/dz = d(P - D)/dz = -30$ mm/100 m·a

The requirement $dP/dz = dD/dz + dE/dz$ must be observed.

The altitude gradients are in many areas of the world not linear, inversions are frequent especially in the tropics.

8. CHARACTERISTICS OF HYDROLOGICAL CYCLES

8.1 World Water Balance (WWB)

There are many investigations of the water balance of the globe in the history, see Baumgartner and Reichel (1970). Two very actual WWB are on the desk since the International Hydrological Decade (IHD), namely the "World Water Balance and Water Resources of the Earth" of UdSSR-Researchers (1974, English translation 1978) and the "World Water Balance" of Baumgartner and Reichel (1975). Both WWB are independent and therefore a comparison of both results is a very good illustration of the accuracy of such an attempt. The results of both investigations can, on the other hand, supplement each other. The USSR-WWB gives a world-wide description of the water resources and regional balances. The WWB of Baumgartner and Reichel presents in the appendix a set of tables for $5°$ latitudinal averages and basic maps of P, D and E for the world, which are iteratively balanced at any point of the globe; these are 41,840 grid net values for $1°$ width on land and $2.5°$ on sea. The results are given in Table 3.

Table 3: World Water Balances

	Area 10^6km^2	P	E	D	P	E	D
		Water-depths mm			Water-volumes 10^3km^3		
		UdSSR-WWB (1974)					
Land	149	800	485	315	119	72	47
Sea	361	1,270	1,400	-130	458	505	-47
Globe	510	1,130	1,130	.	577	577	.
		Baumgartner-Reichel (1975)					
Land	148.9	746	480	266	111.1	71.4	39.7
Sea	361.1	1,066	1,176	-110	385.0	424.7	-39.7
Globe	510.0	973	973	.	496.1	496.1	.

The balanced world maps for P, E and D by Baumgartner and Reichel (1975) are designated as Figure 3 a-c.

The differences are very pronounced in precipitation over sea, in evaporation at sea surfaces and in D over land. The crucial requirement, however, of $D_S = D_L$ is in both cases given, but with more than 10% differing quantities. The $47 \times 10^3 km^3$ include $44.7 \times 10^3 km^3$ surface water and $2.2 \times 10^3 km^3$ groundwater. Glacial runoff of Antarctica and of Arctic glaciers is about 3,000 km^3/a. The higher P_L and P_S in the UdSSR-WWB is a consequence of the rigorous corrections of precipitation data. The evaporation on sea is gained by fully different methods: UdSSR with parameterization, Baumgartner from balance requirements. The global total of water volume in circulation is 577 and $491 \times 10^3 km^3$; the average water depth 1,130 or 973mm respectively. These are big differences.

In both balances the importance of a better knowledge of the balance terms of the sea is striking. It is illustrated in Figure 2. The ratio of the volumes $P_L/P_S = 111/385 \simeq 1/4$ and the ratio of $E_L/E_S = 71/425 \simeq 1/6$. Only the discharges D are per condition equal. The requirement for the balance terms P and E over the sea $P_S - E_S = D_S = D_L$ is not very sensitive to errors in $D_S = D_L$ due to the ratio of E_S or P_S to D_S or $D_L \simeq 1/10$. Therefore it cannot be decided at the moment which of the WWB should be given more credit. For this reason the author of this review paper can follow for more detail his own WWB.

8.2   Water balances of hemispheres

The different area of land and water surfaces of the two hemispheres leads to inequalities of the water balances of north and south hemispheres (Table 4, Figure 4).

Table 4:   Water Balance of Hemispheres by Baumgartner and Reichel (1975)

	North		South		Globe	
\multicolumn{7}{c}{Water depth (mm, %)}
P	970	(100)	975	(100)	973	(100)
E	897	(93)	1,048	(107)	973	(100)
D	73	(7)	-73	(7)	.	0
\multicolumn{7}{c}{Water volumes ($10^3 km^3$, %)}
P	247.4	(50)	248.7	(50)	496.1	(100)
E	228.9	(46)	267.2	(54)	496.1	(100)
D	18.5	(4)	-18.5	(-4)	.	0

The precipitation maximum of zonal averages lies in the northern hemisphere in the ITC. The evaporation maximum, however, is situated south of the equator. The southern hemisphere produces more water vapour through evaporation than the northern hemisphere. The cause is the larger water surface in the south and the reduced evaporating potential of the land surfaces in the north. The excess in evaporation of the southern hemisphere corresponds to a depth of 151 mm (or 7% of global mean) and to a volume of $8 \cdot 10^3 km^3$, or 8%.

The surplus water vapour in the order of $18.5 \times 10^3 km^3/a$ will be transported across the equator to the northern hemisphere, precipitated there and transported back with the ocean currents to the south. The vapour transport may occur primarily in the monsoon circulation and with the annual variations of the ITC. It may vary considerably from year to year. The withdrawal of latent energy in the south and the release of the latent heat in the northern hemisphere has a magnitude of 1% of the annual sun radiation. The transport of latent heat is a supplementary energy source of the northern hemisphere for atmospheric motion and climate. It must be considered or simulated in global or hemispherical motion and climate models.

Baumgartner and Reichel (1975) investigated the dependence of E on land and sea to precipitation, temperature and latitude. The result is shown in Figure 5. Temperature and latitude are well correlated on land. The evaporation from sea surfaces is in depth limited in the tropical area by the energy flux. Therefore direct or linear parameterization of E_S with Q, P or T is not allowed.

8.3 Water balances of oceans

As explained in 4.4 and mentioned in 8.2, the balance terms have individual ratio for each of the oceans. Quantifications are presented in the World Water Balances and illustrated in Figure 6 by Baumgartner and Reichel (1975), see pp. 97-103.

The Arctic Ocean has an excess of water $ds_o = 3 \cdot 10^3 km^3$. It is derived from $2.6 \cdot 10^3 km^3$ inflow and $D_o = P_o - E_o = 0.4 \cdot 10^3 km^3$ of precipitation surplus. In comparison the ocean current through the Bering Strait transports $35 \cdot 10^3 km^3$ of water. The great loss $D_o = 36.5 \cdot 10^3 km^3$ of water vapour by the Atlantic Ocean is partly reduced by the inflow of water from land $D_C = 19.3 \cdot 10^3 km^3$. Equilibrium currents from other oceans of a volume of $17.2 \cdot 10^3 km^3$ are included. At the surface of the Indian Ocean evaporation causes a water loss of $19.5 \cdot 10^3 km^3$. This is partly compensated by inflow from the Asiatic rivers of $5.6 \cdot 10^3 km^3$. The permanent deficit of $ds_o = 13.9 \cdot 10^3 km^3$ must be eliminated by currents from the connected Atlantic, Pacific and South Polar Ocean. The Pacific is the great water distributor where $P_S - E_S$ is positive and amounts to $15.9 \cdot 10^3 km^3$. It has to be added to an inflow $D_C = 12.2 \cdot 10^3 km^3$ from surrounding continents. The surplus $ds_o = 28.1 \cdot 10^3 km^3$ is delivered to the Atlantic, to the Indian and North Polar Ocean.

A quantification of the water cycles for the sea of the northern and southern hemisphere is given at the right side of Figure 6. Naturally the greater part of water flow from the continents to sea occur in the northern hemisphere, where greater land surfaces are situated. D_C there is $24.9 \cdot 10^3 km^3$. The vapour loss of the sea in the north is $6.3 \cdot 10^3 km^3$. The sea in the south has a runoff gain $D_C = 14.6 \cdot 10^3 km^3$ and a vapour loss of $33.7 \cdot 10^3 km^3$. The net loss of $18.9 \times 10^3 km^3$ of water vapour will be transported to the sea of the northern hemisphere, as described above, and the loss of vapour will be compensated by a corresponding ocean water stream of $18.9 \times 10^3 km^3$.

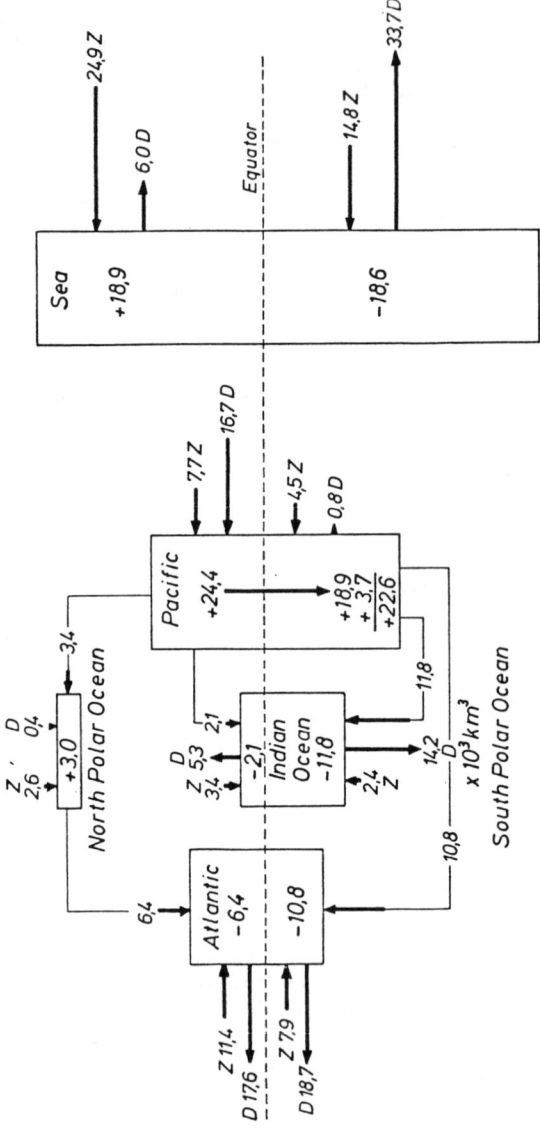

Figure 6 — Water balances of the oceans. Surplus (+) or deficit (-) runoff from land (Z), water vapour-discharge (D) and resulting ocean "Equilibrium Currents".

8.4 Short-term water balances

Worldwide water balances for shorter periods than a year are not available at the moment. Only precipitation distributions for the seasons by Möller (1951) and precipitation maps by Jaeger (1976) are basis for completion with maps for E and D. It shall be proposed to collect worldwide data P, E and D for a distinct year on a daily basis for further investigations.

8.5 Regional and local water balances

Many investigations in the meso- and microscale have developed the principles of regional or local water balances. They can be used as control or as supplement information. For the transformation to the macroscale normalized data as D/P, E/P or D/P are more helpful than the quantities of E, P and D itself. A few examples should explain the fact:

Lysimeter results from Baumgartner (1970) show for humid regions the following averages of E/P-values: Bare soil 30%, crop-land 40%, grass-land 65%, deciduous forest 70%, surface water 75% and wet soil, swamps 100%. Runoff ratio D/P, depending on forest percentage (W) has been quantified by Molchanow (1963).

W	0	20	40	60	100	%
D/P	0.42	0.33	0.26	0.21	0.18	.

Afforestation increases evaporation and reduces runoff. The water content m which can be evaporated by a 1m soil layer is dependent on the soil type and of the order for

Humus	Clay	Loam	Loamy sand	Sand	
32	22	20	16	10	cm/m.

Suction forces in the soil and evapotranspiration of vegetated soil cover are causes of the order of the internal runoff ratio D_i/P, like the following

	Basaltgravel	Sand	Humus	Loamy sand	Loam, clay	
D_i/P	70	65	40	35	25	% .

Therefore vegetation and soil maps may be used for deductions of evaporation, of the water storage in soil or of the runoff terms on land. It has to be noted also, that the increase of the agriculture yields is combined with an increase of the water consumption of the crops. For example a corn yield of 70 kg/ha needed 300mm water depth and a yield of 100 kg/ha about 500mm for evaporation. Changes in the water balance of productive land regions since some decade of years is accompanied by a change of the heat balance.

9. EXISTING ARRANGEMENTS

Though the collection of data for P from the existing networks and the processing to grid net values should not cause problems, the corrections to real data is an open question. The data should be given as depth in mm and controlled under balance requirements.

The meteorological parameters for the evaluation of E and of precipitable water W can be obtained from computerized synoptic data and analysed as grid net values. The water content in soil (m) can be deduced from soil maps with initial saturation or full depletion and processed with the method of Lettau (1969) or Thornthwaite (1958).

Table 5: World Water Balance. Globe, latitudinal Means, N- and S-Hemisphere. After Baumgartner and Reichel (1975).

°	$10^3 km^2$ A	Depth (mm)					
		N			S		
		P	E	D	P	E	D
90 - 85	980	23	26	3	65	10	55
85 - 80	2,933	69	45	24	81	14	67
80 - 75	4,848	145	84	61	172	36	136
75 - 70	6,739	254	167	87	288	71	217
70 - 65	8,567	406	234	172	442	146	296
65 - 60	10,332	608	317	291	656	312	344
60 - 55	12,009	807	422	385	908	480	426
55 - 50	13,600	879	472	407	1,099	625	474
50 - 45	15,077	881	548	333	1,176	762	414
45 - 40	16,479	867	731	136	1,080	962	118
40 - 35	17,653	811	935	-124	943	1,154	-211
35 - 30	18,756	711	1,007	-296	807	1,208	-401
30 - 25	19,705	679	1,052	-373	752	1,266	-514
25 - 20	20,497	671	1,168	-497	802	1,344	-542
20 - 15	21,154	880	1,250	-370	962	1,474	-512
15 - 10	21,634	1,354	1,318	36	1,255	1,540	-285
10 - 5	21,954	2,029	1,275	753	1,450	1,438	12
5 - 0	22,123	1,743	1,225	518	1,420	1,303	117
90 - 0	254,990	970	897	73	975	1,048	-73

A new arrangement is needed for the international exchange of runoff data and for the processing of catchment data to grid net values. The collection of runoff data by Marcinek (1964) and Unesco (1969), as well as the world maps for runoff by Llovich (1964) or Baumgartner and Reichel (1975) could be useful.

While the data sets of the average world water balance are available in grid net values, the water balances of the years in the past have to be investigated; also mean seasonal and mean monthly water balances have to be worked out.

A group of experts should organize for one distinct year, e.g. 1979 or 1980, a set of daily or weekly data as a test material for the models and propose a system of collection processing, archiving and exchange for further actual balances.

10. GAPS AND DEFICIENCIES

In concluding the review, gaps and deficiencies in the present status of knowledge are summarized:

1. Main difficulties arise for the vast area of the sea. P, E and D have to be evaluated by remote sensing. More surface stations, stationary or drifting are needed for calibration and for the following improvements:

    (a) Evaluation of precipitation on the ocean surfaces;

    (b) Evaluation of water vapour flow at the ocean/atmosphere interface and between sea and land;

    (c) Evaluation of runoff and subsurface water flow from land to sea;

    (d) Quantification of vapour transports over the equator from southern to northern hemisphere.

2. The validity of predictions for water balance terms in the present models should be controlled under balance requirements.

3. A data bank for water balances for past years and for actual daily data of a particular year should be established under the guidance of a group of experts.

4. Wanted predictions of circulation models are: floods, droughts, precipitation intensities, regional distributions of P, E, D and integrations for river catchments or streams.

11. SELECTED BIBLIOGRAPHY

References marked with an asterisk (*) are supplementary publications of importance, but not mentioned in the text. See also the extended lists of Literature in the USSR and Baumgartner/Reichel "World Water Balance".

Albrecht, F. (1950): Die Methoden zur Bestimmung der Verdunstung der natürlichen Erdoberfläche. Archiv. Meteor. Geophys. u. Biokl., Ser. A, Vol. 2, 1-38.

_____ (1962): Die Berechnung der natürlichen Verdunstung (Evapotranspiration) der Erdoberfläche aus klimatologischen Messungen. Berichte Deutscher Wetterd. Nr. 83, Offenbach.

_____ (1965): Untersuchungen des Wärme- und Wasserhaushaltes der südlichen Kontinente. Berichte Deutscher Wetterdienst Nr. 99, Offenbach.

*Baumgartner, A. and E. Reichel (1970): Preliminary results of new investigations of world's water balance. Bull. Int. Soc. Hydrol., 65-78, Gentbrügge.

_____ (1970): Energy and water balance of different vegetation covers. IASH, Proc. Reading, Vol. 3, 65-78, Gentbrügge.

_____ and E. Reichel (1975): World Water Balance. Oldenbourg Munich.

Brockamp and Wenner (1963): Verdunstungsmessungen auf dem Steiner See bei Münster. Deutsche Gewässerkdl. Mitt. 7, 149-154.

Budyko, M. I. (1963): Atlas of heat balance of the earth. Moscow.

*_____ (1970): The water balance of the oceans. Int. Ass. Sci. Hydrol. Publ. No. 92, 22-33, Gentbrügge.

Clayton, H. (1927 ff.): World Weather Records. Smithson, Miscell., Coll., Washington.

Dorman, C. E. (1980): Comments on the relationship between the amount and frequency of precipitation over the ocean. J. Appl. Met., 19, 1131-1133.

Eagleson, P. D. (1978): Climate, soil and vegetation. Introduction to water balance dynamics. Water Resources Res., 14, 705-775.

*Hylckama, v. T. E. A. (1956): The water balance of the earth. Publ. in Climatology, Vol. 9, 8, 57-117, Centerton.

*IASH/Unesco (1970): Symposium on World Water Balance, Reading, IASH Publ. No. 92 and 93, Brüssel.

*Jacobs, W. C. (1968): The seasonal apportionment of precipitation over the ocean. Ass. of Pac. Coast, Geogr. Yearbook 30, 63-78.

Jaeger, L. (1976): Monatskarten des Niederschlages für die ganze Erde. Berichte Deutscher Wetterd, Nr. 139, Offenbach.

Kalweit, H. (1953): Der Wasserhaushalt, 2 Vol., Berlin.

Kessler, A. (1965): Über die Luftdruckbilanz der Erde, Meteor. Rundschau 18, 166-169.

*Knapp, R. M. et al. (1975): Development and field testing of a basin hydrology simulator. Water Resources Res. 11, 879-888 (see also comment by Lettau H. H., Water Res. 13 (1977) 691-701).

Kohler, M. A. (1955): Lake Hefner Studies. Weather Bureau Res., Paper Nr. 38.

Korzum, V. I. (ed.) (1974): Mirovoi vodnyi balans i vodnye resurcy zemli. Moskau.

*Landsberg, H. E. (1969): World Survey of Climatology. Elsevier.

Lettau, H. H. (1969, 1973): Evapotranspiration Climatonomy. Monthly Weather Rev. 97, 691-699 and 101, 636-649.

───── and C. L. B. Molion (1979): Amazonas Hydrological Cycle and the Role of Atmospheric Recycling in Assessing Deforestation Effects. Monthly Weather Rev. 107, 227-238.

Liebscher, H.-J. (1974): Wasserhaushalt in den Oberharzer Versuchsgebieten, 171-198, Hannover.

Llvovitch, M. J. (1964): River runoff. Water regime types of rivers of the world. Atlas Mira. Moskau.

Marcinek, J. (1964): Der Abfluss von den Landflächen der Erde und seine Verteilung auf 5°-zonen. Mitt. Inst. f. Wasserforschung. Berlin and Suppl. 1965.

*Mendel, G. (ed.) (1980): Operationale Wasserstands- und Abflussvorhersagen. Schriftenreihe DVWK, Nr. 51, Parey Verl. Hamburg.

*Miller, D. (1965): The heat and water budget of the earth's surface. Adv. in Geophys. 11.

Molchanow, A. A. (1963): The hydrological role of forests. Israel Trans. Jerusalem.

*Molion, C. L. B. (1976): A climatonomy study of the energy and moisture fluxes of the Amazonas Basin with considerations of deforestation effects. Thesis Met. Dpt. Univ. Wisc. Madison.

Möller, F. (1951): Vierteljahreskarten des Niederschlages. Petermanns Geogr. Mitt. 95, 1-7.

*Nace, R. L. (1970): World-hydrology: status and prospects. Int. Ass. Sci. Hydrol. Publ. No. 92, Gentbrügge.

Penman, H. L. (1956): Estimating Evaporation. Transact. Geophys. Union 37, 43-46.

Richter, N. (1973): A comparison of various methods used for the determination of evaporation from free water surfaces. IASH, Publ. No. 109, 235-238.

Sellers, W. D. (1965): Physical Climatology, Chicago.

Siebinger, W. (1950): Statistik aller bestehenden Wasserkraftquellen. IV, Rep. 9, Schr. Hl. World-Power-Conf. London.

SSARR-Modell (1976): Development and application of the SSAR-Modell. US Army Corps of Eng. North Pac. Div., Oregon, U.S.A., Portland.

Thornthwaite, C. W. (1948): An approach toward a rational classification of climate. Monthly Weather Rev. 67, 4-11.

───── (1958): Three water balance maps of SW Asia. Publ. in Climatology, Vol. 11, 1, Centerton.

*───── (1962 ff.): Average Climatic Water Balance Data of the Continents. Centerton.

Trabert (1896): Neuere Beobachtungen über die Verdampfungsge-schwindigkeit. Meteorol. Zeitschr. 13, 7, 261-263.

Unesco (1969): Discharge of selected rivers of the world. Vol. I, Paris.

US Navy (1955-69): Marine climatic atlas of the world. Vol. 1-7. US Off. of Naval Operations.

USSR Conc. IHD (1978): World water balance and water resources of the earth. With: Atlas of World Water Balance. Unesco Press, Paris.

*Van der Leeden, F. (1975): Water resources of the world. Water Info. Center, New York.

*WMO (1962 ff.): Climatological Normals (Clino) for Climate and Climate Ship Stations for the period 1931-1960.

─────── (1966): Measurement and estimation of evaporation and evapotranspiration (= CIMO-Rep.) Techn. Rep. No. 83, Geneva.

*───────/Unesco (1970 ff.) Climatic Atlases of the Continents, Paris.

*─────── (1974): Guide to Hydrological Practices, 3 ed. No. 168, Geneva.

*Wüst, G. (1936): Oberflächensalzgehalt, Verdunstung und Niederschlag auf den Weeltmeeren, nebst Bemerkungen zum Wasserhaushalt der Erde. Festschrift N. Krebs, Stuttgart.

Wundt, W. (1939): Die Verdunstung von den Landflächen der Erde im Zusammenhang mit der Temperatur und dem Niederschlag. Z. Angew. Meteorol. 56, 1-9.

*Wetterdienst, Deutscher (1961 ff.): Marine Climatological Summary. Area $20° - 0°N$, $0 - 50°S$, $50°W - 10°E$, $70°W - 20°E$. Seewetteramt Hamburg.

SESSION V

ACQUISITION OF LAND SURFACE DATA

POSSIBILITIES FOR REMOTE SENSING
OF SURFACE CHARACTERISTICS

Klaus I. Itten

Department of Geography, University of Zurich, Switzerland

ABSTRACT

The scope of this paper is to provide a summary of the current status of remote sensing of several quantities of primary interest in climate modelling. It also gives a short introduction to the basic physical principles involved. Where applicable, estimates of the accuracies and resolutions currently achieved are given.

Some basic requirements for climate modelling in terms of data needed from remote sensing are identified. Extensive ground support during aircraft or satellite remote sensing missions in the micro- and macro-scale, with high resolution devices, can lead to local and regional applicable results. These form a basis for large scale parameterization of factors needed in climate models.

Some factors may require satellite techniques from the beginning, and may already be operationally available on a global scale. Other factors have to be investigated carefully by airborne means.

Many measuring devices and platforms do not yield data in the form needed for direct use in the models, but they can at least partly provide useful information.

1. INTRODUCTION

1.1 The philosophy of remote sensing

In remote sensing it is sought to image or locate, to characterize, and to identify objects versus a certain background, and eventually to measure quantitatively their reflected or emitted radiant energy.

The term "remote sensing" is broadly defined as measuring the properties of objects without direct contact between measuring device and object.

Measurements are made in the electromagnetic spectrum from gamma-rays to audio-waves. The regions within that spectrum commonly used in remote sensing extend from ultraviolet to microwave wavelengths (see Fig. 1).

1.2 The difference between reconnaissance and remote sensing

While much of the development of imaging systems up to today was initiated by military demands, and many of today's commonly used instrumentation are civil versions of more sophisticated military equipment, the identification philosophy is not necessarily, but may be quite different.

From many failed applications we learned, that in a proper approch the whole context has to be studied in view of remote sensing possibilities. Initially this involves theoretical analysis, literature search and laboratory research on spectral properties and behaviour of the targets followed by field measurements, experiments from towers, low flying helicopter or aircraft which lead to final decisions on the appropriateness of a certain method, in applying a specific remote sensing technique from space.

1.3 The ground truth essential

The term "ground truth" may not be taken literally, but it is commonly referred to as the collection of data on objects through alternative direct or indirect (objective) measurements or subjective statements on what we think they are !

Stating the expression "ground truth essential" we believe that ground truth in some form has to accompany every remote sensing mission. Although very sophisticated unsupervised clustering techniques are used in multidimensional image processing approaches, for instance the result of a land use classification cannot be labels as cover type 1, 2, 3 etc. without clearly saying what they stand for. Missing ground information is often substituted by the so-called scientific guesses. Let us hope, that the interpreter states confidence levels of his results. Supervised techniques (including supervised clustering) which contain point information of samples taken on the ground or very close to the object during the remote sensing mission, are to be favoured strongly. There is still much interpolation and uncertainty to those approaches; but error probabilities can be better assessed and confidence intervals defined.

There are only few earth oriented remote sensing missions where absolutely no ground truth exists (be it in the form of direct or indirect measurements, maps, photographs, descriptions, checked interpretation keys etc.). Some more static applications like geologic mapping in arid regions need a less dense ground truth network, whereas for instance vegetation dynamics monitoring may necessitate detailed timely investigations on the ground. The planning of ground data collection is therefore a task specific duty, and depends on the basic requirements of the study.

Essentially it can be stated, that the remote sensing indentification process be it analog or digital, can be greatly simplified and the results improved through the use of proper ground truth data.

1.4 The basic requirements

A structured decision process leads from a certain task, within a given budget and time frame, over user specified requirements, over object specific, analysis, and schedule requirements to the design of an application or a system.

Object specific requirements are spectrally, temporally and spatially induced. The analysis requirements result from timeliness, products (data form), analysis, support data and result evaluation. Schedule requirements are given through the demands of research, development and implementation.

2. BASIC PRINCIPLES OF REMOTE SENSING

In sensing electromagnetic radiation which is either reflected or emitted from an object, against a certain background or standard, some basic principles have to be observed.

Wavelength dependent radiation parameters from sun, space, the atmosphere and earth objects impose limitations on certain problems' solutions.

Referring to global radiation laws and to Fig. 1 "the electromagnetic spectrum" we can summarize:

- the peak of solar shortwavelength radiation reaches the earth at approx. $0.520 \mu m$.
- 30 % of this radiation is reflected through a part or all of the atmosphere (depending on platform altitude).
- 70 % of the energy is emitted from the earth as longwavelength radiation with its peak at $10.2 \mu m$.
- atmospheric windows allowing the earth to be viewed lie in the visible and near infrared part of the spectrum, in the infrared between $3-5 \mu m$ and $8-14 \mu m$, and in the radar and microwave wavelengths.
- inbetween numerous absorption bands caused by atmospheric gases, constituents and pollutants do lead to specific applications (example: watervapor imaging between $5.7-7.1 \mu m$).

2.1 The measurement context - the basic elements

In measuring we have to observe the object context, the atmospheric composition, and the sensor environment. If we ask ourselves what a measurement represents, then we may explain it through:

- the object;   its spectral, spatial, contextual, temporal or textural features and/or variations.
- the atmosphere;   its aerosol, layering, pollutants, clouds, precipitation etc.
- the sensor;   its mechanical and electrical stability and performance, sensitivity, resolution etc.

- the platform;   its stability, orbit characteristics, locational precision etc.

- the treatment of the measured data;   their storage, handling, telemetry, preprocessing, and processing.

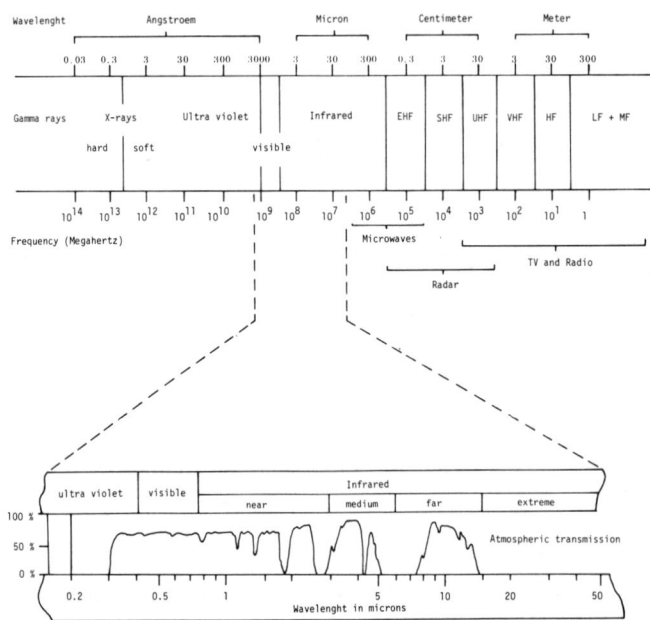

Fig.1: The electromagnetic spectrum, and the atmospheric transmission between ultra violet and infrared.

2.2   The measuring content - what is measured

The content of the measurement is wavelenght or frequency specific. If we observe measurements from ultraviolet to microwave wavelengths we may roughly summarize the following characteristics and indicate some direct applications:

Sensing in:

- ultra violet: Only low level aircraft applications due to atmospheric absorption (for instance ultra violet sensing of oil pollution on water). Reflected solar energy measured.

- visible: Reflection properties measured, and indirect assessment of factors influencing reflection. High atmospheric influence in blue and green bands.

- near infrared: Powerful tool to investigate plant stress. Water has very low reflectivity and is therefore easy to map. Near infrared measurements are less affected by haze than the visible range. Solely reflection measured.

- thermal infrared: Emitted radiation which is proportional to physical temperature and emissivity is sensed. Many factors influence an object's radiant temperature, e.g., physical and chemical composition of material, surface factors, neighbouring effects, atmospheric variables and radiation history etc. Day and night sensing and some haze penetration capabilities enhance application possibilities. Some subsurface features are measurable through their thermal effects.

- passive microwave: Measurements of emitted radiation which depends on physical temperature and emissivity. Day and night sensing capabilities almost unaffected by atmospheric conditions, but even more complex emissivity due to wavelength dependence, partial "penetration" of object, stratification, composition, conductivity, complex scattering etc. Little energy available, e.g. signal to noise problems, and low resolution.

- active microwave (radar): Day and night sensing of reflected energy with similar penetration properties and problems as in passive microwave range. Additional directional flexibility; polarization effects and reflecting material-properties are prominent. High resolution through synthetic aperture, but very complex interpretation of data. Doppler effects enable moving target detection.

2.3 The measuring device - how is the object measured

Photographic and photogrammetric cameras and multiband cameras were used mainly in manned space missions, because the films had to be retrieved whereas the orbiter photo system allowed high precision photographic products to be transmitted line by line to earth.

TV-vidicons were used in satellites since the first missions. A tremendous technological development has been observed from for example the first TIROS vidicons to today's high resolution Return Beam Vidicons of Landsat-3. Weather satellite vidicons allowed direct transmission to local ground stations.

Infrared scanners first, and then multispectral scanners, radiometers and scanning radiometers were developed and installed in satellites also because the continuity of data flow reduces handling problems.

In spaceborne applications of today's radar technology, systems with synthetic aperture (for example Seasat-1 SAR[1]) are favoured due to their great potential in all-weather, day-and-night high resolution imaging. Passive microwave radiometers and imaging scanners offer similar advantages but with far less spatial resolution compared to active systems.

New developments besides the great effort in synthetic aperture radars are found in solid state scanners, silicon detector arrays, charge coupled devices, the so-called push broom scanner, the multispectral linear arrays as well as two dimensional arrays.

Table 1 gives an overview of spectral ranges and resolutions obtained from space with different systems.

System	Spectral Coverage	Groundresolution Range (in m)
Photographic Cameras	uv - nir	10^0
Vidicons	vis - nir	$10^1 - 10^2$
Multispectral Scanner	uv - nir ir	$10^1 - 10^2$ 10^2
Radiometer	vis - ir mw	10^3 $10^4 - 10^5$
SAR	rad	$10^1 - 10^2$
Solid State Scanner	vis - nir	$10^1 - 10^2$

Table 1: Spectral ranges and resolutions obtained from space

2.4 The measuring platform - aircraft and satellites

Despite the general objective of providing information on remote sensing possibilities in obtaining measurements on a global scale, and the economic necessity to solve the problem with spaceborne means, some of the needed quantities in global circulation modelling can only be obtained from airborne missions. Therefore Table 2 gives an overview on standard ranges covered by aircraft measurements.

1) See list in appendix for abbreviations

Type	photogr.	ir	MSS	mw-rad	SLAR	serv.ceiling (m)
single engine light aircraft	x					5 000
twin engine light aircraft	x	x	x	(x)	(x)	6 000
2-4 eng.large aircraft	x	x	x	x	x	8 000
2-4 engin jet	x	x	x	x	x	20 000
best resolution in typical application in m	10^{-2}	10^0	10^0	10^1	10^0	low

Table 2: Standard ranges covered by aircraft measurements

In Tables 3 and 4 some characteristics of selected weather satellites and specific earth resources satellites are summarized.

Country	Name	Duration	Applications / Spectral Coverage
USA	Vanguard 2 and Explorer 6/7	1959	first experimental weathersatellites (vis + ir)
USA	TIROS 1-10	1960-66	experimental weathersatellites with APT + AVCS (vis - nir)
USA	ESSA 1-9	1966-69	operational weathersatellites, simult.1 APT and 1 AVCS-type in orbit (vis - nir)
USA	ATS 1-6	1966-74	geostationary experimental weather- and comm. satellites (vis - ir)
USA	NOAA 1-5	1970-77	second generation of operational weathersatellites, APT (vis - nir + ir) VHRR
USA	TIROS-N and NOAA 6,7..	1978..	third generation of op. weathersatellites,APT (vis - nir + ir) AVHRR
USA	Nimbus 1-7..	1964-80..	experimental satellites for sensor testing (uv + vis - nir + ir + mw) and data relay
USA	SMS 1,2 GOES 1-4	1974-77 1975-80..	geostationary synchronous met.satellites, (vis + ir), data coll.and transm.system, US GARP satellites
USSR	Cosmos 14,23,.. 149 .. 320 ..	1963.. cont.	R&D satellites for sensor evaluation (vis - nir + ir)
USSR	Molniya 3-10	1966..	comm. satellites with tv systems (vis - nir)
USSR	Meteor 1-20..	1969-74..	operational weathersatellites (vis + ir + mw) tv + scanning radiometers
Japan	GMS 1..	1979..	GARP satellite (vis + ir),see SMS/GOES
ESA	Meteosat 1..	1977-79..	GARP satellite (vis + ir),see SMS/GOES

Table 3: Meteorological satellites

Country	Name	Duration	Equipment
USA	Landsat 1	1972-77	3 channel RBV (vis - nir)resol.approx.45 m 4 channel MSS (vis - nir)resol. 79 m DCS (Data Collection System)
USA	Landsat 2	1975-80..	same as Landsat 1
USA	Landsat 3	1978..	2x 1 channel RBV(vis - nir) res.approx.25 m 5 channel MSS (vis - nir + ir) 79 m res.,DCS
USA	HCMM	1978-80	2 channel scann.radiometer (vis - nir + ir) resolution 500/600m
USA	Seasat 1	1978	2 channel scann.radiometer (vis - nir + ir) 4 channel MMR (mw) SAR with resolution better than 25 m
USA	Landsat D	planned for 1982	4-5 channel MSS (vis - nir + ir) see Lands.3 2x 1 channel RBV (vis - nir) as Landsat 3
USA	Landsat D-Prime	planned for 1983	4-5 channel MSS (vis - nir + ir) see above 7 channel Thematic Mapper (TM) (vis - ir)
F	SPOT A	planned for 1984	2x 4 channel high res. "push broom scanners" approx. 10 m res.(vis - nir)
ESA	ERS 1 (COMSS)	planned for 1985	7 channel Ocean Color Monitor (MSS)(vis + ir) 180 m resolution SAR (C-band) 30/100 m res. Imaging Microwave Radiometer (IMR)
ESA	ERS 2 (LASS)	planned for 1988	6 channel CCD "push broom scanner"(vis) 30 m resolution 5 channel MSS (vis - ir) 60 m res.(120 m ir) SAR (C-band) 30-100 m res.
USA	Landsat follow-on	planned for 1985	Thematic Mapper (vis - ir) 10 m res.(120 ir) Multispectral Linear Array (MLA)

Table 4: Earth resources satellites - US and European experiments

3. STATUS OF REMOTE SENSING OF QUANTITIES NEEDED IN CLIMATE MODELS

The status of remote sensing capabilities for a specific selection of quantities needed in climate models is summarized. Where applicable the measuring technique is proposed and practical past or ongoing experiments indicated. The wavelength bands, obtainable resolutions, and if known, accuracies of these approaches are given. Specific problems in the sensing of some needed quantities are discussed and references listed.

It is believed that through the following strictly schematic approach a maximum of structured information can be given.

At the end a summary table allows a condensed overview of the topic (Tab. 5).

3.1 Shortwave albedo

Albedo: $$A(\lambda,\phi,d) = \frac{R(\lambda,\phi,d)}{S(\lambda,\phi,d)} \times 100$$

where: S = Flux of incoming solar radiation
R = Flux of reflected radiation crossing a horizontal element outside the atmosphere
λ,ϕ = Coordinates,   d = Day

Measuring technique: best from satellites, for instance Nimbus 6/7 ERB) experiment with scanning radiometer. Also with operational NOAA weathersatellites.

Wavelength: shortwave albedo 0.2-4.0 μm approx. (ERB)
longwave emission 4.0-50 μm (ERB)

Resolution: spatial(ERB high resolution) 150 x 150 km
spectral 2 W/cm^2

Literature: BARRETT,E.C.: Climatology from Satellites. Methuen, London, 1974.
GRUBER, A.: Determination of the Earth-Atmosphere Radiation Budget From NOAA Satellite Data, NOAA Tech. Rep. NESS 76, Washington D.C. 1978.

NASA: Nimbus-7 Users' Guide GSFC, Greenbelt, Md, 1978.

3.2 Percentage cloud cover

Measuring technique: best from satellites, types: NOAA
TIROS-N
Nimbus (THIR)
GARP-Sat.
HCMM

Wavelength: vis and ir, with multispectral scanners, and scanning radiometers (also in water absorption band, and nir 1.55-1.75 μm)

Resolution: 500 m to 7 km, best temporal resolution with GARP-Sat.

Problems: high albedo from sand and snow, eventually to be solved using nir-channels

3.3 Surface temperature

Measuring technique:	from aircraft and satellites
	aircraft: usually low altitude mission with light aircraft
	satellites: NOAA / Nimbus / HCMM / GARP-Sat. / Seasat
Wavelength:	usually ir-sensing or scanning in atmospheric "windows" normally 10.5-12.5 μm, better 2 wavelengths because of correction of atmospheric attenuation and emissivity factors.
	typical instruments used: NOAA-SR / NOAA-VHRR / NOAA-AVHRR HCMM-HCMR Nimbus-THIR GARP-GOES-VISSR GARP-Meteosat-HRR etc.
	also mw-sensing to assess brightness temperatures with Nimbus-6/7: Nimbus-6 ESMR / Nimbus-7 SMMR
Problems:	remote sensing monitoring of heat flux into the ground is still a topic under discussion. Some experimental results propose thermal infrared application rather than for instance dual microwave detection (Schanda E., 1976).
Resolution:	from aircraft typically 1-2 m (ir) / 10-100 m (mw) from satellites best resolution 500 m HCMM / 20 km (mw)
Accuracy:	from aircraft approx. 0.1 ^{o}C BT (ir) from satellites 0.2-0.5 ºC BT (ir)
	accuracy to obtain real temperature (EBBT) depends on determination of emissivity of sensed material.
Literature:	American Soc. of Photogrammetry: Manual of Remote Sensing, 2 Vol., Falls Church, Va. 1975.
	SCHANDA,E.: Remote Sensing for Environmental Studies, Berlin/Heidelberg, 1976 (Springer).
	WOLFE, W.L. + ZISSIS, G.J.(Edit.): The Infrared Handbook, ERIM, Off.of Naval Res.,Washington DC 1978.

3.4 Surface soil type

Measuring technique:	from aircraft and satellites (limited)
	from aircraft: photography, multiband/MSS from satellites: Landsat, HCMM extensive "ground truth" necessary
Wavelength:	multispectral scanner in vis, nir and ir
Problems:	vegetation cover often prevents soil investigations, then indirect methods become necessary. Best prospects in arid or semi-arid regions. Thermal inertia measurement can ease identfication.
Resolution:	79 m Landsat MSS, 500 m HCMM
Literature:	GIRARD, C.M.: Application of Photointerpretation Technique to the Classification of Agricultural Soils, Choice of the Sensor, Use of the Results, Proc.Seminar IRC/ISPRA Courses, p.37-52, Balkema, Rotterdam 1980.

3.5 Surface soil moisture

Measuring technique:	from aircraft and satellites (both experimental)
	from aircraft: photography and MSS in indirect methods, mw-sensing for direct measurements
	from satellites: mw-sensing, Nimbus-6/7 / Meteor
Wavelength:	Nimbus-6 ESMR 0.8 cm Nimbus-7 SMMR 4.55/2.81/1.67/1.36/0.81 cm Meteor-SR 0.3-3/3-30/8-12 cm Skylab 2.3 + 21 cm
Resolution:	20-100 km (sat.) / 10 m (aircraft low level)

Problems:	Emissivity is moisture dependent. This BT dependency at wavelength of 1 to a few cm is:
	   3-4°K per % moisture content for sandy and loamy soil,
	   near 2°K per % moisture content for arable areas,
	                     (after Basharinov, 1974)

	Long wavelengths (6-21 cm) have greater sensitivity to soil moisture:
	   below 10-20 % little change in emission from soils,
	   more than 20 % - about linear decrease at 2°K per % soil moisture content

	Short wavelengths are attenuated by vegetation and high soil moisture near the surface.

After Kondratyev/Melentyev/Rabinovich/Shulgina, 1977:
Humidity can be assessed only in the surface layer
(test with wavelengths from 1.6 to 18 cm). For the
rest of the soil a model is needed defining a moisture
profile of the minimum fieldwater content for different soil types (down to 1 m).

After Schmugge,T.,1978/80 the soil moisture sampling
depth at 21 cm wavelength is approx. 2-5 cm.
In going from a wet to a dry stage for a medium rough
field, a 80°K change in BT occurs.
A thermal method of sensing bare soil surface moisture
is proposed through measuring thermal inertia by observing the diurnal range of temperatures (Schmugge 1978)
Sensing soil moisture with active microwave systems
is beeing tested at various institutions.

Literature: BASHARINOW, BORODIN and SHUTKO: Passive Microwave
Sensing of Moist Soils, URSI, Berne, 1974.

KONDRATYEV, MELENTYEV, RABINOVICH and SHULGINA:
Passive Microwave Remote Sensing of Soil Moisture,
Proc.Int.Symp.on Remote Sensing, Ann Arbor,
p. 1641-1662, 1977.

SCHMUGGE, T.: Remote Sensing of Surface Soil Moisture,
J.of Appl.Meteor., Vol.17,No.10, p.1549-1557, 1978.

SCHMUGGE, T.: Microwave Approaches in Hydrology,
Photogramm.Engin. Vol.46,No.4, p.495-507, 1980.

3.6 Areal extent of water

Measuring technique: aircraft and satellites

from aircraft: photography, vis/nir, multiband
MSS, ir, SLAR
from satellites: Landsat, Seasat

Wavelength: from aircraft: best nir photography / color ir
film up to 900 nm, or SLAR for large areas

from satellites: best Landsat band 6 or 7
(0.7-0.8 / 0.8-1.1 μm) or Seasat SAR 21.5 cm

Resolution: 79 m for Landsat / 25 m for Seasat SAR (all-weather / day-night)

Literature: SALOMONSON, V.V.+ HALL, D.K.: A Review of Landsat-D
and other advanced Systems to Improving the Utility
of Space Data in Water Resources Management, NASA
Conf.Publ.2116, p.281-296, NASA-GSFC, Greenbelt,Md.
1980.

3.7 Areal extent of ice cover on water

Measuring technique:	from aircraft and satellites
	from aircraft: vis, nir, ir + SLAR
	from satellites: Seasat-SAR, Nimbus-6/7 THIR/+mw Landsat (limited), NOAA
Wavelength:	from aircraft (see above) best for large areas SLAR
	from satellites: best SAR
	NOAA composits (min. brightness)
Resolution:	from aircraft: cm-range for photography / 1 m range for SLAR
	from satellites: Seasat SAR 25 m
	Landsat 79 m
	Nimbus-6 ESMR 32 km
	Nimbus-7 SMMR 20 - 100 km
Literature:	BUSUEVA, A.V. + VOLKOVA, N.A.(Edit.): Remote Measurement of Ice Cover Parameters, (in russian), Rept.of the Arctic and Antarctic Scientific and Res. Institute, Vol.343, Leningrad 1977.
	KONDRATJEV, K.Ja.: Ice and Snow Cover as Observed from Space,(in russian), Allmion Scientific and Research Institute for Hydrometeorological Information - World Data Center, Obninsk, USSR 1978.

3.8 Thickness of ice cover on water

Measuring technique:	aircraft and satellites (both experimental)
	from aircraft: mw radiometers and scanning radiometers
	from satellites: Nimbus-6 ESMR, Nimbus-7 SMMR
Wavelength:	from aircraft: various experimental radiometers
	from satellites: Nimbus-6 ESMR 0.8 cm
	Nimbus-7 SMMR 0.81/1.36/1.67/ 2.81/4.55 cm
Problems:	- Multi-year ice has lower emissivity than young ice.
	- Open water near $0°C$ has BT of approx. $100°K$ which is about $100°K$ less than temperature of thick ice.
	- Very thin ice (8 cm) has about $30°K$ lower temperature than thick ice.
	- The effect of ice thickness on brightness temperature is relatively small for thicknesses over 50 cm. This means that thickness measurement with passive mw radiometry seems impossible for ice thicker than 50 cm.

- Pressure ridges have BT of about $10°K$ less than normal ice (may be important to guide ice breakers). (after Tiuri, M. + Hallikainen, M. + Kaski, K.,1974, - frequencies used 4.77/4.88/4.65 GHz - and: Ramseier, R.O. + Gloersen, P.+ Campbell,W.J. 1974).

- Freeboard height (FH) can be assessed, because of relationship between

  FH and porosity: $1.5°K$ decrease in BT for 1.0 cm increase in FH (means 10 cm more ice) only for multi-year ice. First year ice has too high salinity, there only applicable when large physical temperature difference.

  (after Meeks,D.C.+ Ramseier,R.O.+Campbell,W.J. 1974).

  Ice - Aidjex 1972. Proc. Int. Symp. on RS, Ann Arbor, 1974)

Literature: MEEKS, D.C. + RAMSEIER, R.O. + Campbell, W.J.: A Study of Microwave Emission Properties of Sea Ice - Aidjex 1972, Proc.Int.Symp.on Remote Sensing, Ann Arbor 1974.

RAMSEIER, R.O. + GLOERSEN, P. + CAMPBELL, W.J.: Variation in the Microwave Emissivity of Sea Ice in the Beaufort and Bering Sea. in: Proc.URSI, Berne 1974.

TIURI, M. + HALLIKAINEN, M. + KASKI, K.: Experiments on Remote Sensing of Sea Ice Using a Microwave Radiometer Proc.URSI, Berne 1974.

3.9 Areal extent of snow cover

Measuring technique: preferably from satellites as: NOAA
TIROS-N
GARP-Sat.
Landsat
HCMM

Wavelength: best: red + $0.8-1.1 \mu m$ + $1.55-1.75 \mu m$ (Skylab MSS Exp.) plus additionally ir and mw

partly useful: NOAA VHRR daily 1 vis + 1 ir
NOAA Composit Minimum Brightness Charts
TIROS-N AVHRR daily 1 vis 2 nir 1 ir
GARP-Sat. 20 min. vis + ir
Landsat MSS vis + nir
HCMM-HCMR vis + ir

Resolution: NOAA VHRR 900 m
TIROS-N AVHRR 1,100 m
GARP-Sat. 2.5 - 4 km
Landsat MSS 79 m
HCMM-HCMR 500 m

Problems:	- vis-red channel would offer good measure of areal extent. - nir 0.8-1.0 μm enables assessment of surface condition. - nir 1.55-1.75 μm for cloud and snow separation. - ir 10.5-12.5 μm for determining portion that is melting. - Major concern is cloudiness preventing useful imaging, therefore radar techniques or pass. mw have to be investigated (for the latter see 3.10). High repetition rates are needed for instance for runoff modeling.
Literature:	ITTEN, K.I.: Snow Mapping from Space Platforms, Paper XXII COSPAR Meeting, Bangalore, India 1979, in: SALOMONSON, V.V.+ BHAVSAR, P.D.(Edit.): The Contributions of Space Observations to Water Resources Management, Pergamon, Oxford/New York, 1980. RANGO, A.(Edit.): Operational Applications of Satellite Snow-Observations, NASA SP-391, Washington, D.C., 1975. RANGO, A.+ PETERSON, R.(Edit.): Operational Applications of Satellite Snow-Cover Observations, NASA Conf.Publ. 2116, Washington, D.C. 1980.

3.10 Liquid water content of snowpack

Measuring technique:	aircraft and satellites (both experimental) from aircraft: mw-sensing from satellite: Nimbus-6 ESMR mw Nimbus-7 SMMR mw Meteor SR mw
Wavelength:	Nimbus-6 ESMR 0.8 cm Nimbus-7 SMMR 0.81/1.36/1.67/2.81/4.55 cm Meteor SR 0.3-3 / 3-30 / 8-12 cm
Resolution:	ESMR 32 km SMMR 20 - 100 km SR 70 km
Problems:	- 0.81 cm measurements reveal BT near the surface, its emissivity, roughness parameters, and a very high influence by the free water content. - Intermediate wavelengths 1.4-1.7 cm (Nimbus-5) show information on the composition of the snowpack. - Long wavelength (21 cm) measurements seem to be affected by moisture conditions of the underlying ground.

- All passive microwave measurements are low resolution data, useful for large area non mountainous investigations.
- Melting causes very distinct increase of BT (about $70°K$ at 2.2 cm with only about 1 % liquid water. Water equivalent and BT are directly related.)

Literature: CHANG, A.T. et al: Microwave Emission from Snow, NASA TM 79671, NASA-GSFC, Greenbelt, Md. 1978.

RANGO, A.(EDIT.): Microwave Remote Sensing of Snowpack Properties, NASA Conf.Publ.2153, Washington,D.C. 1980.

SCHANDA, E.: Passive Microwave Sensing, in: Remote Sensing for Environmental Studies, Berlin/Heidelberg, 1976 (Springer).

STILES, W.H.+ Ulaby, F.T.: Microwave Remote Sensing of Snowpacks, NASA Contractor Peport 3263, Washington, D.C., 1980.

3.11 Density and species of vegetation

Measuring technique:　aircraft and satellites
ground truth is essential

from aircraft: photography, MSS, vis + nir + ir
from satellites: same as above with: Landsat MSS+RBV
　　　　　　　　　　　　　　　　　　　　HCMM (partly)

Wavelength:　aircraft: vis + nir phot. up to $0.9 \mu m$
　　　　　　　vis + nir + ir MSS from $0.3-14 \mu m$

satellites: Landsat MSS 0.5-0.6 / 0.6-0.7 / 0.7-0.8 / 0.8-1.1 μm
RBV 0.5-0.75 μm
HCMM HCMR 0.5-0.9 / 10.5-12.5 μm

Resolution:　aircraft: phot. cm range / MSS best 1 m
satellite: Landsat MSS 79 m / RBV 25 m / HCMM 500 m

Problems:
- The determination of vegetation species is a problem of spectral, spatial and temporal variations overlain by atmospheric signature attenuation according to the wavelengths used.
- Careful supervised classification procedures supported by extensive ground investigations are essential.
- Unsupervised multispectral clustering often leads to problems in interpretation.
- Density and vigour of vegetation can best be investigated in nir bands, eventually supported by mw soil moisture determination.
- Best approach is multistage surveying according to the following model:

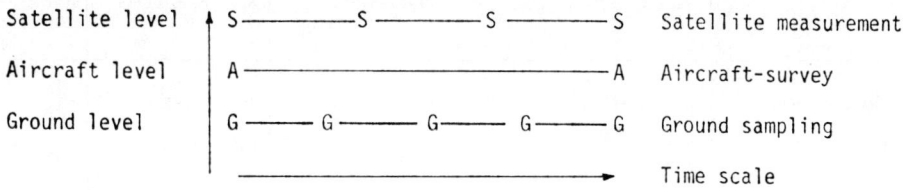

Extensive ground research in test plots (G) + "long"-periodic detailed aircraft survey (A) + "short" periodic large area satellite classification (S).

Literature: BILLINGSLEY, F.C.: Remote Sensing for Vegetation Monitoring, ICSU/SCOPE, Woods Hole, Mass. 1979.

TUCKER, C.J.: A Comparison of Satellite Sensor Bands for Vegetation Monitoring, Photogramm. Engin. Vol.XLIV, No.11, p.1369-1384, 1978.

4. Concluding remarks

In summary table No.5 it is shown, which of the treated topics have reached an operational applications status in remote sensing. This means however not operationality in the sense of having operational satellite systems available, that would allow a continous data transfer. We rather mean that the technical questions of solving the given problem can today be answered.

For those of the topics, where dynamic changes have to be observed and implemented in the models, active and passive microwave techniques will have to be applied. Recent developments in these fields make believe, that in the near future real operationality can be reached for the topics concerned.

Where an experimental status of application is given, the microwave technology mentioned is used presently to resolve most of the remaining problems.

Whether global satellite data will be operationally available to the scientific community is a question of international cooperation and coordination. The experience in the GARP experiments has set the way in which to proceed.

Topic	Measuring technique	Wavelenghts	Best resolution	Status of application (o = operational e = experimental)
Shortwave albedo	Nimbus ERB	vis - ir	12.8 x 267 km	o
Percentage cloudcover	Weathersatellites + HCMM	vis - ir	0.5 - 1 km	o
Surface temperature	Weathersatellites + HCMM	ir, mw	0.5 - 1 km	o
Surface soil type	Aircraft + earth resources sat.	vis - ir, MSS	79 m (sat.)	e
Surface soil moisture	Aircraft + earth res. satellites + weather satellites	mw	10 m (aircraft) 20 - 100 km (sat.)	e
Areal extent of water	Aircraft, earth res. satellites (Landsat, Seasat)	vis, nir, rad	25-79 m (sat.)	o
Areal extent of ice cover on water	Aircraft, weather-satellites, Seasat	vis - ir, mw, rad	25 m (sat, SAR)	o
Thickness of ice cover on water	Aircraft + weather-satellites	mw	20 - 100 km (sat.)	e
Areal extent of snow cover	Weathersatellites, Landsat, HCMM	vis - ir	79 m	o
Liquid water content of snowpack	Aircraft + weather-satellites	mw	20 m (aircraft) 20 km (sat.)	e
Density and species of vegetation	Aircraft + earth res. satellites	vis - ir, MSS	25 m (sat,)	e-o

Table 5: Summary table - Status of remote sensing of quantities needed in climate models

REFERENCES

American Society of Photogrammetry: Manual of Remote Sensing, 2 Vol., Falls Church, Va. 1975.

Barrett, E.C.: Climatology from Satellites, Methuen, London 1974.

Barrett, E.C.+ Curtis, L.F.: Environmental Remote Sensing, Chapman and Hall, London 1976.

Basharinow, A.E.+ Borodin, L.F.+ Shutko, A.M.: Passive Microwave Sensing of Moist Soils, Proc.URSI, Berne 1974.

Billingsley, F.C.: Remote Sensing for Vegetation Monitoring, ICSU/SCOPE, Woods Hole, Mass. 1979.

Busueva, A.V.+ Volkova, N.A.(Edit.): Remote Measurement of Ice Cover Parameters, (in russian), Report of the Arctic and Antarctic Scientific and Research Institute, Vol.343, Leningrad 1977.

Chang, A.T. et al.: Microwave Emission from Snow, NASA TM 79671, NASA-GSFC, Greenbelt, Md. 1978.

European Space Agency: Use of Data from Meteorological Satellites, Proc. Tech. Conf., Lannion 1979.

Girard, C.M.: Application of Photointerpretation Technique to the Classification of Agricultural Soils, Choice of Sensor, Use of the Results, Proc.Seminar IRC/ISPRA Courses, p.37-52, Balkema, Rotterdam 1980.

Gruber, A.: Determination of the Earth-Atmosphere Radiation Budget From NOAA Satellite Data, NOAA Tech.Rep. NESS 76, Washington D.C. 1978.

Holz, R.K.: The Surveillant Science, Houghton Mifflin, Boston 1973.

Itten, K.I.: Snow Mapping from Space Platforms, Paper XXII COSPAR, Bangalore, India 1979, in: Salomonson, V.V.+ Bhavsar, P.D.(Edit.): Contributions of Space Observations to Water Resources Management, Pergamon, Oxford/New York 1980.

Kondratyev, K.Ya.+ Melentyev, V.V.+ Rabinovich, Yu.I.+ Shulgina, E.M.: Passive Microwave Remote Sensing of Soil Moisture, Proc. Int.Symp.on Remote Sensing, p.1641-1662, AnnArbor 1977.

Kondratyev, K.Ya.: Ice and Snow Cover as Observed from Space, (in russian), Allmion Scientific and Research Institute for Hydrometeorological Information, World Data Center, Obninsk, USSR 1978.

Kondratyev, K.Ya.(Edit.): Meteorological Sounding of the Underlying Surface from Space, (in russian), Hydromet. Izdad, Leningrad 1979.

Meeks, D.C.+ Ramseier, R.O.+ Campbell, W.J.: A Study of Microwave Emission Properties of Sea Ice - Aidjex 1972, Proc.Int. Symp.on Remote Sensing, Ann Arbor 1974.

NASA: Nimbus 6/ Nimbus 7 Users Guide's, NASA-GSFC, Greenbelt, Md. 1975/1978.

Ramseier, R.O. + Gloersen, P.+ Campbell, W.J.: Variation in the Microwave Emissivity of the Sea Ice in the Beaufort and Bering Sea, in: Proc. USRSI, Berne 1974.

Rango, A.(Edit.): Operational Applications of Satellite Snow Observations, NASA SP-391, Washington, D.C. 1975.

Rango, A.+ Peterson, R.(Edit.): Operational Applications of Satellite Snow Cover Observations, NASA Conf.Publ. 2116, Washington, D.C. 1980.

Rango, A.+ Chang, A.T.C.+ Foster, J.L.: The Utilization of Spaceborne Microwave Radiometers for Monitoring Snowpack Properties, Nordic Hydrology, 10, p.25-40, 1979.

Rango, A.(Edit.): Microwave Remote Sensing of Snowpack Properties, NASA Conf. Publ. 2153, Washington,D.C. 1980.

Salomonson, V.V.+ Hall, D.K.: A Review of Landsat-D and other Adavanced Systems to Improving the Utility of Space Data in Water Resources Management, NASA Conf.Publ.2116, p.281-296, Washington D.C. 1980.

Schanda, E.: Remote Sensing for Environmental Studies, Berlin/Heidelberg 1976 (Springer).

Schmugge, T.: Remote Sensing of Surface Soil Moisture, Journal of Appl. Meteor., Vol.17,No.10, p.1549-1557, 1978.

Schmugge, T.: Microwave Approaches In Hydrology, Photogramm.Engin.,Vol.46, No.4, p.495-507, 1980.

Stiles, W.H.+ Ulaby, F.T.: Microwave Remote Sensing of Snowpacks, NASA Contracto Report 3263, Washington D.C.,1980.

Tiuri, M.+ Hallikainen, M.+ Kaski, K.: Experiments on Remote Sensing of Sea Ice Using a Microwave Radiometer, Proc.URSI, Berne 1974.

Tucker, C.J.: A Comparison of Satellite Sensor Bands for Vegetation Monitoring, Photogramm. Engin.,Vol.XLIV, No.11, p.1369-1384, 1978.

Wolfe, W.L.+ Zissis, G.J.(Edit.): The Infrared Handbook, ERIM, Off.of Naval Res., Washington D.C. 1978.

World Data Center: Report on Active and Planned Spacecraft and Experiments, NASA-GSFC, Greenbelt, Md. 1979.

APPENDIX

List of Abbreviations

API	=	Automatic Picture Transmission
AVCS	=	Advanced Vidicon Camera System
AVHRR	=	Advanced Very High Resolution Radiometer
BT	=	Brightness Temperature
CCD	=	Charge Coupled Device
ERB	=	Earth Radiation Budget
ESA	=	European Space Agency
ESMR	=	Electrically Scanning Microwave Radiometer
GARP-Sat.	=	geostationary Satellites used in the GARP Experiment
HCMM	=	Heat Capacity Mapping Mission
HCMR	=	Heat Capacity Mapping Radiometer
HRR	=	High Resolution Radiometer
MLA	=	Multispectral Linear Array
MMR	=	Multifrequency Microwave Radiometer
MSS	=	Multispectral Scanner
RBV	=	Return Beam Vidicon
SAR	=	Synthetic Aperture Radar
SLAR	=	Side Looking Airborne Radar
SMMR	=	Scanning Multispectral Microwave Radiometer
SR	=	Scanning Radiometer
THIR	=	Temperature Humidity Infrared Radiometer
VHRR	=	Very High Resolution Radiometer

uv	= ultra violet	10 – 400	nm
vis	= visible	400 – 700	nm
nir	= near infrared (reflected)	0.7 – 3	μm
ir	= infrared (emitted)	3 – 1000	μm
mw	= microwaves	3 mm – 300	mm
rad	= radar	1 mm – 10	m